Electroceramics

Second Edition

Materials · Properties · Applications

A. J. Moulson
and
J. M. Herbert

WILEY

Electroceramics
Materials • Properties • Applications
Second Edition

Authorized reprint by Wiley India Pvt. Ltd., 4435-36/7, Ansari Road, Daryaganj, New Delhi – 110002.

Reprint: 2015

Printed at: *Ritomate International, Delhi*

ISBN: 978-81-265-5394-5

Electroceramics

Second Edition

CONTENTS

Electroceramics: Materials, Properties, Applications. 2nd Edition. Edited by A. J. Moulson and J. M. Herbert.
 ISBN 0 471 49747 9 (hardback) 0 471 49748 7 (paperback)

PREFACE

'Ceramics' describes an engineering activity embracing the design and fabrication of ceramic components. Because the optimum physical and chemical properties of a ceramic are defined by the specific requirements of the end use, the pursuit is, of necessity, interdisciplinary. For example, the design and manufacture of refractories is a challenging technology, spanning physical chemistry and metallurgical and chemical engineering. Even greater challenges confront the electroceramist involved in, for example, the development of the ceramic battery or fuel cell. Here a combination of a good understanding of solid state chemical physics with expert knowledge and experience of advanced ceramics fabrication technologies is essential. In addition there should be a sound appreciation of the many considerations, usually complex, concerned with 'end use'. The same is true of the very diverse range of types of electroceramic component discussed in the text.

In the UK, the necessary basic disciplines, solid state chemical physics and electrical and electronic engineering, have, in the main, attracted into higher education those students who at school displayed strengths in mathematics, physics and chemistry. Materials science undergraduates have tended to be more qualitative in their approach to learning, an approach now difficult to justify and sustain:

> *I often say that when you can measure what you are speaking about, and express it in numbers you know something about it; but when you cannot measure it, when you cannot express it in numbers, your knowledge is of a meagre and unsatisfactory kind: it may be the beginning of knowledge, but you have scarcely, in your thoughts advanced to the stage of science, whatever the matter may be.*

(Sir William Thomson – Lord Kelvin (1883) *Electrical units of measurement* 'Popular Lectures and Addresses' **1**, Macmillan & Co. London 1989).

The electroceramist must cultivate, at an appropriate level, a quantitative understanding of the basic science of a wide range of physical properties of solids, including conductive, dielectric, optical, piezoelectric and magnetic. The understanding must embrace how the science of ceramics can be exploited to optimise properties, not only through the design of material composition, but also through the tailoring of microstructure and texture. Because the objective

Electroceramics: Materials, Properties, Applications. 2nd Edition. Edited by A. J. Moulson and J. M. Herbert.
 ISBN 0 471 49747 9 (hardback) 0 471 49748 7 (paperback)

is an improved component for some particular function – a capacitor, thermistor, fuel cell, battery, microwave filter, chemical sensor, actuator, etc. – there has to be an intelligent appreciation of the significance of the various relevant properties to the particular application, and how to 'engineer' the material to optimise them. This is well illustrated by the piezoceramic–polymer composites for ultrasound transducers, pyroelectric materials for infrared detectors and imaging systems, and thin film ceramics for random access memories. Their development demands an interplay between the basic sciences, electronic engineering and materials science, or better, 'materials engineering' – a term increasingly encountered.

Not surprisingly, most of the available texts concentrate on one or other of the relevant basic solid state science, the ceramics science and technology, or on component applications; the other two aspects receiving only superficial coverage. The nearest to what might be seen as offering interdisciplinary treatments are edited contributions from specialists in various topics. Whilst these are valuable they may present difficulties to the undergraduate and newcomer to the field. There are plenty of specialist papers but they are mostly published for the benefit of those well grounded in their subjects and capable of a balanced and critical appreciation.

In the UK, the teaching of electroceramics passed through its formative years as a natural development of 'traditional ceramics' and then became absorbed into the framework of 'materials science' courses. It now seems that the very interdisciplinary nature of the topics embraced, together with changing fashions as far as the aspirations of many young people in the West are concerned, are having their impact and the next decade may well see the basic science, materials and engineering communities in higher education merging into interdisciplinary institutes of one sort or another. The authors have very much in mind the teaching and postgraduate research personnel in higher education and also the large community of physicists, chemists and engineers who enter industry without the benefit of specialized training.

A great deal has happened since the first edition was published and there is no reason to believe that the rate of technological progress will diminish. However, principles do not change. The authors' objective is not to present up-to-the-minute descriptions encompassing all of what comprises the science and technology of electroceramics, but to concentrate on the most significant advances, which encompass what seem to be the unchanging principles underpinning the science, fabrication and applications of electroceramics. However where significant developments are occurring at the subject frontiers and about which the authors feel the well-informed electroceramist should be aware, the coverage is sufficient to serve as a 'lead-in' for a more in-depth study. This is the case with, for example, ceramics in photonics and ferroelectric random access memories.

The past decade has seen significant advances in fabricating electroceramics. For example the demand for higher volumetric efficiency multilayer capacitors has had its impact on the technologies concerned with the preparation of powders having closely defined chemistry and physical characteristics, and on their processing into components. The (often overriding) need to reduce costs has led to the widespread adoption of base metal electrodes in multilayer technology which, in turn, has stimulated the design of special chemically modified powders. Multilayer technology itself has penetrated many sectors of electroceramics technology. The decade has also seen significant developments in electroceramics associated with microwave telecommunications and the emergence of LTCC (low temperature co-fired ceramic) technology.

Since the appearance of the first edition there has been a growing awareness of the 'global environment' and of the negative impact our apparently insatiable demand for energy is having upon it. The strong growth of interest in fuel cells and batteries is one very important response to this. High temperature superconductor technologies are maturing with 'current leads' and 'fault current limiters' now commercial products. Impressive progress is being made with regard to 'trapped field' magnets and these, together with superconducting cables, are set to play important roles in the more efficient generation and distribution of electrical energy.

To an ever-increasing extent the functioning of manufactured products from the 'air-bus' and motor-car through to the 'fridge' and washing-machine, depends upon 'sensor-actuator' technology where electroceramics play essential roles. The technology of sensing harmful gases and of chemical sensing in general has seen significant developments over the past decade.

In addition to covering these developments, the revision has presented the opportunity to rectify what were identified as shortcomings of one sort or another. Minor errors have been corrected and the discussions on many topics modified to bring them up-to-date. The cross-referencing throughout the text has been extended and so also has the bibliography. The review papers and texts referenced have been carefully selected with the objective of easing entry into the specialised literature; at the same time every effort has been taken to maintain the essentially 'free-standing' character of the book. With very few exceptions, the bibliography is restricted to 'readily accessible' texts and papers which are also judged to be essential information sources for the electroceramist. The web contains a wealth of information – some excellent in quality and some less so. In general the authors have resisted the temptation to reference web-sites in the belief that these are better located and consulted by the adequately informed reader.

The exercises have been extended. As for the first edition they are designed to assist in deepening understanding. It is only when one tackles an illustrative problem that one's deficiencies as far as understanding is concerned are revealed, and this equally applies to the design of problems! It is hoped that the answers given are sensibly correct; most, but not all, have been checked by kind colleagues. If errors of any kind, here or elsewhere, are identified then the authors would be grateful to be informed of them.

The text is balanced as it is because of the interdisciplinary nature of the authorship. JMH, trained as a chemist, spent most of his working life engineering electroceramics into existence for specific purposes for a major electronics company. AJM, trained as a physicist, spent the greater part of his working life attempting to teach ceramics and to keep a reasonably balanced postgraduate research activity ongoing in one of the major university centres for materials science. Both have learnt a great deal in putting the text together, and hope that others will benefit from their not inconsiderable effort.

Finally, although the usual place to acknowledge assistance is under 'Acknowledgements' (and this has been done), it seems appropriate to put on record here that the revision could not have been completed without the generous support of so many colleagues around the world. They have read drafts e-mailed to them, suggested modifications and re-read them. Our awareness of just how burdensome giving such help is makes us all the more appreciative of it. It has given the authors the confidence so necessary when attempting to cover reasonably comprehensively the many and diverse topics embraced by 'electroceramics', and contributed immeasurably to making this second edition what we trust is a fitting successor to the first.

A. J. Moulson
J. M. Herbert

ACKNOWLEDGEMENTS

The authors extend their sincere thanks to the many colleagues who have helped, sometimes unwittingly, including:

Prof. Andrew Bell, Mr. Torsten Bove, Prof. Robert Freer, Dr. Ian Hunter, Dr. Reinhardt Kulke, Prof. Vilho Lantto, Mr. Kazuhisa Niwano, Prof. David Payne, Dr. David Pearce, Mr. Mike Rendell, Mr. Adrian Smith, Prof. Katsuhisa Tanaka, Dr. Mike Thomas, Prof. Noel Thomas, Prof. Heiko Thust, Mr. Rob. Twiney, Dr. Eva Vogel, Prof. Roger Whatmore, Dr. Wolfram Wersing, Dr. Liang Xue, Mr. Noriaki Yamana.

We are especially indebted to the following for the kindness shown in reading and constructively criticizing various parts of the revision:

Prof. Alan Atkinson, Mr. Jake Beatson, Dr. Les Bowen, Dr. John Bultitude, Dr. Tim Button, Dr. David Cardwell, Prof. Archie Campbell, Prof. Stephen Evans, Dr. David Hall, Prof. Derek Fray, Dr. David Iddles, Dr. Heli Jantunen, Prof. Tom H. Johansen, Dr. Charles King, Dr. Ian McAuley, Dr. Mira Naftali, Mr. Derek Nicker, Dr. Elvin Nix, Mr. Bill Phillips, Prof. Götz Reinhardt, Prof. Jim Scott, Dr. Brian Shaw, Dr. Subhash Singhal, Dr. Wallace Smith, Prof. Brian Steele, Dr. Jim Sudworth, Dr. Michael Todd, Dr. Pieter van der Zaag, Prof. Alan Williams, Prof. David Williams, Prof. Rainer Waser.

Of course we again thank all those colleagues who helped with the first edition, especially Prof. Denis Greig, Dr George Johnson, Dr. Chris Groves-Kirkby, Mr. Peter Knott, Mr. John McStay, Prof. Don Smyth and Mr. Rex Watton.

We also thank Rachael Ballard, Robert Hambrook and Andrew Slade, and others of John Wiley & Sons, Ltd. for their patience and help and Alison Woodhouse for her helpful copy-editing.

Those we may have inadvertently missed have our sincere apologies.

Finally we thank our families for their patience.

Electroceramics: Materials, Properties, Applications. 2nd Edition. Edited by A. J. Moulson and J. M. Herbert.
 ISBN 0 471 49747 9 (hardback) 0 471 49748 7 (paperback)

GLOSSARY

In expressing quantities throughout the text the authors have been guided by the recommendations in 'Quantitites, Units and Symbols in Physical Chemistry', prepared for publication by Ian Mills, International Union of Pure and Applied Chemistry, Blackwell Scientific Publications, London 1988. (ISBN 0 632 01773 2)

Fundamental constants

c	speed of light in vacuum	$2.998 \times 10^{8}\,\mathrm{m\,s^{-1}}$
e	elementary charge	$1.602 \times 10^{-19}\,\mathrm{C}$
F	Faraday constant	$9.649 \times 10^{4}\,\mathrm{C\,mol^{-1}}$
h	Planck constant	$6.626 \times 10^{-34}\,\mathrm{J\,s}$
$\hbar$	$= h/2\pi$	$1.055 \times 10^{-34}\,\mathrm{J\,s}$
k	Boltzmann constant	$1.381 \times 10^{-23}\,\mathrm{J\,K^{-1}}$
m_e	electron rest mass	$9.109 \times 10^{-31}\,\mathrm{kg}$
N_A	Avogadro's constant	$6.022 \times 10^{23}\,\mathrm{mol^{-1}}$
R_0	gas constant	$8.315 \times \mathrm{J\,K^{-1}\,mol^{-1}}$
ε_0	permittivity of a vacuum	$8.854 \times 10^{-12}\,\mathrm{F\,m^{-1}}$
μ_B	Bohr magneton	$9.274 \times 10^{-24}\,\mathrm{J\,T^{-1}}$
μ_0	permeability of a vacuum	$4\pi \times 10^{-7}\,\mathrm{H\,m^{-1}}$ (exactly)
σ	Stefan–Boltzmann constant	$5.671 \times 10^{-8}\,\mathrm{W\,m^{-2}\,K^{-4}}$
$\mathrm{M_u}$	atomic mass unit	$1.66 \times 10^{-27}\,\mathrm{kg}$

Symbols with the same meaning in all chapters (unless the text makes clear an alternative meaning)

A	area
C	capacitance
E	electric field
e	electronic charge
f	frequency
I	electric current
J	current density
j	$(-1)^{1/2}$
k	Boltzmann constant
R	electrical resistance
R_0	gas constant
S	entropy
T	temperature
t	time
U	electric potential difference
u	mobility
V	volume
$\mathbf{v}$	linear velocity
$\mathcal{E}$	energy, work
ε	absolute permittivity
μ	absolute permeability
μ_r	relative permeability
ε_r	relative permittivity
λ	wavelength
χ_e	electric susceptibility
χ_m	magnetic susceptibility
ω	angular velocity

Electroceramics: Materials, Properties, Applications. 2nd Edition. Edited by A. J. Moulson and J. M. Herbert.

1

INTRODUCTION

The word ceramic is derived from *keramos*, the Greek work for potter's clay or ware made from clay and fired, and can simply be interpreted as 'pottery'. Pottery is based on clay and other siliceous minerals that can be conveniently fired in the 900–1200 °C temperature range. The clays have the property that on mixing with water they form a mouldable paste, and articles made from this paste retain their shape while wet, on drying and on firing. Pottery owes its usefulness to its shapability by numerous methods and its chemical stability after firing. It can be used to store water and food, and closely related materials form the walls of ovens and vessels for holding molten metals. It survives almost indefinitely with normal usage although its brittleness renders it susceptible to mechanical and thermal shock.

The evolution from pottery to advanced ceramics has broadened the meaning of the word 'ceramics' so that it now describes '...solid articles which have as their essential component, and are composed in large part of, inorganic non-metallic materials' [1]. Here the term will be restricted to polycrystalline, inorganic, non-metallic materials that acquire their mechanical strength through a firing or sintering process. However, because glass and single crystals are components of many polycrystalline and multiphase ceramics, and because single crystals of some compositions are grown for special applications, discussion of them is included as appropriate.

The first use of ceramics in the electrical industry took advantage of their stability when exposed to extremes of weather and to their high electrical resistivity, a feature of many siliceous materials. The methods developed over several millennia for domestic pottery were refined for the production of the insulating bodies needed to carry and isolate electrical conductors in applications ranging from power lines to the cores bearing wire-wound resistors and electrical fire elements.

Whilst the obvious characteristic of ceramics in electrical use in the first half of the twentieth century was that of chemical stability and high resistivity, it was

Electroceramics: Materials, Properties, Applications. 2nd Edition. Edited by A. J. Moulson and J. M. Herbert.

evident that the possible range of properties was extremely wide. For example, the ceramic form of the mineral magnetite, known to the early navigators as 'lodestone', was recognized as having a useful electrical conductivity in addition to its magnetic properties. This, combined with its chemical inertness, made it of use as an anode in the extraction of halogens from nitrate minerals. Also, zirconia, combined with small amounts of lanthanide oxides (the so called 'rare earths') could be raised to high temperatures by the passage of a current and so formed, as the Nernst filament, an effective source of white light. It was recognized that some ceramics, the 'fast-ion conductors', conduct electricity well, and predominantly by the transport of ions, and over the last two decades interest in them has intensified because of their crucial roles in fuel cell, battery and sensor technologies.

The development from 1910 onwards of electronics accompanying the widespread use of radio receivers and of telephone cables carrying a multiplicity of speech channels led to research into ferrites in the period 1930–1950. Nickel–zinc and manganese–zinc ferrites, closely allied in structure to magnetite, were used as choke and transformer core materials for applications at frequencies up to and beyond 1 MHz because of their high resistivity and consequently low susceptibility to eddy currents. Barium ferrite provided permanent magnets at low cost and in shapes not then achievable with ferromagnetic metals. From 1940 onwards magnetic ceramic powders formed the basis of recording tapes and then, as toroids of diameter down to 0.5 mm, were for some years the elements upon which the mainframe memories of computers were based. Ferrites, and similar ceramics with garnet-type structures, remain valuable components in microwave technology.

From the 1920s onwards conductive ceramics found use, for instance, as silicon carbide rods for heating furnaces up to 1500 °C in air. Ceramics with higher resistivities also had high negative temperature coefficients of resistivity, contrasting with the very much lower and positive temperature coefficients characteristic of metals. They were therefore developed as temperature indicators and for a wide range of associated applications. Also, it was noticed at a very early stage that the resistivity of porous specimens of certain compositions was strongly affected by the local atmosphere, particularly by its moisture content and oxidation potential. Latterly this sensitivity has been controlled and put to use in detectors for toxic or flammable components.

It was also found that the electrical resistivity of ceramics based on silicon carbide, and, more recently, zinc oxide could be made sensitive to the applied field strength. This has allowed the development of components that absorb transient surges in power lines and suppress sparking between relay contacts. The non-linearity in resistivity is now known to arise because of potential barriers between the crystals in the ceramic.

Ceramics as dielectrics for capacitors have the disadvantage that they are not easily prepared as self-supporting thin plates and, if this is achieved, are extremely fragile. However, mica (a single-crystal mineral silicate) has been widely used in capacitors and gives very stable units. Thin-walled (0.1–0.5 mm) steatite tubes have been extruded for use in low-capacitance units. The low relative permittivity of steatite ($\varepsilon_r \approx 6$) has limited its use but the introduction of titania ($\varepsilon_r \approx 100$) in the 1930s led to the development of capacitors having values in the 1000 pF range in convenient sizes but with a high negative temperature coefficient. Relative permittivities near to 30 with low temperature coefficients have since been obtained from titanate and zirconate compositions.

The situation was altered in the late 1940s with the emergence of high-permittivity dielectrics based on barium titanate ($\varepsilon_r \approx 2000$–$10\,000$). For a wide range of applications small plates or tubes with thicknesses of 0.2–1 mm gave useful combinations of capacitance and size. The development of transistors and integrated circuits led to a demand for higher capacitance and small size which was met by monolithic multilayer structures. In these, thin films of organic polymer filled with ceramic powder are formed. Patterns of metallic inks are deposited as required for electrodes and pieces of film are stacked and pressed together to form closely adhering blocks. After burning out the organic matter and sintering, robust multilayer units with dielectrics of thicknesses down to $< 5\,\mu m$ have been obtained. Such units fulfil the bypass, coupling and decoupling functions between semiconductor integrated circuits in thick-film semiconductor circuitry. The monolithic multilayer structure can be applied to any ceramic dielectric, and multilayer structures for a variety of applications are the subject of continuous development effort. In particular ‘low temperature co-fired ceramic’ (LTCC) technology is intensively pursued for electronics packaging, especially for mainframe computer and telecommunications systems.

The basis for the high permittivity of barium titanate lies in its ferroelectric character which is shared by many titanates, niobates and tantalates of similar crystal structure. A ferroelectric possesses a unique polar axis that can be switched in direction by an external field. The extent of alignment of the polar axes of the crystallites in a ceramic is limited by the randomness in orientation of the crystallites themselves but is sufficient to convert a polycrystalline isotropic body into a polar body. This polarity results in piezoelectric, pyroelectric and electro-optic behaviour that can be utilized in sonar, ultrasonic cleaners, infrared detectors and light processors. Ceramics have the advantage, over the single crystals that preceded them in such applications, of greater ease of manufacture. Ferroelectrics in thin film form are now becoming established as one type of digital memory element.

Barium titanate can be made conductive by suitable substitutions and/or by sintering in reducing atmospheres, which has led to two developments: firstly, high-capacitance units made by reoxidizing the surface layers of conductive plates and using the thin insulating layers so formed; secondly, high positive temperature coefficient (PTC) resistors since the resistivity of suitably doped and fired bodies increases by several orders of magnitude over a narrow temperature range close to the transition from the ferroelectric to the paraelectric states. Uses for PTC resistors include thermostatic heaters, current controllers, degaussing devices in television receivers and fuel-level indicators. As with voltage-sensitive resistors, the phenomenon is based on electrical potential barriers at the grain boundaries. Finally, superconducting ceramics with transition temperatures of over 100 K have been discovered. This enables the development of devices operable at liquid nitrogen temperatures, in particular cables for electric power distribution and permanent magnets capable of producing exceptionally high magnetic field strengths for a variety of applications, including magnetically levitated transport systems.

The evolution of ferrimagnetic, ferroelectric and conductive ceramics has required the development of compositions almost entirely free from natural plasticizers such as clays. They require organic plasticizers to enable the 'green' shapes to be formed prior to sintering. Densification is no longer dependent on the presence of large amounts of fusible phases (fluxes) as is the case with the siliceous porcelains. Instead it depends on small quantities of a liquid phase to promote 'liquid phase sintering' or on solid state diffusional sintering or on a combination of these mechanisms. Crystal size and very small amounts of secondary phases present at grain boundaries may have a significant effect on properties so that close control of both starting materials and preparation conditions is essential. This has led to very considerable research effort devoted to the development of so-called 'wet chemical' routes for the preparation of starting powders.

Ceramics comprise crystallites that may vary in structure, perfection and composition as well as in size, shape and the internal stresses to which they are subjected. In addition, the interfaces between crystallites are regions in which changes in lattice orientation occur, often accompanied by differences in composition and attendant electrical effects. As a consequence it is very difficult, if not impossible, to account precisely for the behaviour of ceramics. The study of single-crystal properties of the principal components has resulted in valuable insights into the behaviour of ceramics. However, the growth of single crystals is usually a difficult and time-consuming business while the complexities of ceramic microstructures renders the prediction of properties of the ceramic from those of the corresponding single crystal very uncertain. Consequently, empirical observation has usually led to the establishment of new devices based on ceramics before there is more than a partial understanding of the underlying physical mechanisms.

In the following chapters the elementary physics of material behaviour has been combined with an account of the preparation and properties of a wide range of ceramics. The physical models proposed as explanations of the observed phenomena are often tentative and have been simplified to avoid mathematical difficulties but should provide a useful background to a study of papers in contemporary journals.

Bibliography

1. Kingery, W.D., Bowen, H.K. and Uhlmann, D.R. (1976) *Introduction to Ceramics*, 2nd edn, Wiley, New York.

2

ELEMENTARY SOLID STATE SCIENCE

2.1 Atoms

The atomic model used as a basis for understanding the properties of matter has its origins in the α-particle scattering experiments of Ernest Rutherford (1871–1937). These confirmed the atom to be a positively charged nucleus, of radius of order 10^{-14} m, in which the mass of the atom is largely concentrated and around which negatively charged electrons are distributed. The radius of the atom is of order 10^{-10} m and thus much of it is empty space. The total negative charge of the electrons compensates the positive charge of the nucleus so that the atom is electrically neutral.

A dynamic model of the atom has to be adopted, as a static model would be unstable because the electrons would fall into the nucleus under the electrostatic attraction force. Niels Bohr (1885–1962) developed a dynamic model for the simplest of atoms, the hydrogen atom, using a blend of classical and quantum theory. In this context the term 'classical' is usually taken as meaning pre-quantum theory.

The essentials of the Bohr theory are that the electron orbits the nucleus just as a planet orbits the sun. The problem is that an orbiting particle is constantly accelerating towards the centre about which it is rotating and, since the electron is a charged particle, according to classical electromagnetic theory it should radiate electromagnetic energy. Again there would be instability, with the electron quickly spiralling into the nucleus. To circumvent this problem Bohr introduced the novel idea that the electron moved in certain allowed orbits without radiating energy. Changes in energy occurred only when the electron made a transition from one of these 'stationary' states to another. In a stationary state the electron moves so that its angular momentum is an integral multiple of

Electroceramics: Materials, Properties, Applications. 2nd Edition. Edited by A. J. Moulson and J. M. Herbert.
 ISBN 0 471 49747 9 (hardback) 0 471 49748 7 (paperback)

$\hbar = h/2\pi$ where h is the Planck constant ($\hbar = 1.055 \times 10^{-34}$ J s), i.e. the angular momentum is quantized.

If the electron is moving in a circular orbit of radius r then the electrostatic attraction force it experiences is balanced by the centripetal force:

$$\frac{e^2}{4\pi\varepsilon_0 r^2} = m_e \frac{\upsilon^2}{r} = m_e \omega^2 r \tag{2.1}$$

in which υ is the linear velocity of the electron and ω is its angular frequency.

The total energy $\mathcal{E}$ of the electron is made up of its kinetic energy $\mathcal{E}_k$ and its electrostatic potential energy $\mathcal{E}_p$. The value of $\mathcal{E}_p$ is taken as zero when the electron is so far removed from the nucleus, i.e. at 'infinity', that interaction is negligible. Hence

$$\mathcal{E} = \mathcal{E}_p + \mathcal{E}_k \tag{2.2}$$

$$\mathcal{E} = -\frac{e^2}{4\pi\varepsilon_0 r} + \frac{1}{2} m_e \upsilon^2$$

Substituting from Eq. (2.1) gives

$$\mathcal{E} = -\frac{e^2}{8\pi\varepsilon_0 r} \tag{2.3}$$

From the quantum condition for angular momentum,

$$m_e \omega r^2 = n\hbar \tag{2.4}$$

which, together with Eq. (2.1), leads to

$$\omega^2 = \frac{e^2}{4\pi m_e \varepsilon_0 r^3} = \frac{n^2 \hbar^2}{m_e^2 r^4}$$

or

$$\frac{1}{r} = \frac{m_e e^2}{4\pi\varepsilon_0 n^2 \hbar^2} \tag{2.5}$$

which substitutes into Eq. (2.3) to give

$$\mathcal{E} = -\frac{m_e e^4}{32\pi^2 \varepsilon_0^2 \hbar^2} \frac{1}{n^2} \tag{2.6}$$

The integer n (1, 2, 3 etc.) is called the 'principal quantum number' and defines the energy of the particular electron state. Although the situation for multielectron atoms is complicated by the repulsive interaction between electrons, the energy of a particular electron is still defined by its principal quantum number. In general, the smaller the value of n, the lower is the energy, with the energy differences between states defined by successive n values decreasing with increasing n.

The Bohr theory of the atom was further developed with great ingenuity to explain the complexities of atomic line spectra, but the significant advance came with the formulation of wave mechanics.

It was accepted that light has wave-like character, as evidenced by diffraction and interference effects; the evidence that light has momentum, and the photoelectric effect, suggested that it also has particle-like properties. A ray of light propagating through free space can be considered to consist of a stream of 'photons' moving with velocity $2.998 \times 10^8\ \mathrm{m\,s^{-1}}$. The kinetic energy of a photon is given by $\mathcal{E}_k = h\nu$ where ν is the frequency of the light according to the wave model. The converse idea that particles such as electrons exhibit wave-like properties (the electron microscope is testimony to its correctness) was the first step in the development of wave mechanics. It turns out that the 'de Broglie wavelength' of a particle moving with momentum mv is given by

$$\lambda = \frac{h}{mv} \tag{2.7}$$

Electron states are described by the solutions of the following equation which was developed by Erwin Schrödinger (1887–1961) and which bears his name:

$$\nabla^2\psi + \frac{2m}{\hbar^2}(\mathcal{E} - \mathcal{E}_p)\psi = 0 \tag{2.8}$$

This form of the Schrödinger equation is independent of time and so is applicable to steady state situations. The symbol ∇^2 denotes the operator

$$\frac{\partial^2}{\partial x^2} + \frac{\partial^2}{\partial y^2} + \frac{\partial^2}{\partial z^2}$$

$\psi(x,y,z)$ is the wave function, and $\mathcal{E}(x,y,z)$ and $\mathcal{E}_p(x,y,z)$ are respectively the total energy and the potential energy of the electron. The value of $|\psi|^2\,\mathrm{d}V$ is a measure of the probability of finding an electron in a given volume element $\mathrm{d}V$.

To apply Eq. (2.8) to the hydrogen atom it is first transformed into polar coordinates (r,θ,ϕ) and then solved by the method of separation of the variables. This involves writing the solution in the form

$$\psi(r,\theta,\phi) = R(r)\Theta(\theta)\Phi(\phi) \tag{2.9}$$

in which $R(r)$, $\Theta(\theta)$ and $\Phi(\phi)$ are respectively functions of r, θ and ϕ only.

Solution of these equations leads naturally to the principal quantum number n and to two more quantum numbers, l and m_l. The total energy of the electron is determined by n, and its orbital angular momentum by the 'azimuthal' quantum number l. The value of the total angular momentum is $\{l(l+1)\}^{1/2}\hbar$. The angular momentum vector can be oriented in space in only certain allowed directions with respect to that of an applied magnetic field, such that the components along the field direction are multiples of $\hbar$; the multiplying factors are the m_l quantum numbers which can take values $-l, -l+1, -l+2, \ldots, +l$. This effect is known as 'space quantization'.

Experiments have demonstrated that the electron behaves rather like a spinning top and so has an intrinsic angular momentum, the value of which is $\{s(s+1)\}^{1/2}\hbar = (\sqrt{3}/2)\hbar$, in which $s(=\frac{1}{2})$ is the spin quantum number. Again, there is space quantization, and the components of angular momentum in a direction defined by an 'internal' or applied magnetic field are $\pm\frac{1}{2}\hbar$.

The set of quantum numbers n, l, m_l and s define the state of an electron in an atom. From an examination of spectra, Wolfgang Pauli (1900–1958) enunciated what has become known as the Pauli Exclusion Principle. This states that there cannot be more than one electron in a given state defined by a particular set of values for n, l, m_l and s. For a given principal quantum number n there are a total of $2n^2$ available electronic states.

The order in which the electrons occupy the various n and l states as atomic number increases through the Periodic Table is illustrated in Table 2.1. The prefix number specifies the principal quantum number, the letters s, p, d and f respectively† specify the orbitals for which $l = 0$, 1, 2 and 3, and the superscript specifies the number of electrons in the particular orbital. For brevity the electron configurations for the inert gases are denoted [Ar] for example.

An important question that arises is how the orbital and spin angular momenta of the individual electrons in a shell are coupled. One possibility is that the spin and orbital momenta for an individual electron couple into a resultant and that, in turn, the resultants for each electron in the shell couple. The other extreme possibility is that the spin momenta for individual electrons couple together to give a resultant spin quantum number S, as do the orbital momenta to give a resultant quantum number L; the resultants S and L then couple to give a final resultant quantum number J. In the brief discussion that follows the latter coupling is assumed to occur.

In most cases the 'ground' state (lowest energy) of the electron configuration of an atom is given by the Hund rules, according to which electrons occupy states fulfilling the following conditions.

1. The S value is the maximum allowed by the Pauli Exclusion Principle, i.e. the number of unpaired spins is a maximum.
2. The L value is the maximum allowed consistent with rule 1.
3. The value of J is $|L - S|$ when the shell is less than half-full and $L + S$ when it is more than half-full. When the shell is just half-full, the rules require $L = 0$, so that $J = S$.

The way in which the rules operate can be illustrated by applying them to, in turn, an isolated Fe atom and isolated Fe^{2+} and Fe^{3+} ions.

† The letters s, p, d and f are relics of early spectroscopic studies when certain series were designated 'sharp', 'principal', 'diffuse' or 'fundamental'.

Table 2.1 Electronic structures of the elements

H	**He**						
$1s^1$	$1s^2 = [He]$						
Li	**Be**	**B**	**C**	**N**	**O**	**F**	**Ne**
$[He]2s^1$	$[He]2s^2$	$[He]2s^2p^1$	$[He]2s^2p^2$	. . .	. . .	. . .	$[He]2s^2p^6 = [Ne]$
Na	**Mg**	**Al**	**Si**	**P**	**S**	**Cl**	**Ar**
$[Ne]3s^1$	$[Ne]3s^2$	$[Ne]3s^2p^1$	$[Ne]3s^2p^2$	. . .	. . .	. . .	$[Ne]3s^2p^6 = [Ar]$
K	**Ca**						
$[Ar]4s^1$	$[Ar]4s^2$ . . .	to Sc, Ti etc. and filling of 3d states					
Sc	**Ti**	**V**	**Cr**	**Mn**	**Fe**	**Co**	**Ni**
$[Ar]3d^14s^2$ etc.	$[Ar]3d^24s^2$	$[Ar]3d^34s^2$	$[Ar]3d^54s^1$	$[Ar]3d^54s^2$	$[Ar]3d^64s^2$	$[Ar]3d^74s^2$	$[Ar]3d^84s^2$ . . .

The electron configuration for an Fe atom is $[Ar]3d^6 4s^2$. Because completed shells ($[Ar]$ and $4s^2$) have a zero resultant angular momentum, only the partially filled 3d shell need be considered. A d shell can accommodate a total of 10 electrons and application of the rules to the six electrons can be illustrated.

Rule 1.	Spin:	$\frac{1}{2}$	$\frac{1}{2}$	$\frac{1}{2}$	$\frac{1}{2}$	$\frac{1}{2}$;	$S=2$
		$-\frac{1}{2}$					
Rule 2.	Orbit:	2	1	0	-1	-2;	$L=2$
		2					
Rule 3.							$J=4$

The total angular momentum of the electrons is given by

$$\{J(J+1)\}^{1/2}\hbar = \sqrt{(20)}\hbar$$

In the case of Fe^{2+} the result is the same since all that has changed is that two 4s electrons which contributed nothing have been lost.

For an isolated Fe^{3+} ion with configuration $[Ar]3d^5$ the situation is different.

Rule 1.	Spin:	$\frac{1}{2}$	$\frac{1}{2}$	$\frac{1}{2}$	$\frac{1}{2}$	$\frac{1}{2}$;	$S=5/2$
Rule 2.	Orbit:	2	1	0	-1	-2;	$L=0$
Rule 3.							$J=L+S=5/2$

Hence the total angular momentum of the electrons is $\{J(J+1)\}^{1/2}\hbar = \frac{1}{2}\sqrt{35}\hbar$.

It must be stressed that the above examples serve only to illustrate the way in which the Hund rules are applied. When ions are in crystal lattices the basic coupling between spin momenta and orbital momenta differs from what has been assumed above, and under certain conditions the rules are not obeyed. For instance six-coordinated Fe^{2+} and Co^{3+} ions contain six d electrons which most commonly have one pair with opposed spins and four with parallel spins but, in exceptional circumstances, have three pairs with opposed spins and two unfilled states. These differences have a marked effect on their magnetic properties (cf. Section 9.1) and also alter their ionic radii (Table 2.2). The possibility of similar behaviour exists for six-coordinated Cr^{2+}, Mn^{3+}, Mn^{2+} and Co^{2+} ions but occurs only rarely [1].

2.2 The Arrangement of Ions in Ceramics

When atoms combine to form solids their outer electrons enter new states whilst the inner shells remain in low-energy configurations round the positively charged nuclei.

The relative positions of the atoms are determined by the forces between them. In ionic materials, with which we are mainly concerned, the strongest influence is the ionic charge which leads to ions of similar electrical sign having their lowest energies when as far apart as possible and preferably with an ion of opposite sign between them. The second influence is the relative effective size of the ions which governs the way they pack together and largely determines the crystal structure. Finally there are quantum-mechanical 'exchange' forces, fundamentally electrostatic in origin, between the outer electrons of neighbouring ions which may have a significant influence on configuration. In covalently bonded crystals the outer electrons are shared between neighbouring atoms and the exchange forces are the main determinants of the crystal structure. There are many intermediate states between covalent and ionic bonding and combinations of both forms are common among ceramics, particularly in the silicates. The following discussion of crystal structure is mainly devoted to oxides, and it is assumed that the ionic effects are dominant.

Ionic size is determined from the distances between the centres of ions in different compounds and is found to be approximately constant for a given element in a wide range of compounds provided that account is taken of the charge on the ion and the number of oppositely charged nearest neighbours (the coordination number). Widely accepted values, mostly as assessed by R.D. Shannon and C.T. Prewitt [2], are given in Table 2.2.

It must be realized that the concept of ions in solids as rigid spheres is no more than a useful approximation to a complex quantum wave-mechanical reality. For instance, strong interactions between the outer electrons of neighbouring ions, i.e. covalent effects, reduce the ionic radius while the motion of ions in ionic conduction in solids often requires that they should pass through gaps in the structure that are too small for the passage of rigid spheres. Nevertheless, the concept allows a systematic approach to the relation of crystal structure to composition. For convenience radii are given the symbol r_j, where j is the coordination number.

The effect of atomic number on r_j can be seen by comparing Sr^{2+} ($Z=38$, $r_6 = 116\,pm$) with Ca^{2+} ($z=20$, $r_6 = 100\,pm$). However, for $Z > 56$, the 'lanthanide contraction' greatly reduces the effect of nuclear charge; for instance both Nb^{5+} ($Z=41$) and Ta^{5+} ($Z=73$) have $r_6 = 64\,pm$. The contraction is due to the filling of the 4f levels in the lanthanides which reduces their radii as their atomic numbers increase. The radius increases with coordination number; for example, for Ca^{2+}, $r_8 = 112\,pm$ and $r_{12} = 135\,pm$. The radius decreases when the positive charge increases; for example, Pb^{4+} has $r_6 = 78\,pm$ and Pb^{2+} has $r_6 = 118\,pm$. Values of the radii not included in Table 2.2 can be evaluated roughly by comparison with ions of approximately the same atomic number, charge and coordination. Some of the transition elements can have various electronic configurations in their d shells owing to variations in the numbers of unpaired and paired electrons. These result in changes in ionic radius with the larger number of unpaired electrons (high spin state indicated by the superscript h) giving larger ions.

Table 2.2 Ionic radii

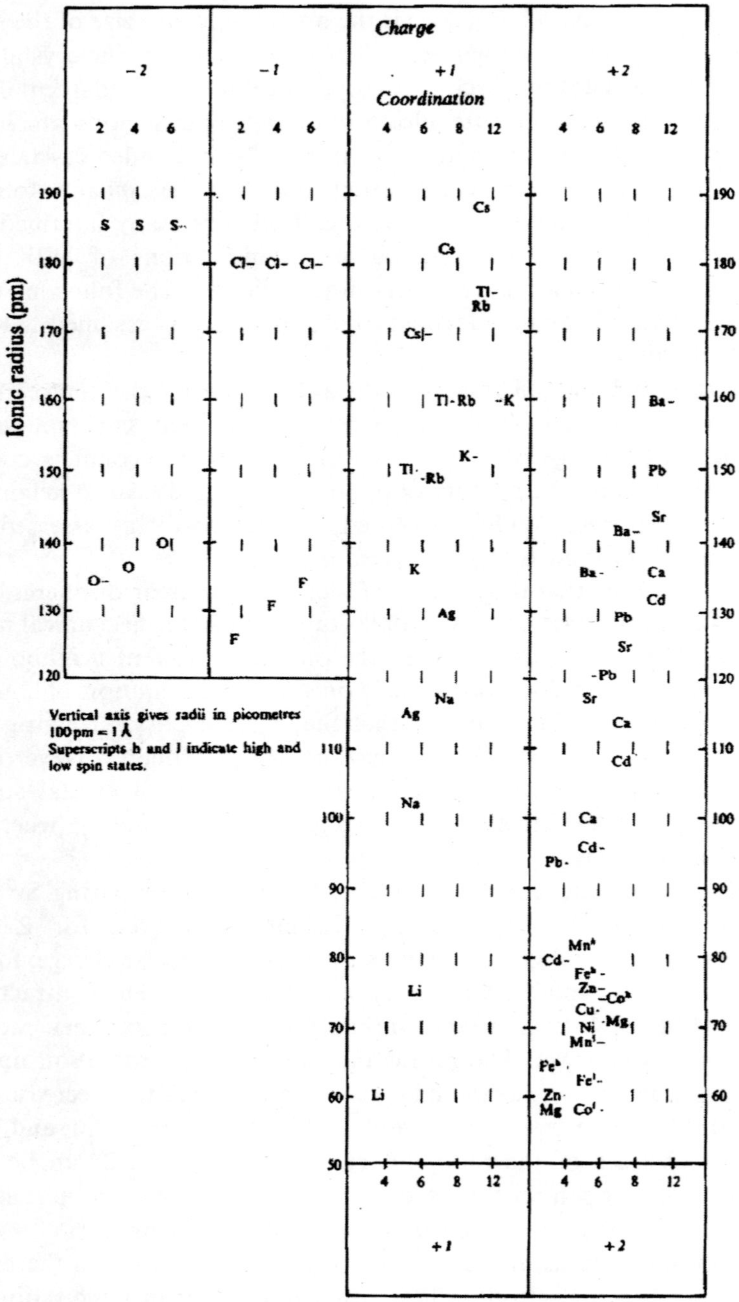

Vertical axis gives radii in picometres
100 pm = 1 Å
Superscripts h and l indicate high and low spin states.

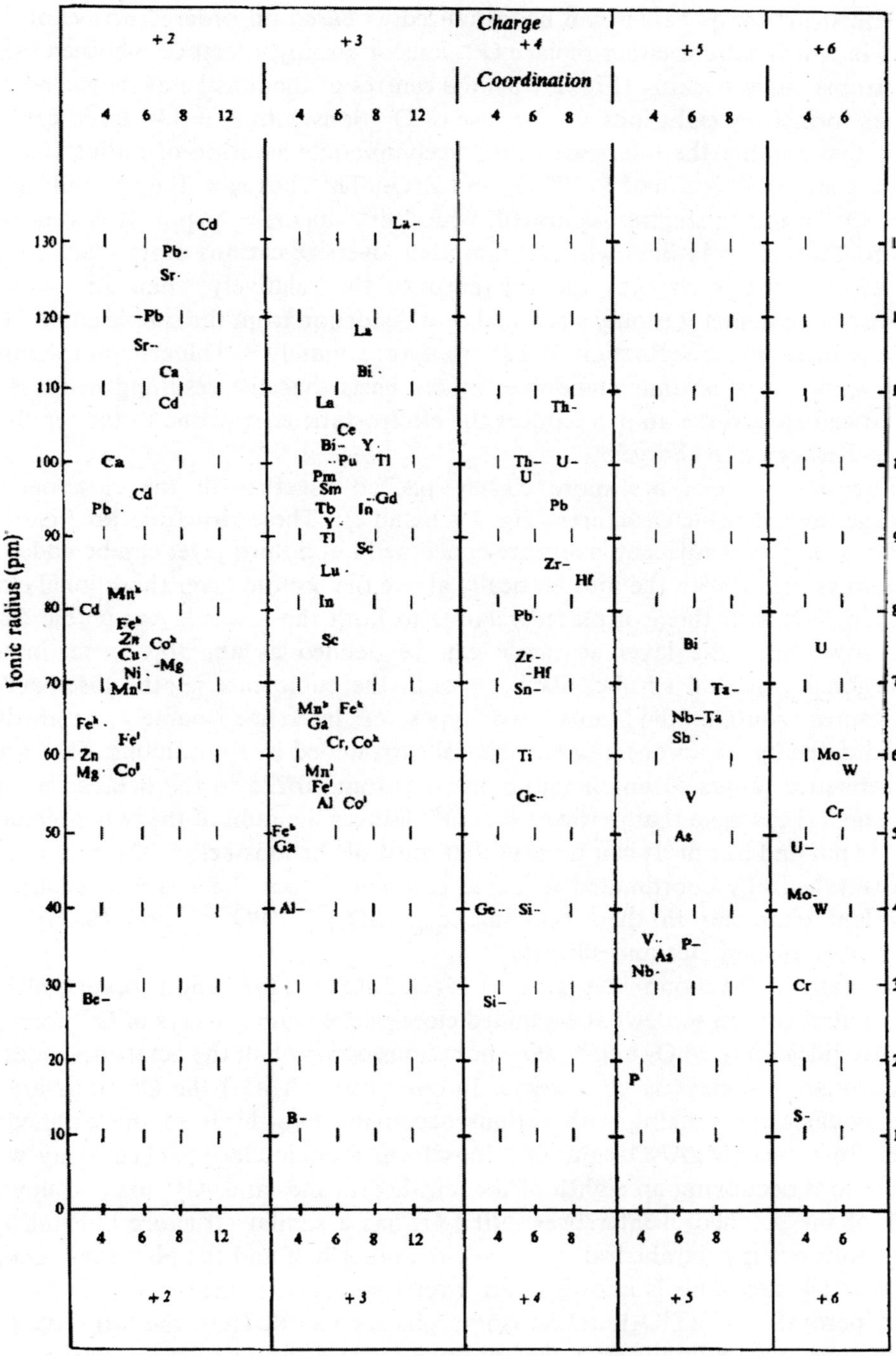
Charge
Coordination
Ionic radius (pm)
+2
+3
+4
+5
+6

The structure of oxides can be visualized as based on ordered arrays of O^{2-} ions in which cations either replace O^{2-} ions or occupy interstices between them. In simple cubic packing (Fig. 2.1(a)) the centres of the ions lie at the corners of cubes formed by eight ions. In the case of O^{2-} ions with $r_6 = 140$ pm in contact with one another the interstice would accommodate a cation of radius 103 pm. Such a structure is found for ThO_2 and ZrO_2. Th^{4+} has $r_8 = 106$ pm, indicating that O^{2-} must be slightly separated, whilst Zr^{4+} has $r_8 = 84$ pm. It is generally found that anion lattices will accommodate oversize cations more readily than undersize cations so that the tolerance to the relatively small Zr^{4+} ion is exceptional; in fact it is only sustained by a distortion from the simple cubic form that reduces the coordination of Zr^{4+} to approximately 7. The general tolerance to oversize ions is understandable on the basis that the resulting increase in distance between the anions reduces the electrostatic energy due to the repulsive force between like charges.

The oxygen ions are more closely packed together in the close-packed hexagonal and cubic structures (Fig. 2.1(b and c)). These structures are identical as far as any two adjacent layers are concerned but a third layer can be added in two ways, either with the ions vertically above the bottom layer (hexagonal close packing) or with them displaced relative to both the lower layers (cubic close packing). Thus the layer sequence can be defined as ab, ab,... etc. in the hexagonal case and as abc, abc,... etc. in the cubic case. Both close-packed structures contain the same two types of interstice, namely octahedral surrounded by six anions and tetrahedral surrounded by four anions. The ratios of interstice radius to anion radius are 0.414 and 0.225 in the octahedral and tetrahedral cases, so that in the case of O^{2-} lattices the radii of the two interstices are 58 pm and 32 pm. It can be seen that most of the ions below 32 pm in radius are tetrahedrally coordinated in oxide compounds but there is a considerable covalent character in their bonding, e.g. $(SO_4)^{2-}$, $(PO_4)^{3-}$ and $(SiO_4)^{4-}$ in sulphates, phosphates and silicates.

In many of the monoxides, such as MgO, NiO etc., the cations occupy all the octahedral sites in somewhat expanded close-packed cubic arrays of O^{2-} ions. In the dioxides TiO_2, SnO_2 and MnO_2 the cations occupy half the octahedral sites in hexagonal close-packed O^{2-} arrays. In corundum (Al_2O_3) the O^{2-} ions are in hexagonal close packing with cations occupying two-thirds of the octahedral sites. In spinel ($MgAl_2O_4$) the O^{2-} ions form a cubic close-packed array with Mg^{2-} ions occupying an eighth of the tetrahedral sites and Al^{3+} ions occupying half of the octahedral interstices. $NiFe_2O_4$ has a similar structure but half the Fe^{3+} ions occupy tetrahedral sites while the other half and the Ni^{2+} ions occupy octahedral sites. This is known as an inverse spinel structure.

In perovskite ($CaTiO_3$) and its isomorphs such as $BaTiO_3$, the large alkaline earth ions replace O^{2-} ions in the anion lattice and the Ti^{4+} ions occupy all the octahedral interstices that are surrounded only by O^{2-} ions, i.e. no Ti^{4+} ions are immediately adjacent to divalent cations.

(a)

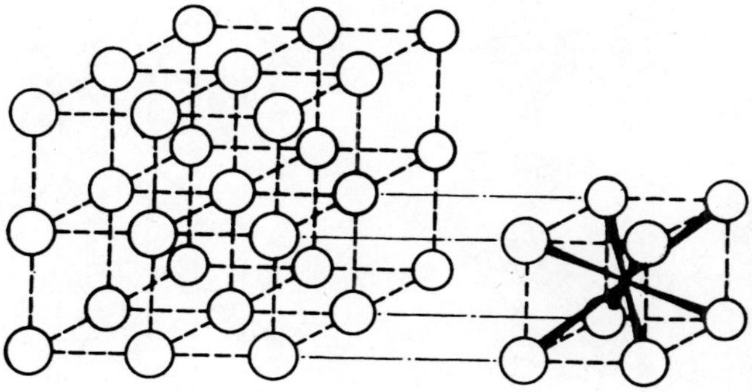

(b)

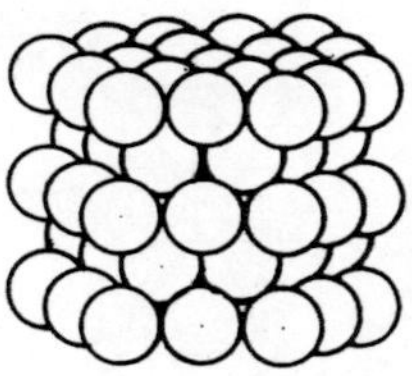

(c)

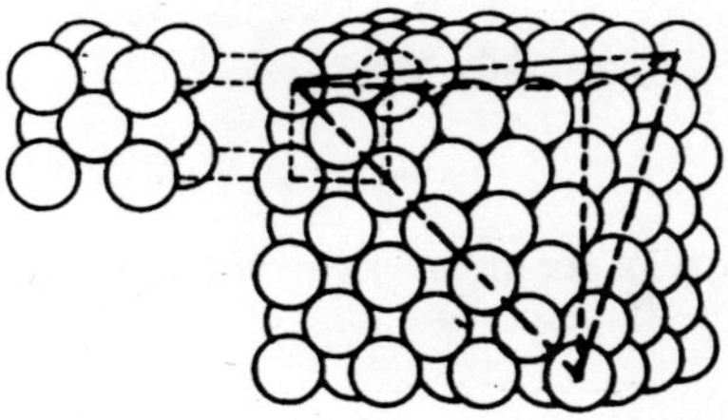

Fig. 2.1 Packing of ions: (a) simple cubic packing showing an interstice with eightfold coordination; (b) hexagonal close packing; (c) cubic close packing showing a face-centred cubic cell.

In many cases the arrangement of structural units gives a more enlightening view of crystals than do considerations based on close packing. Thus perovskite-type crystals can be viewed as consisting of a simple cubic array of corner-sharing octahedral MO_6 groups with all the interstices filled by divalent ions (Fig. 2.2(a)). On this basis the rutile form of TiO_2 consists of columns of edge-sharing TiO_6 octahedra linked by shared corners of the TiO_6 units (Fig. 2.2(b)). A hexagonal form of $BaTiO_3$, where the BaO_3 lattice is hexagonal close packed, contains layers of two face-sharing TiO_6 groups linked by single layers of corner-sharing TiO_6 groups (Fig. 2.2(c)).

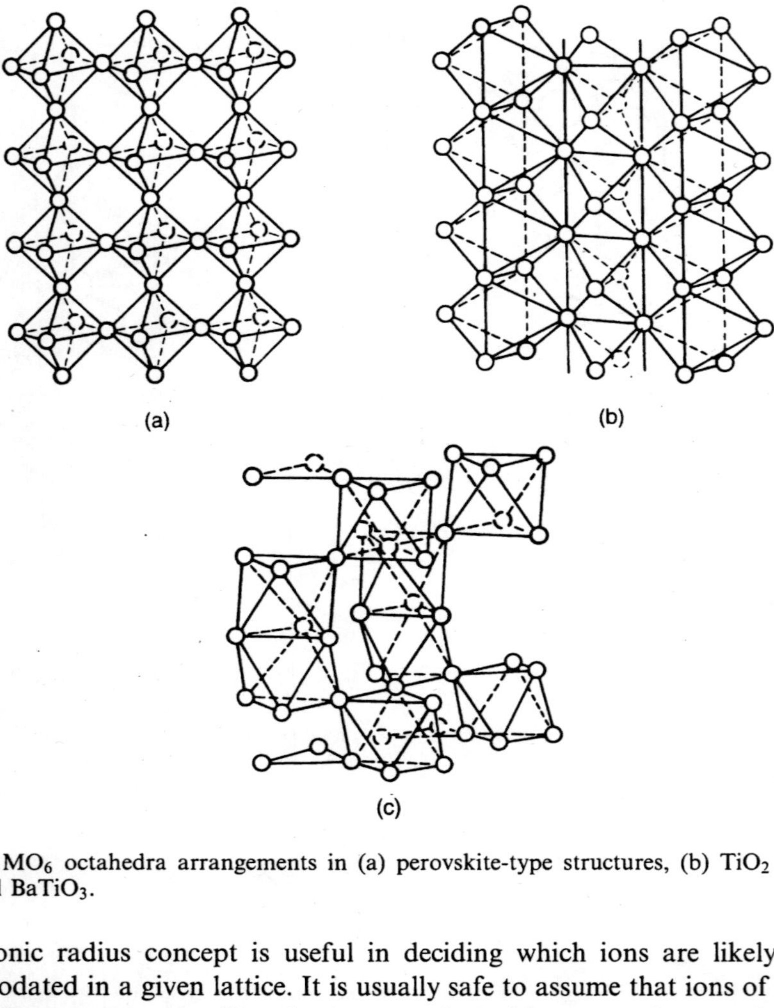

(a) (b) (c)

Fig. 2.2 MO_6 octahedra arrangements in (a) perovskite-type structures, (b) TiO_2 and (c) hexagonal $BaTiO_3$.

The ionic radius concept is useful in deciding which ions are likely to be accommodated in a given lattice. It is usually safe to assume that ions of similar size and the same charge will replace one another without any change other than in the size of the unit cell of the parent compound. Limitations arise because there is always some exchange interaction between the electrons of neighbouring ions.

In the case of the crystalline silicates an approach which takes account of the partly covalent character of the Si–O bond is helpful. The $[SiO_4]^{4-}$ tetrahedron is taken as a basic building unit, and in most of the silicates these tetrahedra are linked together in an ordered fashion to form strings as in diopside ($MgCa(SiO_3)_2$), sheet structures as in clay minerals, or three-dimensional frameworks as in quartz and the feldspars. Within these frameworks isomorphic replacement of one cation type for another is extensive. For example, the replacement of Si^{4+} by Al^{3+} is common, with the necessary lattice charge balance being maintained either by the incorporation of interstitial cations such as Na^+

and K^+, as in the case of the feldspars, or by 'exchangeable' cations such as Ca^{2+}, which are a feature of clays. The exchangeable ions are held on the surfaces of the small (typically 10^{-6} m) clay particles and can be easily exchanged for other ions.

Silicates readily form glasses which are vitreous materials in which the atoms do not have the long-range order characteristic of the crystalline state. Thus vitreous silica consists of a three-dimensional network of $(SiO_4)^{4-}$ tetrahedra joined at their corners, in which the Si–O–Si bond angles vary randomly throughout the structure. Alkali and alkaline earth ions can be introduced into silica in variable amounts, up to a certain limit, without a crystalline phase forming. One effect of these ions is to cause breaks in the Si–O–Si network according to the following reaction:

$$\rangle Si—O—Si\langle + Na_2O \rightarrow 2Na^+ + \rangle Si—O^- + O^-—Si\langle$$

Although there are important exceptions, a characteristic feature of the crystalline state is that compositions are stoichiometric, i.e. the various types of ion are present in numbers which bear simple ratios one to the other. In contrast, glass compositions are not thus restricted, the only requirement being overall electrical neutrality.

Vitreous materials do not have the planes of easy cleavage which are a feature of crystals, and they do not have well-defined melting points because of the variable bond strengths that result from lack of long-range order.

A wide variety of substances, including some metals, can be prepared in the vitreous state by cooling their liquid phases very rapidly to a low temperature. In many cases the glasses so formed are unstable and can be converted to the crystalline state by annealing at a moderate temperature. An important class of material, the 'glass-ceramics', can be prepared by annealing a silicate glass of suitable composition so that a large fraction of it becomes crystalline. Strong materials with good thermal shock resistance can be prepared by this method.

2.3 Spontaneous Polarization

In general, because the value of a crystal property depends on the direction of measurement, the crystal is described as anisotropic with respect to that property. There are exceptions; for example, crystals having cubic symmetry are optically isotropic although they are anisotropic with respect to elasticity. For these reasons, a description of the physical behaviour of a material has to be based on a knowledge of crystal structure. Full descriptions of crystal systems are available in many texts and here we shall note only those aspects of particular relevance to piezoelectric, pyroelectric and electro-optical ceramics. The monograph by R.E. Newnham [3] is recommended for further study.

For the present purpose it is only necessary to distinguish polar crystals, i.e. those that are spontaneously polarized and so possess a unique polar axis, from the non-polar variety. Of the 32 crystal classes, 11 are centrosymmetric and consequently, non-piezoelectric. Of the remaining 21 non-centrosymmetric classes, 20 are piezoelectric and of these 10 are polar. An idea of the distinction between polar and non-polar structures can be gained from Fig. 2.3 and Eqs (2.70) and (2.71).

The piezoelectric crystals are those that become polarized or undergo a change in polarization when they are stressed; conversely, when an electric field is applied they become strained. The 10 polar crystal types are pyroelectric as well as piezoelectric because of the polarization inherent in their structure. In a pyroelectric crystal a change in temperature produces a change in polarization.

A limited number of pyroelectric materials have the additional property that the direction of the polarization can be changed by an applied electric field or mechanical stress. Where the change is primarily due to an electric field the material is said to be ferroelectric; when it is primarily due to a stress it is said to be ferroelastic. These additional features of a pyroelectric material cannot be predicted from crystal structure and have to be established by experiment.

Because a ceramic is composed of a large number of randomly oriented crystallites it would normally be expected to be isotropic in its properties. The possibility of altering the direction of the polarization in the crystallites of a ferroelectric ceramic (a process called 'poling') makes it capable of piezoelectric, pyroelectric and electro-optic behaviour. The poling process – the application of a static electric field under appropriate conditions of temperature and time – aligns the polar axis as near to the field direction as the local environment and the crystal structure allow.

The changes in direction of the polarization require small ionic movements in specific crystallographic directions. It follows that the greater is the number of possible directions the more closely the polar axes of the crystallite in a ceramic can be brought to the direction of the poling field. The tetragonal (4mm) structure allows six directions, while the rhombohedral (3m) allows eight and so should permit greater alignment. If both tetragonal and rhombohedral

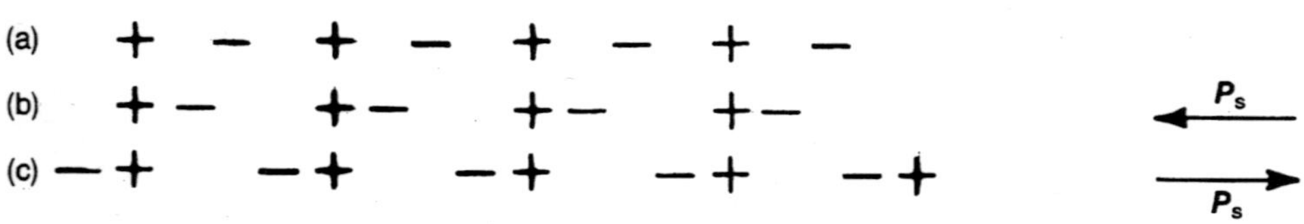

Fig. 2.3 (a) Non-polar array; (b), (c) polar arrays. The arrows indicate the direction of spontaneous polarization $\boldsymbol{P}_s$.

crystallites are present at a transition point, where they can be transformed from one to the other by a field, the number of alternative crystallographic directions rises to 14 and the extra alignment attained becomes of practical significance (cf. Section 6.3.1).

2.4 Phase Transitions

Effective ionic sizes and the forces that govern the arrangement of ions in a crystal are both temperature dependent and may change sufficiently for a particular structure to become unstable and to transform to a new one. The temperature at which both forms are in equilibrium is called a transition temperature. Although only small ionic movements are involved, there may be marked changes in properties. Crystal dimensions alter and result in internal stresses, particularly at the crystallite boundaries in a ceramic. These may be large enough to result in internal cracks and a reduction in strength. Electrical conductivity may change by several orders of magnitude. In some respects crystal structure transitions are similar to the more familiar phase transitions, melting, vaporization and sublimation when, with the temperature and pressure constant, there are changes in entropy and volume.

If a system is described in terms of the Gibbs function G then a change in G can be written

$$dG = -S\mathrm{d}T + V\mathrm{d}p \tag{2.10}$$

where S, V and P are respectively entropy, volume and pressure. During an isothermal structural change G is continuous, but there may be discontinuities in the derivatives of G. It follows from Eq. (2.10) that

$$S = -\left(\frac{\partial G}{\partial T}\right)_p \tag{2.11}$$

and so if there is a discontinuity in the first derivative of $G(T)$ there is a change in entropy at constant temperature, which implies latent heat. This is a characteristic of a 'first-order' transition. A 'second-order' transition occurs when the first derivative of the Gibbs function is continuous but the second derivative is discontinuous. The definitions are illustrated in Fig. 2.4.

An important transition which will be discussed later is that between the ferroelectric and paraelectric states which involves changes in crystal symmetry. In the case of magnetic materials the transition between the spontaneously magnetized and magnetically disordered states that occurs at the Curie or Néel temperatures does not involve changes in crystal structure but only small dimensional changes that result from changes in the coupling forces between the outer electrons of neighbouring magnetic ions.

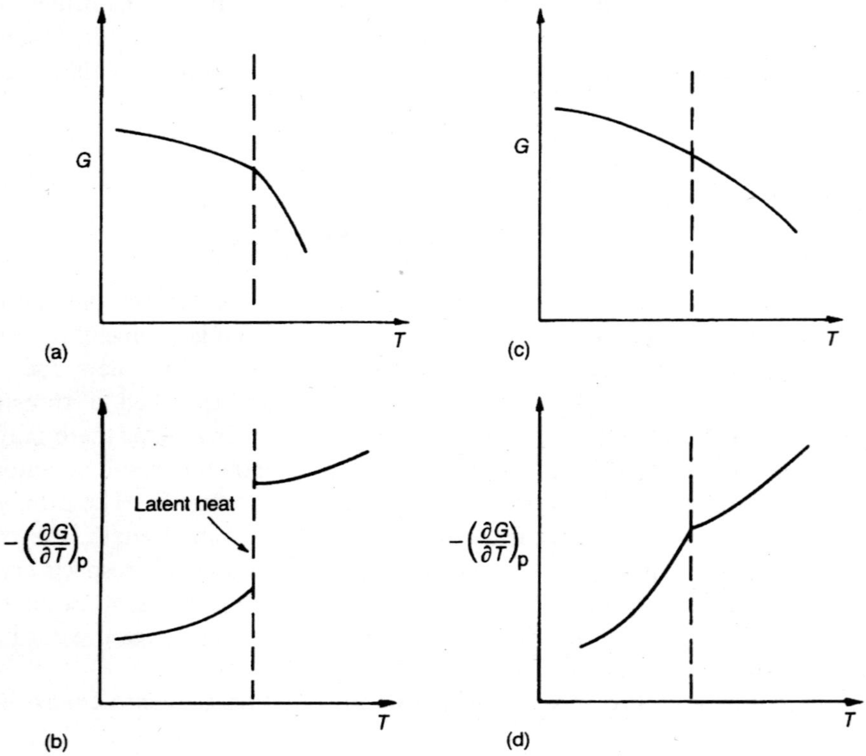

Fig. 2.4 Free-energy changes at transitions: (a) first-order transition; (b) change in S at constant T and, consequently, latent heat; (c) second-order transition; (d) continuous change in entropy and so no latent heat (discontinuity in $\partial^2 G/\partial T^2$).

2.5 Defects in Crystals

Early chemists believed that inorganic compounds obeyed the law of definite proportions under which they had invariable compositions determined by the valence of the constituent atoms. From the early part of the twentieth century views began to change when many compounds were found experimentally to be non-stoichiometric, and theoretical predictions by Wagner and Schottky demonstrated that exact stoichiometric compositions are the exception rather than the rule. The literature contains many treatments of the topic; the text by D.M. Smyth [4] is recommended.

2.5.1 Non-stoichiometry

The ratio between anions and cations can vary from a simple integral value because of the variable valence of a cation. Thus manganese dioxide is a well-established compound but it always contains less than the stoichiometric amount of oxygen. Iron monoxide, however, always contains an excess of oxygen. Such deviations can be accounted for by the presence of Mn^{3+} in the first case and Fe^{3+} in the second case. The positive charge deficiency in the first case can be balanced by vacant oxygen sites, and the charge excess in the second case can be balanced by cation vacancies.

Where there are two species of cation present in a compound the ratio between them may vary. This occurs in $LiNbO_3$ which has a structure based on face-sharing MO_6 octahedra. The O^{2-} ions are hexagonally close packed with a third of the octahedral sites occupied by Nb^{5+}, a third by Li^+ and a third empty. This compound can be deficient in lithium down to the level $Li_{0.94}Nb_{1.012}O_3$. There is no corresponding creation of oxygen vacancies; instead the Nb^{5+} content increases sufficiently to preserve neutrality. The withdrawal of five Li^+ ions is compensated by the introduction of one Nb^{5+} ion and so leaves four additional electrically neutral empty octahedral sites. $LiNbO_3$ can only tolerate a very small excess of lithium (see Section 6.4.5). The $LiNbO_3$ type of non-stoichiometry can be expressed as limited solid solubility; it could be said that $LiNbO_3$ and Nb_2O_5 form solid solutions containing up to 3.8 mol.% Nb_2O_5.

In most compounds containing two cation types some variability in composition is possible. This varies from less than 0.1% in compounds such as $BaTiO_3$, in which there is a marked difference in charge and size between the two cations corresponding to differences between their lattice sites, to complete solid solutions over the whole possible range where the ions are identical in charge and close in size and can only occupy one type of available lattice site such as, for instance, Zr^{4+} and Ti^{4+} in $Pb(Zr_xTi_{1-x})O_3$.

2.5.2 Point defects

Crystals contain two major categories of defect: 'point' defects and 'line' defects. Point defects occur where atoms are missing (vacancies) or occupy the interstices between normal sites (interstitials); 'foreign' atoms are also point defects. Line defects, or dislocations, are spatially extensive and involve disturbance of the periodicity of the lattice.

Although dislocations have a significant effect on some of the important properties of electroceramics, especially those depending on matter transport, our understanding of them is, at best, qualitative. In contrast, there is a sound basis for understanding the effects of point defects and the relevant literature is extensive. It is for these reasons that the following discussion is confined to point defects and, because of the context, to those occurring in oxides.

Schottky defects, named after W. Schottky, consist of unoccupied anion and cation sites. A stoichiometric crystalline oxide having Schottky disorder alone contains charge-equivalent numbers of anion and cation vacancies. A Frenkel defect, named after Y. Frenkel, is a misplaced ion, and so a crystal having only Frenkel disorder contains the same concentrations of interstitial ions and corresponding vacant sites. Frenkel defects depend on the existence in a crystal lattice of empty spaces that can accommodate displaced ions. Ti^{4+} ions occur interstitially in rutile (see Section 5.6.1) and F^- ions can occupy interstitial sites in the fluorite structure (see Fig. 4.29).

The equilibrium concentrations of point defects can be derived on the basis of statistical mechanics and the results are identical to those obtained by a less fundamental quasi-chemical approach in which the defects are treated as reacting chemical species obeying the law of mass action. The latter, and simpler, approach is the one widely followed.

The equilibrium concentrations of defects in a simple binary oxide MO are given by

$$n_S \approx N \exp\left(-\frac{\Delta H_S}{2kT}\right) \tag{2.12}$$

$$n_F \approx (NN')^{1/2} \exp\left(-\frac{\Delta H_F}{2kT}\right) \tag{2.13}$$

where n_S and n_F are the Schottky and Frenkel defect concentrations respectively and ΔH_S and ΔH_F are the enthalpy changes accompanying the formation of the associated defects (cation vacancy + anion vacancy and ion vacancy + interstitial ion); N is the concentration of anions or cations and N' is the concentration of available interstitial sites.

If $\Delta H_S \approx 2$ eV (0.32 aJ) then, at 1000 K, $n_S/N \approx 10^{-6}$, i.e. 1 ppm. High defect concentrations can be retained at room temperature if cooling is rapid and the rate at which the defect is eliminated is slow.

The notation of Kröger and Vink is convenient for describing a defect and the effective electrical charge it carries relative to the surrounding lattice. A defect that carries an effective single positive electronic charge bears a superscript dot ($^\bullet$), and a defect that carries an effective negative charge bears a superscript prime ($'$). Neutral defects have no superscript. These effective charges are to be distinguished from the real charges on an ion, e.g. Al^{3+}, O^{2-}. An atom or ion A occupying a site normally occupied by an atom or ion B is written A_B. An interstitial ion is denoted A_I.

The effective charge on a defect is always balanced by other effective or real charges so as to preserve electrical neutrality. The notation and ideas are conveniently illustrated by considering the ionization of an anion and cation vacancy in the metal oxide MO.

Consider first an oxygen vacancy. Its effective charge of $2e$ can be neutralized by a cation vacancy with an effective charge $-2e$; an example of such 'vacancy compensation' is an associated Schottky pair. Alternatively, an oxygen vacancy might be electron-compensated by being associated with two electrons. Similarly a divalent cation vacancy might be compensated by association with two positive 'holes'.

In the case of electron compensation, the neutral defect can be progressively ionized according to

$$\begin{aligned} V_O &\rightarrow V_O^{\cdot} + e' \\ V_O^{\cdot} &\rightarrow V_O^{\cdot\cdot} + e' \\ V_M &\rightarrow V_M' + h^{\cdot} \text{ etc.} \end{aligned} \tag{2.14}$$

Whereas the defect chemistry of pure stoichiometric compounds is largely of academic interest, the effects of the introduction of foreign ions are of crucial significance to electroceramics. The defect chemistry of barium titanate itself and, in particular, the effect of lanthanum doping are of such importance that they are discussed in detail in Section 2.6.2. It is for these reasons that the system is chosen here to illustrate basic ideas relating to the aliovalent substitution of one ion for another.

When a small amount (< 0.5 cation % (cat. %)) of La_2O_3 is added to $BaTiO_3$ and fired under normal oxidizing conditions, the La^{3+} ions substitute for Ba^{2+} and the defect $La_{Ba}^{\cdot}$ is compensated by an electron in the conduction band derived from the Ti 3d states (see Eq. (2.19)).

$La_{Ba}^{\cdot}$ and, in general, any substituent ion with a higher positive charge than the ion it replaces is termed a 'donor'. In some circumstances $La_{Ba}^{\cdot}$ can be compensated by V_{Ba}'' species, two dopant ions to every vacancy, or by one V_{Ti}'''' to every four $La_{Ba}^{\cdot}$.

An ion of lower charge than the one it replaces is called an 'acceptor', e.g. Ga^{3+} on a Ti^{4+} site. Ga_{Ti}' will have an effective negative charge, which can be compensated by a positively charged oxygen vacancy or an interstitial positively charged cation or a 'hole' in the valence band.

In summary, a chemical equation involving defects must balance in three respects:

4. the total charge must be zero;
5. there must be equal numbers of each chemical species on both sides;
6. the available lattice sites must be filled, if necessary by the introduction of vacant sites.

Vacancies do not have to balance since as chemical species they equate to zero, but account must be taken of their electrical charges. Thus the introduction of an acceptor Mn^{3+} on a Ti^{4+} site in $BaTiO_3$ can be expressed as

$$Mn_2O_3 + 2BaO \rightleftharpoons 2Ba_{Ba} + 2Mn_{Ti}' + 5O_O + V_O^{\cdot\cdot} \tag{2.15}$$

which replaces the equilibrium equation for the pure crystal:

$$2TiO_2 + 2BaO \rightleftharpoons 2Ba_{Ba} + 2Ti_{Ti} + 6O_O$$

Since $BaO = Ba_{Ba} + O_O$, Eq. (2.15) simplifies to

$$Mn_2O_3 \rightleftharpoons 2Mn'_{Ti} + 3O_O + V_O^{\cdot\cdot} \tag{2.16}$$

where it is understood that only Ti and the corresponding O sites are under consideration.

The equilibrium constant for Eq. (2.16) is

$$K_A = \frac{[Mn'_{Ti}]^2[V_O^{\cdot\cdot}]}{[Mn_2O_3]} \tag{2.17}$$

since the activity of O_O, and any other constituent of the major phase, can be taken as unity.

K_A is expressed as a function of temperature by

$$K_A = K'_A \exp\left(-\frac{\Delta H_A}{kT}\right) \tag{2.18}$$

where ΔH_A is the change in enthalpy for the reaction and K'_A is a temperature-insensitive constant.

The replacement of Ba^{2+} in $BaTiO_3$ by the donor La^{3+} is represented by

$$La_2O_3 \rightleftharpoons 2La_{Ba}^{\cdot} + 2O_O + \tfrac{1}{2}O_2(g) + 2e' \tag{2.19}$$

and the equilibrium constant K_D is

$$K_D = K'_D \exp - \left(\frac{\Delta H_D}{kT}\right) = \frac{[La_{Ba}^{\cdot}]^2 n^2 p_{O_2}{}^{1/2}}{[La_2O_3]} \tag{2.20}$$

where n is the electron concentration.

2.6 Electrical Conduction

The electrical conduction characteristics of ceramics can range from those of superconductors through those of metals to those of the most resistive of materials; in between the extremes are characteristics of semi-conductors and semi-insulators. It is the purpose of this section to provide a framework for an understanding of this very diverse behaviour of apparently basically similar materials. The monographs by C. Kittel [5] and B.I. Bleaney and B. Bleaney [6] are recommended to supplement the discussion.

2.6.1 Charge transport parameters

If a material containing a density, n, of mobile charge carriers, each carrying a charge Q, is situated in an electric field $\boldsymbol{E}$, the charge carriers experience a force

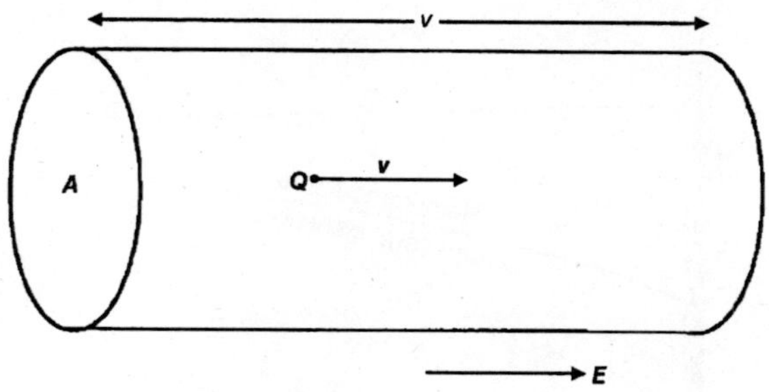

Fig. 2.5 Flow of charge in a prism.

causing them to accelerate but, because of the interaction with the lattice owing to thermal motion of the atoms or to defects, they quickly reach a terminal velocity, referred to as their drift velocity $\boldsymbol{v}$. All the carriers contained in a prism of cross section A, and length v (Fig. 2.5) will move through its end face in unit time. The current density $\boldsymbol{j}$ will therefore be given by

$$\boldsymbol{j} = nQ\boldsymbol{v} \tag{2.21}$$

If the drift velocity of the charges is proportional to the force acting on them, then

$$\boldsymbol{v} = u\boldsymbol{E} \tag{2.22}$$

where u is the mobility, which is defined as the magnitude of the drift velocity per unit electric field $\boldsymbol{E}$. It follows from Eqs (2.21) and (2.22) that

$$\boldsymbol{j} = nQu\boldsymbol{E} \tag{2.23}$$

For materials for which nQu is constant at constant temperature, this is a statement of Ohm's law:

$$\boldsymbol{j} = \sigma\boldsymbol{E} \tag{2.24}$$

where

$$\sigma = nQu \tag{2.25}$$

is the *conductivity* of the material.

The resistivity ρ, like the conductivity, is a material property and the two are related by

$$\rho = 1/\sigma \tag{2.26}$$

In practice it is often the conductive or resistive characteristics of a specimen of uniform section A and length l which are relevant. The resistance R, conductance G and specimen dimensions are related as follows:

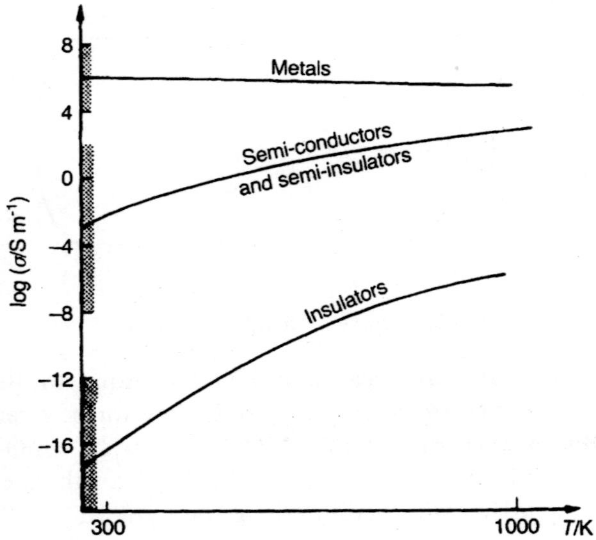

Fig. 2.6 Conductivities of the various classes of material: shading indicates the range of values at room temperature.

Table 2.3 Conductivity characteristics of the various classes of material

Material class	*Example*	*Conductivity level*	*dσ/dT*	*Carrier type*
Metals	Ag, Cu	High	Small, negative	Electrons
Semiconductors	Si, Ge	Intermediate	Large, positive	Electrons
Semi-insulators	ZrO_2	Intermediate	Large, positive	Ions or electrons
Insulators	Al_2O_3	Very low	Very large, positive	Ions or electrons; frequently 'mixed'

Table 2.4 Electrical quantities introduced so far

Quantity	*Symbol*	*Unit*
Electric charge	Q	coulomb (C)
Electric field	E	volt per metre ($\mathrm{V\,m^{-1}}$)
Current density	j	ampere per square metre ($\mathrm{A\,m^{-2}}$)
Mobility	u	drift velocity/electric field ($\mathrm{m^2\,V^{-1}\,s^{-1}}$)
Conductivity	σ	siemen per metre ($\mathrm{S\,m^{-1}}$)
Resistivity	ρ	reciprocal conductivity (Ω m)
Conductance	G	siemen (S)
Resistance	R	ohm (Ω)

$$R = G^{-1} = \rho l/A$$
$$G = R^{-1} = \sigma A/l \tag{2.27}$$

It should be emphasised that from Eq. (2.24) σ is in general a tensor of the second rank. Unless otherwise stated it will be assumed in the discussions that follow that materials are isotropic, so that j and E are collinear and σ is a scalar.

It follows from Eq. (2.25) that, to understand the behaviour of σ for a given material, it is necessary to enquire into what determines n, Q and u separately; in particular, the variation of σ with temperature T is determined by the manner in which these quantities depend on T.

No reference has been made to the type of charge carrier and the equations developed so far in no way depend upon this. However, the electrical behaviour of solids depends very much on whether the charge carriers are electrons, ions or a combination of both.

At this point it will be helpful to summarize the charge transport characteristics of the various types of material so that those of the ceramics can be seen in proper perspective. Figure 2.6 shows the room temperature values of conductivity characteristic of the broad categories of material together with typical dependencies of conductivity on temperature. What is immediately striking is the large difference between the room temperature values of conductivity for the metallic and insulating classes of material, which span about 30 orders of magnitude.

Table 2.3, which should be considered in conjunction with Fig. 2.6, completes the general picture. Table 2.4 summarizes the quantities introduced so far together with the units in which they are measured.

In the following sections closer attention is given to the two principal mechanisms whereby charge is transported in a solid, i.e. 'electronic' and 'ionic' conduction.

2.6.2 Electronic conduction

Band conduction

So far we have considered atoms in isolation and in their ground state. There exist a large number of higher energy states into which electrons can be promoted thermally or by interaction with external energy sources such as photons.

When a large number of atoms condense to form a crystal, quantum mechanics dictates that the discrete sharply defined electron energy levels associated with a free atom broaden into bands of discrete levels situated close together in the energy spectrum. The multiplication of possible energy states as atoms approach one another is shown diagrammatically in Fig. 2.7. In general each band is separated by a 'forbidden zone' – a region in the energy spectrum forbidden to electrons – although there are instances where the bands overlap. The problem of determining the electronic properties of a solid becomes a matter of finding the distribution of electrons among the available energy states making up the bands, and this is achieved with the help of Fermi statistics. It turns out that, at the absolute zero of temperature, some bands are completely filled with electrons and some are completely empty. It is also possible for a band to be only partly filled. Fortunately it is only bands associated with the outer electronic shells, or valence electrons, which need be considered since the deeper core electrons play no part in the conduction process. The possible situations outlined above are shown schematically in Fig. 2.8.

If an energy band is only partly filled the electron population is able to move into the higher energy states available to it. Therefore in an electric field the electrons are able to acquire additional kinetic energy and, as in the case of a metal, a current will flow. However, if the band is full, then at zero temperature the electrons cannot normally acquire energy from an electric field and so no current flows. However, if the energy gap (e.g. Figs 2.7 and 2.8(b)) is not too wide (say about 1 eV or 0.16 aJ), then at around room temperature (where $kT \approx 0.025$ eV) some electrons can be thermally excited across the gap into an empty band where they can conduct. In addition to electrons excited into the so-called 'conduction band', there are also empty electronic states in the previously full 'valence band', allowing the valence electron population as a whole to accept energy from the field. The description of this redistribution of energy in the valence band is simply accomplished with the aid of the concept of the 'positive hole'; attention is then focused on the behaviour of the empty state – the positive hole – rather than on that of the electron population as a whole. (This is analogous to focusing attention on a rising air bubble in a liquid in preference to the downwards motion of the water as a whole.)

Fig. 2.7 (a) Atomic levels having identical energies merging to a broad band of levels differing slightly in energy as free atoms condense to form a crystal; (b) band structure at equilibrium interatomic spacing in a crystal.

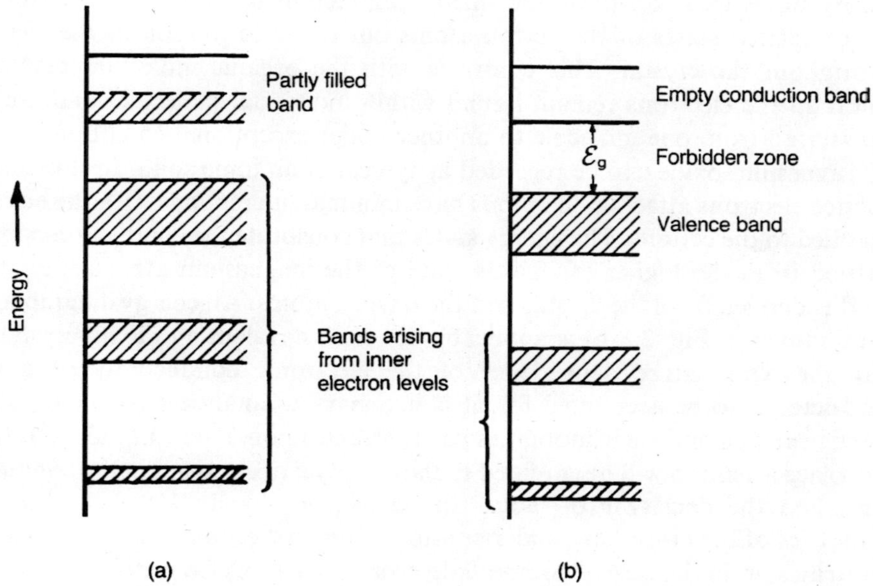

Fig. 2.8 Schematic electron energy band structures for (a) a metallic crystal and (b) a semiconducting or insulating crystal.

In metals the conduction band is partly filled with electrons derived from the outer quantum levels of the atoms. The atoms become positive ions with the remaining electrons in low energy states, e.g. as complete eight-electron shells in the case of the alkaline and alkaline earth elements. Since the valence electrons are in a partly filled band they can acquire kinetic energy from an electric field which can therefore cause a current to flow. Conduction is not limited by paucity of carriers but only by the interaction of the conducting electrons with the crystal lattice. The thermal motion of the lattice increases with temperature and so intensifies electron–phonon interactions and consequently reduces the conductivity. At temperatures above 100–200 °C the resistivity of most metals is approximately proportional to the absolute temperature. Silver is the most conductive metal with $\sigma = 68 \times 10^6\,\mathrm{S\,m^{-1}}$ at 0 °C; manganese, one of the least conductive, has $\sigma = 72 \times 10^4\,\mathrm{S\,m^{-1}}$. Magnetite ($Fe_3O_4$) is one of the most conductive oxides with $\sigma \approx 10^4\,\mathrm{S\,m^{-1}}$, but it does not have the positive temperature coefficient of resistivity typical of metals.

In covalently bonded non-polar semiconductors the higher levels of the valence band are formed by electrons that are shared between neighbouring atoms and which have ground state energy levels similar to those in isolated atoms. In silicon, for instance, each silicon atom has four sp^3 electrons which it shares with four similar atoms at the corners of a surrounding tetrahedron. As a result each silicon atom has, effectively, an outer shell of eight electrons. The energy states that constitute the silicon conduction band are derived from the higher excited states of the silicon atoms but relate to motion of the electrons throughout the crystal. This contrasts with the organic molecular crystals in which all the electrons remain bound within individual molecules and are only transferred from one molecule to another under exceptional conditions.

Magnesium oxide can be regarded as typical of an ionic solid. In this case the valence electrons attached to anions have minimal interactions with the electrons attached to the cations. The energy states that constitute the conduction band are derived from the higher excited 3s state of the magnesium atoms; the valence band is derived from the 2p states of the oxygen atoms. An energy diagram of the form shown in Fig. 2.8(b) is applicable for both covalent and ionic crystals.

If the temperature dependence of the electronic conductivity of a semiconductor is to be accounted for, it is necessary to analyse how the density of charge carriers and their mobilities each depend upon T (see Eq. (2.25)). In the first place attention will be confined to the density n of electrons in the conduction band and the density p of 'holes' in the valence band. When the 'intrinsic' properties of the crystal are under consideration, rather than effects arising from impurities or, in the case of compounds, from departures from stoichiometry, the corresponding conductivity is referred to as 'intrinsic conductivity'. The approach to the calculation of n and p in this instance is as follows.

Figure 2.9 illustrates the situation in which a small fraction of the valence electrons in an intrinsic semiconductor have been thermally excited into the conduction band, with the system in thermal equilibrium. Since the only source of electrons is the valence band $p_i = n_i$, where the subscript indicates intrinsic.

Formally, the electron density in the conduction band can be written

$$n_i = \int_{\mathcal{E}_c}^{\mathcal{E}_{top}} Z(\mathcal{E})F(\mathcal{E})\,d\mathcal{E} \tag{2.28}$$

in which $Z(\mathcal{E})d\mathcal{E}$ represents the total number of states in the energy range $d\mathcal{E}$ around $\mathcal{E}$ per unit volume of the solid, and the Fermi–Dirac function $F(\mathcal{E})$ represents the fraction of states occupied by an electron. $F(\mathcal{E})$ has the form

$$F(\mathcal{E}) = \left\{\exp\left(\frac{\mathcal{E}-\mathcal{E}_F}{kT}\right)+1\right\}^{-1} \tag{2.29}$$

where $\mathcal{E}_F$ is the Fermi energy, which is a characteristic of the particular system under consideration.

The evaluation of n_i is readily accomplished under certain simplifying assumptions. The first is that $\mathcal{E}-\mathcal{E}_F \gg kT$, which is often the case since kT is approximately 0.025 eV at room temperature and $\mathcal{E}-\mathcal{E}_F$ is commonly greater than 0.2 eV. If this condition is met the term +1 can be omitted from Eq. (2.29); if it is not met then the electron distribution is said to be *degenerate* and the full Fermi function must be used. The second assumption is that the excited electrons

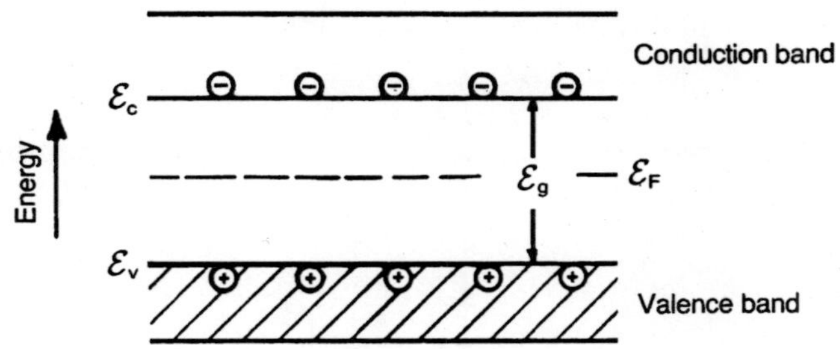

Fig. 2.9 Band structure with electrons promoted from the valence to the conduction band.

and holes occupy states near the bottom of the conduction band and the top of the valence band respectively. Under these circumstances the electrons and holes behave as free particles for which the state distribution function is known. Thirdly, the upper limit of the integration in Eq. (2.28) is taken as infinity since the probability of occupancy of a state by an electron rapidly approaches zero as the energy increases through the band. Under these assumptions it is readily shown that

$$n_i = N_c \exp\left(-\frac{\mathcal{E}_c - \mathcal{E}_F}{kT}\right) \tag{2.30}$$

$$p_i = N_v \exp\left(-\frac{\mathcal{E}_F - \mathcal{E}_v}{kT}\right) \tag{2.31}$$

The form of Eqs (2.30) and (2.31) suggests that N_c and N_v are effective state densities for electrons in the conduction band and holes in the valence band respectively. It turns out that $N_c \approx N_v \approx 10^{25}\,\mathrm{m}^{-3}$. If we put $n_i = p_i$,

$$\mathcal{E}_F \approx \frac{\mathcal{E}_c + \mathcal{E}_v}{2} \tag{2.32}$$

A more rigorous treatment shows that

$$\mathcal{E}_F = \frac{\mathcal{E}_c + \mathcal{E}_v}{2} + \frac{3kT}{4}\ln\left(\frac{m_e^*}{m_h^*}\right) \tag{2.33}$$

in which m_e^* and m_h^* are respectively the *effective* electron and hole masses. The electrons and holes move in the periodic potential of the crystal as though they are charged particles with masses which can differ markedly from the free electron mass m_e. Under conditions in which $m_e^* \approx m_h^*$, $\mathcal{E}_F$ is approximately at the centre of the band gap. Therefore it follows that

$$n_i = p_i \approx 10^{25}\exp\left(-\frac{\mathcal{E}_c - \mathcal{E}_F}{kT}\right) \approx 10^{25}\exp\left(-\frac{\mathcal{E}_g}{2kT}\right) \tag{2.34}$$

From Eq. (2.25) the conductivity can be written

$$\sigma = nu_e e + pu_h e \tag{2.35}$$

in which u_e and u_h are the electron and hole mobilities respectively and e is the electronic charge. Therefore

$$\begin{aligned}\sigma &= n_i e(u_e + u_h) \\ &\approx 10^{25} e(u_e + u_h)\exp\left(-\frac{\mathcal{E}_g}{2kT}\right)\end{aligned} \tag{2.36}$$

Both theory and experiment show a temperature dependence for u lying typically in the range $T^{-1.5}$–$T^{-2.5}$, which is so weak compared with that for n (and p) that for most purposes it can be ignored. Therefore the conductivity is written

$$\sigma = B\exp\left(-\frac{\mathcal{E}_g}{2kT}\right) \tag{2.37}$$

It is usual to show $\sigma(T)$ data as a plot of log σ versus T^{-1} – the Arrhenius plot – where the slope of the straight line is proportional to the band gap.

The theory outlined above was developed for group IV semiconducting elements such as silicon and germanium; some of the compounds of group III and V elements, the III–V compounds, are also covalently bonded and have similar electrical properties which can be described in terms of a band model. The best known semiconducting III–V compound is GaAs, which is exploited for both its photonic and semiconducting properties.

The same model can be applied to an ionic solid. In this case, for the example of MgO, Fig. 2.9 represents the transfer of electrons from anions to cations resulting in an electron in the conduction band derived from the Mg^{2+} 3s states and a hole in the valence band derived from the 2p states of the O^{2-} ion. Because the width of the energy gap is estimated to be approximately 8 eV, the concentration of thermally excited electrons in the conduction band of MgO is low at temperatures up to its melting point at 2800 °C. It is therefore an excellent high-temperature insulator.

Apart from the wider band gaps, electrons and holes in ionic solids have mobilities several orders lower than those in the covalent semiconductors. This is due to the variation in potential that a carrier experiences in an ionic lattice.

The effect of dopants

The addition of small quantities of impurity atoms to a semiconductor has a dramatic effect on conductivity. It is, of course, such extrinsic effects that are the basis on which silicon semiconductor technology has developed. Their origins can be understood by considering a silicon crystal in which a small fraction of the

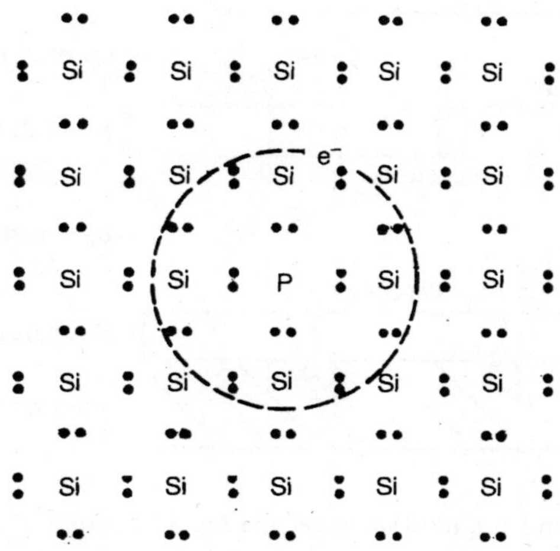

Fig. 2.10 Planar representation of a silicon crystal doped with P^{5+} giving rise to a P_{Si} defect.

Si atoms is replaced by, say, P atoms. A P atom has five valence electrons of which only four are required to form the four electron-pair bonds with neighbouring Si atoms. The 'extra' electron is not as strongly bound as the others and an estimate of its binding energy can be arrived at as follows. The electron can be regarded as bound to an effective single positive charge ($+e$), which is the P^{5+} ion on a Si^{4+} lattice site, as shown in Fig. 2.10. The configuration therefore resembles a hydrogen atom for which the ground state ($n = 1$) energy is

$$\mathcal{E} = -\frac{m_e e^4}{32\pi^2 \varepsilon_0^2 \hbar^2} \tag{2.38}$$

and has a value of about 13.5 eV (see Eq. (2.6)).

In the case when the electron is bound to the P^{5+} ion, Eq. (2.38) needs modifying to take account of the relative permittivity of the material separating the two charges. For a crude approximation the relative permittivity of bulk silicon (about 12) is used, leading to an ionization energy of approximately 0.09 eV. Another modification is necessary to allow for the effective mass of the electron being approximately $0.2\,m_e$, further reducing the estimate of the ionization energy $\mathcal{E}_i$ to about 0.01 eV, a value consistent with experiment. The doping of silicon with phosphorus therefore leads to the introduction of localized donor states about 0.01 eV below the conduction band, as shown in Fig. 2.11, and to n-type semiconductivity.

The addition of a trivalent atom (e.g. boron) to silicon leads to an empty electron state, or positive hole, which can be ionized from the effective single negative charge $-e$ on the B atom. The ionization energy is again about 0.01 eV, as might be expected. Therefore the doping of silicon with boron leads to the

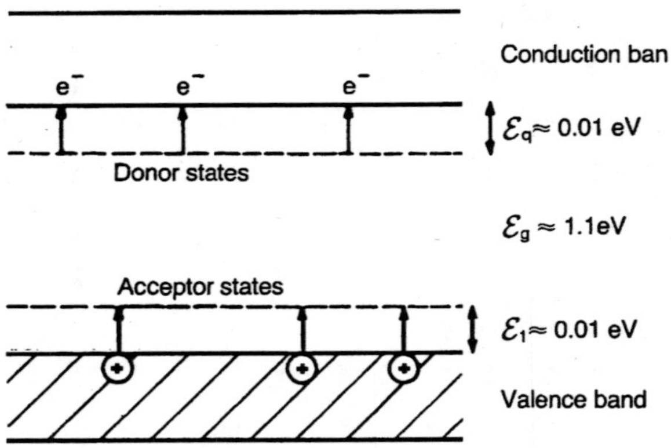

Fig. 2.11 Effect of n- and p-type doping on the band structure of a semiconductor (e.g. silicon).

introduction of acceptor states about 0.01 eV above the top of the valence band, as shown in Fig. 2.11, and to p-type semiconductivity. In the case of n- or p-type semiconductivity the temperature dependence of the conductivity is similar to that in Eq. (2.37) with $\mathcal{E}_g$ replaced by $\mathcal{E}_i$.

In practice, doping concentrations in silicon technology range from 1 in 10^9 to 1 in 10^3; B and Al are the usual acceptor atoms and P, As and Sb the usual donor atoms, the choice depending on the particular function that the doped silicon has to perform.

Because of doping $n \neq p$, but the equilibrium relation

$$e' + h^{\cdot} \rightleftharpoons \text{nil} \tag{2.39}$$

still holds, where 'nil' indicates a perfect crystal with all electrons in their lowest energy states. From Eq. (2.39) it follows that

$$[e'][h^{\cdot}] = np = K(T) = K' \exp\left(-\frac{\mathcal{E}_g}{kT}\right) \tag{2.40}$$

where $\mathcal{E}_g$ is the band gap (at 0 K) and K' is independent of temperature.

Semiconductivity in oxides

The discussion draws on the extensive studies by Philips' researchers [7] and [8] and by D.M. Smyth and co-workers [4]. Several cases of oxide systems in which the conductivity is controlled by the substitution of aliovalent cations are given in Chapter 4. For instance, Sb^{5+} can replace Sn^{4+} in SnO_2 and be compensated by an electron in the conduction band conferring n-type conductivity (see Section 4.1.4). However, models for oxide systems are generally more complex than for silicon and have been studied far less intensively. An important limitation to present research is the non-availability of oxides that approach the parts in 10^9 purity of available silicon crystals. The term 'high purity' applied to oxides usually implies less than 1 in 10^4 atoms of impurities which consist mainly of the more generally abundant elements magnesium, aluminium, silicon, phosphorus, calcium and iron, and a much smaller content of the less abundant elements such as niobium, tantalum, cerium, lanthanum etc. The bulk of the predominant impurities in $BaTiO_3$ are cations such as Mg^{2+}, Al^{3+}, Fe^{3+} and Ca^{2+} that form acceptors when substituted on Ti^{4+} sites. The resulting deficit in charge is compensated by oxygen vacancies which may therefore be present in concentrations of order 1 in 10^4, a far greater concentration than would be expected from Schottky defects in intrinsic material (cf. Section 2.5.2).

One consequence of the high impurity levels is the use of high dopant concentrations to control the behaviour of oxides. The dopant level is seldom below 1 in 10^3 moles and may be as high as 1 in 10 moles so that defects may interact with one another to a far greater extent than in the covalent semiconductors silicon, GaAs etc.

The study of semiconduction in oxides has necessarily been carried out at high temperatures ($>500\,°C$) because of the difficulties of making measurements when they have become highly resistive at room temperature. However, the form and magnitude of conductivity at room temperature will depend on the difference in energy between the sources of the electronic current carriers from the conduction and valence bands. Thus while n- and p-type conduction can be observed in $BaTiO_3$ at high temperatures, p-type $BaTiO_3$ is a good insulator at room temperature whereas n-type is often conductive. The cause lies in the structures of the orbital electrons in Ti^{4+} and O^{2-} which correspond to those of the inert elements argon and neon. The transfer of an electron from the stable valence bands of the ions to a defect requires energy of over 1 eV, which is available only at high temperatures. Recombination occurs at room temperature and only a very low level of p-type conductivity remains. However, the Ti^{4+} ion possesses empty 3d orbitals from which a conduction band is derived which allows occupancy by electrons transferred from defects at low energy levels so that appreciable n-type conductivity can persist at room temperature.

There are also oxides in which p-type conduction persists at lower temperatures than n-type does. For instance, Cr^{4+} in $LaCrO_3$ has two electrons in its d levels and one of these can be promoted with a relatively small expenditure of energy to give p-type conduction. The addition of an electron to the d levels requires greater energy so that n-type material is less conductive than p-type at room temperature.

***$BaTiO_3$*: the effects of oxygen pressure** One of the most important features of oxide semiconductors is the effect on their behaviour of the external oxygen pressure. This has only been examined at high temperatures because the

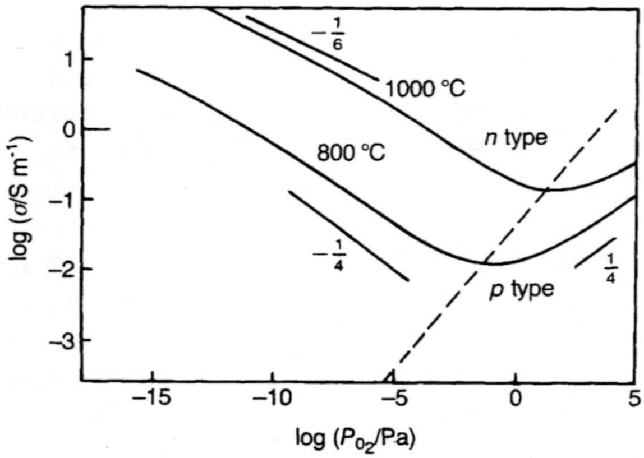

Fig. 2.12 Conductivity of undoped $BaTiO_3$ as a function of p_{O_2} and T (adapted from Smyth [4]).

establishment of equilibrium at low temperatures is very slow. It will be discussed here for the case of $BaTiO_3$ with acceptor and donor substituents since this is one of the most important systems in terms of applications and so has been well researched (see Section 5.6.1 for discussion of TiO_2).

Figure 2.12 shows the electrical conductivity of $BaTiO_3$ containing only dopants, predominantly acceptors present as natural impurities, as a function of oxygen pressure p_{O_2} at high temperatures. The conductivity is n type at low p_{O_2} and p type at high p_{O_2}. The general shape of the curves in Fig. 2.12 can be explained under the assumption that the observed conductivity is determined by the electron and hole concentrations, and that the electron and hole mobilities are independent of p_{O_2}. Under these assumptions information concerning the relative concentrations of electrons and holes under given conditions and an estimate of $K(T)$ can be arrived at as follows.

Combining Eqs (2.35) and (2.40) leads to

$$\frac{\sigma}{e} = u_e n + \frac{u_h K(T)}{n} \tag{2.41}$$

It follows from differentiating Eq. (2.41) that the value n_m of n corresponding to a minimum σ_m in σ is given by

$$n_m^2 = \frac{u_h}{u_e} K(T) \tag{2.42}$$

which, on substituting in Eq. (2.41), gives

$$\left(\frac{\sigma_m}{e}\right)^2 = 4u_e u_h K(T) \tag{2.43}$$

Combining Eqs (2.35) and (2.43) gives

$$\frac{\sigma}{\sigma_m} = \frac{1+\alpha}{2\alpha^{1/2}} \tag{2.44}$$

where $\alpha = u_h p / u_e n$.

Eq. (2.44) enables the relative contributions of electrons and holes to the conductivity to be estimated from the ratio of the conductivity to its minimum value, without having to determine $K(T)$. In estimating σ and σ_m an allowance must be made for contributions from current carriers other than e' and $h^{\cdot}$, such as $V_O^{\cdot\cdot}$.

It can be seen from Eq. (2.44) that, when $\sigma = \sigma_m$, $\alpha = 1$ and

$$u_h p_m = u_e n_m \tag{2.45}$$

It is also clear from Eq. (2.41) that when n is large

$$\frac{\sigma_e}{e} = u_e n \tag{2.46}$$

and, using Eq. (2.40), that when p is large

$$\frac{\sigma_h}{e} = u_h p \tag{2.47}$$

Eq. (2.43) shows that $K(T)$ can be estimated from the minima in the conductivity isotherms and a knowledge of the mobilities. u_e has been estimated to be $0.808T^{-3/2}\exp(-\mathcal{E}_u/kT)\ \mathrm{m^2V^{-1}s^{-1}}$, where $\mathcal{E}_u = 2.02\ \mathrm{kJ\,mol^{-1}}$ (0.021 eV). This gives $u_e = 15 \times 10^{-6}\ \mathrm{m^2V^{-1}s^{-1}}$ at 1000 °C and $24 \times 10^{-6}\ \mathrm{m^2V^{-1}s^{-1}}$ at 600 °C. There are few data on u_h but it is likely to be about $0.5u_e$.

The further analysis of the dependence of σ on p_{O_2} for $BaTiO_3$ is mainly based on work by Smyth [4]. It is assumed that the conductive behaviour is controlled by the equilibrium between p_{O_2}, $V_O^{\cdot\cdot}$, n, p and cation vacancies, most probably V_{Ti}'''' rather than V_{Ba}''. All the vacancies are assumed to be fully ionized at high temperatures. Under these assumptions, a schematic diagram of the dependence of $[V_O^{\cdot\cdot}]$, $[V_{Ti}'''']$, n and p on p_{O_2} at constant T can be deduced (Fig. 2.13(a)). The various p_{O_2} regions are now considered separately for the 1000 °C isotherm of acceptor-doped (ie nominally pure) $BaTiO_3$.

$p_{O_2} < 10^{-10}$ Pa (AB in Fig. 2.13(a)) The equilibrium reduction equation is

$$O_O \rightleftharpoons \tfrac{1}{2}O_2(g) + V_O^{\cdot\cdot} + 2e' \tag{2.48}$$

which, by the law of mass action, leads to

$$K_n = n^2[V_O^{\cdot\cdot}]p_{O_2}{}^{1/2} \tag{2.49}$$

where K_n is the equilibrium constant.

At these low oxygen pressures the acceptor-compensating oxygen vacancy concentration is regarded as insignificant compared with that arising through loss of oxygen according to Eq. (2.48). Therefore, since $n \approx 2[V_O^{\cdot\cdot}]$,

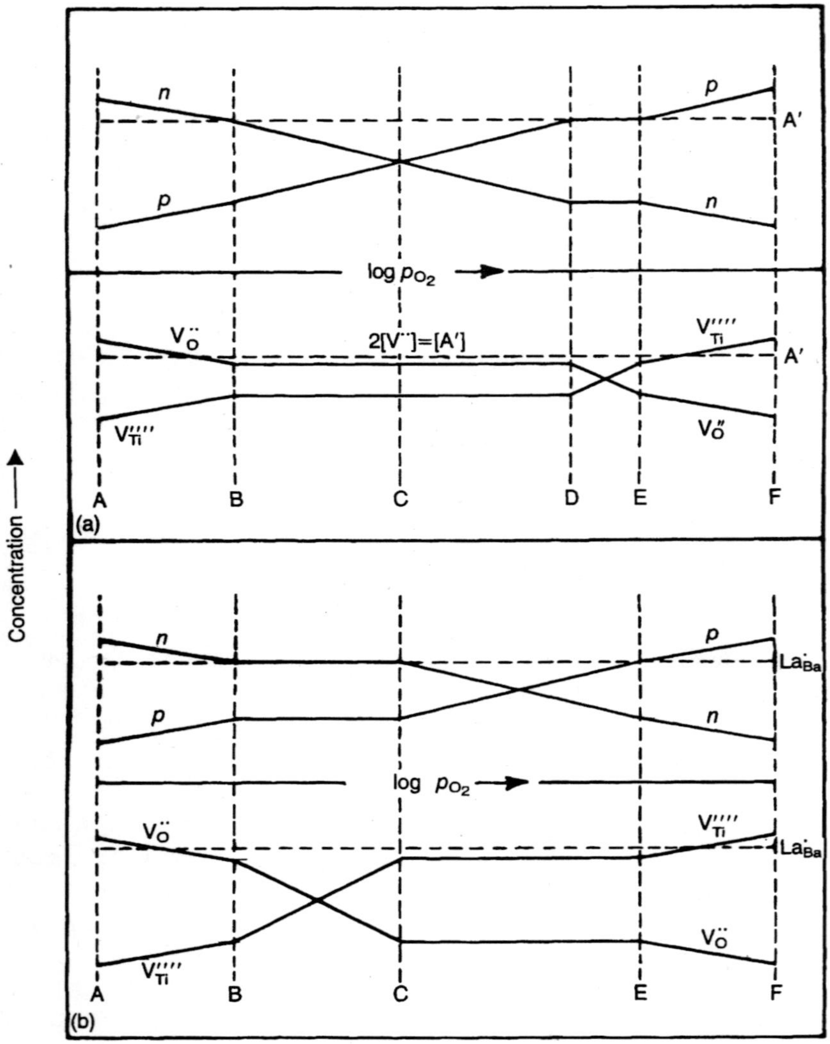

Fig. 2.13 Schematic representation of the dependence of n, p, $[V_{Ti}'''']$ and $[V_O^{\cdot\cdot}]$ on p_{O_2} for (a) acceptor-doped and (b) donor-doped $BaTiO_3$.

$$n \approx (2K_n)^{1/3} p_{O_2}{}^{-1/6} \tag{2.50}$$

10^{-10} *Pa* < p_{O_2} < 10^5 *Pa* (*BD*) The oxygen vacancy concentration now determined by the acceptor impurity concentration [A'], is little affected by changes in p_{O_2} and remains sensibly constant. It follows from Eq. (2.49) that

$$n = \left(\frac{K_n}{[V_O^{\cdot\cdot}]}\right)^{1/2} p_{O_2}{}^{-1/4} \tag{2.51}$$

The p-type contribution to semiconductivity arises through the oxidation reaction involving take-up of atmospheric oxygen by the oxygen vacancies according to

$$V_O^{\cdot\cdot} + \tfrac{1}{2}O_2(g) \rightleftharpoons O_O + 2h^{\cdot} \qquad (2.52)$$

leading to

$$p = [V_O^{\cdot\cdot}]^{1/2} K_p^{1/2} p_{O_2}{}^{1/4} \qquad (2.53)$$

At $p_{O_2} \approx 100\,\text{Pa}$, $n = p$, $\sigma = \sigma_m$ and the material behaves as an intrinsic semiconductor.

$p_{O_2} \gtrsim 10^5$ Pa (DF) Over this p_{O_2} regime the discussion is more speculative since measurements against which the model can be checked have not been made.

In the region DE the dominating defect changes from $V_O^{\cdot\cdot}$ to V_{Ti}'''' since the oxygen vacancies due to the acceptors are now filled. The conductivity is largely governed by acceptor concentration and may be independent of p_{O_2} over a small pressure range.

In the EF region the equilibrium is

$$O_2(g) \rightleftharpoons V_{Ti}'''' + 2O_O + 4h^{\cdot} \qquad (2.54)$$

so that

$$K_p' = p^4 [V_{Ti}''''] p_{O_2}{}^{-1} \qquad (2.55)$$

which, because

$$p \approx 4[V_{Ti}'''']$$

leads to

$$p = (4K_p')^{1/5} p_{O_2}{}^{1/5} \qquad (2.56)$$

Measurements in the region $10^{-17}\,\text{Pa} < p_{O_2} < 10^5\,\text{Pa}$ as shown in Fig. 2.12 show good agreement between the σ–p_{O_2} slopes and the calculated n–p_{O_2} and p–p_{O_2} relations given above. Increased acceptor doping moves the minimum in the σ–p_{O_2} towards lower pressures (see Fig. 5.49).

Intentionally donor-doped $BaTiO_3$ The effect of p_{O_2} on the conductivity of a donor-doped system has been studied for lanthanum-doped $BaTiO_3$ as shown in Fig. 2.14 for 1200 °C. The behaviour differs from that shown in Fig. 2.12 for acceptor-doped material. Firstly, there is a shift of the curves towards higher oxygen pressures. Secondly, at intermediate p_{O_2} there is a region, particularly at higher lanthanum contents, where the conductivity becomes independent of p_{O_2}. At sufficiently low pressures the curves coincide with those of the 'pure' ceramic.

Figure 2.13(b) is a schematic diagram of the proposed changes in charge carrier and ionic defect concentrations as p_{O_2} is increased from very low values

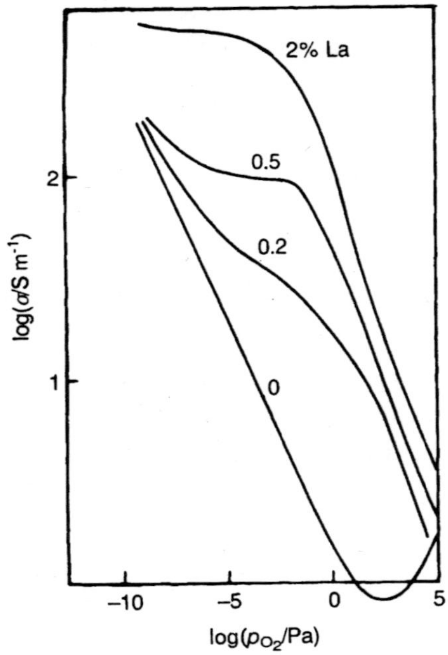

Fig. 2.14 Dependence of σ on p_{O_2} for lanthanum-doped BT ceramics at 1200 °C [8].

(below about 10^{-10} Pa). At the lowest p_{O_2} values (AB) loss of oxygen from the crystal is accompanied by the formation of $V_O^{\cdot\cdot}$ and electrons according to Eq. (2.48) and (2.50). As p_{O_2} is increased, n falls to the level controlled by the donor concentration so that $n \approx [La_{Ba}^{\cdot}]$ as shown in the following equation:

$$La_2O_3 \rightleftharpoons 2La_{Ba}^{\cdot} + 2O_O + \tfrac{1}{2}O_2(g) + 2e' \qquad (2.57)$$

When n is constant over BC, corresponding to the plateau in the curves of Fig. 2.14, there are changes in the energetically favoured Schottky disorder so that $[V_O^{\cdot\cdot}] \propto p_{O_2}{}^{-1/2}$, according to Eq. (2.49), and $[V_{Ti}''''] \propto p_{O_2}{}^{1/2}$. At C the condition

$$4[V_{Ti}''''] = [La_{Ba}^{\cdot}] \qquad (2.58)$$

is established from the equilibrium

$$2La_2O_3 + 4TiO_2 \rightleftharpoons 4La_{Ba}^{\cdot} + 3Ti_{Ti} + V_{Ti}'''' + 12O_O + \text{'}TiO_2\text{'} \qquad (2.59)$$

where 'TiO₂' indicates incorporation in a separate phase.

Both $[V_{Ti}'''']$ and $[V_O^{\cdot\cdot}]$ remain sensibly constant over the range CE so that, according to Eq. (2.55),

$$p = K_p'' p_{O_2}{}^{1/4} \text{ and } n = K_p'' p_{O_2}{}^{-1/4} \qquad (2.60)$$

At still higher values of p_{O_2} (EF), the dependence of p on p_{O_2} would be expected to follow Eq. (2.56) for the same reasons.

This model can be applied to $BaTiO_3$ containing Nb^{5+} or Sb^{5+} on the Ti site or trivalent ions of similar ionic radius to La^{3+} on the Ba site. Because the donors shift the change to p-type conductivity to pressures above atmospheric the n-type conductivity may be high at room temperature after sintering in air; this is accompanied by a dark coloration of the ceramic. The conductivity diminishes as the donor concentration is increased beyond 0.5 mol.% and at levels in the range 2–10 mol.% the ceramics are insulators at room temperature and cream in colour. This is an instance of a change in regime that may occur at high dopant concentrations. The structural cause has not yet been determined but it is notable that the grain size is diminished.

Band model for $BaTiO_3$ A primary objective of studies of semiconductors is the development of appropriate band models. In the case of elemental semiconductors such as silicon or germanium, and for the covalently bonded compound semiconductors such as GaAs and GaP, there is confidence that the essential features of the band models are correct. There is less confidence in the band models for the oxide semiconductors because sufficiently precise physical and chemical characterization of the materials is often extremely difficult. In addition, measurements are necessarily made at high temperatures where knowledge of stoichiometry, impurity levels, dislocation content, defect association and other characteristics is poor. Figure 2.15 shows a tentative band model diagram for doped barium titanate.

Doping and barium titanate technology The n- and p-type substituents, at low concentrations, have important effects on the room temperature behaviour of $BaTiO_3$. Acceptor-doped material can be fired at low oxygen pressures without losing its high resistivity at room temperature because of the shift of the σ–p_{O_2} characteristic to low pressures (Fig. 5.49) which makes it possible to co-fire the ceramic with base metal electrodes (see Section 5.7.3). The acceptors also cause the formation of oxygen vacancies which affect the changes in properties with time ('ageing', see Section 2.7.3) and the retention of piezoelectric properties under high compressive stress (see Section 6.4.1). Oxygen vacancies are also associated with the fall in resistance that occurs at temperatures above 85 °C under high d.c. fields ('degradation', see Section 5.6.2).

Donor-doped $BaTiO_3$ is the basis of positive temperature coefficient (PTC) resistors (see Section 4.4.2). The insulating dielectrics formed with high donor concentrations have a low oxygen vacancy content and are therefore less prone to ageing and degradation.

The effects of aliovalent substituents in $PbTiO_3$ and $Pb(Ti,Zr)O_3$ are, broadly speaking, similar to those in $BaTiO_3$.

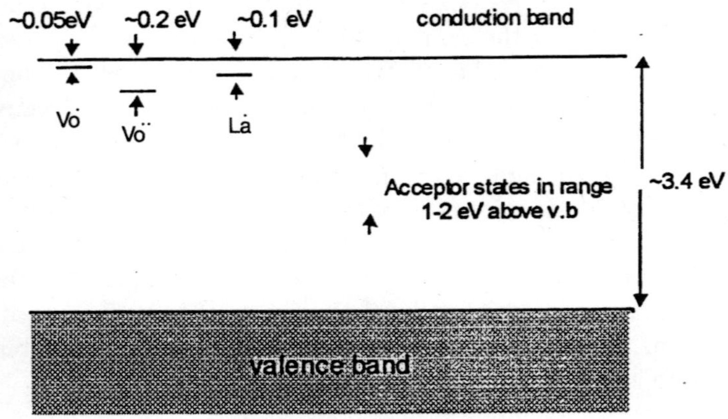

Fig. 2.15 Tentative band diagram for doped $BaTiO_3$.

Polaron conduction

The band model is not always appropriate for some oxides and the electron or hole is regarded as 'hopping' from site to site. 'Hopping' conduction occurs when ions of the same type but with oxidation states differing by unity occur on equivalent lattice sites and is therefore likely to be observed in transition metal oxides. Doping of transition metal oxides to tailor electrical properties (valency-controlled semiconduction) is extensively exploited in ceramics technology, and an understanding of the principles is important. These can be outlined by reference to NiO, the detailed study of which has taught physicists much concerning the hopping mechanism.

The addition of Li_2O to NiO leads to an increase in conductivity, as illustrated in Fig. 2.16. The lithium ion Li^+ (74 pm) substitutes for the nickel ion Ni^{2+} (69 pm) and, if the mixture is fired under oxidizing conditions, for every added Li^+ one Ni^{2+} is promoted to the Ni^{3+} state, the lost electron filling a state in the oxygen 2p valence band. The lattice now contains Ni^{2+} and Ni^{3+} ions on equivalent sites and is the model situation for conduction by 'polaron hopping', which is more often referred to simply as 'electron hopping'.

The Ni^{3+} ion behaves like a perturbing positive charge and polarizes the lattice in its immediate surroundings; the polaron comprises the Ni^{3+} ion together with the polarized surrounding regions of the ionic lattice. Polarons can be thermally excited from Ni^{3+} ions to Ni^{2+} ions; the equivalent electron transfer is from Ni^{2+} ions to Ni^{3+} ions. This conduction mechanism contrasts with the band conduction observed in silicon, for example, in two respects. In the case of 'hopping' the concentration of carriers is determined solely by the doping level and is therefore temperature independent, whereas the carrier mobility is temperature activated:

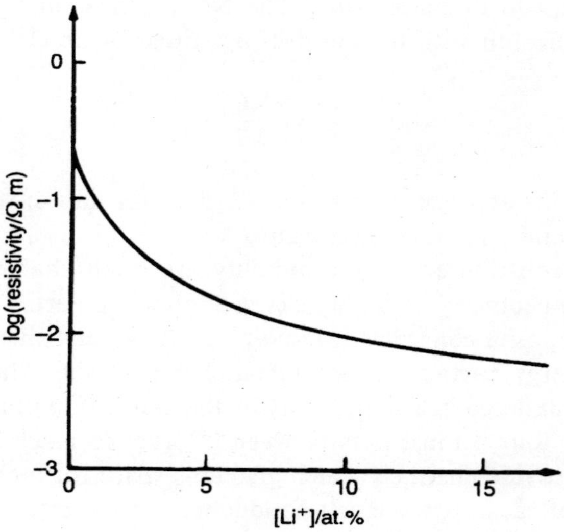

Fig. 2.16 Resistivity of NiO as a function of lithium content.

$$u \propto \exp\left(-\frac{\mathcal{E}_A}{kT}\right) \tag{2.61}$$

Thus it follows that the temperature dependence of conductivity is similar to that for band conduction (Eq. (2.37)), but for different reasons.

Because the room temperature hopping mobility is low ($< 10^{-5}\,m^2\,V^{-1}\,s^{-1}$) in contrast to that typical for band conduction ($\approx 10^{-1}\,m^2\,V^{-1}\,s^{-1}$), hopping conductors are sometimes referred to as 'low-mobility semiconductors'. Another important distinction between band and hopping conductors is the very different doping levels encountered. Whereas doping levels for silicon are usually in the parts per million range, in the case of hopping conductors they are more typically parts per hundred.

Although the mechanism of conduction in lithium-doped NiO and other 'low-mobility semiconductors' is a controversial matter, the simple polaron hopping model outlined above serves well as a basis for understanding conduction processes in many of the systems discussed later (eg Section 4.4.1).

2.6.3 Ionic conduction

Crystals

Ionic conduction depends on the presence of vacant sites into which ions can move. In the absence of a field, thermal vibrations proportional to kT may cause ions and vacancies to exchange sites. The Nernst–Einstein equation links this process of self-diffusion with the ion drift σ_i caused by an electric field:

$$\frac{\sigma_i}{D_i} = \frac{N_i Q_i}{kT} \tag{2.62}$$

where D_i is the self- or tracer-diffusion coefficient for an ion species i, Q_i is the charge it carries and N_i is its concentration.

Features that contribute to ionic mobility are small charge, small size and favourable lattice geometry. A highly charged ion will polarize, and be polarized by the ions of opposite charge as it moves past them, and this will increase the height of the energy barrier that inhibits a change of site. The movement of a large ion will be hindered in a similar way by the interaction of its outer electrons with those of the ions it must pass between in order to reach a new site. Some structures may provide channels which give ions space for movement.

The presence of vacant sites assists conduction since it offers the possibility of ions moving from neighbouring sites into a vacancy which, in consequence, moves in the opposite direction to the ions (Fig. 2.17). This is particularly likely in the case of the oxygen lattice since the smaller cations do not present large energy barriers impeding the process. However, the cations usually have to pass through the relatively small gap between three O^{2-} ions to reach any neighbouring cation vacancy.

NaCl is a suitable material for further discussion since it has been extensively investigated. The lattice contains Schottky defects but the V_{Na} (Na^+ has $r_6 = 102$ pm) moves more readily than the V_{Cl} (Cl^- has $r_6 = 181$ pm) so that charge transport can be taken as almost wholly due to movement of V'_{Na}.

In the absence of an electric field the charged vacancy migrates randomly, and its mobility depends on temperature since this determines the ease with which the Na^+ surmounts the energy barrier to movement. Because the crystal is highly ionic in character the barrier is electrostatic in origin, and the ion in its normal lattice position is in an electrostatic potential energy 'well' (Fig. 2.17).

It is clear from Fig. 2.17 that in the simplified one-dimensional representation and in the absence of an electric field the vacancy would have equal probability of jumping to the right or to the left, because the barrier height $\mathcal{E}_j$ is the same in both directions. However, when an electric field $\boldsymbol{E}$ is imposed the barrier heights are no longer equal, and the jump probability is higher for the jump across the lower barrier (in the illustrated case, to the right) of height $\mathcal{E}_j - \Delta\mathcal{E}_j$ where

$$\Delta\mathcal{E}_j = e\boldsymbol{E}\frac{a}{2} \tag{2.63}$$

Since we know the bias in jump probability in one direction, it is not difficult to arrive at the following expression for the current density:

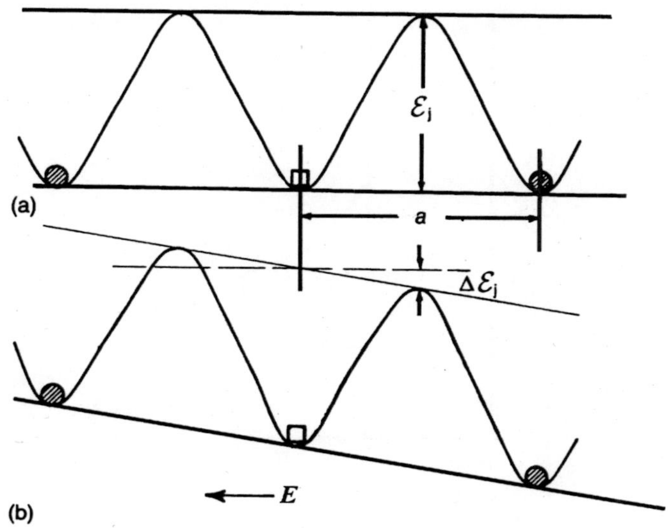

Fig. 2.17 Energy barriers to ionic transport in a crystal (a) in the absence of a field and (b) with applied field $\boldsymbol{E}$.

$$j = \frac{n_v}{N}\frac{A}{T}\boldsymbol{E}\exp\left(-\frac{\mathcal{E}_j}{kT}\right) \tag{2.64}$$

in which n_v/N is the fraction of Na^+ sites that are vacant and A is a constant describing the vibrational state of the crystal. Since it is assumed that the vacancy is part of the Schottky defect, then $n_v = n_S$ and hence, using Eq. (2.12), we obtain

$$j = \frac{A}{T}\boldsymbol{E}\exp\left(-\frac{\Delta H_S}{2kT}\right)\exp\left(-\frac{\mathcal{E}_j}{kT}\right) = \sigma\boldsymbol{E} \tag{2.65}$$

or

$$\sigma = \frac{A}{T}\exp\left\{-\frac{1}{kT}\left(\mathcal{E}_j + \frac{\Delta H_S}{2}\right)\right\} \tag{2.66}$$

Because the temperature dependence of σ is dominated by the exponential term, the expression for conductivity is frequently written

$$\sigma = \sigma_0 \exp\left(-\frac{\mathcal{E}_i}{kT}\right) \tag{2.67}$$

in which $\mathcal{E}_i = \mathcal{E}_j + \Delta H_S/2$ is an activation energy and σ_0 is regarded as approximately temperature independent.

Vacancies might also be introduced into the crystal extrinsically by the addition of impurities. For example, the addition of $SrCl_2$ to NaCl would

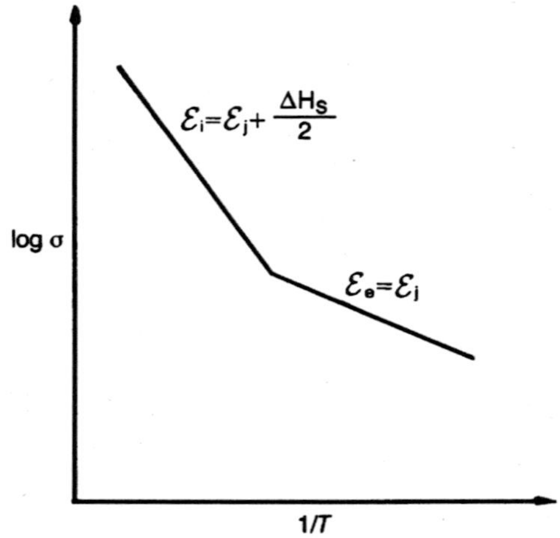

Fig. 2.18 Extrinsic and intrinsic regimes in the log σ versus $1/T$ relation.

introduce an Na vacancy for every Sr ion added. Under these circumstances, and depending upon the dopant level and the temperature, the concentration of extrinsic defects might be orders of magnitude greater than that of the intrinsic vacancies and independent of temperature. The $\sigma(T)$ relationship would therefore be as shown in Fig. 2.18 where the steep slope in the high-temperature (intrinsic) regime reflects the energy $\mathcal{E}_i$ required both to create and to move defects, and the shallower slope in the lower-temperature (extrinsic) regime reflects the energy $\mathcal{E}_j$ required only to move defects.

Glasses

The glass formers SiO_2, B_2O_3 and Al_2O_3 provide a fixed random three-dimensional network in the interstices of which are located the modifier ions such as Li^+, Na^+, K^+, Ca^{2+} and Mg^{2+}. Some of these ions, particularly Li^+ and Na^+, are very mobile whereas others, such as Ca^{2+} and Mg^{2+}, serve only to block the network. A little reflection leads to the following reasonable expectations which are borne out by observation:

1. Conductivity σ depends upon temperature through an exponential term, as in Eq. (2.67), because mobile ions need to be 'activated' to squeeze their way past oxygen ions in moving from one site to the next.

2. For a given temperature and alkali ion concentration, σ decreases as the size of the mobile ion increases (e.g. $\sigma_{Li^+} > \sigma_{Na^+} > \sigma_{K^+}$, where the corresponding sizes of the three ion types are in the ratio 1:1.5:2).

3. For a given temperature and mobile ion content, σ decreases as the concentration of blocking ions (Ca^{2+}, Mg^{2+}) increases.

Mixed-phase materials

In practice ceramics are usually multiphase, consisting of crystalline phases, glasses and porosity. The overall behaviour depends on the distribution as well as the properties of these constituents. A minor phase that forms a layer round each crystallite of the major phases, and therefore results in a 3–0 connectivity system (see Section 2.7.4), can have a major effect. If the minor phase is conductive it can greatly reduce the resistivity of the composite or, if insulating, it can reduce its conductivity. Also, an abrupt change in the mode of conduction at the main phase–intercrystalline phase boundary may introduce barriers to conduction that dominate the overall electrical behaviour. In contrast, minor phases present as small discrete particles, or porosity present as empty cavities, can only modify properties to a minor extent as indicated by one of the mixture relations such as Lichtenecker's rule (see Section 2.7.4).

2.6.4 Summary

In the technology of ceramics, electronic conductors (semiconductors), ionic conductors (solid electrolytes) and mixed electronic–ionic conductors are encountered. In all cases the conductivity is likely to vary with temperature according to

$$\sigma = \sigma_0 \exp\left(-\frac{\mathcal{E}_A}{kT}\right) \tag{2.68}$$

In the case of intrinsic band conduction the 'experimental activation energy' $\mathcal{E}_A$ is identified with half the band gap (Eq. (2.37)); in the case of 'extrinsic' or 'impurity' semiconductivity, $\mathcal{E}_A$ is either half the gap between the donor level and the bottom of the conduction band or half the gap between the acceptor level and the top of the valence band, depending upon whether the material is n or p type. In such cases the temperature dependence is determined by the concentration of electronic carriers in the appropriate band, and not by electron or hole mobility.

In ceramics containing transition metal ions the possibility of hopping arises, where the electron transfer is visualized as occurring between ions of the same element in different oxidation states. The concentration of charge carriers remains fixed, determined by the doping level and the relative concentrations in the different oxidation states, and it is the temperature-activated mobility, which is very much lower than in band conduction, that determines σ.

In crystalline ionic conductors charge transport occurs via lattice defects, frequently vacancies, and again the same dependence of conductivity on temperature is observed. For pure compounds $\mathcal{E}_A$ is identified with the energy to form defects together with the energy to move them; if defects are introduced by doping, then the thermal energy is required only to move them and $\mathcal{E}_A$ is correspondingly lower.

The common glassy or vitreous materials encountered, either as window glass or as the glass phase in ceramics, conduct by the migration of ions, often Na^+, through the random glass network. (The chalcogenide glasses, which are based on arsenic, selenium or tellurium, conduct electronically by a hopping mechanism.) Again conductivity depends upon temperature through the familiar exponential term, but the experimental activation energy $\mathcal{E}_A$ is interpreted differently depending on the mechanism. For ionic conduction, it is identified with the energy to activate an ion to move from one lattice site to an adjacent site, together with, possibly, the formation energy of the defect which facilitates the move. (In the case of chalcogenide glasses it is identified with the energy to activate the electron hopping process.)

Finally, it is important to appreciate that, for most ceramics encountered, the conduction mechanism is far from fully understood. Probably it will involve a combination of ionic and electronic charge carriers, and the balance will depend upon temperature and ambient atmosphere. The effects of impurity atoms may well dominate the conductivity and there is also the complication of contributions, perhaps overriding, from grain boundaries and other phases – glass, crystalline or both. Only through long and painstaking study can a true understanding of the conduction mechanisms emerge, and advances in technology can seldom wait for this. Such advances are therefore made through a combination of systematic research and intuitive development work, based on an appreciation of underlying principles.

2.6.5 Schottky barriers to conduction

As discussed in Section 2.6.2 electrons in a solid in thermal equilibrium obey Fermi–Dirac statistics in which the probability $F(\mathcal{E})$ that a state of energy $\mathcal{E}$ is occupied is given by the Fermi–Dirac function

$$F(\mathcal{E}) - \left\{\exp\left(\frac{\mathcal{E}-\mathcal{E}_F}{kT}\right)+1\right\}^{-1} \qquad (2.69)$$

where $\mathcal{E}_F$ is the Fermi energy. An important property of $\mathcal{E}_F$ is that, for a system in thermal equilibrium, it is constant throughout the system.

In a metal at 0 K the electrons occupy states up to the Fermi energy and so the most energetic electrons have kinetic energy $\mathcal{E}_F$. The energy ϕ_m required to remove an electron with the Fermi energy to a point outside the metal with zero

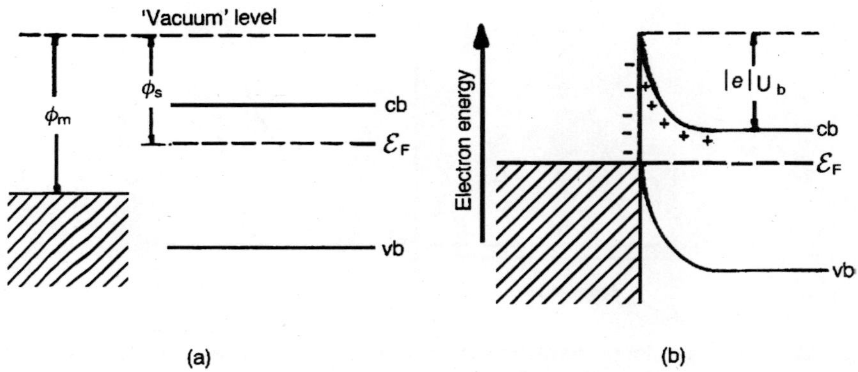

Fig. 2.19 Metal–n-type semiconductor junction ($\phi_m > \phi_s$) (cb, bottom of the conduction band; vb, top of the valence band; $\mathcal{E}_F$, Fermi energy): (a) before contact and (b) after contact.

kinetic energy (the 'vacuum' level) is called the 'work function' of the metal. When electrons are thermally excited out of a semiconductor, the effective work function ϕ_s of the semiconductor is the energy difference between the Fermi energy and the vacuum level.

Important ideas are well illustrated by considering the consequences of making a junction between a metal and an n-type semiconductor and, for the sake of argument, it is assumed that $\phi_m > \phi_s$. The situation before and after contact is illustrated in Fig. 2.19. When contact is made electrons flow from the semiconductor into the metal until the Fermi energy is constant throughout the system. It should be noted that, since the ordinate on the diagrams represents electron energies, the more negative the higher the location on the energy diagram. In the vicinity of the junction, typically within 10^{-6}–10^{-8} m depending on the concentration of n dopant, the donors are ionized. This gives rise to a positive space charge and consequently to a difference between the potential of cb near the junction and the value at a point well removed from it. Electrons moving up to the junction from the semiconductor then encounter an energy barrier – a Schottky barrier – of height $|e|U_b$. The electrons have a greater barrier to overcome in moving from the metal to the semiconductor. At equilibrium the thermally excited electron currents from metal to semiconductor and from semiconductor to metal are equal: the combination of a very large population of electrons in the metal trying to cross a large barrier matches the relatively small electron population in the semiconductor attempting to cross a smaller barrier.

If a voltage difference is established across the junctions by a battery in the sense that the semiconductor potential is made less positive, the barrier height is lowered and electrons are more readily excited over it. The rate of passage of

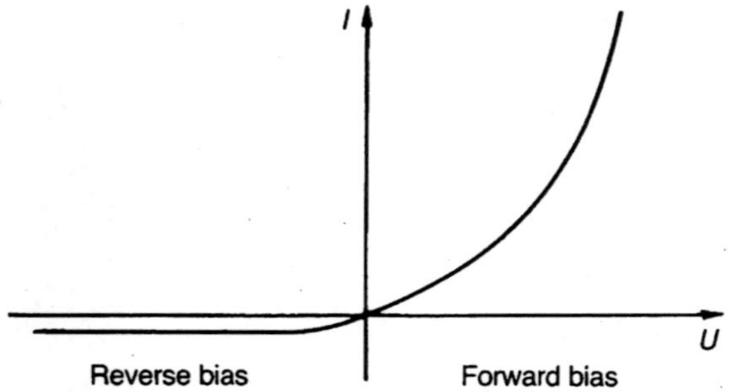

Fig. 2.20 Current–voltage characteristic for a metal–semiconductor rectifying junction.

electrons crossing the barrier, i.e. the current, depends exponentially on the barrier height. If, however, the semiconductor is positively biased, the barrier height is increased and the junction is effectively insulating. The two cases are referred to as 'forward' and 'reverse' biasing respectively, and the current–voltage characteristic is shown in Fig. 2.20.

A sandwich comprising a semiconductor between two metallic electrodes presents the same effective barrier irrespective of the sense of an applied voltage. The situation is illustrated in Fig. 2.21. The resistance will be high at low voltages, because few electrons cross the barriers, but will be low once a voltage is reached which enables electrons to cross the metal–semiconductor barrier at a significant rate. The resulting characteristic is shown in Fig. 2.22 and is similar to that for two rectifiers back to back.

In the case of wide band gap insulators the current will be low at all voltages but will initially exceed that to be expected under equilibrium conditions because of a redistribution of charges in the vicinity of the junctions. This makes it

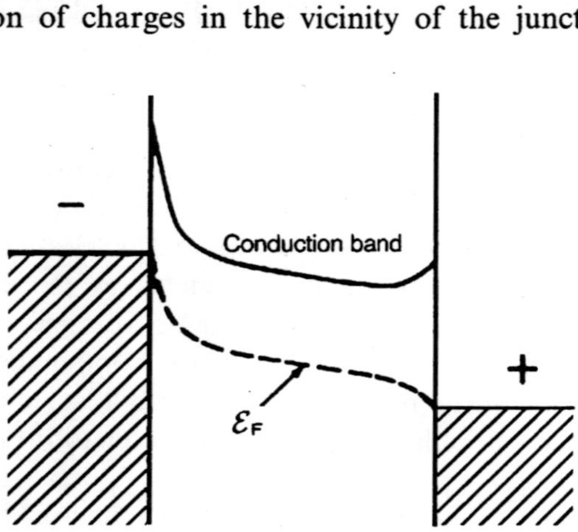

Fig. 2.21 n-type semiconductor sandwiched between two metal electrodes.

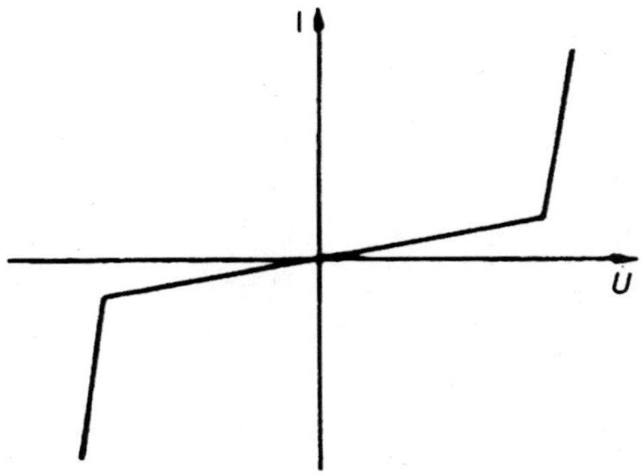

Fig. 2.22 Current–voltage characteristic for back-to-back Schottky barriers.

difficult to measure the resistivity of insulators at temperatures below 150–200 °C because of the prolonged period over which the discharge of the space charges may mask the true conduction current.

A complicating matter disregarded in the discussion concerns 'surface states'. At a semiconductor surface, because of the discontinuity, the atomic arrangement is quite different from that in the interior of the crystal. In consequence the electrons on the surface atoms occupy localized energy states quite different from those in the interior. As an added complication impurity atoms carrying their own localized energy states will e adsorbed on the semiconductor surface.

Barriers of the Schottky type control the behaviour of voltage-dependent resistors (VDRs), PTC resistors and barrier-layer capacitors. Their behaviour is by no means as well understood as that occurring in semiconductors such as silicon but, where appropriate in the text, simplified models will be presented to indicate the principles involved.

Another important type of junction is that between n and p types of the same semiconductor. The situation before and after contact is illustrated in Fig. 2.23. The n-type material has a higher concentration of electrons than the p-type material so that electrons will diffuse down the concentration gradient into the p-type material. Similarly 'holes' will diffuse into the n-type material. These diffusions result in a positive space charge on the n side and a negative space charge on the p side. The space charge generates a field that transfers carriers in the opposite direction to the diffusion currents and, at equilibrium, of equal magnitude to them. When a voltage is applied to the junction the barrier is either raised or lowered depending on the polarity. The former case is 'reversed biased' and the latter is 'forward biased'. They are illustrated in Fig. 2.24. The *U–I* characteristic is very similar to that shown in Fig. 2.20. The junction has a rectifying action which is the basis of the bipolar (p–n–p or n–p–n) transistor.

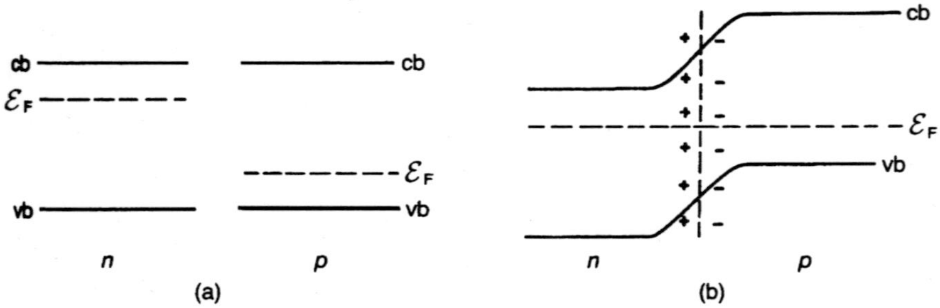

Fig. 2.23 Junction between an n-type and a p-type semiconductor (a) before contact and (b) after contact.

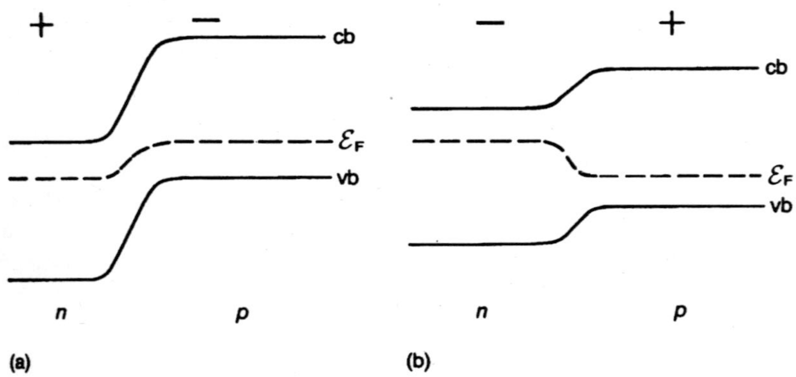

Fig. 2.24 n–p junction: (a) reverse biased; (b) forward biased.

2.7 Charge Displacement Processes

2.7.1 Dielectrics in static electric fields

Macroscopic parameters

When an electric field is applied to an ideal dielectric material there is no long-range transport of charge but only a limited rearrangement such that the dielectric acquires a dipole moment and is said to be *polarized*. Atomic polarization, which occurs in all materials, is a small displacement of the electrons in an atom relative to the nucleus; in ionic materials there is, in addition, ionic polarization involving the relative displacement of cation and anion sublattices. Dipolar materials, such as water, can become polarized because the applied electric field orients the molecules. Finally, space charge polarization involves a limited transport of charge carriers until they are stopped

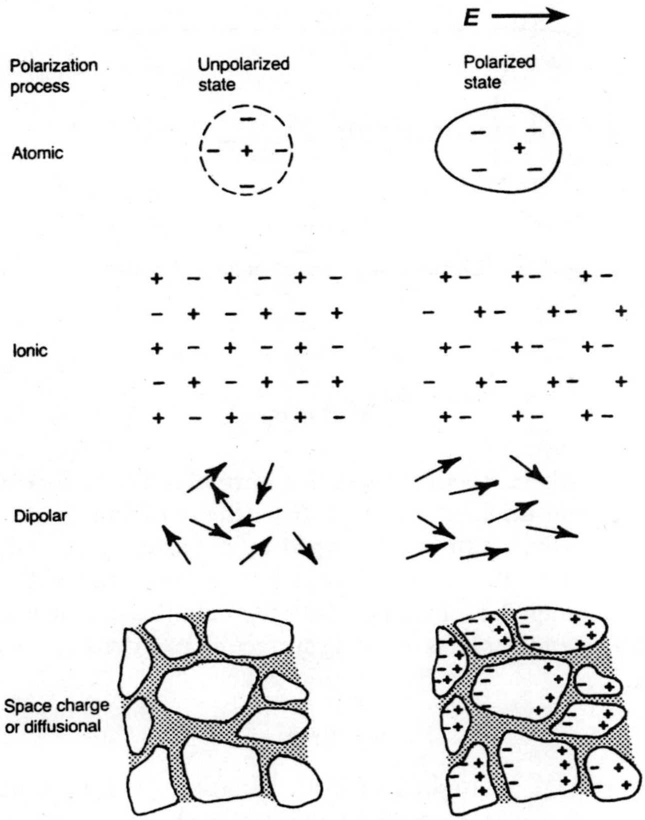

Fig. 2.25 Various polarization processes.

at a potential barrier, possibly a grain boundary or phase boundary. The various polarization processes are illustrated in Fig. 2.25.

In its most elementary form an electric dipole comprises two equal and opposite point charges separated by a distance δx. The dipole moment $\boldsymbol{p}$ of the dipole, defined as

$$\boldsymbol{p} = Q\delta\boldsymbol{x} \tag{2.70}$$

is a vector with its positive sense directed from the negative to the positive charge.

A polarized material can be regarded as made up of elementary dipolar prisms, the end faces of which carry surface charge densities of $+\sigma_p$ and $-\sigma_p$ as shown in Fig. 2.26. The dipole moment per unit volume of material is termed the *polarization* $\boldsymbol{P}$ and can vary from region to region. From Fig. 2.26 the magnitudes of the vectors are given by

$$\delta p = \sigma_p \delta A \delta x = \sigma_p \delta V$$

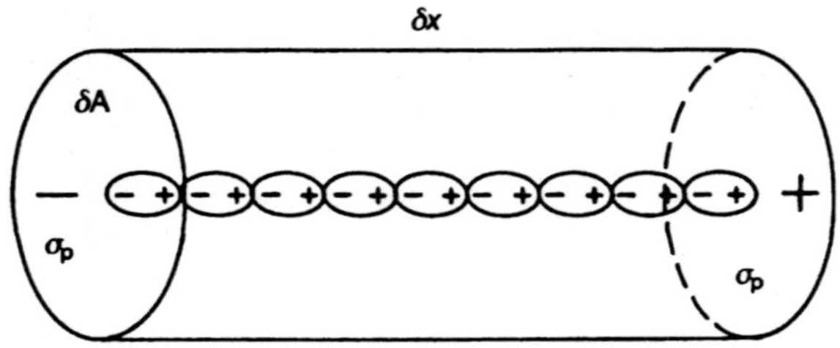

Fig. 2.26 Elementary prism of polarized material.

or

$$\frac{\delta p}{\delta V} = P = \sigma_p \tag{2.71}$$

In general $\sigma_p = \boldsymbol{n}.\boldsymbol{P}$ where $\boldsymbol{n}$ is the unit vector normal to the surface enclosing the polarized material and directed outwards from the material.

Important relationships can be developed by considering the effect of filling the space between the plates of a parallel-plate capacitor with a dielectric material, as shown in fig. 2.27. From Gauss's theorem the electric field E between and normal to two parallel plates carrying surface charge density σ and separated by a vacuum is

$$E = \sigma/\varepsilon_0 \tag{2.72}$$

Since the same voltage is applied in both situation (a) and situation (b), E remains the same. However, in (b) the polarization charge density σ_p appearing on the surfaces of the dielectric compensates part of the total charge density σ_T

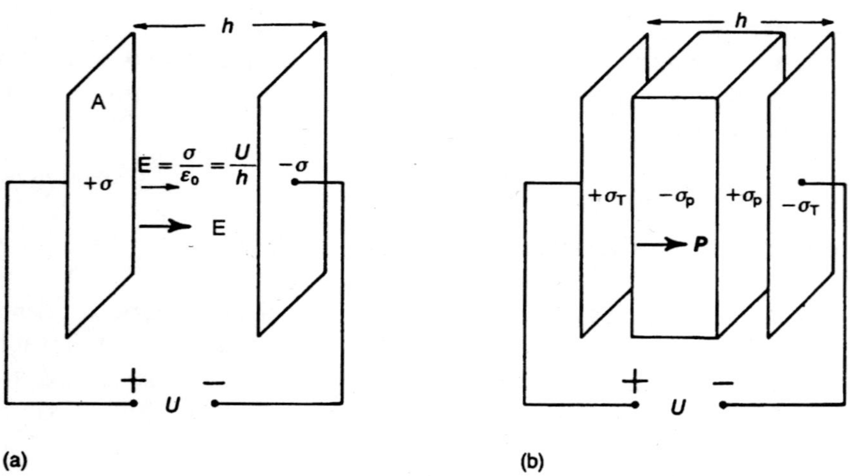

Fig. 2.27 The role of the dielectric in a capacitor.

carried by the plates. Thus the effective charge density giving rise to E is reduced to $\sigma_T - \sigma_p$ so that

$$E = \frac{\sigma_T - \sigma_p}{\varepsilon_0} \tag{2.73}$$

The total charge density σ_T is equivalent to the magnitude of the electric displacement vector $\boldsymbol{D}$, so that

$$\boldsymbol{D} = \varepsilon_0 \boldsymbol{E} + \boldsymbol{P} \tag{2.74}$$

If the dielectric is 'linear', so that polarization is proportional to the electric field within the material, which is commonly the case,

$$\boldsymbol{P} = \chi_e \varepsilon_0 \boldsymbol{E} \tag{2.75}$$

where the dimensionless constant χ_e (chi, pronounced 'ky' as in 'sky') is the *electric susceptibility*. In general χ_e is a tensor of the second rank. Unless otherwise stated it will be assumed in the following discussions that $\boldsymbol{P}$ and $\boldsymbol{E}$ are collinear, in which case χ_e is simply a scalar.

It follows from Eq. (2.74) and Eq. (2.75) that

$$\boldsymbol{D} = \varepsilon_0 \boldsymbol{E} + \chi_e \varepsilon_0 \boldsymbol{E} = (1 + \chi_e)\varepsilon_0 \boldsymbol{E} \tag{2.76}$$

and, since $D = \sigma_T$,

$$\frac{Q_T}{A} = (1 + \chi_e)\varepsilon_0 \frac{U}{h} \tag{2.77}$$

in which Q_T is the total charge on the capacitor plate. Therefore the capacitance is

$$C = \frac{Q_T}{U} = (1 + \chi_e)\varepsilon_0 \frac{A}{h} \tag{2.78}$$

Since vacuum has zero susceptibility, the capacitance C_0 of an empty parallel-plate capacitor is

$$C_0 = \varepsilon_0 \frac{A}{h} \tag{2.79}$$

If the space between the plates is filled with a dielectric of susceptibility χ_e, the capacitance is increased by a factor $1 + \chi_e$.
The *permittivity* ε of the dielectric is defined by

$$\varepsilon = \varepsilon_0(1 + \chi_e) \tag{2.80}$$

where

$$\frac{\varepsilon}{\varepsilon_0} = 1 + \chi_e = \varepsilon_r \tag{2.81}$$

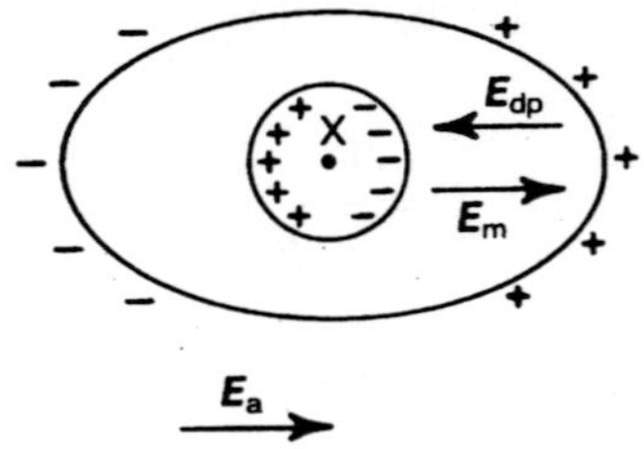

Fig. 2.28 The 'local' field in a dielectric.

and ε_r is the *relative permittivity* (often, unfortunately, referred to as the 'dielectric constant') of the dielectric.

From induced elementary dipoles to macroscopic properties

An individual atom or ion in a dielectric is not subjected directly to an applied field but to a *local field* which has a very different value. Insight into this rather complex matter can be gained from the following analysis of an ellipsoidal solid located in an applied external field $\boldsymbol{E}_a$, as shown in Fig. 2.28. The ellipsoid is chosen since it allows the depolarizing field $\boldsymbol{E}_{dp}$ arising from the polarization charges on the external surfaces of the ellipsoid to be calculated exactly. The internal *macroscopic* field $\boldsymbol{E}_m$ is the resultant of $\boldsymbol{E}_a$ and $\boldsymbol{E}_{dp}$, i.e. $\boldsymbol{E}_a - \boldsymbol{E}_{dp}$.

It is assumed that the solid can be regarded as consisting of identifiable polarizable entities on the atomic scale. It is then necessary to know the electric field experienced by an entity; this is termed the *local* field $\boldsymbol{E}_L$. It was recognized in the early part of the last century that $\boldsymbol{E}_L$ differs from $\boldsymbol{E}_m$ since the latter is arrived at by considering the dielectric as a continuum. In reality the atomic nature of matter dictates that the local field, which is also known as the Lorentz field, must include contributions from the adjacent, individual dipoles. Furthermore, the local field arises from the charges in their *displaced* positions, and because it is also doing the displacing, calculation of it is by no means straightforward!

Lorentz calculated $\boldsymbol{E}_L$ in the following way. A spherical region within the dielectric, centred on the point X at which $\boldsymbol{E}_L$ is required, is selected. The radius is chosen so that, as viewed from X, the region external to the spherical boundary can be regarded as a continuum, whereas within the boundary the discontinuous atomic nature of the dielectric must be taken into consideration. $\boldsymbol{E}_L$ can then be written

$$\boldsymbol{E}_L = \boldsymbol{E}_m + \boldsymbol{E}_p + \boldsymbol{E}_d \tag{2.82}$$

in which $\boldsymbol{E}_p$ is the contribution from the charges at the surface of the spherical cavity (imagining for the moment that the sphere of material is removed) and $\boldsymbol{E}_d$

is due to the dipoles within the boundary. $\boldsymbol{E}_p$ can be shown to be $\boldsymbol{P}/3\varepsilon_0$, but $\boldsymbol{E}_d$ must be calculated for each particular site chosen and for each type of dielectric material. However, for certain crystals of high symmetry and glasses it can be shown that $\boldsymbol{E}_d = 0$, and so for these cases

$$\boldsymbol{E}_L = \boldsymbol{E}_m + \frac{\boldsymbol{P}}{3\varepsilon_0} \tag{2.83}$$

In the more general case it is assumed that

$$\boldsymbol{E}_L = \boldsymbol{E}_m + \gamma \boldsymbol{P} \tag{2.84}$$

in which γ is the 'internal field constant'.

The dipole moment $\boldsymbol{p}$ induced in the entity can now be written

$$\boldsymbol{p} = \alpha \boldsymbol{E}_L \tag{2.85}$$

in which α is the *polarizability* of the entity, i.e. the dipole moment induced per unit applied field. If it is assumed that all entities are of the same type and have a density N, then

$$\boldsymbol{P} = N\boldsymbol{p} = N\alpha(\boldsymbol{E}_m + \gamma \boldsymbol{P}) \tag{2.86}$$

or

$$\frac{\boldsymbol{P}}{\varepsilon_0 \boldsymbol{E}_m} = \chi_e = \frac{N\alpha/\varepsilon_0}{1 - N\alpha\gamma} \tag{2.87}$$

In the particular cases for which $\gamma = 1/3\varepsilon_0$ rearrangement of Eq. (2.87) leads to the Clausius–Mosotti relationship, here in SI units

$$\frac{\varepsilon_r - 1}{\varepsilon_r + 2} = \frac{N\alpha}{3\varepsilon_0} \tag{2.88}$$

Using the cgs system the Clausius–Mossotti relation becomes

$$(\varepsilon_r - 1)/(\varepsilon_r + 2) = 4\pi N\alpha_{cgs}/3 \tag{2.89}$$

and this form, with α_{cgs} measured in $Å^3$ and N the number of 'molecules' per $Å^3$, has been widely used in calculating polarizabilities.

Eqs (2.88) and (2.89) cannot be applied to all solids but they are valid for the many high symmetry ionic structures that are non-polar and non-conductive. Also it has to be borne in mind that in the case of ceramics, grain boundaries can give rise to anomalies in the applied field distribution, and occluded layers of water can contribute to increased permittivities. R.D. Shannon [9] and others have calculated polarizabilities using Eq. (2.89) with the established values of molecular volume ($V_m = 1/N$) and permittivity. They find that each constituent ion can be assigned a unique polarizability which is the same whatever other ions they are associated with. Table 2.5 gives the polarizabilities of a wide selection of ions and using these it is possible to calculate the permittivity of any combination

Table 2.5 Ion dielectric polarizabilities (in units of Å^3)

Li 1.20	Be 0.19											B 0.05	C	N	O 2.01	F 1.62	Ne
Na 1.80	Mg 1.32											Al 0.79	Si 0.87	P (5+) 1.22	S	Cl	
K 3.83	Ca 3.16	Sc 2.81	Ti (4+) 2.93	V (5+) 2.92	Cr (3+) 1.45	Mn (2+) 2.64	Fe (2+)2.23 (3+)2.29	Co (2+) 1.65	Ni (2+) 1.23	Cu (2+) 2.11	Zn 2.04	Ga 1.50	Ge 1.63	As (5+) 1.72	Se	Br	
Rb 5.29	Sr 4.24	Y 3.81	Zr 3.25	Nb 3.97	Mo	Tc				Ag	Cd 3.40	In 2.62	Sn 2.83	Sb (3+) 4.27	Te (4+) 5.23	I	
Cs 7.43	Ba 6.40	La 6.07	Hf	Ta 4.73	W	Re				Au	Hg	Tl (1+) 7.28	Pb (2+) 6.58	Bi 6.12			
				Ce (3+)6.15 (4+)3.94	Pr 5.32	Nd 5.01	Pm	Sm 4.74	Eu (2+)4.83 (3+)4.53	Gd 4.37	Tb 4.25	Dy 4.07	Ho 3.97	Er 3.81	Tm 3.82	Yb 3.58	Lu 3.64
				Th 4.92	Pa	U (4+) 4.45											

After R.D. Shannon [9].

of ion that yields compounds with structures within the limitations stated above. Such compounds are said to be 'well behaved'. If reliable experimental values do not agree with prediction then the cause may be that the solid is not 'well behaved' because of one or more of the reasons stated above. It may of course be that 'extrinsic' factors such as pores, microcracks and chemical impurities are responsible for lack of agreement; it may also be that poor experimental procedures are partly responsible.

Measurements must be made at frequencies in the 10 kHz to 1 MHz range so as to confine the response to ionic and electronic polarizations.

Plots of ionic polarizability against ionic volume are approximately linear with the slope α/V_m increasing with cation charge.

Eq. (2.87) also suggests the possibility of 'spontaneous polarization', i.e. lattice polarization in the absence of an applied field. Considering Eq. (2.87), $\chi_e \rightarrow \infty$ as $N\alpha\gamma \rightarrow 1$, implying that under certain conditions lattice polarization produces a local field which tends to further enhance the polarization – a 'feedback' mechanism. Such spontaneously polarized materials do exist and, as mentioned in Section 2.3, 'ferroelectrics' constitute an important class among them.

Ferroelectric behaviour is limited to certain materials and to particular temperature ranges for a given material. As shown for barium titanate in Section 2.7.3, Fig. 2.40(c), they have a Curie point T_c, i.e. a temperature at which the spontaneous polarization falls to zero and above which the properties change to those of a 'paraelectric' (i.e. a normal dielectric). A few ferroelectrics, notably Rochelle Salt (sodium potassium tartrate tetrahydrate ($NaKC_4O_6.4H_2O$)) which was the material in which ferroelectric behaviour was first recognized by J. Valasek in 1920, also have lower transitions below which ferroelectric properties disappear.

Many ferroelectrics possess very high permittivity values which vary considerably with both applied field strength and temperature. The permittivity reaches a peak at the Curie point and falls off at higher temperatures in accordance with the Curie–Weiss law

$$\varepsilon_r = \frac{A}{T - \theta_c} \tag{2.90}$$

where A is a constant for a given material and θ_c is a temperature near to but not identical with the Curie point T_c. This behaviour is illustrated for barium titanate ceramic in Section 2.7.3, Fig. 2.48.

The reason for coining the term 'ferroelectric' is that the relation between field and polarization for a ferroelectric material bearing electrodes takes the form of a hysteresis loop similar to that relating magnetization and magnetic field for a ferromagnetic body (see Figs 2.46 and 9.10). There are some other analogies between ferroelectric and ferromagnetic behaviour, but the two phenomena are so fundamentally different that a comparison does not greatly assist understanding.

Various models have been suggested to explain why some materials are ferroelectric. The most recent and successful involves a consideration of the vibrational states of the crystal lattice and does not lend itself to a simple description.

Lattice vibrations may be acoustic or optical: in the former case the motion involves all the ions, in volumes down to that of a unit cell, moving in unison, while in the optical mode cations and anions move in opposite senses. Both acoustic and optical modes can occur as transverse or longitudinal waves.

From the lattice dynamics viewpoint a transition to the ferroelectric state is seen as a limiting case of a transverse optical mode, the frequency of which is temperature dependent. If, as the temperature falls, the force constant controlling a transverse optical mode decreases, a temperature may be reached when the frequency of the mode approaches zero. The transition to the ferroelectric state occurs at the temperature at which the frequency is zero. Such a vibrational mode is referred to as a 'soft mode'.

The study of ferroelectrics has been greatly assisted by so-called 'phenomenological' theories which use thermodynamic principles to describe observed behaviour in terms of changes in free-energy functions with temperature. Such theories have nothing to say about mechanisms but they provide an invaluable framework around which mechanistic theories can be constructed. A.F. Devonshire was responsible for much of this development between 1949 and 1954 at Bristol University.

There is as yet no general basis for deciding whether or not a particular material will be ferroelectric. Progress in discovering new materials has been made by analogy with existing structures or by utilizing simple tests that allow the rapid study of large numbers of materials.

The detailed discussion of the prototype ferroelectric ceramic barium titanate in Section 2.7.3 provides the essential background to an understanding of the later discussion in the text.

2.7.2 Dielectrics in alternating electric fields

Power dissipation in a dielectric

The discussion so far has been concerned with dielectrics in steady electric fields; more commonly they are in fields that change with time, usually sinusoidally. This is clearly the case for capacitors in most ordinary circuit applications, but there are less obvious instances. For example, because electromagnetic waves have an electric field component it would be the case for dielectric resonators in microwave devices and also for light passing through a transparent material. Fortunately, no matter how the field may vary with time, the variation can be

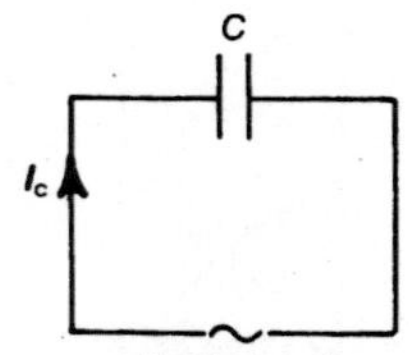

Fig. 2.29 Sinusoidal voltage applied to a perfect capacitor.

synthesized from its Fourier components and therefore there is no loss of generality if the response of a dielectric to a field changing sinusoidally with time is discussed in detail.

The earlier discussion can be extended by considering a capacitor to which a sinusoidal voltage is applied (Fig. 2.29). At a time when the voltage is U, the charge on C is $Q = UC$ and, since the current $I_c = \dot{Q}$, it follows that

$$I_c = C\dot{U} \tag{2.91}$$

Thus, with the usual notation, if the voltage is described by $U_0 \sin(\omega t)$ then the current is $U_0 C\omega\cos(\omega t)$ and leads U by 90°. Since the instantaneous power drawn from the voltage source is $I_c U$, the time average power dissipated is $\bar{P}$ where

$$\bar{P} = \frac{1}{T}\int_0^T I_c U \mathrm{d}t \tag{2.92}$$

or

$$\bar{P} = \frac{1}{T}\int_0^T U_0 I_0 \sin(\omega t)\cos(\omega t)\mathrm{d}t = 0 \tag{2.93}$$

where $T = 2\pi/\omega$ is the time period.

The average power drawn from the voltage source is zero because during one half-cycle the capacitor is charged and during the next it releases its charge reversibly (i.e. without loss of energy) back into the source. Put another way, during one half-cycle the source does work on the capacitor and during the next the discharging capacitor does work on the source. The mechanical analogy is a mass oscillating under gravity on a perfect spring for which there is no loss of energy from the system but only an exchange between elastic energy in the spring and gravitational potential energy of the mass.

The current–voltage relationship for the charging and discharging capacitor can be described with the help of a 'phasor' diagram (Fig. 2.30) in which the applied voltage at a given time is represented by a horizontal line and the instantaneous current by a vertical line since it leads the voltage by 90°. The 'phasor' diagram is an instantaneous snapshot of the voltage and current vectors

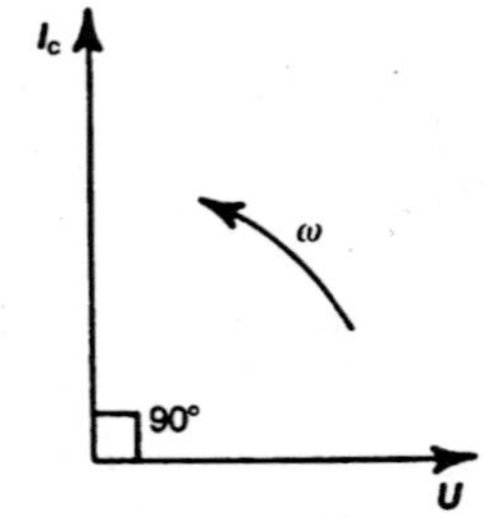

Fig. 2.30 Phasor diagram for a perfect capacitor.

as they rotate in an anticlockwise sense with angular frequency ω, maintaining a constant phase difference, in this case 90°.

If there is to be a net extraction of power from the source, there must be a component I_l of $\boldsymbol{I}$ in phase with $\boldsymbol{U}$, as shown in Fig. 2.31; I_l leads to power loss, whereas the capacitative component I_c does not. Therefore the time average power loss is

$$\bar{P} = \frac{1}{T}\int_0^T UI\mathrm{d}t$$
$$= \frac{1}{T}\int_0^T U_0\sin(\omega t)I_0\cos(\omega t - \delta)\mathrm{d}t \quad (2.94)$$

Integrating Eq. (2.94) gives

$$\bar{P} = \tfrac{1}{2}U_0 I_0 \sin\delta \quad (2.95)$$

Since $I_0 = I_c/\cos\delta$ and $I_c = \omega U_0 C$,

$$\bar{P} = \tfrac{1}{2}U_0 I_c \tan\delta = \tfrac{1}{2}U_0^2\omega C\tan\delta \quad (2.96)$$

$U_0/\sqrt{2}$ and $I_0/\sqrt{2}$ being respectively the rms voltage and rms current.

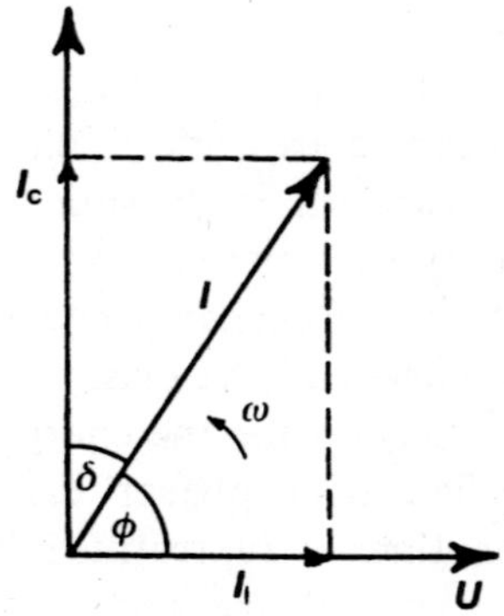

Fig. 2.31 Phasor diagram for a real capacitor.

It can be seen from Eq. (2.95) that $\sin\delta$ (or $\cos\phi$) represents the fraction of the current–voltage product that is dissipated as heat. It is termed the 'power factor'. From Eq. (2.96) the 'dissipation factor' $\tan\delta$ is the fraction of the product of the capacitive current (the component 90° out of phase with the voltage) and the voltage, dissipated as heat. In most cases of interest δ is small enough for $\sin\delta \approx \tan\delta$.

Eq. (2.96) can be put in terms of dielectric material parameters by substituting $E_0 h$ for U_0, $\varepsilon_r\varepsilon_0 A/h$ for C and V for Ah (see Fig. 2.27), leading to the equation for the dissipated power density in the dielectric:

$$\frac{\bar{P}}{V} = \tfrac{1}{2}E_0^2\omega\varepsilon_0\varepsilon_r \tan\delta \qquad (2.97)$$

$\varepsilon_r \tan\delta$ is termed the 'loss factor' of the dielectric and $\omega\varepsilon_0\varepsilon_r \tan\delta$ is the 'dielectric (or a.c.) conductivity':

$$\sigma_{\text{a.c.}} = \omega\varepsilon_0\varepsilon_r \tan\delta \qquad (2.98)$$

The complex permittivity

The behaviour of a.c. circuits can conveniently be analysed using complex quantities. The method makes use of the identity

$$\exp(\mathrm{j}\theta) = \cos\theta + \mathrm{j}\sin\theta \qquad \text{where } \mathrm{j} = \sqrt{(-1)} \qquad (2.99)$$

Thus $U = U_0 \exp(\mathrm{j}\omega t)$ can represent a complex sinusoidal voltage. The time differential of U is given by

$$\dot{U} = \mathrm{j}\omega U_0 \exp(\mathrm{j}\omega t) = \mathrm{j}\omega U$$

where the j multiplier indicates that $\dot{U}$ is 90° in advance of U. In linear equations $U = U_0 \exp(\mathrm{j}\omega t)$ can be used to represent $U_0 \cos(\omega t)$, although more correctly it should be represented by $U_0\mathrm{Re}\{\exp(\mathrm{j}\omega t)\}$ where Re stands for 'real part of'. The latter form must be used in higher-order equations.

The method can be demonstrated by repeating the preceding analysis for a lossy dielectric. The instantaneous charge on a 'lossless' vacuum capacitor C_0 is

$$Q = UC_0$$

and

$$I = \dot{Q} = \dot{U}C_0 = \mathrm{j}\omega C_0 U \qquad (2.100)$$

If the capacitor is now filled with a lossy dielectric, its effect can be taken into account by introducing a complex relative permittivity $\varepsilon_r^* = \varepsilon_r' - \mathrm{j}\varepsilon_r''$, where ε_r' and ε_r'' are respectively the real and imaginary parts of the relative permittivity. It follows that

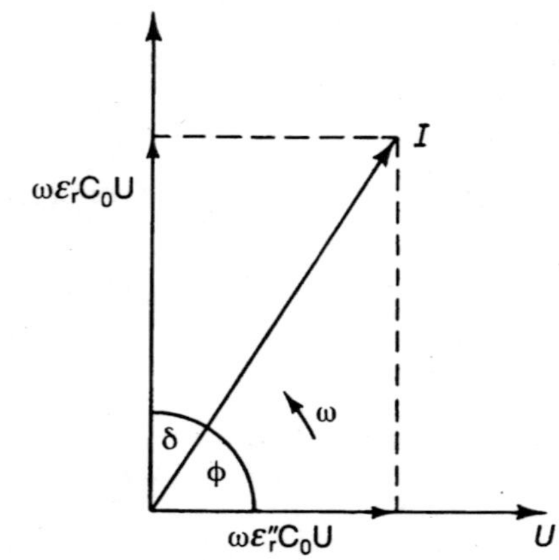

Fig. 2.32 Capacitative and 'loss' components of total current I.

$$\begin{aligned} \boldsymbol{I} &= \mathrm{j}\omega\varepsilon_r^* C_0 \boldsymbol{U} \\ &= \mathrm{j}\omega\varepsilon_r' C_0 \boldsymbol{U} + \omega\varepsilon_r'' C_0 \boldsymbol{U} \end{aligned} \tag{2.101}$$

The current is made up of two components, one capacitive and 'lossless' and the other in phase with $\boldsymbol{U}$ and 'lossy', as indicated in Fig. 2.32. It can be seen from the figure that

$$\varepsilon_r''/\varepsilon_r' = \tan\delta \tag{2.102}$$

The current I_l in phase with $\boldsymbol{U}$ can be written

$$I_l = \omega\varepsilon_r'' \frac{\varepsilon_0 A}{h} U$$

so that the current density is given by

$$\frac{I_l}{A} = \omega\varepsilon_r''\varepsilon_0 E = \omega\varepsilon'' E \tag{2.103}$$

It follows that the average dissipated power density is given by

$$\frac{\bar{P}}{V} = \tfrac{1}{2}E_0^2\sigma_{AC} = \tfrac{1}{2}E_0^2\omega\varepsilon_0\varepsilon_r'' = \tfrac{1}{2}E_0^2\omega\varepsilon_0\varepsilon_r' \tan\delta \tag{2.104}$$

which is identical in meaning with Eq. (2.97).

Frequency and temperature dependence of dielectric properties

Resonance effects Because charges have inertia, polarization does not occur instantaneously with the application of an electric field. In the case of atomic and ionic polarization, the electrons and ions behave, to a first approximation, as though bound to equilibrium positions by linear springs so that the restoring force is proportional to displacement x. Because the polarization process is accompanied by energy dissipation, a damping factor γ is included in the equation of motion. If the damping (resistive) force is assumed to be proportional to the velocity $\dot{x}$ of the moving charged particle, the equation becomes

$$m\ddot{x} + m\gamma\dot{x} + m\omega_0^2 x = QE_0 \exp(j\omega t) \tag{2.105}$$

in which m, Q and ω_0 are respectively the mass, charge and natural angular frequency of the particle and $E_0 \exp(j\omega t)$ is the sinusoidal forcing field. The appropriate field is of course the local field and only in the case of a gas would this be approximately equal to the applied field. For the present, attention is restricted to a gas consisting of N hydrogen-like *atoms* per unit volume, for which $Q = -e$.

Solving Eq. (2.105) and ignoring the transient term yields

$$x(t) = -\frac{eE_0 \exp(j\omega t)}{m\{(\omega_0^2 - \omega^2) + j\gamma\omega\}} \tag{2.106}$$

Since $-ex(t)$ is the induced dipole moment per atom, the complex polarization $\boldsymbol{P}^*$ is given by

$$\boldsymbol{P}^* = N(-e)x(t) \tag{2.107}$$

and

$$\chi^*_{\mathrm{e}\infty} = \frac{Ne^2}{m\varepsilon_0}\left\{\frac{1}{(\omega_0^2 - \omega^2) + j\gamma\omega}\right\} \tag{2.108}$$

so that

$$\varepsilon^*_{\mathrm{r}\infty} = 1 + \frac{Ne^2}{m\varepsilon_0}\left\{\frac{1}{(\omega_0^2 - \omega^2) + j\gamma\omega}\right\} \tag{2.109}$$

where the asterisks indicate complex quantities. Because the resonances of electrons in atoms and of ions in crystals occur at optical frequencies (about $10^{15}\,\mathrm{s}^{-1}$ and $10^{13}\,\mathrm{s}^{-1}$ respectively), the susceptibility and permittivity are written with the additional subscript ∞ to distinguish them from values measured at lower frequencies. It follows from Eq. (2.109), by equating real and imaginary parts, that

$$\varepsilon'_{\mathrm{r}\infty} - 1 = \frac{Ne^2}{m\varepsilon_0}\left\{\frac{{\omega_0}^2 - \omega^2}{(\omega_0^2 - \omega^2)^2 + \gamma^2\omega^2}\right\} \tag{2.110}$$

$$\varepsilon''_{\mathrm{r}\infty} = \frac{Ne^2}{m\varepsilon_0}\left\{\frac{\gamma\omega}{({\omega_0}^2 - \omega^2)^2 + \gamma^2\omega^2}\right\} \tag{2.111}$$

which are shown graphically in Fig. 2.33.

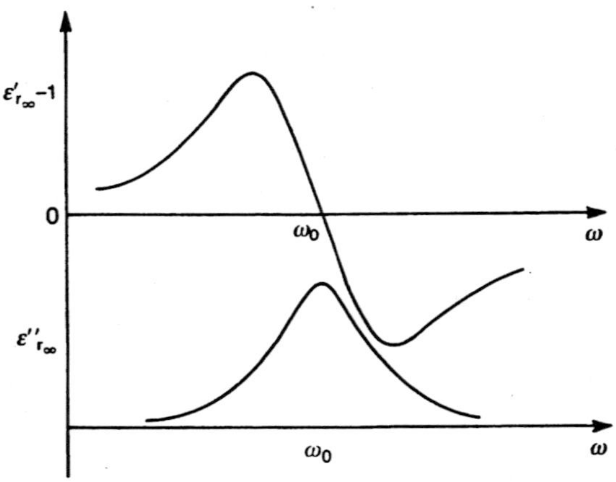

Fig. 2.33 Variation in $\varepsilon'_{r\infty}$ and $\varepsilon''_{r\infty}$ with frequency close to a resonance frequency ω_0.

As already mentioned, the above analysis would be valid for a gas; for a solid a properly calculated local field would have to be used in Eq. (2.105). Fortunately, doing this does not change the general forms of Eqs (2.110) and (2.111) but leads only to a shift in ω_0. Furthermore, because the restoring forces are sensibly independent of temperature, so too are the resonance curves.

Relaxation effects Polarization processes occur in ceramics for which the damped, forced harmonic motion approach is inappropriate. For example, because of the random structure of glass the potential energy of a cation moving through a glass can be shown schematically as in Fig. 2.34. The application of an alternating electric field causes ions to diffuse over several atomic distances, over a length such as b for example, surmounting the smaller energy barriers *en route*. It therefore takes a considerable time for the new charge distribution to establish itself after the application of the field.

In contrast to the atomic and ionic polarization processes, the diffusional polarization and depolarization processes are relatively slow and strongly temperature dependent. Temperature-activated diffusional polarization processes also occur in ionic crystals and can involve ionic migration and changes in the orientation of defect complexes.

Figure 2.35 illustrates how, on the application of a field and following the initial instantaneous atomic and ionic polarization, the slow diffusional polarization P_d approaches its final static value P_{ds}. It is assumed that at time t the polarization $P_d(t)$ develops at a rate proportional to $P_{ds} - P_d(t)$:

$$\dot{P}_d = \frac{1}{\tau}\{P_{ds} - P_d(t)\} \tag{2.112}$$

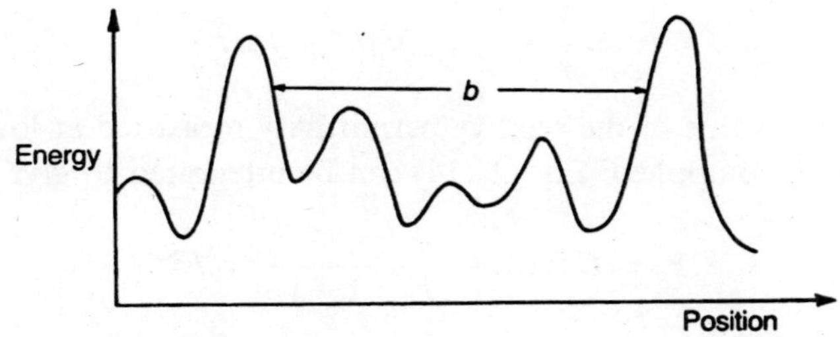

Fig. 2.34 Schematic one-dimensional representation of the variation in the electrostatic potential in a glass.

in which $1/\tau$ is a proportionality constant. Integrating Eq. (2.112) with the initial condition $P_d = 0$ when $t = 0$ gives

$$P_d = P_{ds}\left\{1 - \exp\left(-\frac{t}{\tau}\right)\right\} \tag{2.113}$$

where τ is a *relaxation time.*

The extension of this analysis to account for alternating applied fields is not straightforward. P_{ds} will now depend upon the instantaneous value of the applied field and so will be time dependent. Furthermore, because the local field is a function of both position and time, its estimation would be extremely difficult. However, to make progress, if it is assumed that the polarizing field is $E^* = E_0 \exp(j\omega t)$, it can be shown that Eq. (2.112) is modified to

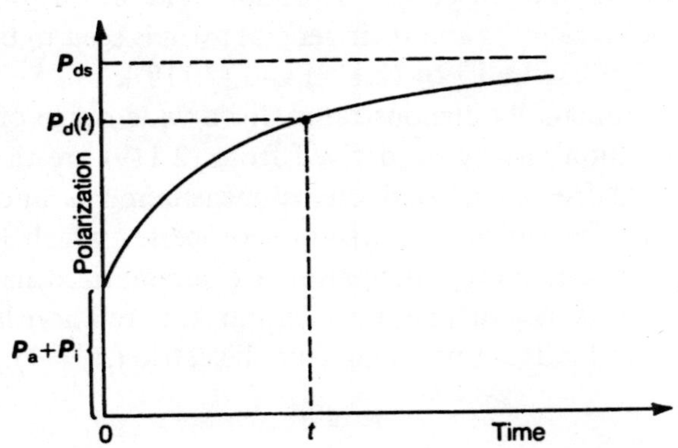

Fig. 2.35 Development of polarization by slow diffusional processes; P_a and P_i are the 'instantaneous' atomic and ionic polarization processes capable of responding to very high frequency (∞) fields.

$$\dot{P}_{\mathrm{d}} = \frac{1}{\tau}\{(\varepsilon'_{\mathrm{rs}} - \varepsilon'_{\mathrm{r}\infty})\varepsilon_0 E^* - P_{\mathrm{d}}(t)\} \tag{2.114}$$

in which $\varepsilon'_{\mathrm{rs}}$ is the value of the relative permittivity measured at low frequencies or with a static field applied. Eq. (2.114) can be integrated to give

$$P_{\mathrm{d}} = C\exp\left(-\frac{t}{\tau}\right) + \frac{\varepsilon'_{\mathrm{rs}} - \varepsilon'_{\mathrm{r}\infty}}{1 + \mathrm{j}\omega\tau}\varepsilon_0 E^* \tag{2.115}$$

If the transient $C\exp(-t/\tau)$ is neglected, Eq. (2.115) leads to

$$\varepsilon^*_{\mathrm{r}} = \varepsilon'_{\mathrm{r}\infty} + \frac{\varepsilon'_{\mathrm{rs}} - \varepsilon'_{\mathrm{r}\infty}}{1 + \mathrm{j}\omega\tau} \tag{2.116}$$

or, equating the real and imaginary parts, to

$$\varepsilon'_{\mathrm{r}} - \varepsilon'_{\mathrm{r}\infty} = \frac{\varepsilon'_{\mathrm{rs}} - \varepsilon'_{\mathrm{r}\infty}}{1 + \omega^2\tau^2} \tag{2.117}$$

and

$$\varepsilon''_{\mathrm{r}} = (\varepsilon'_{\mathrm{rs}} - \varepsilon'_{\mathrm{r}\infty})\frac{\omega\tau}{1 + \omega^2\tau^2} \tag{2.118}$$

Eqs (2.117) and (2.118), which are known as the Debye equations, are shown graphically in Fig. 2.36; the relaxation frequency is $\omega_{\mathrm{r}} = 1/\tau$. Because the polarization occurs by the same temperature-activated diffusional processes which give rise to d.c. conductivity (see Eq. (2.68)), τ depends on temperature through an exponential factor:

$$\tau = \tau_0 \exp\left(\frac{\mathcal{E}_{\mathrm{A}}}{kT}\right) \tag{2.119}$$

Figure 2.37 shows the dielectric dispersion and absorption curves for a common soda lime silica glass and their general form is seen to be consistent with the predictions (Fig. 2.36 and Eqs (2.117) and (2.118)).

It has been experimentally demonstrated, for a wide range of glass types that values of the activation energy $\mathcal{E}_{\mathrm{A}}$ derived from (2.119) are in close agreement with those obtained from d.c. conductivity measurements and so support the model as outlined. The various polarization processes which lead to dielectric dispersion and attendant energy dissipation are summarized in Fig. 2.38.

In conclusion, it is opportune to mention the relationship between the refractive index n and ε_{r} for a non-magnetic dielectric ($\mu_{\mathrm{r}} = 1$), i.e.

$$\varepsilon_{\mathrm{r}} = n^2 \tag{2.120}$$

which is a consequence of Maxwell's electromagnetic theory. It is only valid, of course, when the same polarization processes occur during the measurement of both ε_{r} and n. This would be so for elemental solids such as diamond and silicon

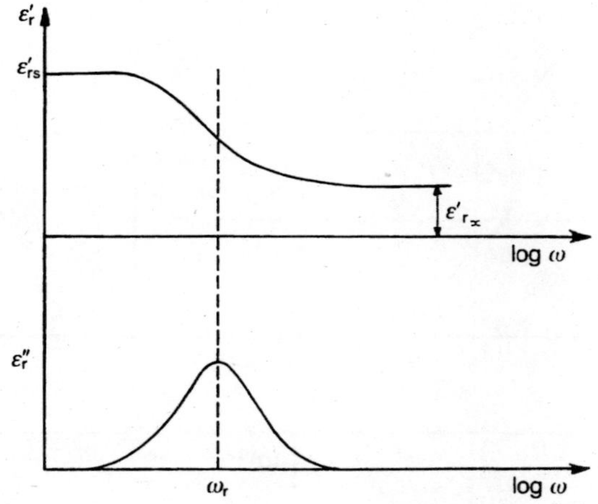

Fig. 2.36 Variation in permittivity with frequency for a dielectric showing 'Debye' relaxation.

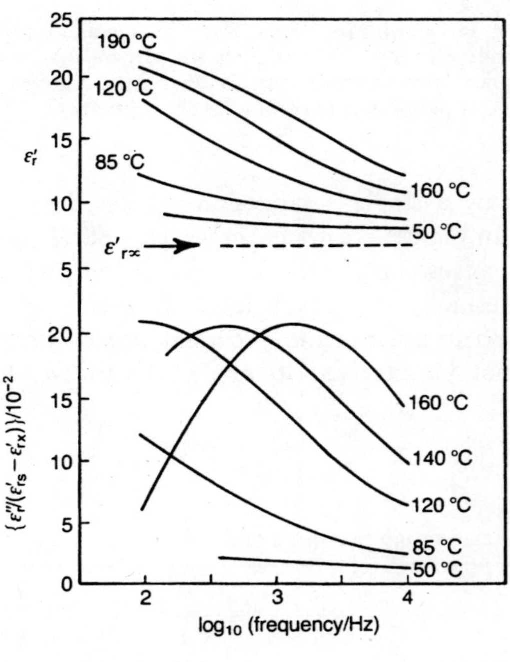

Fig. 2.37 Permittivity dispersion and dielectric loss for a glass: 18 $Na_2O \cdot 10CaO \cdot 72SiO_2$ (after Taylor, H.E., *J. Soc. Glass Tech.*, **43**, 124T, 1959. Also see Rawson, H. *Properties and Applications of Glass*. Elsevier, Amsterdam, p. 266, 1980).

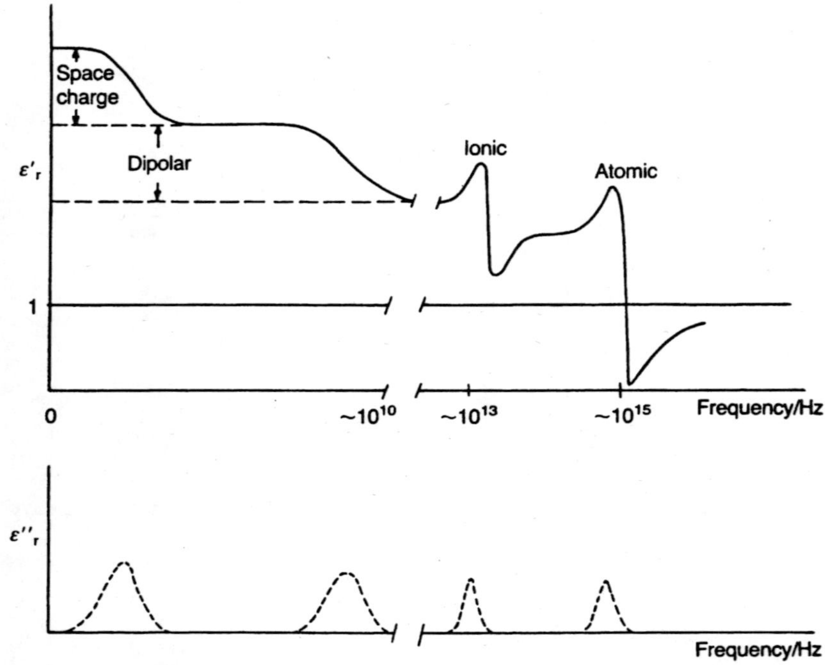

Fig. 2.38 Variation of ε_r' and ε_r'' with frequency. Space charge and dipolar polarizations are relaxation processes and are strongly temperature dependent; ionic and electronic polarizations are resonance processes and sensibly temperature independent. Over critical frequency ranges energy dissipation is a maximum as shown by peaks in $\varepsilon_r''(\omega)$.

for which there is only a single polarization process, i.e. atomic, across the frequency spectrum. In the case of water, however, optical refraction measured at 10^{15} Hz, say, involves only atomic polarization, whereas during the measurement of ε_r, usually at a much lower frequency of typically 10^5 Hz, both atomic and dipolar polarization processes are occurring and therefore Eq. (2.120) would not be expected to apply. Illustrative data are given in Table 2.6.

Table 2.6 Relationship between ε_r and n^2

	ε_r	n	n^2
Diamond	5.68	2.42	5.85
Germanium	~16	4.09	16.73
NaCl	5.9	1.54	2.37
H_2O	~80	1.33	1.77

2.7.3 Barium titanate – the prototype ferroelectric ceramic

Barium titanate, the first ceramic material in which ferroelectric behaviour was observed, is the ideal model for a discussion of the phenomenon from the point of view of crystal structure and microstructure; see also [10] and [11].

$BaTiO_3$ is isostructural with the mineral perovskite ($CaTiO_3$) and so is referred to as 'a perovskite'. The generalized perovskite structure ABO_3, is visualized as based on a cubic close-packed assembly (see Fig. 2.1) of composition AO_3 with the A-ion cordinated with 12 oxygen ions and the B-ion in the octahedral interstices (see Fig. 2.39). A consideration of the geometry shows that for a prefect fit the following relationship between the ionic radii holds [12].

$$R_A + R_O = 2(R_B + R_O) \tag{2.121}$$

For the many compounds having the perovskite structure the relationship will not hold exactly because of small variations in the sizes of the A and B ions. Therefore, to allow for this Eq. (2.121) is written

$$R_A + R_O = t\,2(R_B + R_O) \tag{2.122}$$

in which 't' is termed the 'tolerance factor' with a value typically in the range $0.95 < t < 1.06$. In the case of $SrTiO_3$ $t = 1$. When $t \neq 1$ then small lattice distortions (the octahedra tilt) occur in order to minimize lattice energy. These distortions have a significant effect on dielectric properties.

Above its Curie point (approximately 130 °C) the unit cell is cubic with the ions arranged as in Fig. 2.39. Below the Curie point the structure is slightly distorted to the tetragonal form with a dipole moment along the c direction. Other transformations occur at temperatures close to 0 °C and −80 °C: below 0 °C the unit cell is orthorhombic with the polar axis parallel to a face diagonal and below −80 °C it is rhombohedral with the polar axis along a body diagonal.

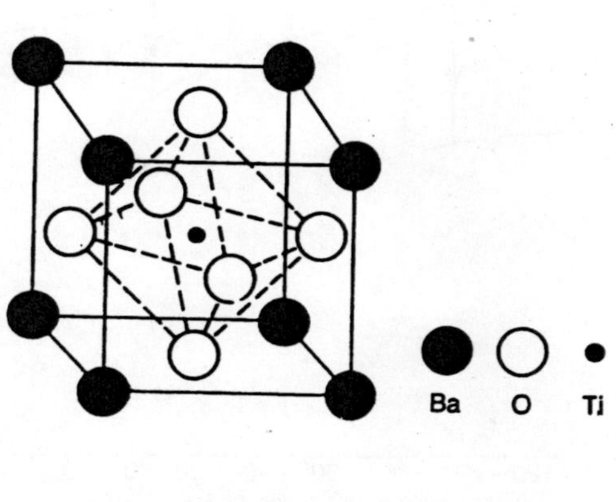

Fig. 2.39 The unit cell of $BaTiO_3$.

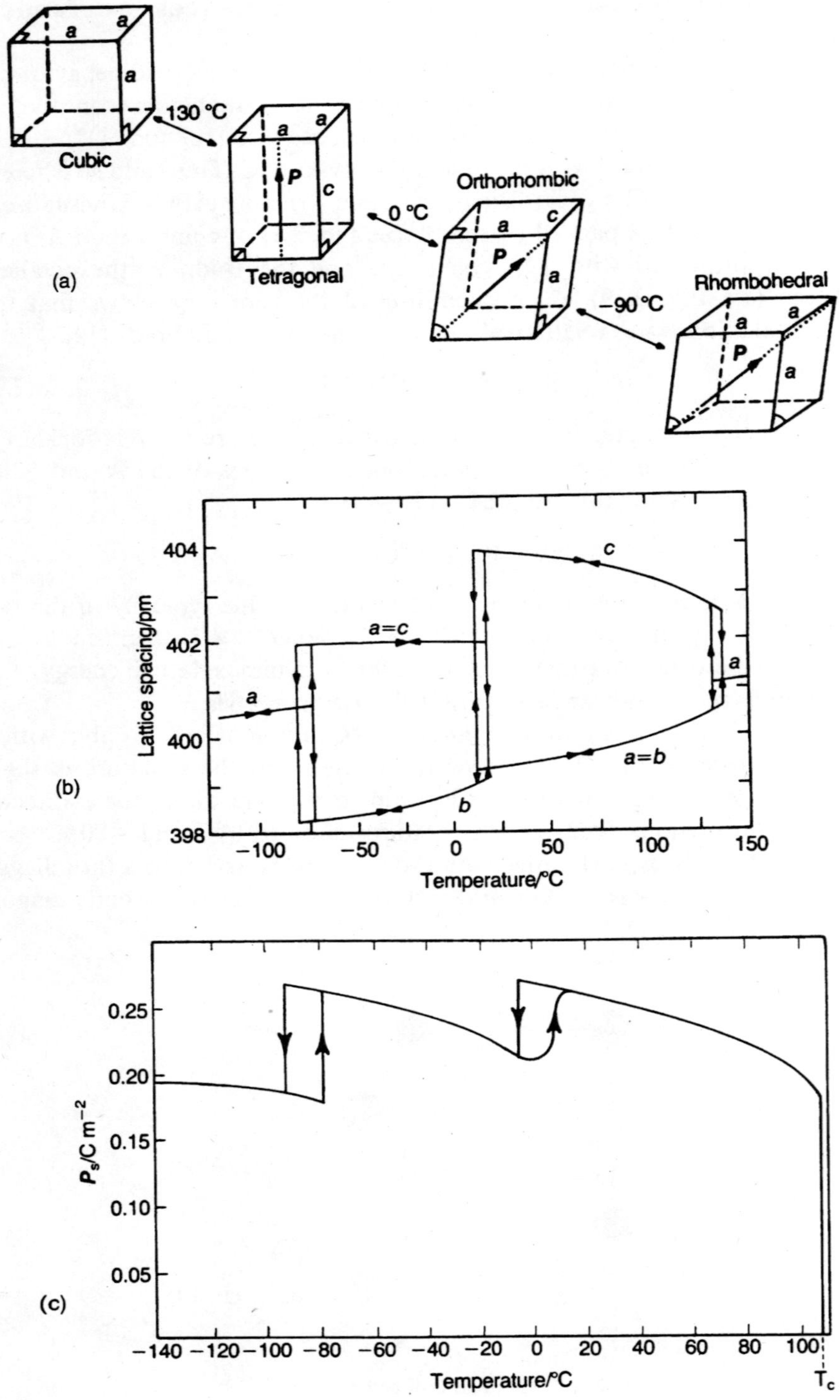

Cubic
130 °C
Tetragonal
0 °C
Orthorhombic
−90 °C
Rhombohedral
(a)
a
c
P
(b)
Lattice spacing/pm
Temperature/°C
a=c
a=b
404
402
400
398
−100
−50
0
50
100
150
(c)
P_s/C m^{-2}
0.25
0.20
0.15
0.10
0.05
−140 −120 −100 −80 −60 −40 −20 0 20 40 60 80 100
Temperature/°C
T_c

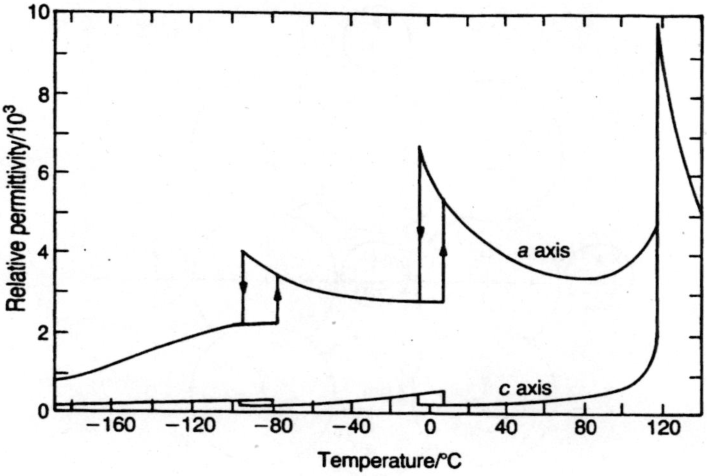

(d)

Fig. 2.40 Properties of single-crystal $BaTiO_3$: (a) unit-cell distortions of the polymorphs; (b) lattice dimensions versus temperature (after R. Clarke *J. Appl. Cryst.* **9**, 335, 1976); (c) spontaneous polarization versus temperature; (d) relative permittivities measured in the *a* and *c* directions versus temperature (after W.J. Merz *Phys. Rev.* **76**, 1221, 1949).

The transformations are illustrated in Fig. 2.40(a), and the corresponding changes in the values of the lattice parameters, the spontaneous polarization and the relative permittivity are shown in Fig. 2.40(b–d).

A consideration of the ion displacements accompanying the cubic–tetragonal transformation can give insight into how the spontaneous polarization might be coupled from unit cell to unit cell. X-ray studies have established that in the tetragonal form, taking the four central (B) oxygen ions in the cubic phase as origin, the other ions are slightly shifted as shown in Fig. 2.41. It is evident that if the central Ti^{4+} ion is closer to one of the O^{2-} ions marked A, it will be energetically favourable for the Ti^{4+} ion on the opposite side of A to be located more distantly from that O^{2-} ion, thus engendering a similar displacement of all the Ti^{4+} ions in a particular column in the same direction. Coupling between neighbouring columns occurs in $BaTiO_3$ so that all the Ti^{4+} ions are displaced in the same direction. In contrast, in the orthorhombic perovskite $PbZrO_3$ the Zr^{4+} ions in neighbouring columns are displaced in opposite senses so that the overall dipole moment is zero. Such a structure is termed *antiferroelectric* if the material shows a Curie point.

In tetragonal $BaTiO_3$ the energy of the Ti^{4+} ion in terms of its position along the *c* axis takes the form of two wells (Fig. 2.42). An applied field in the opposite direction to the polarization may enable a Ti^{4+} ion to pass over the energy barrier between the two states and so reverse the direction of the polarity at that point. When this happens the energy barriers for neighbouring ions are reduced and the entire region affected by the field will eventually switch into the new

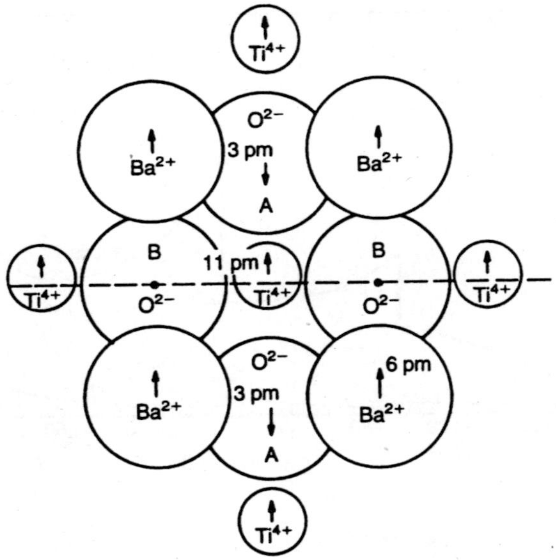

Fig. 2.41 Approximate ion displacements in the cubic–tetragonal distortion in $BaTiO_3$.

direction. A similar mechanism is available for changes of polarity through 90° but in this case there is an accompanying dimensional change because the polar *c* axis is longer than the non-polar *a* axis (Fig. 2.40(b)). Switching through 90° can be induced through the ferroelastic effect by applying a compressive stress along the polar axis without an accompanying electric field. Mechanical stress does not induce 180° switching.

An immediate consequence of the onset of spontaneous polarization in a body is the appearance of an apparent surface charge density and an accompanying depolarizing field $\boldsymbol{E}_D$ as shown in Fig. 2.43(a). The energy associated with the

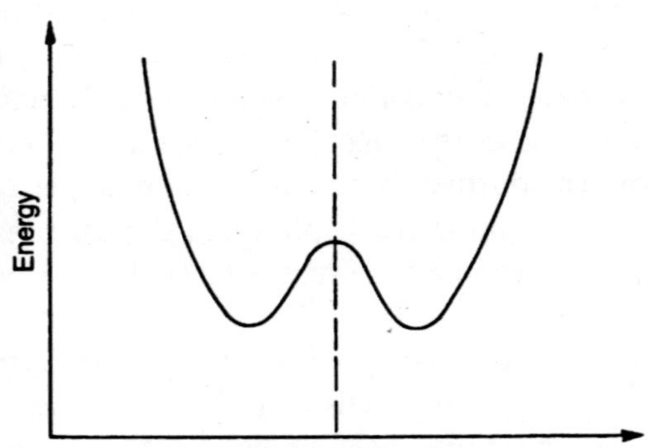

Fig. 2.42 Variation in the potential energy of Ti^{4+} along the *c* axis.

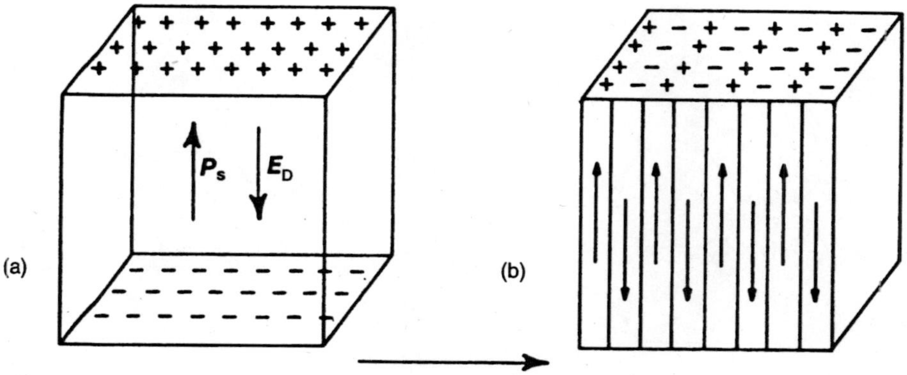

Fig. 2.43 (a) Surface charge associated with spontaneous polarization; (b) formation of 180° domains to minimize electrostatic energy.

polarization in the depolarizing field is minimized by twinning, a process in which the crystal is divided into many oppositely polarized regions, as shown in Fig. 2.43(b). These regions are called *domains* and the whole configuration shown comprises 180° domains. Thus the surface consists of a mosaic of areas carrying apparent charges of opposite sign, resulting in a reduction in E_D and in energy. This multidomain state can usually be transformed into a single domain by applying a field parallel to one of the polar directions. The domains with their polar moment in the field direction grow at the expense of those directed oppositely until only a single domain remains. The presence of mechanical stress in a crystal results in the development of 90° domains configured so as to minimize the strain. For example, as ceramic $BaTiO_3$ cools through the Curie temperature individual crystallites are subjected to large mechanical stresses leading to the development of 90° domains. The configurations can be modified by imposing either an electric or a mechanical stress. A polycrystalline ceramic that has not been subjected to a static field behaves as a non-polar material even though the crystals comprising it are polar. One of the most valuable features of ferroelectric behaviour is that ferroelectric ceramics can be transformed into polar materials by applying a static field. This process is called 'poling'. The ceramic can be depoled by the application of appropriate electric fields or mechanical stresses. These poling and depoling processes are illustrated in Fig. 2.44.

The random directions of the crystallographic axes of the crystallites of a ceramic limit the extent to which spontaneous polarization can be developed. It has been calculated that the fractions of the single-crystal polarization value that can be attained in a ceramic in which the polar axes take all possible alignments

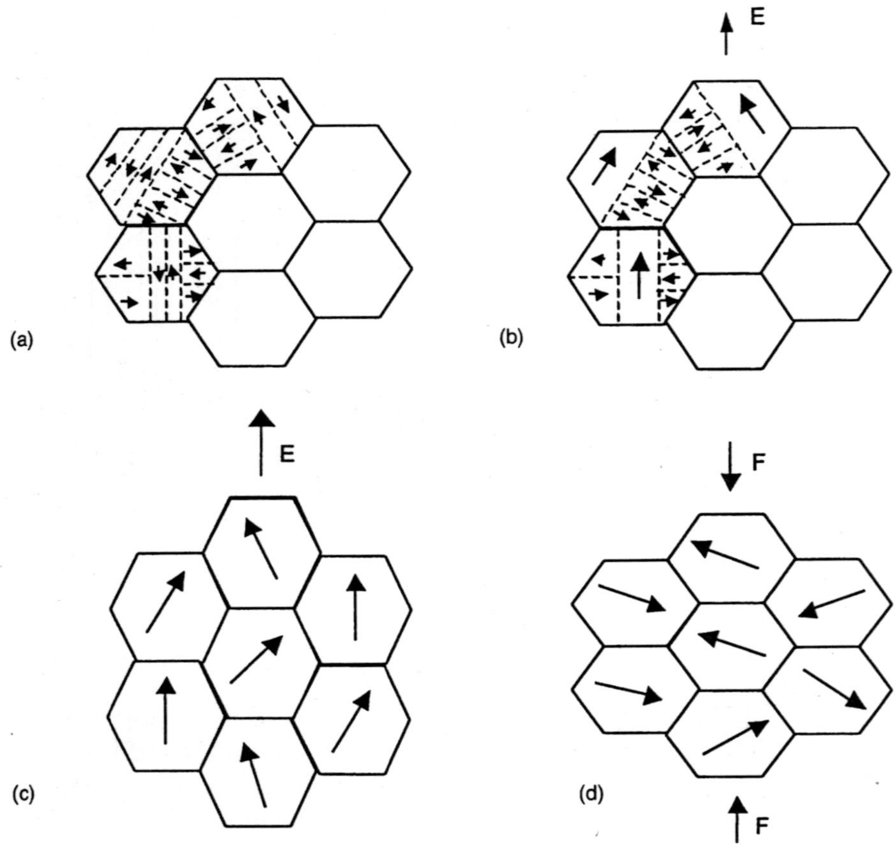

Fig. 2.44 Schematic illustrating the changes accompanying the application of electrical and mechanical stresses to a polycrystalline ferroelectric ceramic: (a) stress-free – each grain is non-polar because of the cancellation of both 180° and 90° domains; (b) with applied electric field – 180° domains switch producing net overall polarity but no dimensional change; (c) with increase in electric field 90° domains switch accompanied by small (~1%) elongation; (d) domains disorientated by application of mechanical stress. (Note the blank grains in (a) and (b) would contain similar domain structures.)

are 0.83, 0.91 and 0.87 for perovskites with tetragonal, orthorhombic or rhombohedral structures respectively. In ceramic tetragonal $BaTiO_3$ the saturation polarization is about half the single-crystal value. The value attainable is limited by the inhibition of 90° switching because of the strains involved, although 180° switching can be almost complete.

The domain structure revealed by polishing and etching an unpoled ceramic specimen is shown in Fig. 2.45(a). The principal features in the form of parallel lines are due to 90° changes in the polar direction. The orientations occurring in a simple domain structure are shown schematically in Fig. 2.45(b). The thickness

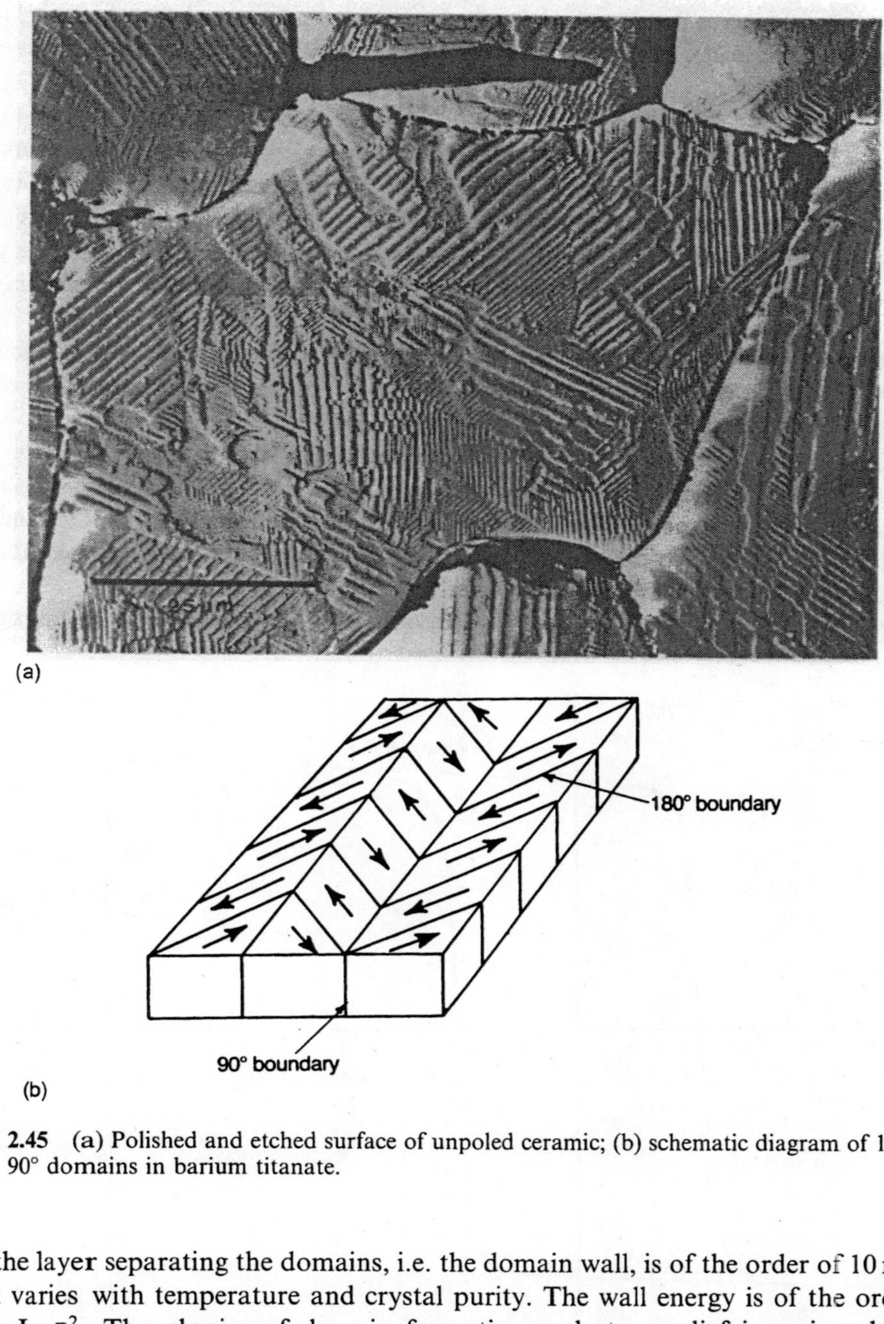

Fig. 2.45 (a) Polished and etched surface of unpoled ceramic; (b) schematic diagram of 180° and 90° domains in barium titanate.

of the layer separating the domains, i.e. the domain wall, is of the order of 10 nm but varies with temperature and crystal purity. The wall energy is of the order $10\,\mathrm{mJ\,m^{-2}}$. The physics of domain formation and stress-relief is reviewed by G. Arlt [13].

The detailed geometry and dynamics of changes in domain configuration in a single crystal accompanying changes in applied field are complex and there is marked hysteresis between induced polarization and an applied field of sufficient strength. Conditions in a crystallite clamped within a ceramic are even more complex.

The hysteresis loop of a single-domain single crystal of $BaTiO_3$ is shown in Fig. 2.46(a). The almost vertical portions of the loop are due to the reversal of the spontaneous polarization as reverse 180° domains nucleate and grow. The almost horizontal portions represent saturated states in which the crystal is single domain with a permittivity ε_r of 160 (see Fig. 2.40(d)) measured in the polar direction. The coercive field at room temperature when the loop is developed by a 50 Hz supply is $0.1\,\mathrm{MV\,m^{-1}}$ and the saturation polarization is $0.27\,\mathrm{C\,m^{-2}}$. For fields in the approximate range 10 to $100\,\mathrm{V\,mm^{-1}}$ the hysteresis loop takes the form of a narrow ellipse, a Rayleigh loop, with its major axis parallel to the almost horizontal part of the fully developed loop.

The hysteresis loop of a ceramic varies according to composition and ceramic structure but is typically of the form shown in Fig. 2.46(b). The coercive field is higher and the remanent polarization is lower than for a single crystal. Changes in both 180° and 90° domain configurations take place during a cycle and are impeded by the defects and internal strains within the crystallites.

In discussing dielectric losses in ferroelectrics it is necessary to distinguish

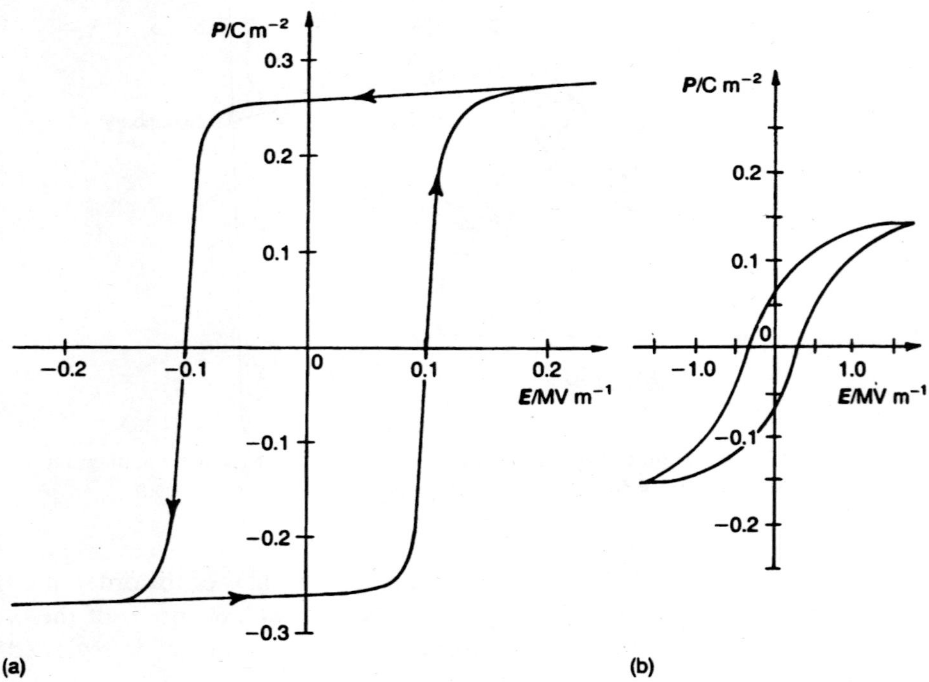

Fig. 2.46 Hysteresis loops for (a) a single-domain single crystal of $BaTiO_3$ and (b) $BaTiO_3$ ceramic.

between three mechanisms. The first involves the vibrating domain wall, the second a limited translation of the wall and the third the switching of the polarization direction of an entire domain. These three mechanisms are now discussed in a little more detail when it is assumed that the driving electric field is sinusoidal and that when field strengths are referred to it is the amplitude which is the relevant parameter.

Considering first the vibrating domain wall, the losses have their origin in the emission of acoustic shear waves resulting from small changes in domain shape induced by the applied field. These losses are present at all frequencies extending up to the GHz range. At around 1 GHz there is a marked Debye-like relaxation effect with the losses reaching a maximum. (The process bears a formal similarity to that discussed earlier (see Section 2.7.2 in the context of dielectric losses in a glass). At this frequency the wavelength of an acoustic wave is of the same order of size as that of the domains (i.e. $\sim 1 \mu m$) and there will be strong scattering. The topic is discussed by G.Arlt and co-workers (e.g. [14]).

This loss mechanism is dominant up to a 'threshold field' (E_{th}) the strength of which depends upon the 'softness' or 'hardness' of the ferroelectric. Anticipating the later discussion (see Section 6.3.2) of the family of piezoelectric ceramics ($PbZrO_3 - PbTiO_3$ ['PZT']) 'hardness' is engineered through specific doping which has the effect of 'pinning' the domain walls. The losses ($\tan \delta$) of a 'hard' and 'soft' PZT in the low field region (below E_{th}) are typically 0.003 and 0.02 respectively.

Above E_{th} the field is sufficiently strong to cause limited translation of the domain wall without disturbing to any significant extent the overall domain structure. This process is described as 'reversible' (more correctly as 'nearly reversible') to distinguish it from the very hysteretic and clearly irreversible process evidenced by the hysteresis loop (Fig. 2.46). In this regime the P-E characteristic is a narrow loop, the Rayleigh loop referred to above (c.f. Fig. 6.9).

When a critical field (E_c) is reached, which is near to the coercive field, the domains switch direction as a whole involving considerable hysteresis loss. This loss is proportional to the area of the loop, so that for the single crystal in Fig 2.46(a) it amounts to about $0.1\,\mathrm{MJ\,m^{-3}}$. At 100 Hz the power dissipated as heat would be $100\,\mathrm{MW\,m^{-3}}$, which would result in a very rapid rise in temperature. The dissipation factor ($\tan \delta$) is also very high at high field strengths, but becomes small at low field strengths, as described above. Modifications to the composition diminish the loss still further.

Dielectric hysteresis in the commercially important ferrolectric piezoceramics (e.g. 'PZT'; c.f. Section 6.3.1) is comprehensively reviewed by D. Hall [15].

A further unusual characteristic of ferroelectric materials is that their properties change with time in the absence of either external mechanical or electrical stresses or temperature changes. This ageing is due to a diminution of domain wall mobility through the gradual build-up of inhibiting causes. These may be internal fields due to the alignment of dipoles formed from lattice defects and impurity ions, the redistribution of internal strains due to crystal anisotropy or the accumulation of defects in the domain walls.

The rate of change $\partial p/\partial t$ of a property with time is approximately inversely proportional to the time after a specimen has cooled from above its Curie point to room temperature:

$$\frac{\partial p}{\partial t} = \frac{a'}{t} \tag{2.123}$$

On integration Eq. (2.123) yields

$$p = a \log_{10}\left(\frac{t}{t_0}\right) \tag{2.124}$$

where $a = a' \log_e 10$ and t_0 is an arbitrary zero for the time. a is usually less than 0.05 for most properties of commercial materials. It varies in sign according to the property and, for instance, is negative for permittivity and positive for Young's modulus. The value of a is often given as a percentage per decade, which implies that the change in a property will be the same between 1 and 10 days as between 100 and 1000 days. The effect therefore becomes negligible after a sufficient lapse of time provided that the component is not subjected to high mechanical or electrical stresses or to temperatures near to or exceeding the Curie point. Such stresses will disturb the domain structure and consequently the ageing rate will increase to a value above that expected from Eq. (2.123).

One very significant advantage of ceramic ferroelectrics is the ease with which their properties can be modified by adjusting the composition and the ceramic microstructure. Additions and the substitution of alternative cations can have the following effects:

1. shift the Curie point and other transition temperatures;
2. restrict domain wall motion;
3. introduce second phases or compositional heterogeneity;
4. control crystallite size;
5. control the oxygen content and the valency of the Ti ion.

The effects are important for the following reasons.

1. Changing the Curie point enables the peak permittivity to be put in a temperature range in which it can be exploited. The substitution of Sr^{2+} for Ba^{2+} in $BaTiO_3$ lowers T_c whilst the substitution of Pb^{+} increases it, as shown in Fig. 2.47.

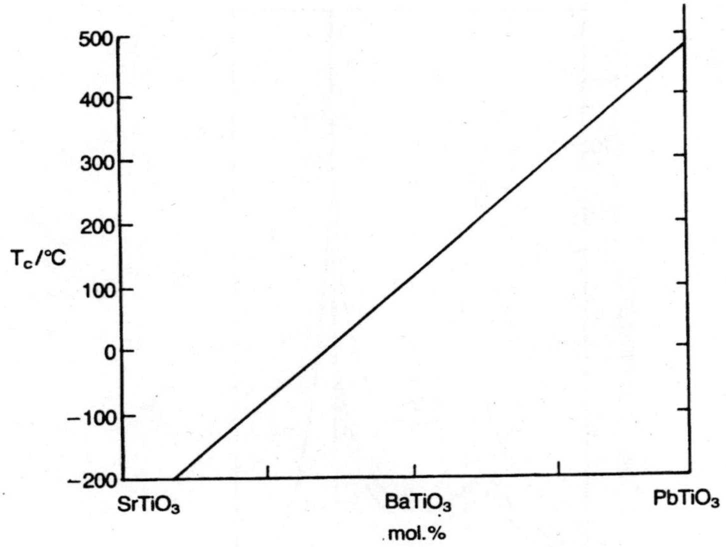

Fig. 2.47 The approximate effect on the Curie point of the substitution of either strontium or lead for barium in $BaTiO_3$.

2. A number of transition ions (Fe^{3+}, Ni^{2+}, Co^{3+}) that can occupy Ti^{4+} sites reduce that part of the dissipation factor due to domain wall motion.
3. Broadening of the permittivity–temperature peak can be effected by making additions, such as $CaZrO_3$ to $BaTiO_3$. The resultant materials may contain regions of variable composition that contribute a range of Curie points so that the high permittivity is spread over a wider temperature range, although at a somewhat lower level than that of the single peak.
4. Cations that have a higher valency than those they replace, when present at levels exceeding about 0.5 cation percent, e.g. La^{3+} in place of Ba^{2+} or Nb^{5+} in place of Ti^{4+}, generally inhibit crystal growth. This has the effect of raising the permittivity level below the Curie point as shown in Fig. 2.48. Crystal size is also controlled by sintering conditions. It has important effects on the electro-optical behaviour.
5. Higher-valency substituents at low concentrations (<0.2 cation percent) in $BaTiO_3$ lead to low resistivity. However, lower-valency substituents, such as Mn^{3+} on Ti^{4+} sites, act as acceptors and enable high-resistivity dielectrics to be sintered in atmospheres with low oxygen contents.

Ferroelectric ceramics find applications in capacitors, infrared detection, sound detection in air and water, the generation of ultrasonic energy, light switches, current controllers and small thermostatic devices. In all these cases

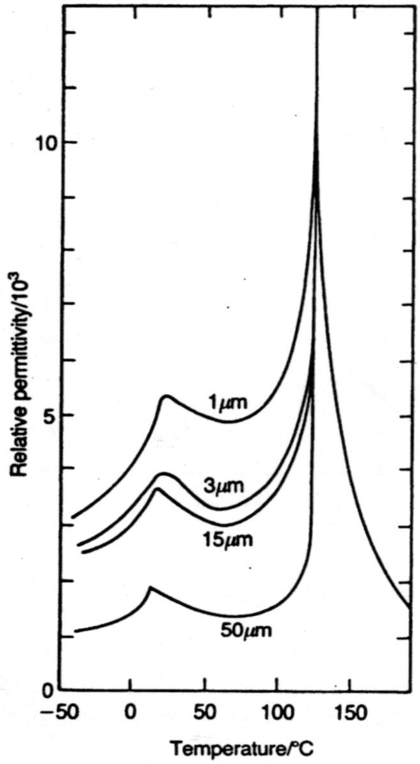

Fig. 2.48 The effect of grain size on the permittivity of a $BaTiO_3$ ceramic. (After K. Kinoshita and A. Yamaji (1976) *J Appl. Phys.* **47**, 371.)

some aspects of their ferroelectric activity have to be suppressed and others enhanced. A suitable compromise is achieved by a combination of composition and ceramic structure as described in the relevant sections.

2.7.4 Mixtures of dielectrics

The properties of mixtures of phases depend on the distribution of the components. The concept of 'connectivity' is useful in classifying different types of mixture. The basis of this concept is that any phase in a mixture may be self-connected in zero, one, two or three dimensions. Thus randomly dispersed and separated particles have a connectivity of 0 while the medium surrounding them has a connectivity of 3. A disc containing a rod-shaped phase extending between its major surfaces has a connectivity of 1 with respect to the rods and of 3 with respect to the intervening phase. A mixture consisting of a stack of plates of two different phases extending over the entire area of the body has a connectivity of 2–2. In all, 10 different connectivities are possible for mixtures of two phases (0–0, 1–0, 2–0, 3–0, 1–1, 2–1, 3–1, 2–2, 3–2, 3–3). There are 20 possibilities for mixtures of three phases and 35 for mixtures of four phases.

The commonest case to have been analysed is that of 3–0 systems. James Clerk Maxwell (1831–1879) deduced that the permittivity ε_m of a random dispersion of spheres of permittivity ε_1 in a matrix of relative permittivity ε_2 is given by

$$\varepsilon_m = \varepsilon_2\left\{1 + \frac{3V_f(\varepsilon_1 - \varepsilon_2)}{\varepsilon_1 + 2\varepsilon_2 - V_f(\varepsilon_1 - \varepsilon_2)}\right\} \tag{2.125}$$

where V_f is the volume fraction occupied by the dispersed particles. The result is independent of the size of the dispersed particles. For $\varepsilon_2 \gg \varepsilon_1$ and $V_f \leqslant 0.1$ Eq. (2.125) reduces to

$$\varepsilon_m = \varepsilon_2\left(1 - \frac{3V_f}{2}\right) \tag{2.126}$$

This formula can be used to convert the values of permittivity found for porous bodies to the value expected for fully dense bodies.

Eq. (2.125) gives results in good agreement with measurement when V_f is less than about 0.1. One reason for this limitation lies in the connectivity of the dispersed phase. At higher volume concentrations the disperse phase forms continuous structures that have a connectivity greater than zero. This is seen very clearly in the resistance–volume concentration relations for dispersions of conductive particles in insulating media (Fig. 2.49). The resistivity remains high until a critical concentration in the neighbourhood of 0.05–0.2 is reached when it drops by several orders of magnitude. There is a transition from a dispersion of separated particles to one of connected aggregates.

A capacitor containing a two-phase 1–3 dielectric consisting of rods of permittivity ε_1 extending from one electrode to the other in a medium of

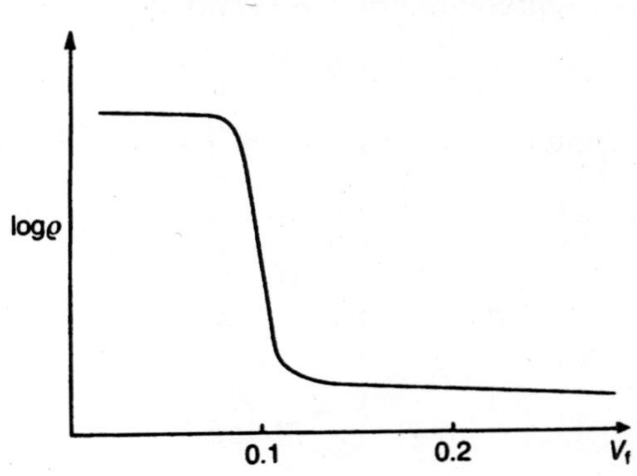

Fig. 2.49 Log resistivity versus volume fraction of conductive particles in an insulating matrix.

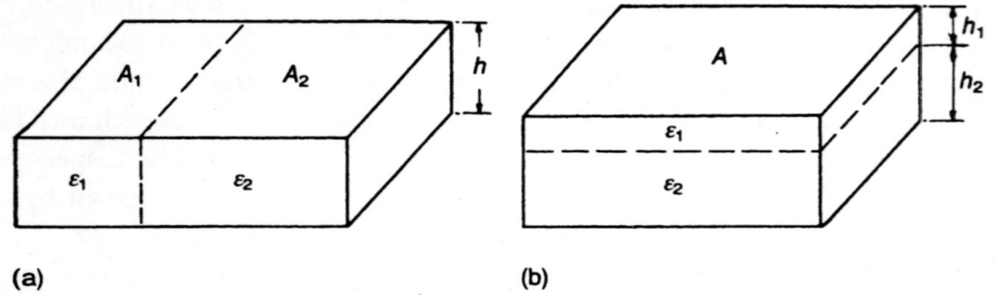

Fig. 2.50 Equivalent structures for dielectrics with (a) 1–3 and (b) 2–2 connectivity.

permittivity ε_2 is equivalent in behaviour to the simple composite shown in Fig. 2.50(a). The structure in Fig. 2.50(a) consists of two capacitors in parallel so that

$$\frac{\varepsilon_m A}{h} = \frac{\varepsilon_1 A_1}{h} + \frac{\varepsilon_2 A_2}{h}$$

that is,

$$\varepsilon_m = (1 - V_f)\varepsilon_2 + V_f \varepsilon_1 \tag{2.127}$$

A 2–2 connectivity dielectric with the main planes of the phases parallel to the electrodes is equivalent to the structure shown in Fig. 2.50(b). In this case there are effectively two capacitors in series so that

$$\frac{h}{A\varepsilon_m} = \frac{h_1}{A\varepsilon_1} + \frac{h_2}{A\varepsilon_2}$$

and

$$\varepsilon_m^{-1} = (1 - V_f)\varepsilon_2^{-1} + V_f \varepsilon_1^{-1} \tag{2.128}$$

Eqs (2.127) and (2.128) represent two extreme connectivity structures for dielectrics. Both can be represented by the formula

$$\varepsilon_m^n = (1 - V_f)\varepsilon_2^n + V_f \varepsilon_1^n \tag{2.129}$$

where $n = \pm 1$, or for a multiplicity of phases of partial volumes $V_{f_1}, V_{f_2}, \ldots, V_{f_i}$

$$\varepsilon_m^n = \sum_i V_{f_i} \varepsilon_i^n \tag{2.130}$$

The approximation $x^n = 1 + n \ln x$ is valid for small values of n and applying it to Eq. (2.130) gives

$$\ln \varepsilon_m = \sum_i V_{f_i} \ln \varepsilon_i \tag{2.131}$$

Although the theoretical backing for Eq (2.131) is perhaps questionable, it serves as a useful rule. It was first proposed by K. Lichtenecker [16] and can also

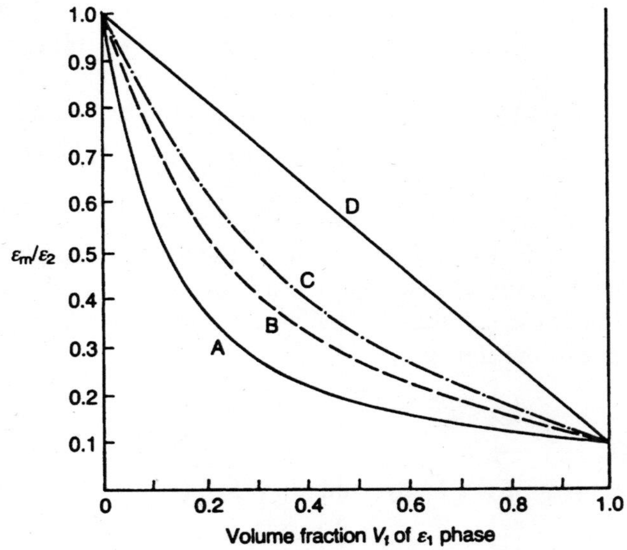

Fig. 2.51 Mixture relations for $\varepsilon_2/\varepsilon_1 = 10$: curve A, equation (2.128) (phases in series); curve B, equation (2.125) (Maxwell); curve C, equation (2.131) (Lichtenecker); curve D, equation (2.127) (phases in parallel).

be applied to the electrical and thermal conductivity and the magnetic permeability of mixed phases. Plots of Eqs (2.125), (2.127), (2.128) and (2.131) are shown in Fig. 2.51 where it can be seen that both Lichtenecker's and Maxwell's equations give predictions intermediate between those of the series and parallel models.

Differentiation of Eq. (2.131) with respect to temperature leads to

$$\frac{1}{\varepsilon_{\mathrm{m}}}\frac{\partial\varepsilon_{\mathrm{m}}}{\partial T} = \sum_i \frac{V_{\mathrm{f}_i}}{\varepsilon_i}\frac{\partial\varepsilon_i}{\partial T} \tag{2.132}$$

which gives the temperature coefficient of permittivity for a mixture of phases and, although not in exact agreement with observation, is a useful approximation.

2.7.5 Impedance spectroscopy

Impedance spectroscopy is discussed in depth in the monograph edited by J.Ross Macdonald [17]. It has its origins in the classical work of K.S. Cole and R.H. Cole, published more than 60 years ago, concerned with methods of plotting the response of a dielectric material to applied voltages as a function of frequency. The method assists in identifying observed relaxation effects with processes at the atomic and microstructural levels. For a system having a single well-defined relaxation time the response would be in accordance with the Debye equations, Eqs (2.117) and (2.118). A plot of $\varepsilon_r''(\omega)$ against $\varepsilon_r'(\omega)$ takes the form of a semicircle.

This can be verified by eliminating $(\omega\tau)$ between the two equations, leading to

$$\{\varepsilon_r'(\omega) - \tfrac{1}{2}(\varepsilon_{rs}' + \varepsilon_{r\infty}')\}^2 + \varepsilon_r''(\omega)^2 = \tfrac{1}{4}(\varepsilon_{rs}' - \varepsilon_{r\infty}')^2, \tag{2.133}$$

the equation for the semi-circle shown in Fig. 2.52.

It is instructive, and more elegant, to prove the form of the plot as follows. In the complex plane vectors $\boldsymbol{\varepsilon}_r^*$, $\boldsymbol{\varepsilon}_{r\infty}$, $\boldsymbol{\varepsilon}_{rs}$, **u** and **v** are defined as shown in Fig. 2.53.

A point on the curve ($\varepsilon'(\omega)$ and $\varepsilon''(\omega)$ at a particular frequency) is defined by the vector $\boldsymbol{\varepsilon}_r^*$. From the figure

$$\mathbf{v} = \mathbf{u} - (\boldsymbol{\varepsilon}_{rs} - \boldsymbol{\varepsilon}_{r\infty}), \tag{2.134}$$

$(\boldsymbol{\varepsilon}_{rs} - \boldsymbol{\varepsilon}_r)$ being a constant length.

The complex form of the Debye equation (Eq. (2.116)) is

$$(\boldsymbol{\varepsilon}_r^* - \boldsymbol{\varepsilon}_{r\infty})(1 + j\omega t) = (\boldsymbol{\varepsilon}_{rs} - \boldsymbol{\varepsilon}_{r\infty}) \tag{2.135}$$

From the figure $\boldsymbol{\varepsilon}_r^* = \boldsymbol{\varepsilon}_{r\infty} +$ **u** and substituting this, and the expression for $(\boldsymbol{\varepsilon}_{rs} - \boldsymbol{\varepsilon}_{r\infty})$ given by Eq. (2.134), into Eq. (2.135) leads to $\mathbf{v} = -j\omega\tau\,\mathbf{u}$. This shows that in the complex plane **u** and **v** are at right angles to each other; that is they define points on a semicircle of diameter $(\varepsilon_{rs} - \varepsilon_{r\infty})$.

When a voltage step is applied to the simple RC parallel circuit shown in Fig. 2.54 the response current decays to zero in a manner describable by a single relaxation time. The frequency response of the impedance also yields a semicircle as shown below. Such a circuit can represent a 'lossy' capacitor, and more elaborate combinations of resistors and capacitors correspondingly more electrically complex materials and systems. It is this rather more general approach which is described by 'impedance spectroscopy'.

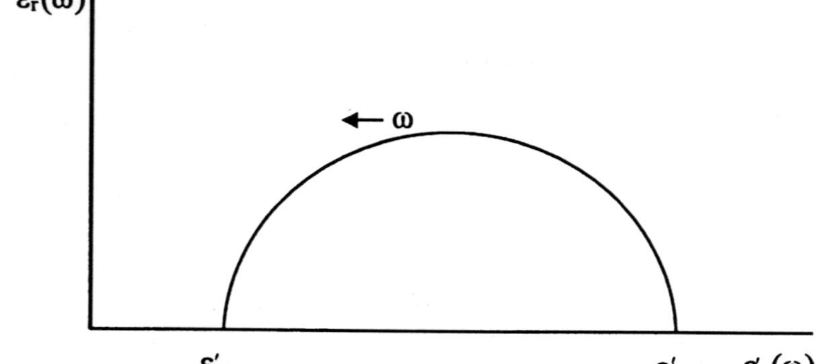

Fig. 2.52 Plot of the frequency response of the real and imaginary components of the relative permittivity of a dielectric satisfying the Debye equations.

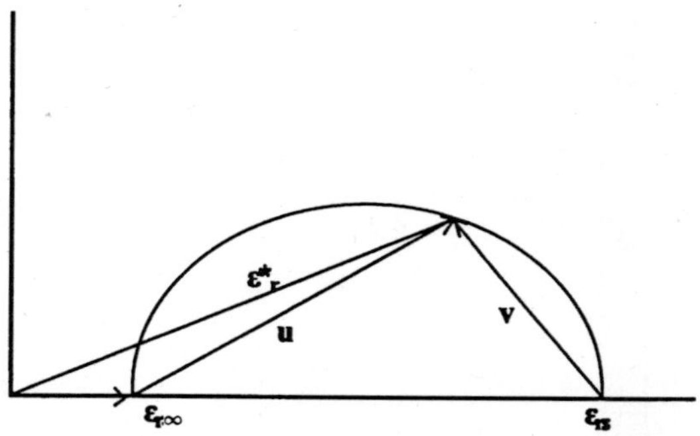

Fig. 2.53 Permittivity as $f(\omega)$ plotted in the complex plane.

The impedance of the circuit Fig. 2.54 is

$$Z^* = R(1 - j\omega RC)/(1 + \omega^2R^2C^2) \tag{2.136}$$

so that the real and imaginary components are

$$Z' = R/(1 + \omega^2R^2C^2) \tag{2.137}$$

and

$$-Z'' = \omega R^2C/(1 + \omega^2R^2C^2) \tag{2.138}$$

Since the time constant of the circuit is τ (= RC) the equations can be written

$$Z' = R/(1 + \omega^2\tau^2) \tag{2.139}$$

and

$$-Z'' = R\omega\tau/(1 + \omega^2\tau^2) \tag{2.140}$$

which are of the same form as Eqs (2.117) and (2.118), that is Debye-like, with R in the place of $(\varepsilon'_{rs} - \varepsilon'_{r\infty})$.

Two such circuits having different relaxation time constants and connected in series lead to two semicircles as shown in Fig. 2.55(b). As in the case of any other spectroscopic analysis the separate responses may overlap and the experimental curve must then be resolved into its separate constituent semicircles. Impedance spectroscopy makes use of other electrical parameters, including the 'admittance' $(Y = Z^{-1})$, to assist in quantifying the circuit parameters.

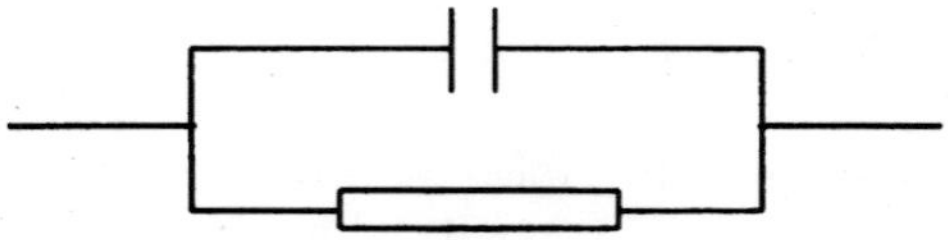

Fig. 2.54 The equivalent circuit simulating a 'lossy' capacitor.

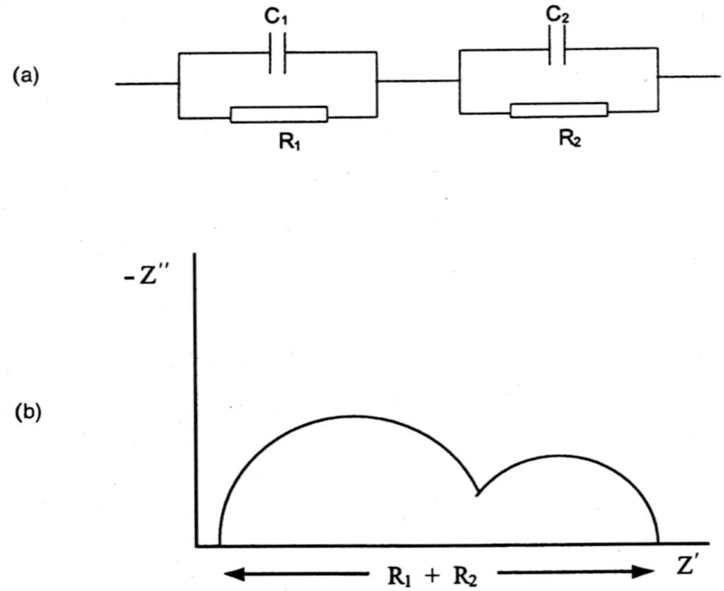

Fig. 2.55 (a) Series-parallel RC circuit which could, depending on component values, produce the frequency response shown in (b).

The frequency response of a real polycrystalline ceramic carrying metallic electrodes may yield three well defined semicircles representing respectively polarization processes associated with the interior of the grains, with the grain boundary regions and with the electrode–ceramic interfacial region. Such a clear-cut situation is illustrated in Fig. 2.56.

Impedance spectroscopy is a valuable addition to the range of techniques for the electrical characterisation of materials and components, especially where interfaces (e.g. metallic electrode/cathode or anode materials) are involved. This is particularly so in the case of batteries and fuel cells, the subjects of the most intensive research and development stimulated by the need to develop power sources which reduce dependence on fossil fuels and their damage to the environment (see Section 4.5.1).

In conclusion, the following extract from the Introduction to a text (V.V. Daniel (1967) *Dielectric Relaxation*, Academic Press, 1967) dealing with dielectric relaxation is thought-provoking.

> Relaxation is a process. The term applies, strictly speaking, to linear systems where a response and stimulus are proportional to one another in equilibrium. Relaxation is a delayed response to a changing stimulus in such a system. Dielectric relaxation occurs in dielectrics, that is in insulating materials with negligible or small electrical conductivity. The stimulus is almost always an

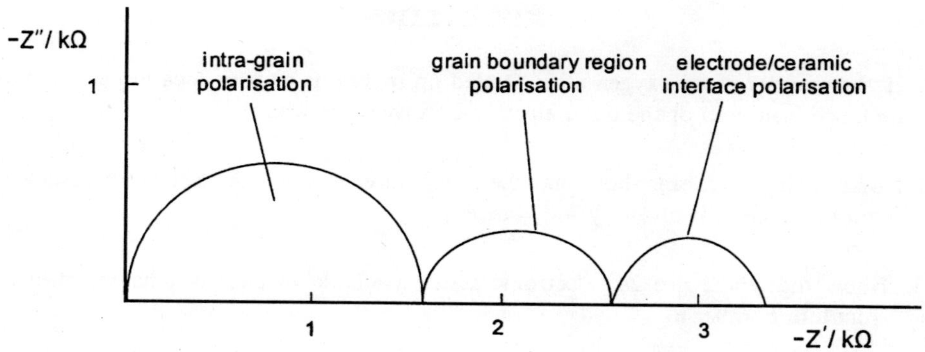

Fig. 2.56 Form of impedance plot for an yttria-stabilized zirconia (see Section 4.5.3.1(b)) at 490°C ($p_{O_2} = 10^{-2}$ atm.) based on data by E.J.L Schouler *et al.* (1983) *Solid State Ionics* **9** and **10**, 989.

electrical field, the response a polarization. The time lag between field and polarization implies an irreversible degradation of free energy to heat.

The objective of this monograph is to describe and interpret the time dependence of the electrical response of dielectrics. Interpretation is difficult because the observable relationship between polarization and field is simple in the cases relevant for dielectric relaxation and because the measurements have relatively little information content. The response of the dielectric can be described by a set of linear differential equations and many models can be described which correspond to the same differential equations. When the dielectric relaxation of a given material has been measured the investigator is in the position of a man presented with a black box which has two terminals. He may apply alternating fields of various kinds and he may heat the box but he is not allowed to look inside. And he finds that the box behaves as if it contained a combination of capacitors and resistors.

The reaction of our investigator to the puzzle presented by the black box will differ according to whether he is a mathematician, electrical engineer, physicist or chemist. The mathematician will be satisfied by a description in terms of differential equations and the engineer by an equivalent circuit. However the physicist or chemist will want an interpretation in terms of the structure of the material whose response can be represented by the black box. The materials scientists will often be disappointed.

The relaxation of the polarization in response to a change of the field applied to a material is not due directly to the pull of the field, as suggested by the naïve imagination. It is brought about by thermal motion, and fields of the magnitude relevant to dielectric relaxation perturb the thermal motion only slightly.

The structural interpretation of dielectric relaxation is a difficult problem in statistical thermodynamics. It can for many materials be approached by considering dipoles of molecular size whose orientation or magnitude fluctuates spontaneously, in thermal motion. The dielectric constant of the material as a whole is arrived at by way of these fluctuations but the theory is very difficult because of the electrostatic interaction between dipoles. In some ionic crystals the analysis in terms of dipoles is less fruitful than an analysis in terms of thermal vibrations. This also is a theoretically difficult task forming part of lattice dynamics. In still other materials relaxation is due to electrical conduction over paths of limited length. Here dielectric relaxation borders on semiconductor physics.

Problems

1. If the nucleus of an oxygen atom, scaled up to 1 m, is at your location estimate the distance from you of the outer electrons. [Answer: ≈ 50 km]

2. For a hydrogen atom, show that the potential energy of the electron at distance R from the nucleus is given by $-e^2/4\pi\varepsilon_0 R$.

3. Show that there are $2n^2$ electronic states available to electrons having principal quantum number n.

4. Estimate the wavelength of the electrons in a 300 kV electron microscope. [Answer: 2.24 pm; 1.97 pm (relativistic)]

5. An electron beam is accelerated through a potential difference of 12 kV and falls onto a copper target. Calculate the shortest-wavelength X-rays that can be generated. [Answer: 103 pm]

6. In a metal oxide the oxygen ions are in a hexagonally close-packed array. Calculate the size of an ion that would just fit (into (i) an octahedral interstice and (ii) a tetrahedral interstice. [Answer: $(\sqrt{2}-1)r_0 = 58$ pm; $(\sqrt{(3/2)}-1)r_0 = 31.5$ pm]

7. Calculate the crystal density of $BaTiO_3$. The dimensions of the tetragonal unit cell are: $a = 399.2$ pm; $c = 403.6$ pm. Relative atomic masses: O, 16.00; Ti, 47.90; Ba, 137.3. [Answer: 6022 kg m^{-3}]

8. Estimate the molar fraction of Schottky defects in a crystal of metal oxide MO at 2000 °C given that the formation enthalpy of a single defect is 2 eV. [Answer: 6.08×10^{-3}]

9. Using the Fermi function, estimate the temperature at which there is a 1% probability that an electron in a solid will have an energy 0.5 eV above the Fermi energy. [Answer: 1300 K]

10. In an experiment concerning the resistivity of 'pure' silicon carbide the specimen was irradiated with electromagnetic radiation the frequency of which was steadily increased. A large decrease in resistivity was observed at a wavelength of 412 nm. If in the absence of radiation the resistivity of the specimen was 0.1 Ω m at room temperature, estimate the intrinsic resistivity at 400 °C.
    Comment on the realism of the estimate.

11. The electrical conductivity of KCl containing a small amount of $SrCl_2$ in solid solution is measured over a temperature range where both intrinsic and extrinsic behaviour is observed. It is assumed that the dominant intrinsic disorder is of the Schottky type. The data fit the relationships

$$\sigma_i = 6.5 \times 10^7 \exp\left(-\frac{2.3 \times 10^4}{T}\right) \mathrm{S\,m^{-1}}$$

$$\sigma_e = 5.3 \exp\left(-\frac{8.2 \times 10^3}{T}\right) \mathrm{S\,m^{-1}}$$

in the intrinsic and extrinsic ranges respectively. Assuming that the dopant has led to the formation of cation vacancies, estimate the enthalpy change associated with the creation of a single Schottky defect. [Answer: 2.55 eV]

12. Derive the Nernst–Einstein relationship.

    The electrical conductivity of sapphire in a particular crystallographic direction was found to be $1.25\,\mathrm{mS\,m^{-1}}$ at 1773 K. An independent experiment on the same material at the same temperature determined the oxygen tracer diffusion coefficient to be $0.4\,\mathrm{nm^2\,s^{-1}}$, the diffusion occurring by a vacancy mechanism. Do these data favour oxygen ion movement as the dominant charge transport mechanism? (Relative atomic masses, Al = 27 and O = 16; density of sapphire, $3980\,\mathrm{kg\,m^{-3}}$.)

13. A sample of potassium ferrite with the chemical formula $K^+_{1.25}Fe^{2+}_{0.25}Fe^{3+}_{10.75}O^{2-}_{17}$ is a mixed ionic/electronic conductor with the β-alumina structure (see Section 4.5). It contains 4.07×10^{27} potassium ions per cubic metre located in the (001) planes. For this material, the total electrical conductivity at 573 K is $1.53 \times 10^{-2}\,\mathrm{S\,m^{-1}}$ and the diffusion coefficient at 573 K for potassium ions is $1.89 \times 10^{-14}\,\mathrm{m^2\,s^{-1}}$. Calculate the transport number for potassium ions at 573 K. If the energy of migration of potassium ions is $23\,\mathrm{kJ\,mol^{-1}}$, what will be the ionic conductivity of the sample at 298 K?

    (Reproduced, in slightly modified form, by kind permission of the University of Cambridge) [Answers: 0.016; $5.57 \times 10^{-6}\,\mathrm{S\,m^{-1}}$]

14. From the data given in Fig. 2.12, together with other data given in the text, estimate the ratio of electron concentration to hole concentration for 'pure' $BaTiO_3$ at 800 °C and $p_{O_2} = 10^{-15}$ Pa. [Answer: of order 10^5]

15. Estimate the band gap for $BaTiO_3$ from the minima in the σ versus p_{O_2} isotherms in Fig. 2.12, together with other data drawn from the text. [Answer: ≈ 3.0 eV]

16. A square parallel-plate capacitor of side 10 mm and thickness 1 mm contains a dielectric of relative permittivity 2000. Calculate the capacitance.

    When 200 V is applied across the plates calculate (i) the electric field in the dielectric, (ii) the total charge on one of the plates, (iii) the polarization in the dielectric and (iv) the energy stored in the dielectric. [Answers: 1.77 nF; (i) $200\,\mathrm{kV\,m^{-1}}$, (ii) 354 nC, (iii) $3.54\,\mathrm{mC\,m^{-2}}$ and (iv) $35\,\mu$J]

17. Using the data given in Table 2.5 together with Eq. (2.89) calculate the relative permittivity of the mineral forsterite (see Section 5.5.1). The orthorhombic cell

dimensions are $a = 475.6$ pm, $b = 102.1$ pm, $c = 598.0$ pm and the cell contains 4 formula units, i.e. $4(Mg_2SiO_4)$. [Answer: $\varepsilon_r = 7.03$]

18. A disc capacitor of thickness 1 mm carries circular electrodes of diameter 1 cm. The real and imaginary parts of the relative permittivity of the dielectric are 3000 and 45 respectively. Calculate the capacitance and the power dissipated in the dielectric when a sinusoidal voltage of amplitude 50 V and frequency 1 MHz is applied to the capacitor. [Answer: 245 mW]

19. A capacitor in the form of a ceramic disc 10 mm diameter and 1 mm thick is electroded to its edges. It has a capacitance of 2000 pF and a dissipation factor of 0.02. Calculate the relative permittivity of the dielectric, the dielectric loss factor, a.c. conductance at 50 Hz and at 50 MHz, and the loss current flowing for an applied sinusoidal voltage of 10 V (r.m.s.) at 1 kHz. [Answers: 2877; 57.5; 1.27×10^{-8} S; 1.27×10^{-2} S; 2.54 μA]

20. A dielectric is a fine interdispersion of two phases, one having a relative permittivity of 20 and a TCC of -500 MK^{-1} and the other a relative permittivity of 8 and TCC of 100 MK^{-1} In what proportion should the two phases be mixed to give zero TCC, and what is the relative permittivity of the mixture? [Answers: 16.7 vol% of phase with TCC $= -500$ MK^{-1}; $\varepsilon_r = 9.3$]

21. A simple circuit comprises a capacitor C shunted by a resistor R. Show that the frequency response of the circuit impedance, plotted in the complex plane, is represented by a semicircle of diameter R and with center at $Z^1 = \frac{1}{2}R$.
    Two such circuits are connected in series. In one the resistor and capacitor values are respectively 3 kΩ and 3.3 nF and in the other 10 kΩ and 1 μF. Sketch the expected frequency response of the impedance in the complex plane, marking the main features of the curves with numerical values.

22. The quote from the monograph 'Dielectric Relaxation' (see p. 88) was written nearly 40 years ago. White a brief critical commentary, about the same length as the quoted passage, giving your opinions on the aptness, or otherwise, of the views expressed to present day materials' research and development in the electroceramics context.

Bibliography

1. Cotton, F.A. and Wilkinson, G. (1980) *Advanced Inorganic Chemistry*, 4th edn, J. Wiley and Sons, New York.
2. Shannon, R.D. and Prewitt, C.T. (1969) Effective Ionic Radii in Oxides and Fluorides. *Acta. Cryst.* **B. 25**, 925–46 (Revised values in *Acta. Cryst.* B.26, 1970, 1046–8).
3. Newnham, R.E. (1975) *Structure-Property Relations*, Springer-Verlag, Berlin.
4. Smyth, D.M. (2000) *The Defect Chemistry of Metal Oxides*, Oxford University Press, Oxford.

5. Kittel, C. (1986) *Introduction to Solid State Physics*, 6th edn, John Wiley and Sons Inc., New York.
6. Bleaney, B.I. and Bleaney, B. (1976) *Electricity and Magnetism*, 3rd edn, Oxford University Press, Oxford.
7. Seuter, A.M.J.H. (1974) *Defect chemistry and electrical transport properties of barium titanate*. Philips Research Reports, Supplement No. 3.
8. Daniels, J., Hardtl, K.H., Hennings, D. and Wernicke, R. (1976) *Defect chemistry and electrical conductivity of doped barium titanate ceramics*. Philips Research Reports **31**, 487–559.
9. Shannon, R.D. (1993) Dielectric polarizabilities of ions in oxides and fluorides. *J Appl. Phys*. **73(1)**, 348–66.
10. Lines, M.E. and Glass, A.M. (1977) *Principles and Applications of Ferroelectrics and Related Materials*, Clarendon Press, Oxford.
11. Herbert, J.M. (1985) *Ceramic Dielectrics and Capacitors*, Gordon and Breach, New York.
12. Megaw, H.D. (1957) *Ferroelectricity in Crystals*, Methuen & Co. Ltd, London, p. 85.
13. Arlt, G. (1990) Twinning in ferroelectric and ferroelastic ceramics: stress relief. *J. Mater. Sci*. **25**, 2655–66.
14. Arlt, G. (1998) Strong ultrasonic microwaves in ferroelectric ceramics. *IEEE Trans. Ultrason. Ferroelectr. Freq. Control* **45**, 4–10.
15. Hall, D. (2001) Nonlinearity in piezoelectric ceramics. *J. Mater. Sci*. **36**, 4575–601.
16. Lichtenecker, K. (1926) Die dielektrizitätskonstante natürlicher und Künstlicher Mischkörper. *Physikal Z*. **27**, 115–158.
17. Macdonald, J.R. (Ed.) (1987) *Impedance Spectroscopy*, J Wiley and Sons, New York.

3

PROCESSING OF CERAMICS

The following summary of ceramics' processing technology can be supplemented by reference to the monographs by J.S. Reed [1] and M.N. Rahaman [2]. Many of the important aspects of present-day ceramics processing outlined below are treated in depth in [3]. Access to a text concerned with elementary colloid science, for example that by D.J. Shaw [4], may be advisable.

3.1 General

The objectives of fabrication are to produce

1. a material with specific properties,
2. a body of a required shape and size within specified dimensional tolerances and
3. the required component at an economic cost.

The material properties are basically controlled by the composition but will also be affected by the grain size and porosity of the sintered ceramic, and the latter features are affected by the method of fabrication.

Key stages in the fabrication of ceramics are calcination and sintering, which are sometimes combined. During these processes the constituent atoms redistribute themselves in such a way as to minimize the free energy of the system. This involves a considerable movement of ions, their interdiffusion to form new phases, the minimization of the internal surface area and an increase in grain size. Usually the overall dimensions shrink and, from a practical viewpoint, it is important that the shrinkage is reproducible so that pieces of specified dimensions can be made. It is also necessary that the shrinkage should be uniform within a given piece, otherwise the shape will be distorted. For this there

Electroceramics: Materials, Properties, Applications. 2nd Edition. Edited by A. J. Moulson and J. M. Herbert.
 ISBN 0 471 49747 9 (hardback) 0 471 49748 7 (paperback)

must be a minimum variation in the density throughout the 'green' body. (The term 'green' is widely used to describe the unfired ceramic.)

The fabrication process comprises five stages:

1. the specification, purchase and storage of raw materials;
2. the preparation of a composition in powder form;
3. forming the powder into a shape;
4. densification;
5. finishing.

3.2 Cost

A detailed consideration of the cost aspect is beyond the scope of this book but it is vitally important. Factory space and plant cost money in rent and maintenance so that, the longer they are occupied in producing a given number of items, the higher is the cost of each. The number of people involved in a process, especially if they are skilled, often accounts for a major part of the cost. The price of raw materials may only contribute a small fraction to the ultimate cost but must be kept to a minimum. An unnecessarily close specification for purity, dimensional tolerances etc. may result in a large increase in price.

Scale of production can greatly affect cost since, if production levels warrant it, the process can be highly automated; it must be borne in mind, however, that automatic machinery is only economic when fully utilized.

Some indication of the relative costs of processes is given, but it has to be emphasized that these are based on the present state of the art and that future developments or changed circumstances may drastically modify them.

3.3 Raw Materials

The simple label on a laboratory container, e.g. 'calcium carbonate', gives only an approximate indication of the composition of its contents. The labels on analytical reagents may also give the maximum levels of certain impurities, which is an adequate description of the contents for many purposes although it needs checking. One aspect of the composition that is difficult to specify exactly is the moisture content. It depends on the ambient humidity, the method of storage and the particle size. It may vary from less than 0.1% to more than 1% for non-hygroscopic materials and, of course, much more widely if they are hygroscopic.

However, it is usually quite easy to determine as a loss in weight after heating to a suitable temperature.

Chemical composition is only one aspect of the specification of a raw material. Particle size and degree of aggregation are often important, and are more difficult to specify and to agree on with a supplier. It is even more difficult to specify the 'reactivity' of a material but it can be of primary importance. Reactivity depends on particle size and the perfection of the crystals comprising the particles. Many raw materials are subjected to heat treatment during their preparation, and variations in the temperature and duration of this may affect reactivity. These raw material parameters are modified by the subsequent processing so that corrections can be made for some variations in them provided that the magnitude of the variations is known. The more detailed is the specification the more expensive is the material, so that a balance has to be kept between the elaboration of the specification and the cost of testing and correcting for variations during processing.

The ultimate test for a starting material is that when it is put through a standard process it yields a product with specified properties. The assumption is often made that the process used in production is sufficiently standard to be used as a basis for raw material testing. This assumption needs careful confirmation since wear and tear causes changes in the behaviour of plant and machinery. Frequently a model production line is set up in the laboratory with a closely controlled and consistent performance and bearing a well-established relationship to the large-scale production line. This can serve as a check on the production process. Such a line is a necessity when introducing a new process but may be expensive to maintain. Most commercial operations employ a mixture of analytical and physical tests coupled with laboratory and manufacturing trials in an endeavour to achieve an economical compromise. The difficult factor to estimate is the possible cost of a breakdown in production leading to a high scrap level or a loss of orders.

The storage of materials is important. The cheapest method is a heap in the open, which may be adequate for coal for a furnace. The most expensive is a room in which the humidity and temperature are controlled. The importance of controlling moisture content is the most relevant factor determining the amount spent on storage space. The quantity of material kept in store has significant economic consequences. On the one hand it represents idle capital; on the other each of a succession of small batches needs handling and testing, both involving expense. It may seem superfluous to mention that every batch of every material should be clearly identified and accessible, but such details can lead to much trouble if neglected.

3.4 Powder Preparation – Mixing and Grinding

The first step is to weigh out the raw material with due allowance for impurity and moisture content. Many modern weighing machines record each weighing;

otherwise the only safeguard against normal human fallibility is to have every weighing checked by a second person.

The next step is mixing, eliminating aggregates and/or reducing the particle size. If compound formation is to occur during calcining or firing, the matter of neighbouring particles must interdiffuse and the time taken to complete the process will be proportional to the square of the particle size. The process will clearly be considerably slower if the particles consist of aggregates of crystals rather than individual crystals. Even where there is only a single component, aggregates that are present during sintering usually densify more rapidly internally than with neighbouring aggregates, with a resulting residue of pores in the spaces originally between the aggregates. Apart from breaking up agglomerates and forming an intimate mixture of the constituents, a milling process introduces defects into the crystals which may enhance diffusion and accelerate sintering.

The most commonly used method of initial mixing is wet ball-milling. A ball-mill is a barrel (usually of ceramic material) that rotates on its axis and is partially filled with a grinding medium in the form of spheres, cylinders or rods. The grinding medium is in such quantity that the rotation of the mill causes it to cascade, rather like the breaking of a wave on the shore, so that both a shearing and a crushing action is applied to any material lying between the milling elements. For efficient action the real volume of material to be milled should be about a third of that of the milling media. The volume of liquid must be sufficient to form a freely flowing cream, usually between 100 and 200% of the volume of the powder. The addition of surface-active agents reduces the amount of water needed for a given fluidity and assists in the dispersion of the softer aggregates. The ball charge normally contains a range of sizes with the largest diameter being of the order of a tenth of the diameter of the barrel. The optimum quantity of balls, their sizes and composition, the quantity of 'slip' and the rate of rotation are normally defined by the supplier of the plant but require adjustment for particular applications. Ball-mills are slow but mechanically simple and robust. They mix and eliminate aggregates and can reduce the particle size to around 10 μm.

The milling media are inevitably abraded by the charge and therefore contaminate it. Contamination is minimized by adhering to well-established practice which will optimize the time of milling, by using hard materials for the grinding media, e.g. flint, alumina or tungsten carbide, or by using materials of the same composition as one of the constituents of the charge, e.g. alumina for aluminous porcelains or steel for ferrites. The contamination may be less than 0.1% but can rise to 1 or 2% under adverse conditions. It is important that the media are inspected periodically and undersized pieces discarded since they will eventually be smashed by impact with larger pieces and contribute excessive contamination. The barrel is normally made of aluminous porcelain and will also be abraded. It can be coated internally with an abrasion-resistant polymer or formed entirely of a polymer, and any contamination from this will be burned away in subsequent firing stages; however, larger pieces of polymer that survive up to the final sintering may result in large pores.

Wet ball-milling is faster than dry-milling and facilitates the separation of the milling media from the charge. Its disadvantage is that the liquid must be removed, which is most economically achieved by filtration, but any soluble constituent will be removed with the liquid. If this is undesirable the alternative is to evaporate the liquid which can be effected by spray-drying if the liquid is water. Evaporation can also take place in shallow trays in ovens; in this case soluble material may appear as a skin on the surface and heavy coarse particles may settle out.

The liquid used is usually water since it is available with adequate purity at low cost and is non-flammable and non-toxic. However, because it results in a powder with a very thoroughly hydrated surface which may have a significantly reduced reactivity during the sintering process, dry-milling is sometimes preferred. In this case it is usually necessary to add a surface-active material such as a stearate containing an inorganic cation to prevent the powder forming a cake on the mill walls. Dry-milling can be combined with the continuous removal of fine particles by forcing air over the surface of the rotating charge and filtering out the entrained material.

Alternatives to ball-mills include vessels with rotating paddles and rotating barrels that contain baffles that ensure that the powder is tossed about. These have no grinding action so that the particle size is unaltered. They also have little effect on aggregates.

So-called 'vibro-energy' mills are machines for vibrating vessels filled with grinding media at amplitudes up to approximately 5 cm. The interstices are filled with a slip containing the material to be ground. Size reduction is far more rapid than in ball-mills and the particle size can be regulated to some extent by the size of the media elements; 3-mm alumina balls give a size range of 1–3 μm. Similar results are obtained by having a paddle rotating through the grinding media in a stationary vessel – the 'attritor' mill. In this case the heat generated must be removed by water cooling. Such methods can yield powders in which a high proportion of particles have sizes below 1 μm.

Contamination by the milling media can be largely avoided by the use of fluid energy mills. In these the powder is entrained in two streams of high-velocity air which are made to impinge on one another so that the particles are broken up by impact. The feedstock must already be ground to within a factor of 10 of the final size required and adjustments are necessary to suit particular materials.

Very good mixing on an atomic scale can be achieved by chemical methods [5], and in the context of electroceramics the production of high purity, sub-micron barium titanate-based powders for the manufacture of multilayer capacitors (see Section 5.4.3) is of paramount importance. Tight control over powder chemistry has a direct and significant influence over capacitor failure rates. Also, the strong interest in ferroelectric random access memories (FeRAMs) and in new ceramics for dynamic random access memories DRAMS (see Section 5.7.5) has stimulated demand for precursors for the fabrication of thin films that can be incorporated into integrated circuits.

If the cations of interest, Ba^{2+} and Ti^{4+} in the case of barium titanate, can be obtained in solution in the correct ratio then precursors to the final required product can be formed. Subsequent heating of the precursors leads to the formation of $BaTiO_3$ (BT) involving the minimum of ion diffusion distances. Because of this the oxides can be formed at relatively low temperatures, for example ~500 °C rather than the usual >1000 °C necessary to effect the solid state reaction between mixed oxides.

An example of this approach is the co-precipitation of mixed oxides, hydroxides and carbonates from aqueous solution. Another approach is for the cations of interest to be complexed to form an organometallic compound, an alkoxide, for example. Subsequent hydrolysis and heat treatment of the precursors yields the required oxide. Cations can also often be incorporated into sols and gels and the mixed oxides produced by dehydration.

There are four principal routes to producing barium titanate powders.

3.4.1 The 'mixed oxide' or solid state route

This route is the solid state reaction at temperatures above ~1100 °C of a mixture of $BaCO_3$ and TiO_2 – together with 'modifiers'. The calcination is followed by comminution to reduce the particle size to approximately 1 μm.

Although the route is still widely used for commercial scale production, as the demands for 'modified' powders having precisely controlled physical and chemical characteristics intensifies, so there will be a progressive move away from the 'mixed oxide' toward the 'chemical' routes.

3.4.2 The oxalate route

Well-crystallized particles of barium titanyl oxalate ($BaTiO(C_2O_4)_2\ 4H_2O$; 'BTO') are precipitated from acidic (HCl) solutions of $TiOCl_2$ and $BaCl_2$ with oxalic acid ($C_2O_4H_2$). The chemical precipitation conditions can be adjusted to control the Ba/Ti ratio. Calcination at ~1000 °C leads to the development of sub-micron BT crystallites strongly bonded together in aggregates as large as 10 μm. These have to be milled down to the primary BT sub-micron particle size Dopants can be co-precipitated with the BTO.

The fully developed route is capable of producing powders of closely controlled stoichiometry, composition and particle size on a large scale commercial basis.

3.4.3 The alkoxide route

Essentially this route involves mixing a solution of $Ba(OH)_2$ with Ti $(C_3H_7O)_4$ (titanium isopropoxide). Depending upon the reaction conditions, extremely fine (~50 nm) single crystal particles can be formed. The route is suited to large scale commercial production.

3.4.4 Hydrothermal synthesis

This is a relatively low-temperature, water-based route capable of producing submicron, spherical and uniform sized particles of either high purity or chemically modified BT. Essentially barium, titanium and dopant compounds are reacted in a basic aqueous medium to form hydroxides. Under the hydrothermal conditions, typically in the temperature and pressure ranges 100–250 °C and 100 kPa–3 MPa (~ −30 atm) respectively, sub-micron particles of either pure or modified barium titanate are precipitated. There are many variables which need careful control, especially the reactive areas of the precursors and the degrees of supersaturation of the various species.

Not surprisingly 'water' in the form of hydroxyl groups can become incorporated into the titanate structure and the evolution of this during subsequent processing stages towards the sintered ceramic has to be planned for. In its fundamentals the situation is identical to that described in Section 4.6.1 concerning the development of proton conductors.

Developments of the process permit coatings of a range of dopants to be applied to the surfaces of the particles which, during sintering, produce the 'core-shell' structure leading to X7R characteristics (see Section 5.7.1). The coatings may also consist of sintering aids.

It is also possible to retain the hydrothermally produced barium titanate particles in aqueous suspension to form the basis of a tape-casting slurry capable of producing $< 3\,\mu m$ dielectric layers. This route avoids the risk of the formation of hard agglomerates on drying the precipitates.

Hydrothermal synthesis is being increasingly adopted for the commercial production of compositionally tailored barium titanates for MLCCs.

3.5 Calcination

In some cases (e.g. certain grades of barium ferrite) a powder can immediately be formed into a shape and sintered without prior heat treatment, but usually there is an intermediate calcination at a lower temperature. Calcination causes the constituents to interact by interdiffusion of their ions and so reduces the extent of the diffusion that must occur during sintering in order to obtain a homogeneous

Fig. 3.1 Industrial-scale rotary kiln. (Courtesy of Philips Technical Review.)

body. It can therefore be considered to be part of the mixing process. The calcination conditions are important factors controlling shrinkage during sintering. The required final phases may not be completely formed but the remaining chemical gradients may assist sintering. The main requirement is that calcination should yield a very consistent product.

Calcination can be carried out by placing the mixed powders in shallow saggers in a batch or continuous kiln. The saggers may need to be closed if any of the constituents are volatile, as is the case with lead oxide. The container surfaces in immediate contact with the powder must not react with it both to avoid contamination and to permit reuse of the sagger. The thermal conductivity of powdered materials is always low, so that a sufficiently uniform temperature can only be obtained through a depth of a few centimetres when the period at maximum temperature is, as is usual, only 1 or 2 h.

Calcination can also be carried out continuously in a rotary kiln of the type shown in Fig. 3.1. This consists of a slowly rotating tube mounted with its axis at an angle to the horizontal such that powder fed in at the top emerges at the bottom in the desired state. Heat is usually supplied by the combustion of oil or gas inside the barrel.

The calcined material has usually undergone a limited amount of sintering and must be milled to give a powder or slip suitable for the shaping stage. The machinery and problems are essentially the same as those discussed above in relation to mixing. The calcine is usually coarser and more abrasive than the raw materials so that precautions against contamination are of more importance. The initial size reduction to produce suitable particles for the finer grinding processes can be carried out using any one of the wide range of pieces of equipment, jaw-crushers, roller-mills, hammer-mills, developed for 'ore-dressing'.

3.6 Shaping

The treatment of the milled powders depends on the method of fabricating shapes from it. The forms in which it is required for various shaping processes are given in Table 3.1.

Unless the material concerned contains a substantial quantity of clay (usually 10% or more) it is necessary to incorporate an organic binder. The primary function of the binder is to give the dry shape sufficient strength to survive the handling necessary between shaping and sintering, but it may also be essential to the method of shaping as, for example, in items (iii)–(ix) of Table 3.1.

One of the most important requirements for a binder is that it should be possible to eliminate it from the compact without any disruptive effect. When particles are in high concentrations in a fluid they tend to form a continuous network with points of direct, or almost direct, contact between them. These points of contact remain when the binder is volatilized or burned out and provide, through van der Waals forces, sufficient strength to resist the disintegrating effect of small stresses. It can be seen that in the initial stage the binder at the surface of a shape is removed and the fragile porous outer layer so formed is held together by the bulk of the body which still contains binder. The binder-free layer grows inwards allowing gases from the interior to escape through its pores and, as its bulk increases, contributes to the restraining forces holding the interior particles in place. The same tendency to aggregation that

Table 3.1 Feed materials for various shaping methods and the type of product

Shaping method	*Type of feed material*	*Type of shape*
(i) Dry-pressing	Free-flowing granules	Small simple shapes
(ii) Isostatic pressing	Fragile granules	Larger more intricate shapes
(iii) Calendering; viscous plastic processing	Plastic mass based on an elastic polymer	Thin plates Simple shapes
(iv) Extrusion	Plastic mass using a viscous polymer solution	Elongated shapes of constant cross-section
(v) Jiggering	Stiff mud containing clay	Large simple shapes
(vi) Injection moulding	Organic binder giving fluidity when hot	Complex shapes
(vii) Slip-casting	Free-flowing cream	Mainly hollow shapes
(viii) Band-casting	Free-flowing cream	Thin plates and sheets
(ix) Screen-printing	Printing ink consistency	Thin layers on substrates

may make the dispersion of powders difficult is essential to the mechanical stability of a body from which the binder has been removed.

To achieve adequate binder 'burn-out' is a matter of experience, and size of component, 'green' particle packing, manner of oven loading, temperature/time schedule and atmosphere are all determining parameters. Removal of the final traces of binder residues may be essential especially in the case of some electroceramics, for example superconductors (see Section 4.7.4) where residual carbon could be detrimental to properties.

3.6.1 Dry-pressing

Dry-pressing is carried out in a die with movable top and bottom punches (Fig. 3.2). A cavity is formed with the bottom punch in a low position and this is filled with free-flowing granulated powder (Fig. 3.2(a)) which is then struck off level with the top of the die. The top punch then descends and compresses the powder to a predetermined volume (Fig. 3.2(b)) or, in more elaborate presses, to a set pressure (75–300 MPa or $\sim$750–300 atm). Both punches then move upwards until the bottom punch is level with the top of the die and the top punch is clear of the powder-feeding mechanism (Fig. 3.2(c)). The compact is then removed, the bottom punch is lowered and the cycle is repeated (Fig. 3.2(d)).

The feedstock is usually contained in a hopper attached to the press, and if the granules have a range of sizes they may segregate as the die fills and cause density variations. Also, during pressing the granules must flow between the closing punches so that, finally, the space between them is uniformly filled.

Spherical granules of uniform size can be produced by spray-drying. A spray-drier (Fig. 3.3) consists of a large tank with a cylindrical upper part and a conical lower section. The slip is sprayed in at the top and meets hot air injected into the centre of the tank. The droplets are dried and fall into the conical section from which they can be removed. Any very fine particles are carried out by the hot air and recovered. In order to obtain granules of a suitable size, usually 0.1–1 mm, tanks with diameters of 2–4 m are necessary. Smaller tanks do not allow sufficient residence time for the removal of moisture from the larger descending droplets. The method is therefore best suited to large-scale operations involving 100–1000 kg batches.

Alternatively, a stiff paste of ceramic and binder solution can be forced through a wire mesh and granules of the required size extracted from the dried product by sieving. The flow of the granules can be enhanced and die-wall friction reduced by 'tumbling' with a small quantity of a powdered lubricant such as calcium stearate.

Once the granules are properly distributed in the die they must crush readily into small fragments so that their structure is not apparent in the final pressing, since this would result in the formation of large pores. Compaction must not be

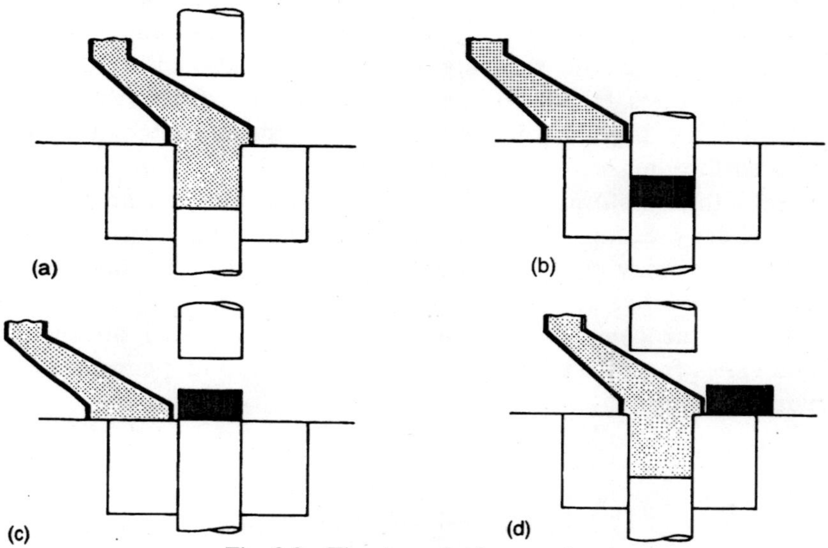

Fig. 3.2 The stages in dry pressing.

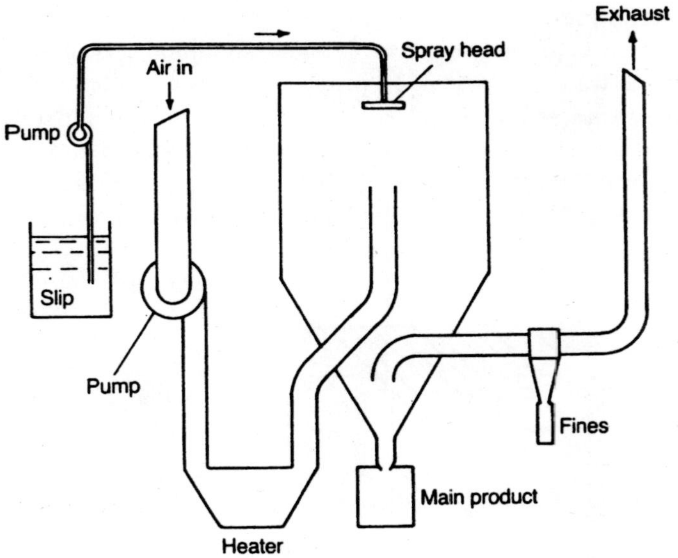

Fig. 3.3 Schematic diagram of a spray-drier.

so rapid that the air entrapped within the powder fails to escape; if this should happen there is likely to be a crack across the compact perpendicular to the pressing direction. Friction between the compact and the die walls must be minimal, since it results in a reduction in the pressure, and therefore in the density, at points remote from the moving punch (or punches if both move within the die). The green density is not usually greatly increased by applying pressures exceeding 74–150 MPa but in practice pressures as high as 300 MPa may be used to minimize density gradients due to die-wall friction and so minimize distortion during sintering. High pressures also ensure the destruction of granule structures. Highly polished die and punch surfaces help to reduce wall friction, and the tools are made of hardened steels to minimize wear and maintain surface finish.

Shapes with a uniform section in the pressing direction are the easiest to produce by dry pressing. Pieces that vary in section require very careful powder preparation and may need special press facilities such as floating dies, i.e. dies free to move relative to the punches, or dies that split open to allow easy extraction of the compact. The time taken by a pressing on an automatic machine varies from 0.2 s for pieces of diameter around 1 mm to 5 s for large complex shapes.

3.6.2 Isostatic-pressing

Many of the difficulties encountered in dry-pressing can be avoided by some form of isostatic-pressing. Ideally, this simply involves the application of hydrostatic pressure to powder in a flexible container. Powder movement is minimal and side-walls are absent. In practice shapes are often produced by the use of rigid mandrels as illustrated in Fig. 3.4. Powder is weighed into a rubber bag with a rigid mouth and a mandrel is then inserted and makes a seal with the mouth. Pressures of 20–280 MPa (~200–2800 atm) can be applied through either liquid or gas media. The pressure must be released slowly since the air originally

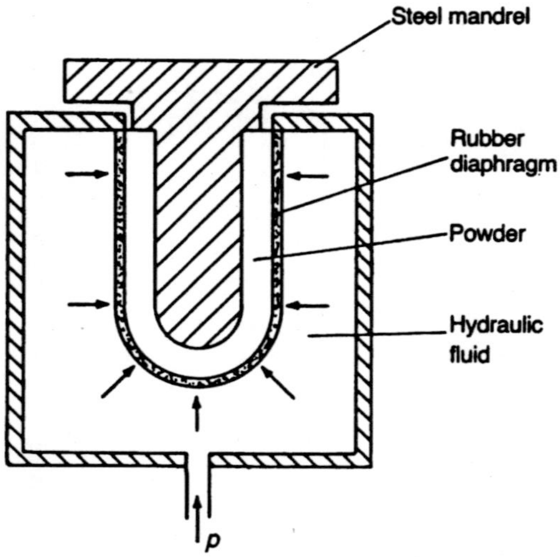

Fig. 3.4 Dry-bag isostatic pressing.

within the powder is compressed within it. Pressure can usefully be applied a second time when there is far less air within the compact.

This method gives highly uniform densities and freedom from laminations. The powder does not have to be in the form of the free-flowing granules required for automated punch and die pressing so that the risks of pores associated with granules are avoided. This method is used in the mass production of spark-plug insulators and ceramics for high-voltage devices. The green compacts can be machined when more complex shapes are required.

3.6.3 'Jolleying'

'Jolleying' is used for the manufacture of domestic crockery, typically cups and other symmetrical hollow shapes. It is also exploited in the manufacture of clay-based porcelain insulators. The plastic body is introduced into a rotating, porous mould, commonly plaster. The mould is shaped to form one surface of the article, for example the outer surface of a cup or insulator. A stationary profiling steel tool is brought close to the inner surface of the rotating mould, spreading the body to the required thickness. The surface in contact with the mould loses some moisture to the plaster and is readily detached. The process is well adapted to producing shapes having simple symmetry but is little used for shaping electroceramics since a high clay content seems essential for the necessary rheological properties.

3.6.4 Extrusion

Components with a uniform section and large length-to-diameter ratios can be produced by extrusion (Fig. 3.5). In the case of bodies containing 10% or more clay, a suitable paste for extrusion can generally be obtained by mixing with water and passing through a de-airing pug mill, which is a screw-fed extruder with a means for extracting air. Clay-free starting materials need the addition of a viscous liquid such as water containing a few per cent of starch, polyvinyl alcohol, methylcellulose etc. An entirely organic soap is also added as a lubricant and wetting agent. Very thorough mixing is essential. The paste is first forced through a die with a large number of small orifices and the resulting 'spaghetti' is loosely packed into the barrel of an extrusion press which is evacuated before pressure is applied from the ram. The binder content has to be adjusted so that the extruded body is sufficiently strong to be broken off and moved to a drying rack without significant distortion. It requires experience to achieve this but, once the correct conditions are established, large quantities of material can be extruded very rapidly (about $10\,\mathrm{cm\,s^{-1}}$). Rods and tubes are readily produced. Dielectric sheets down to a fraction of a millimetre thick and up to a metre wide

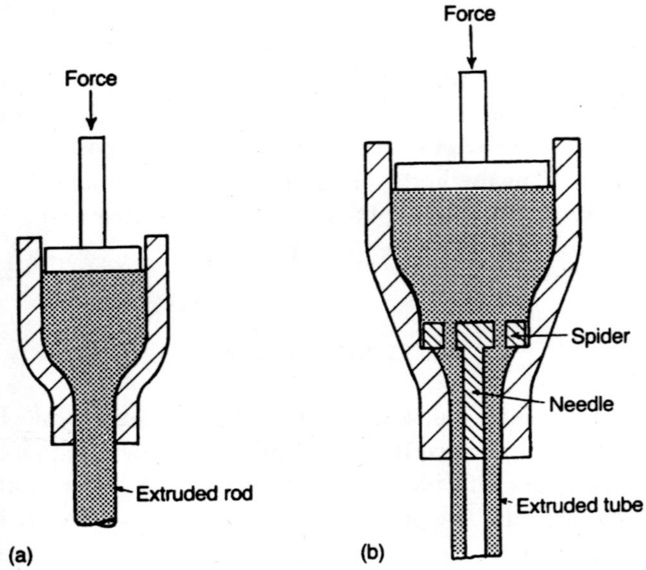

Fig. 3.5 Extrusion of (a) a rod and (b) a tube.

have been produced by extrusion. There is an appreciable shrinkage on drying and generally a greater shrinkage on sintering than for dry or isostatically pressed pieces, so that very precise dimensions are difficult to maintain without machining after firing.

3.6.5 Colloidal processing: slip-casting

Slip-casting has been a major part of traditional ceramics' processing technology for at least 250 years. It was introduced for the casting of alumina ware in the early part of the last century and since then has been extensively exploited for forming advanced ceramics.

In relatively recent years the term 'colloidal processing' has been coined to describe processing technologies in which the colloidal properties of ceramic powders are exploited. J.A. Lewis [6] comprehensively reviews colloidal processing and its potential.

In the slip-casting process the ceramic powder is suspended in a fluid vehicle, usually water. The suspension, or 'slip', has a high solids content, typically 50 vol.%, and the individual particles are fine, usually less than 10 μm. Deflocculants, which modify the electrical environment of each particle so that they repel one another, are added to the slip. The fineness of the powder and consequent high surface area ensure that electrostatic forces dominate gravity forces so that settling does not occur. When exceptionally heavy powder particles are involved the viscosity of the suspending medium can be increased to hinder settling. A plaster of Paris mould is made by casting round a model of the required shape, suitably enlarged to allow for the shrinkage on drying and firing. The inner surface of the plaster mould must have a very smooth finish free from holes originating from air bubbles in the plaster so that the cast article can be removed without damage. The mould is dried and the slip is poured into it. Water passes into the porous plaster leaving a layer of the solid on the wall of the mould. When a sufficient thickness is cast, the surplus slip is poured out and the mould and cast are allowed to dry. Slips containing a high percentage of clay give casts that shrink away from the mould and are easily extracted from it. Most other materials give only a small shrinkage and therefore greater care is needed in mould design and preparation.

The casts are usually sufficiently dense to yield low-porosity (5% or less) bodies on sintering. The relatively slow dewatering process evidently results in close-packed particles and, compared to dry powder compaction routes, with less risk of introducing defects.

This process offers a route for the manufacture of complex shapes and, in the traditional pottery industry, is the accepted method for the production of teapots, milk jugs, figurines and large articles such as wash-hand basins. It may be necessary for the mould to be made up of a number of pieces so that the cast article can be removed.

Slip-casting of technical ceramics has been steadily introduced over the past 60 years or so, and now it is standard practice to cast alumina crucibles and large tubes. The process has been successfully extended to include silica, beryllia, magnesia, zirconia, silicon (to make the preforms for reaction-bonded silicon nitride articles) and mixtures of silicon carbide and carbon (to make the preforms for a variety of self-bonded silicon carbide articles). Many metallics and intermetallics, including tungsten, molybdenum, chromium, WC, ZrC and $MoSi_2$, have also been successfully slip-cast.

3.6.6 Tape-casting

Tape-casting is also called doctor-blading or band-casting. A slip is spread on a moving band (Fig. 3.6), dried, peeled from the band and reeled up prior to further processing. The slip differs from that used for slip-casting because it has to act as a far stronger binder for the ceramic particles when the liquid phase is removed. A water-based slip may contain polyvinyl alcohol as a binder, glycerine as a plasticizer and ammonium polyacrylate as a deflocculant. It is more usually based on a mixture of organic solvents containing, for instance, polyvinyl alcohol, dibutyl phthalate and fatty acid deflocculants. Air bubbles must be removed from the slip by the application of a vacuum and steps taken to exclude foreign particles (e.g. from clothes or skin) that would leave pores on sintering.

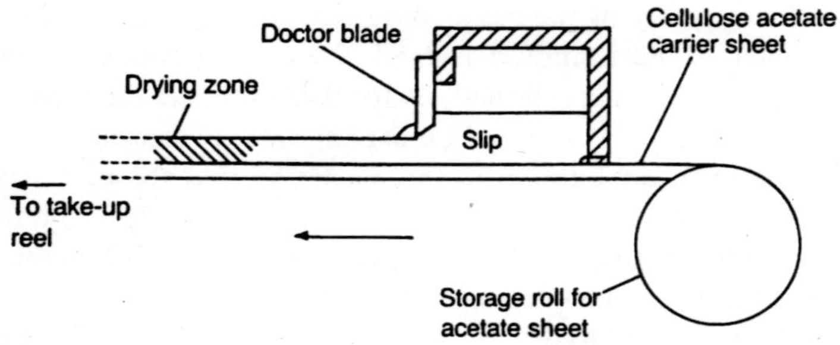

Fig. 3.6 Schematic diagram of the doctor-blade tape-casting process.

The band may consist of highly polished stainless steel or may bear a carrier layer of a suitable polymer. It moves under a hopper containing the slip with one edge raised sufficiently for the slip to be carried out. The evaporation of the solvent may be assisted by heating and by a draught of filtered air. It is difficult to dry the thicker (>0.5 mm) films successfully because the surface film that forms first shrinks laterally and cracks as it is restrained from motion at its edges, which dry more rapidly than the rest. With thinner layers and suitable solvent compositions the film dries out uniformly. Adhesion to the band prevents sideways contraction so that, at this stage, shrinkage occurs entirely in thickness. Hence the lateral shrinkage during sintering is about four times greater than the shrinkage in thickness (e.g. 20 and 5%).

There are variations on the above approach. For instance, the hopper may move over a stationary carrier film. It is also possible to draw a polymer ribbon through a slip so as to pick up thin layers on both surfaces.

The cast film is porous because of the evaporation of solvent and contains up to 30 vol. % of organic solids as binder. It is sufficiently strong for handling and the screen-printing of electrode patterns (see Section 5.4.3). Despite a relatively high porosity after burning out the organic material the cast film can be sintered to a high density. The review by H. Hellebrand [7] is comprehensive.

3.6.7 Calendering and viscous polymer processing

Thin tape can also be prepared by calendering (see Fig. 9.45). In this process a mixture of powder and an elastomeric polymer (e.g. rubber) is fed to a pair of rollers which are rotating at different speeds. As a consequence of this speed differential the mixture is subjected to a strong shearing action as it passes through the gap (pinch) between the rollers. The shearing action disintegrates the agglomerates and forces the elastomer into close contact with the surfaces of the grains. Dispersion is enhanced by reducing the pinch. Considerable energy is dissipated in the process and the rollers are water cooled. The mixture is eventually reduced to the required thickness by running it through a set of rollers rotating at equal speeds.

The process has the advantage that the green tape is free from porosity since there is no solvent loss and the ceramic powder is very well dispersed with the elastomer uniformly distributed. The tape can be further processed for substrates or multilayer capacitors in the same way as band-cast material.

Over the past decade there has been impressive progress in 'viscous polymer processing' (VPP), a technology related to calendering but applied to the fabrication of complex shapes. The technology involves generating very high shear rates in a highly viscous, plastic/ceramic powder mix to form a 'dough' from which the required net shape is formed. B. Su *et al.* [8] describe its application to the fabrication of a variety of electroceramic devices, some having features of size $<150\,\mu m$.

3.6.8 Injection-moulding

Injection-moulding, widely used in the plastics industry, is ideally suited to long production runs of complex shapes. The process has been adapted for forming ceramics and the last 20 years or so have seen it firmly established as an attractive route for forming structural ceramics.

The ceramic powder is blended with a thermoplastic binder which, after forming the shape, has to be removed by careful burning out; a combination of solvent extraction and burning out may also be employed. The volume fraction of ceramic powder (V_c) lies in the range $0.5 < V_c < 0.7$, the optimal loading having to be determined empirically. The high sacrificial polymer loadings lead to high 'green' porosities and in consequence high green-to-sintered shrinkages. Despite linear shrinkages as high as 18% good control over final dimensions can be achieved. For example the dimensions of silicon carbide turbine blades (thickness $\sim$10 mm) can be maintained to within 50 μm. The technology is reviewed by J.R.G. Evans [9].

Recently injection-moulding has been exploited by L.J. Bowen [10] to fabricate a range of piezoceramic–polymer composites (see Section 6.4.6) for the active elements of sonar devices.

3.6.9 Films and layers

The strong research and development activity in fuel cell, sensor and ferroelectric memory technologies (see Sections 4.5, 4.6 and 5.7.5) has stimulated efforts to fabricate films and supported and self-supporting layers of electroceramics spanning the sub-micron to 200 μm thickness range. The essentials of the various technologies are summarized below. They can be classified according to whether the electroceramic constituents are present in the solid, liquid or vapour phases. J. Will *et al.* [11] comprehensively reviews the technologies.

Solid phase

All the techniques in this category are regarded as economically viable so far as very expensive equipment is not a requirement. As with all processes involving a particulate solid ceramic there is shrinkage during the drying and firing stages and the attendant risk of cracking.

Because of their widespread exploitation in the manufacture of electroceramic components in general, the following technologies are discussed elsewhere in the text; *screen-printing* (Section 4.2.2), *slip-casting* (Section 3.6.5), *tape-casting* (Section 3.6.6), and *calendering* (Section 3.6.7).

Slurry–coating: Dense ceramic films have been successfully fabricated starting with low solids content (1–2 vol. %) suspensions of powder in water or ethanol. Multiple coatings may be applied to the particular substrate with controlled drying between each. The coating is sintered in a final co-firing of substrate and deposit. Care has to be taken to tailor the thermal expansivities of both substrate and coating so as to avoid tensile stresses in the layer and consequential cracking. Layer thicknesses in the range 5–30 μm have been successfully fabricated.

Electrophoretic deposition (EPD): In EPD advantage is taken of the electric charge carried by a particle in suspension (see [6]). The charged particles are carried towards a conducting substrate by an applied d.c. electric field. The deposition rates can be fast and typically of order 1 mm min^{-1}. Because little or no organics are involved (in contrast to tape-casting, for example) 'burn-out' procedures are unnecessary.

Transfer-printing: The method consists in screen-printing an ink pattern (see Section 5.4.3 'wet process') onto a water-soluble 'paper' substrate. A temporary 'carrier overcoat' layer (e.g. an acrylic) is then printed over the pattern. The ceramic layer and overcoat are removed from the water-soluble paper and 'squeegeed' on to the chosen substrate. The substrate and transferred layer are then co-fired, the carrier overcoat burning away.

Liquid phase

As in the case of the 'solid phase' approach, these techniques are economically attractive because very expensive equipment is not a requirement.

Sol–gel process: The sol–gel process consists in first forming a 'sol' (particles usually in the nm size range – or large molecules) from which a gel is derived. The gel may be processed to form a powder or a film. A common route starts with a solution of organometallic compounds (for example, isopropoxides of the appropriate cations (e.g. Ba, Ti)) which are hydrolysed (e.g. by exposure to the atmosphere) to form a gel. Films can be made by dripping or spraying the sol onto a spinning substrate (~200 r.p.m.), drying and firing.

Spray pyrolysis: An appropriate metal salt is sprayed from an atomizer onto a hot substrate where decomposition occurs yielding the metal oxide. There are advantages to be gained by electrically charging the droplets using an 'electrostatic atomizer' Deposition rates are quite low, typically in the range 1–10 $\mu m\,h^{-1}$.

Gaseous and vapour phases

Although, generally speaking, these techniques are expensive, some are being actively explored, especially for fuel cell and high temperature superconductor technologies.

Chemical vapour deposition (CVD): In CVD (see Section 4.2.1) process reactant vapours (e.g. metal chlorides) are transported to the substrate where they are adsorbed on the surface, the reaction and subsequent crystal growth occurring on the substrate surface. Deposition rates typically lie in the 1–10 $\mu m\,h^{-1}$.

Electrochemical vapour deposition (EVD): A modification of the CVD process, EVD has been developed specifically to grow a dense coating on a porous substrate. A porous ceramic substrate divides a chamber into two compartments, one containing the metal compound reactant and the other an oxygen-containing reactant. The first step in the process is closure of the pores in the substrate by the normal CVD process. Thereafter a dense layer grows by essentially the Wagner oxidation process by which a coherent and fully dense oxide film grows on a metal. Once a layer has grown by the CVD process and the surface porosity is closed, then the metal oxide layer grows on the metal reactant side, the oxygen ions arriving at the reaction site by solid state diffusion through the layer. The resultant structure is a dense thin layer of oxide on a porous substrate. Such a structure is exploited in solid oxide fuel cell (SOFC) technology (see Fig. 4.33).

Physical vapour deposition (PVD): Evaporation–condensation consists simply in evaporating a target ceramic and condensing the vapour onto a substrate. For example a high power pulsed laser beam is directed onto the target material, the intensity being such that in the outermost layers of the target the atoms are ablated with the compound stoichiometry maintained. The ablated atoms condense on a suitable substrate and a film develops. Currently growth rates are low and typically approximately 1 $\mu m\,h^{-1}$.

Sputtering: Normal sputtering (see Section 4.2.1) is a well established technology characterized by low deposition rates (typically or order 0.1 $\mu m\, h^{-1}$).

In 'magnetron sputtering' a magnet is positioned behind the target. It increases the electron path in the plasma and hence the collision probabilities and deposition rates. In 'reactive d.c. magnetron sputtering' the metal target atoms react with intentionally introduced oxygen (for example) and the oxide is deposited. Various oxide compositions and structures can be grown by varying the target, introduced gas and sputter conditions. Magnetron sputtering deposition rates of ceramics are relatively high, up to ~10 $\mu m\, h^{-1}$ being achieved.

3.7 High-temperature Processing

3.7.1 Densification

The 'driving force' for densification is the reduction in surface energy as the free surfaces of particles disappear and how this is accomplished defines the terms 'firing', solid state sintering and liquid phase sintering.

The densification of tableware and electrical porcelain depends on the formation of a glass phase developed principally from the 'fluxes' – feldspathic minerals – which are a major constituent. The process is commonly referred to as 'firing'. The glass is formed at the firing temperature and wets the surface of the solid phase. The surface tension forces pull the mass of particles together so that a large fraction of the pores is filled with glass. There is always some porosity as trapped gas cannot escape sufficiently quickly through the vitreous phase. Because it forms a continuous network the glass has an important influence on the electrical properties, in particular dielectric loss, so that electrical porcelain, for example, is exploited only in applications where adequate insulation is the prime requirement (see Section 5.5.1).

In the absence of a substantial glassy constituent densification is achieved either by solid state or liquid phase sintering. In both processes the starting compact consists of particles in the 1–10 μm size range held together by a minimum amount of binder. After the binder is volatilized or decomposed and burned out in the early stages of sintering, there is sufficient contact between the particles to maintain the shape of the component. At temperatures of approximately 0.8 to 0.9 T_m (T_m is the melting temperature in K) the constituent ions have sufficient mobility for the solid state sintering process to take place.

In the early stages of solid state sintering the microstructure undergoes changes mainly as a result of the surface diffusion of ions from convex surfaces (where their energy is high) to the concavities at particle contact points (where their energy is lower). This process brings about changes in microstructure but no appreciable densification. Densification necessitates mass transport from grains to pores, a process which commonly occurs by vacancy diffusion from regions close to the pore surface (where the vacancy concentration is high relative to the equilibrium concentration in the bulk) to the grain boundaries which, by virtue of their intrinsic disorder, act as vacancy sinks.

In liquid phase sintering the movement of ions is assisted by the presence of small quantities of liquid in which the crystalline phase has limited solubility. Mass transport occurs by a 'solution precipitation' process involving ions dissolving at high energy sites and precipitating at low energy sites. Probably the most common liquid phase sintered electroceramic component is the alumina substrate for carrying microcircuits (see Section 5.5.5).

In both solid state and liquid phase sintering grain growth accompanies densification, driven by the reduction in overall grain boundary energy. Grain size can be an important determinant of mechanical and electrical properties and therefore efforts are made to control it. Starting particle sizes, heating schedules and chemical composition all influence grain size and the optimum conditions have to be determined empirically.

Grain size can be controlled by exploiting differences in the grain growth and densification kinetics. For example very 'fast firing' at temperatures higher than the normal sintering temperature can effect full densification with minimum grain growth. Hot-pressing (see Section 3.7.2) achieves full densification with little grain growth. Appropriate annealing of a dense ceramic can be used to increase grain size, should this be desirable.

Specific dopants for particular systems are known to control grain sizes but the mechanisms are in the main conjectural. For example, in the solid state sintering of Al_2O_3 the addition of MgO leads to a fully dense, translucent ceramic widely exploited in high pressure Na-vapour street lighting. However, intensive efforts over half a century since the discovery have failed to reveal a definitive explanation of the role of the MgO although it seems clear that control of grain growth kinetics is crucial.

Typical microstructures developed by solid state sintering, liquid phase sintering and firing a porcelain are shown in Fig. 5.20.

3.7.2 Hot-pressing

It is not always possible to obtain a low-porosity body by 'pressureless sintering', i.e. by sintering at atmospheric pressure. For example, difficulties are experienced with silicon nitride and silicon carbide. More commonly it may prove difficult to combine the complete elimination of porosity with the maintenance of small crystal size. These problems can usually by overcome by hot-pressing, i.e. sintering under pressure between punches in a die, as shown in Fig. 8.9. The pressure now provides the major part of the driving force eliminating porosity and the temperature can be kept at a level at which crystal growth is minimized.

Care has to be taken in selecting materials for the die and punches. Metals are of little use above 1000 °C because they become ductile, and the die bulges under pressure so that the compact can only be extracted by destroying the die. However, zinc sulphide (an infrared-transparent material) has been hot pressed at 700 °C in stainless steel moulds. Special alloys, mostly based on molybdenum, can be used up to 1000 °C at pressures of about 80 MPa (5 ton in^{-2}). Alumina, silicon carbide and silicon nitride can be used up to about 1400 °C at similar pressures and are widely applied in the production of transparent electro-optical ceramics based on lead lanthanum zirconate as discussed in Section 8.2.1.

Graphite is widely used at temperatures up to 2200 °C and pressures between 10 and 30 MPa. At 1200 °C it has only a fraction of the strength of alumina, silicon carbide or silicon nitride, but it retains its strength at higher temperatures so that above 1800 °C it is the strongest material available. It has the disadvantage, for processing many electroceramics, of generating a strongly reducing atmosphere and needing some protection against oxidation. Like the metals and silicon carbide it can be used as a susceptor for induction heating.

Although hot-pressing is usually regarded as an expensive process and only simple shapes with a wide tolerance on dimensions can be made, it provides the only route for several valuable materials. Continuous hot-pressing methods have been developed for some magnetic ferrites and piezoelectric niobates. They offer higher production rates but tool wear is very severe.

3.7.3 Isostatic hot-pressing

In isostatic hot-pressing ('HIP'), sintering or a post-sintering operation is carried out under a high gas pressure (typically 30–100 MPa). This method, like most other sintering methods, was first developed for metals and is used routinely for high-performance turbine blades and hip-joint prostheses.

A furnace is constructed within a high-pressure vessel and the objects to be sintered are placed in it. Powders or pieces containing interconnected pores are encapsulated in impervious envelopes of a ductile metal such as platinum or, in the case of metals and some ceramics, of glass. Sintered bodies in which residual pores are isolated need not be enclosed. A neutral gas such as nitrogen or argon is introduced at a suitable pressure while the temperature is raised to the required level.

The method has the advantage of avoiding interaction with die and punch materials and allows the sintering of complex shapes in controlled atmospheres. If the pieces concerned are small a large number can be processed in one operation so that the expense involved is moderate. As with cold isostatic pressing, the method avoids internal flaws and density variations due to the shearing action caused by die-wall friction.

An intermediate form of isostatic pressure sintering, sometimes referred to as pseudo-isostatic hot-pressing, uses uniaxial hot-pressing apparatus but immerses the object to be sintered in a refractory non-interacting powder within the die and punches. This avoids the very considerable expense of building a furnace inside a thick-walled pressure vessel but the results are inferior to those achieved with true isostatic hot-pressing.

3.7.4 Glass-ceramics

Glass-ceramics are one of the relatively new members of the ceramics' family and an important electroceramic type. They were successfully developed during the 1950s and 1960s principally due to the pioneering work of S.D. Stookey at Corning in the United States and are extensively exploited as electrical insulators and in electronics 'packaging' technology (see Section 5.5.5).

In the manufacture of the most familiar glass ceramic articles (e.g. tableware and the cooker hob), the process starts with making the glass shape by any one of the standard glass-forming methods. The shape is then subjected to a two-stage heat treatment. The first stage is at a relatively low temperature when crystal nuclei form, and the second at a higher temperature when the crystals grow by diffusion. The devitrification of the glass is, of course, a manifestation of the free energy of the crystalline phase being lower than that of the randomly structured vitreous phase from which it is derived. The final ceramic is a crystal–glass composite with typically a few vol% residual glass and no porosity. The glass-ceramic route offers close control over the relative amounts and natures of the phases developed and consequently so too over chemical and physical properties.

Glass-ceramics can also be made starting from a powdered glass followed by a single step combined sintering and controlled crystal nucleation and growth steps. This route is the one most relevant to the electroceramics context.

Glass-ceramics based on the Li_2O-Al_2O_3–SiO_2 can be tailored, principally through varying the alumina content, to have linear thermal expansivities in the range from close to zero to approximately $18\,MK^{-1}$. The low expansion materials have excellent resistance to thermal shock whilst those with the higher expansivities can be successfully joined to a range of metals.

The stages in the manufacture of a controlled atmosphere housing for a high current electrical switch are illustrated in Fig. 3.7.

An important attribute in the electroceramics context is that, in contrast to a glass which would soften and deform at high temperatures, the glass-ceramic is relatively dimensionally stable up to temperatures close to the melting points of the crystals, and therefore remains so during subsequent processing stages. Because of this, coupled with the ability to control thermal expansivities, glass-ceramics play an important role in microelectronics 'packaging' technology (see Section 5.5.5).

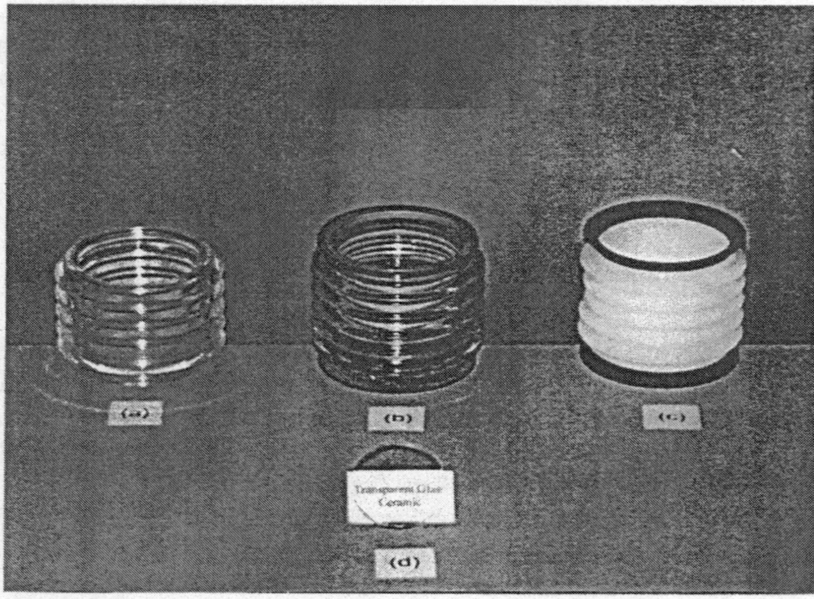

Fig. 3.7 Stages in the fabrication of a glass-ceramic, gas-tight, high current switch housing: (a) the formed glass shape, (b) 18% Cr stainless steel (linear thermal expansivity: ~13 p.p.m. K^{-1}) flanges bonded to glass, (c) finished glass-ceramic component. The transparency of the disc (d) results from a very small grain size. (Courtesy of Advanced Ceramics Limited, UK.)

Electrical characteristics are controlled by the composition and size of the crystals, slightly modified by the thin intergranular layer, the residue of the glass phase. Optical transparency is determined to a large extent by scattering (see Section 8.1.5) and necessitates controlling the crystallite size to about 50 nm (Fig. 3.7).

A properly designed glass-ceramic is pore-free, mechanically strong (modulus of rupture ~250 MPa) and tough (fracture energy ~17 J m^{-2}).

For the reader requiring an in depth discussion of glass-ceramics the classic monograph by P.W. McMillan [12] is recommended, and also the articles very relevant to electroceramics in the text edited by M.H. Lewis [13].

3.8 Finishing

The sintered bodies may require one or more of the machining, glazing and metallizing operations known as 'finishing'. Tool wear during shaping and variations in shrinkage during drying and sintering contribute to a variation of

1–2% in the dimensions of pieces emerging from a furnace. Grinding and lapping with abrasives, such as silicon carbide and diamond powder, can be used to bring component dimensions to within the closest of engineering tolerances, but are expensive operations except in the simplest of cases such as centreless grinding of rods to diameter or adjusting the thickness of slabs.

Glazes are applied when particularly smooth or easily cleaned surfaces are required. The glazes are applied as layers of powder by dipping the component in a suitable slip or by spraying. They are fired on at temperatures sufficient to melt them, generally in the range 600–1000 °C. The thermal expansion of the glaze should be slightly less than that of the material to which it is applied so that it is under compression after cooling, but the mismatch should not exceed 2–3 parts in $10^6\,°C^{-1}$. A surface under compression can increase the strength of a body significantly.

Metallizing may be required in order to provide an electrical contact or as part of the process of joining a ceramic to a metal part. Metals in contact with conductive or semiconductive ceramics must be chosen so that they do not introduce unwanted barriers to the movement of current carriers. Aluminium, silver, gold and Ni–Cr are all readily deposited by evaporation in a vacuum from a heated source or by sputtering, but adhesion is not usually very good. Nickel containing 10% or more of phosphorus (or boron) can be deposited from solutions containing nickel salts and a reducing agent such as sodium hypophosphite, but again with limited adhesion. Ceramic dielectrics are usually coated with a paint containing silver or silver oxide particles mixed with a small amount of a glass. This is fired on at 600–800 °C and gives very good adhesion provided that the glass is matched chemically to the substrate so that, for instance, it does not react strongly and become completely absorbed into the substrate.

Many alumina parts such as klystron microwave windows and lead-throughs (see Fig. 5.25) must form strong vacuum-tight joints with metals. In this case a paint containing molybdenum and manganese powders is applied to the alumina and fired on in wet hydrogen. Enough interaction occurs at the alumina surface to form a strong bond and nickel is then deposited electrolytically on the metallized surface. The ceramic can then be brazed to a metal using Cu–Ag eutectic alloy.

3.9 Porous Materials

In the majority of electrical applications sintered ceramics are required to have minimum porosity. Properties usually reach their optimum values at the highest densities, whilst porosity in excess of 5–10% allows the ingress of moisture leading to many serious problems. However, there are cases where porosity is desirable: for example, in humidity and gas sensors and where thermal shock resistance is of overriding importance. Porous structures can be obtained in the following ways.

1. Calcining at a high temperature so that considerable crystal growth takes place and, after grinding coarsely, separating out particles in a limited size range. Bodies compacted and sintered from such powders will have continuous porosity and cavities as large as 30 μm. The total accessible specific surface area will be low.

2. Underfiring an otherwise normally processed body. The pore structure will be fine and the total accessible specific surface area high.

3. Mixing organic or carbon particles with diameters exceeding 20 μm into a ceramic powder; cavities of corresponding size are left after burning out and sintering. The content of such particles needs to exceed 20 vol.% if porosity is to be continuous. This method allows a control over the final structure that is largely independent of the sintering conditions.

4. High porosities are obtainable by using a high proportion of binder containing a foaming agent. Gas is generated within a fluid binder–powder mixture that subsequently becomes rigid through the polymerization of the binder. The porosity may be either continuous or discontinuous according to the formulation of the binder and is little changed by burn-out and sintering. Materials containing large cavities (several millimetres) can be formed in this way.

5. An existing porous structure, e.g. certain corals, can be reproduced by impregnating it with wax and then dissolving out the original solid (calcium carbonate in the case of coral). The porous wax structure is then impregnated with a concentrated slip containing the ceramic powder. After drying, the wax can be melted out and the ceramic fired. The method becomes more economic if a suitable foam with a continuous porosity is impregnated with a slip and then burned out. In this case the final ceramic structure corresponds to the vacant spaces within the foam.

3.10 Processing and Electroceramics Research and Development

Electroceramics, more so than any other branch of ceramics, relies heavily on research and development for which a range of specialized 'laboratory scale' apparatus has been developed for comminution, spray-drying, pressing, tape-casting, screen-printing and sintering. The development scientist/engineer also requires access to equipment to characterize materials with respect to crystal structure, chemical composition and microstructure and in the case of powders, particle size and size distribution, specific surface area etc. A well-appointed electroceramics laboratory will also have access to X-ray diffraction, X-ray fluorescence, scanning electron microscopy coupled with X-ray analysis, optical microscopy, both reflected and transmitted light, Brunauer–Emmett–Teller (BET)-type apparatus, sedimentometers and optical scattering analysers and, of course, equipment for determining electrical, magnetic and electromechanical properties (e.g. wide frequency band impedance analysers and apparatus for measuring small changes in sample dimensions).

A small-scale, controlled atmosphere, gradient furnace is a valuable addition to the laboratory since it allows optimum sintering conditions to be quickly determined.

Taking precautions to keep inadvertent contamination to a minimum is necessary for reliable interpretation of experimental data and the design of equipment should be such that thorough cleaning between processing batches is facilitated. The introduction of contaminants is particularly likely during comminution and high energy bead attritor-milling offers significant advantages over ball-milling in this regard.

Multilayer technology has arguably become the most important facet of electroceramics technology and there is a constant drive to reduce layer thicknesses, in the case of multilayer capacitors because of the higher volumetric efficiencies achievable (see Section 5.4.3). This, and a general trend to micro-fabrication, has made it essential to remove all risk of airborne contaminating 'dust' consisting of clothing fibres, dry skin flakes etc. Clean-room processing is now commonplace not only on the laboratory scale but to an increasing extent on the commercial production scale.

3.11 The Growth of Single Crystals

The growth of single crystals having special relevance to electroceramics is reviewed by A.L. Gentile and F.W. Ainger [14]. For some purposes materials must be prepared as single crystals which is, in most cases, a more difficult and expensive process than preparing the same compositions in polycrystalline ceramic form. The perfection of a crystal structure is easily upset by impurities or by small changes in conditions during its formation. The growth of crystals usually involves a change of state from liquid or gas to solid, or from liquid solution to solid. The atomic species in a fluid at any instant are arranged randomly; during crystal growth they must take on the ordered structure of the crystalline phase. Too rapid growth results in the trapping of disordered regions in the crystal or in the nucleation of fresh crystals with varying orientations. The growth process must therefore be slow and so requires a precise control over conditions over prolonged periods.

The formation of a solid phase at the freezing point of a liquid only occurs at the surface of existing solids, e.g. dust particles in water at 0 °C. This process is known as heterogeneous nucleation. Liquids free from solid particles can be cooled to some extent below their melting points. Such a supercooled liquid crystallizes through the spontaneous formation of ordered regions closely similar to the solid in structure, and the growth of these nuclei leads to the rapid formation of a large number of randomly oriented crystals. Single-crystal growth can be obtained by providing nuclei in the form of a small single crystal (a 'seed') and ensuring that the conditions confine growth to the immediate vicinity of the seed.

Where a seed crystal is not available or, by reason of the growth method, cannot be utilized, advantage can be taken of the tendency for crystals to grow more rapidly in one crystallographic direction than in others, as for instance in the Bridgman–Stockbarger method outlined later. In the same way, if a thin ceramic rod having the required stoichiometric composition is held vertically and a small region near one end is fused by means of a laser beam and the molten region is moved towards the other end of the rod, those crystals that grow fastest along the axis of the rod will enlarge at the expense of those growing at angles to it. By passing the molten zone up and down the crystal several times relatively large single-crystal regions can be obtained. At the same time there will be a zone-refining effect through which the impurities become concentrated at the ends of the rod, thus leading to the growth of more perfect crystals.

Whilst every material presents its own problems in crystal growth, which must be solved experimentally, there is a general thermodynamic principle that gives an indication of how difficult the process is likely to be. Since crystal growth takes place reversibly, the Gibbs energy G must be constant and so, under isothermal conditions,

$$\Delta G = 0 = \Delta H - T\Delta S \tag{3.1}$$

where H is the enthalpy or heat content of the system, S is its entropy and T is the thermodynamic temperature. Therefore

$$\Delta S = \frac{\Delta H}{T} \tag{3.2}$$

If we take $\Delta H = L_a$, where L_a is the latent heat per atom, Jackson's dimensionless parameter α is defined by

$$\alpha = \frac{\Delta S}{k} = \frac{L_a}{kT_{CR}} \tag{3.3}$$

where k is Boltzmann's constant and T_{CR} is the temperature of the change of state. α represents the decrease in entropy on going from the disordered structure

Table 3.2 Values of Jackson's parameter α

Substance	*Change of state*	*Temperature T_{CR}/K of change of state*	*Latent heat L_m/kJ mol^{-1} at T_{CR}*	$\alpha = L_m/R_0T_{CR}$
Si	Liquid–solid	1680	46.4	3.3
Ge	Liquid–solid	1230	31.8	3.1
Fe	Liquid–solid	1810	15.4	1.02
Pb	Liquid–solid	601	4.77	0.96
H_2O	Liquid–solid	273	6.03	2.7
H_2O	Vapour–solid	273	45	20

in the liquid state to the ordered lattice structure of the solid state. Some values of α are given in Table 3.2.

For $\alpha < 2$ crystals grow without facets and their shape is determined by the isotherms in the melt. Many metals, e.g. iron and lead, are in this category. For $2 < \alpha < 10$, which covers the bulk of materials, facetting may occur during growth as with germanium and silicon. For $\alpha > 10$ crystals nucleate readily so that it is difficult to avoid a polycrystalline structure. The value of α for the growth of ice from water vapour at 0 °C is 20, which accounts for the low-density polycrystalline character of snow. In contrast, $\alpha = 2.7$ for the growth of ice from liquid water, and a dense material containing large crystals is readily formed.

An important consideration in the case of oxides containing several cations is whether the melting point is congruent or incongruent. In the latter case crystals cannot be grown from stoichiometric melts and a method must be found for growth at a temperature at which the required compound is stable. A similar difficulty occurs when the structure of the crystals formed at the melting point differs from that required. For example, silica yields cristobalite on solidifying at 1720 °C; cristobalite transforms into tridymite below 1470 °C and tridymite transforms to β-quartz below 867 °C, while β-quartz transforms to α-quartz below 573 °C. α-quartz must therefore be grown below 573 °C, as explained below.

The following methods have been used to grow crystals of oxides.

1. From an aqueous solution by cooling.
2. From solution in an oxide or fluoride flux by cooling.
3. From the liquid phase by cooling:
   (a) by first freezing at the lowest point of a melt;
   (b) by first freezing at the upper surface of a melt;
   (c) by the flame-fusion or Verneuil method.

4. (a) From the liquid phase kept at constant temperature by dipping a seed in the surface and then withdrawing it into a cooler zone;
   (b) from the liquid phase by deposition on a substrate of differing composition.
5. From the vapour phase by chemical reaction close to the surface of a seed crystal.

These methods are illustrated by the following examples.

1. Quartz is grown from an alkaline solution in water under a pressure of about 150 MPa (1500 atm). Pure mineral quartz fragments are placed at the bottom of a tall cylindrical autoclave that is 80% filled with either a 2.2% solution of NaOH or a 5% solution of Na_2CO_3. Seed crystals are held in wire frames near the top of the autoclave. Because the base of the autoclave is kept at 400 °C and the top is maintained at a temperature some 40 °C cooler, quartz dissolves at the bottom of the vessel and is deposited on the seeds at the top. Crystals weighing over a kilogram can be grown in a few weeks (Fig. 3.8). This hydrothermal method is also of value for such materials as zinc oxide which have high vapour pressures at their melting points.

2. Crystals of many compounds, of sufficient size for scientific purposes, can be obtained by cooling a solution of the required compound in a suitable flux. $BaTiO_3$ crystals have been prepared by using a mixture of KF with 30 wt% $BaTiO_3$ and 0.2 wt% Fe_2O_3 [15]. A temperature of 1150–1200 °C is maintained for 8 h, and the mixture is then cooled slowly to 900 °C when the flux is poured off and the crystals are allowed to cool in the furnace to room temperature. The crystals form as butterfly twins, i.e. as pairs of triangular plates joined at one edge with an angle of about 60° between them. The iron assists the formation of flat plates of relatively large area that are convenient for the investigation of the ferroelectric properties of $BaTiO_3$. A small amount of iron, as Fe^{3+}, is incorporated in the $BaTiO_3$ lattice on Ti^{4+} sites. The charge difference is compensated by the replacement of an equivalent number of O^{2-} ions by F^- ions. This result is typical of flux-growth methods which frequently result in the incorporation of a small proportion of the solvent in the crystal lattice.

   Single crystal rods of a variety of compositions, including complex garnets (see Section 9.5.4) can be grown by the 'floating zone' method. For example, two sintered Bi-substituted garnet (BIG) rods of the chosen composition are placed end to end with a single crystal 'seed' of the same composition sandwiched between them. Halogen lamps are focused onto the join area as the rods are rotated and provide sufficient heat to form a molten zone. If the molten zone is moved into the sintered rod in a suitably controlled way then a

Fig. 3.8 Hydrothermally grown quartz crystals.

high quality single crystal can be grown. The whole operation is carried out in an oxygen atmosphere to ensure the iron is in the fully oxidized state. Very high quality crystals of size 12 mm diameter 120 mm long suited to magneto-optical applications can be grown.

In recent years strong interest has been shown in single crystals of the relaxor ferroelectrics for actuator applications (see Section 6.4.3). Members of this family of electroceramics exhibit exceptionally large electric field-induced strains. One such material has the composition approximately $0.9Pb(Zn_{1/3}Nb_{2/3})O_3$-$0.1PbTiO_3$ abbreviated 0.9PZN–0.1PT. The starting materials are Pb_3O_4, ZnO, Nb_2O_5 and TiO_2. The constituents, in amounts to give the required overall composition, are well mixed and then heated with an excess of Pb_3O_4 which serves as the flux, in a sealed platinum crucible to approximately 1150 °C. The melt is then slowly cooled at approximately 5 °C h^{-1} when crystals with dimensions up to 20 mm grow. When the crucible is cold the crystals are extracted with the help of hot nitric acid.

In order to form manganese zinc ferrite crystals using the Bridgman–Stockbarger method (Fig. 3.9) a charge of the mixed oxides is melted in a Pt–Rh crucible and kept just above its solidification temperature. The furnace is designed such that there is a sharp drop in temperature just below the bottom tip of the crucible in its initial position. The crucible is lowered so that the tip enters the colder zone causing the nucleation of crystals. The crystals grow fastest in particular crystallographic directions and those growing at angles greater than half the cone angle terminate at the walls of the cone; only those crystals oriented so that growth is favoured in the axial direction persist into the bulk of the charge as the crucible is lowered into the cooler zone. As a result the upper part of the crucible finally contains either a single crystal or a few large crystals.

There are a number of problems associated with growing manganese zinc ferrite single crystals by the process outlined above. Temperatures close to 1800 °C are required to melt the oxides, and then there is rapid evaporation of zinc and loss of oxygen from the Fe_2O_3 resulting in the formation of FeO. These difficulties have been largely overcome through the development of a high-pressure version of the apparatus. The crucible, now in a sealed container so that an oxygen pressure of up to 2 MPa (20 atm) can be maintained over the melt, is heated indirectly by a surrounding thick cylindrical Pt–Rh susceptor. The thickness of the inductively heated susceptor is sufficient to shield the melt from the radiofrequency field which would otherwise produce eddy current heating in the melt, resulting in undesirable convective agitation.

A further problem arises owing to differences in composition between the liquid and solid phases. The cation distribution can be optimized and the excess FeO oxidized by annealing the crystal in a suitable atmosphere after cutting the required shapes.

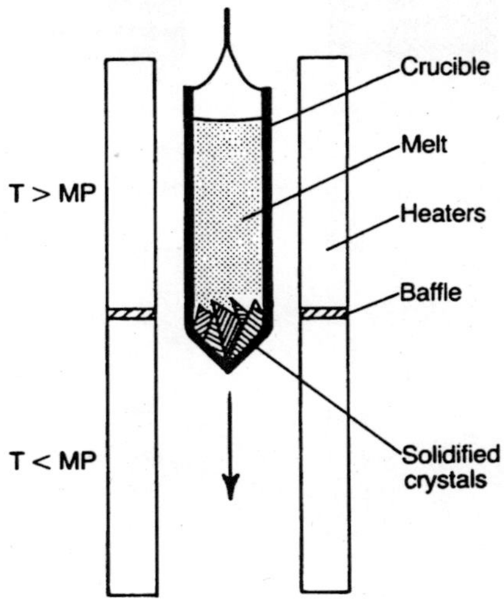

Fig. 3.9 The Bridgman–Stockbarger technique (MP, melting point).

Pure cubic $BaTiO_3$ cannot be grown from a melt of that composition because, as shown in Fig. 5.41, the hexagonal form is in equilibrium with the liquid at the solidification temperature (1618 °C). It can, however, be grown from a composition containing 35 mol.% BaO, 65 mol.% TiO_2, which solidifies below 1460 °C, the temperature below which the cubic form is stable. The melt is held just above its solidification point, a seed crystal is dipped into its surface and the melt is then cooled at between 0.1 and 0.5 °C h^{-1}. The seed crystal is attached to a platinum stirrer which is rotated during growth so that the liquid near the crystal does not become depleted in barium. Only a fraction of the melt can be obtained as single-crystal $BaTiO_3$ since the whole of it solidifies, forming a mixture of cubic $BaTiO_3$ and $Ba_6Ti_{17}O_{40}$, when the BaO content falls to about 32 mol.%.

The Verneuil flame-fusion method illustrated in Fig. 3.10 is a well-established but relatively crude method for growing refractory oxide single crystals, particularly sapphire and ruby. In this case alumina powder is fed at a controlled rate down a tube at the end of which it is melted by an oxyhydrogen flame; it then drips down into a shallow pool of liquid on top of the seed crystal. As the crystal grows, it is lowered into the annealing zone. This method suffers from the disadvantage of poor control over the growth environment, and the quality of crystal produced is inferior to that obtained by crystal pulling. It is adequate, however, for jewelled bearings.

Crystal pulling is one of the most widely exploited and successful processes, and is generally termed the Czochralski method. Silicon, gallium arsenide, alumina, lithium niobate, lithium tantalate and gadolinium gallium garnet are all grown on a large scale by variants of this process which is applicable to most materials that melt congruently. The melt is kept at a temperature just

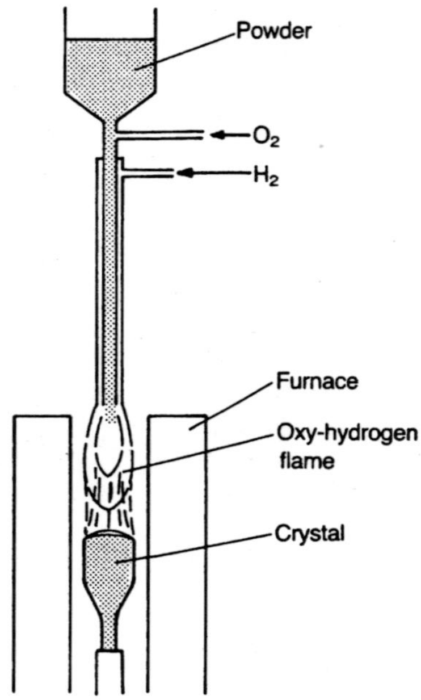

Fig. 3.10 The Verneuil technique.

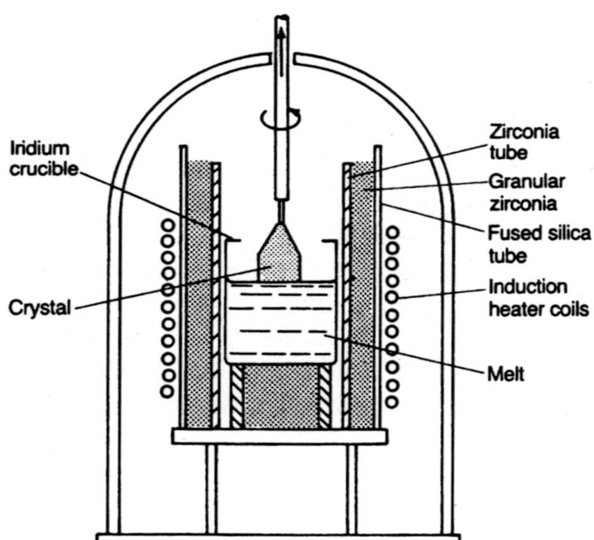

Fig. 3.11 The Czochralski technique for pulling crystals.

above its freezing point and a seed crystal, firmly fixed to a rotating tube, is lowered into the surface and then slowly withdrawn (Fig. 3.11). Air can be blown down the tube to control the extraction of heat necessary to the growth process. The crystal diameter can be varied by changing the rate of withdrawal.

Flaws and misoriented crystals can be eliminated by first enlarging and then diminishing the crystal diameter; misoriented regions are thereby terminated at the crystal surface in much the same way as in a Stockbarger–Bridgman crucible. In favourable cases nearly 90% of a melt can be obtained in single-crystal form.

Smaller crystals can be grown from solutions by a similar method; for example, a mixture of barium and strontium titanates containing a 28% excess of TiO_2 can be melted at 1500 °C and crystals of (Ba, Sr)TiO_3 withdrawn on a seed. The cubic crystals are stoichiometric in their ratio of Sr+Ba to Ti but contain a substantially greater proportion of strontium than the liquid phase. Provided that the mass of the crystals constitutes less than 1% of that of the melt, they have a highly homogeneous composition despite the inevitable change in composition of the liquid phase as the crystal forms.

In some applications only a thin film of single-crystal material is needed on a substrate that differs in composition. In this case advantage can be taken of the possibility of epitaxial deposition. A seed crystal of similar crystal structure but differing composition from the required film can be used in place of a seed of identical composition. The method is of particular value when the deposit has a complex composition and a suitable substrate material is available. The magnetic garnets that were used in the now obsolete 'bubble' memory stores with complex compositions such as $(Y_{0.9}Sm_{0.5}Tm_{1.1}Ca_{0.5})(Fe_2)(Fe_{2.5}Ge_{0.5})O_{12}$, were grown by liquid phase epitaxy (LPE) on gadolinium gallium garnet substrates.

The apparatus for LPE growth of garnet film is shown schematically in Fig. 3.12. The melt composition is typically PbO–B_2O_3 in a 50:1 ratio by weight to which sufficient garnet constituent oxides are added to form a saturated solution between 950 and 1000 °C. A set of circular substrates is held in a platinum holder above the solution which is attached to the rotatable spindle. The melt is allowed to cool a few degrees so that it becomes supersaturated and the substrates are then lowered into it and rotated for a sufficient time (some minutes) for a film of a garnet of the required thickness to grow. They are then raised from the melt and spun rapidly to remove the excess liquid. The thickness for bubble memories was usually about 4 μm, though films of thickness up to 15 μm can be grown.

The growth of epitaxial layers from the vapour phase has been developed extensively for III–V compounds for use in semiconductor and electro-optical devices. The method most applicable to oxide systems is metal–organic chemical vapour phase deposition (MOCVD). Metals form a wide range of volatile organic compounds such as lead tetraethyl, $(Pb(C_2H_5)_4)$, silicon tetraethoxide $(Si(OC_2H_5)_4)$ and titanium isopropoxide $(Ti(C_3H_7O)_4)$, and these are readily converted to oxide by reaction with oxygen or water and by thermal decomposition. Mixtures of such compounds in vapour form can be reacted in close proximity to heated substrates to form oxide deposits that interact to form

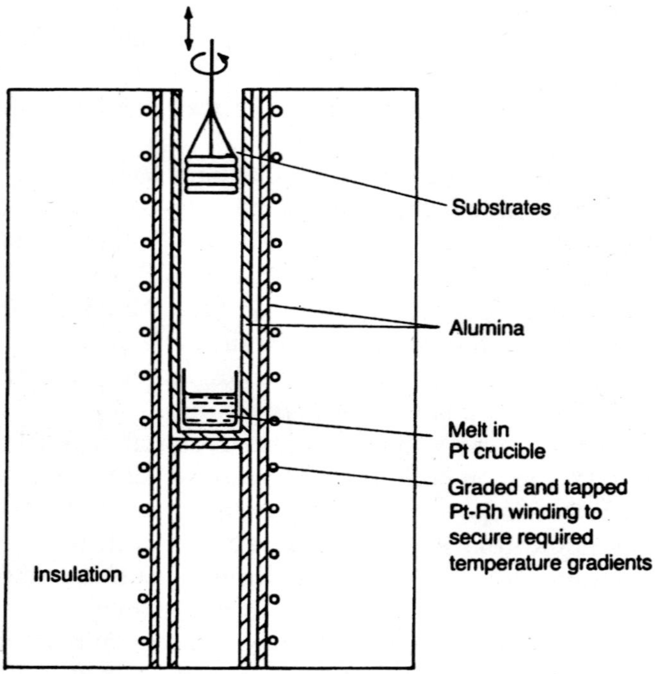

Fig. 3.12 Schematic diagram of a furnace for LPE growth of magnetic garnet films.

titanates, zirconates etc. Polycrystalline layers may be adequate for some purposes, but with suitable substrates single-crystal layers are possible.

Problems

1. Estimate the density of a compact of uniformly close-packed alumina powder, assuming a mean particle diameter of 25 μm. What would the density be if the close-packed particles have a mean diameter of 10 μm? Estimate the compact density if 25 wt% of the powder has a small enough particle diameter for it to occupy the interstices in the close-packed larger fraction (crystal density of Al_2O_3, 4000 kg m^{-3}). [Answers: 2960 kg m^{-3}; 2960 kg m^{-3}; 3730 kg m^{-3}].

2. A sample of alumina powder has a mean 'particle' diameter of 25 μm. What is the apparent specific surface area (SSA) of the powder? If each 'particle' is in reality an agglomerate of ultimate particles of diameter 0.5 μm, estimate (i) the number of ultimate particles per agglomerate and (ii) the true SSA of the powder. [Answers: 0.06 mg^2 g^{-1}; $\sim 10^5$; 3 m^2 g^{-1}].

3. A colloid in suspension in an electrolyte acquires a surface charge and an 'atmosphere' of oppositely charged ions. An inner part of the atmosphere moves with the colloid

and, together with it, constitutes an 'electrokinetic unit'. If the net charge on the unit is $-Q$, and assuming that Stokes' law applies, derive the equation

$$u = \frac{4\varepsilon\zeta}{6\eta}$$

where u is the electrophoretic mobility, ε is the effective permittivity of the electrolyte in the immediate vicinity of the colloid, ζ is the zeta potential and η is the viscosity of the liquid. Write a brief account of the significance of the zeta potential in determining the stability of casting slips. (see [a], or similar)

4. The density of an aqueous alumina slip is 2.5 kg dm^{-3}. Calculate the concentration of dry alumina in the slip on (i) a weight basis and (ii) a volume basis. [Answers: 80 wt%; 50 vol.%]

5. A pore-free polycrystalline solid contains two phases, A of density 4000 kg m^{-3} and B of density 3000 kg m^{-3}. Calculate the density of the solid if it contains (i) 60 wt% of phase A and (ii) 60 vol.% of phase A. [Answers: 3529 kg m^{-3}; 3600 kg m^{-3}]

6. As part of the characterization of an electrical porcelain the following weighings were made at 25 °C using a balance which was accurate to the third decimal place:

single-piece sample of 'body'	15.234 g
sample saturated with water	15.408 g
saturated sample suspended in water	8.948 g
specific gravity bottle	22.432 g
bottle + sample of finely ground dry 'body'	27.362 g
bottle + powder + water	75.379 g
bottle + water	72.429 g

Calculate (i) the true density of the 'body', (ii) the bulk density of the 'body', (iii) the open porosity and (iv) the total porosity (density of water at 25 °C, 997 kg m^{-3}). [Answers: 2.482 Mg m^{-3}; 2.351 Mg m^{-3}; 2.7%; 5.3%]

7. Assuming a reasonable value for the surface energy of an oxide estimate the 'negative pressure' tending to close spherical pores of radius (i) 1 μm and (ii) 0.01 μm during sintering. Express your answer in megapascals and 'atmospheres'. [Answers: 2 MPa ($\approx$ 20 atm); 200 MPa ($\approx$ 2000 atm)]

8. Compounds A and B interdiffuse a distance of approximately 1 μm in 1 h at 1000 °C. If the activation energy for the rate-determining step is 3.2×10^{-19} J, calculate the interdiffusion distance for a 10 h anneal at 1600 °C assuming the same diffusion mechanism. [Answer: 58 μm]

9. A circular hole in an alumina plate at 20 °C is 1 cm in diameter. Calculate the diameter at 1000 °C. If the density of alumina at 20 °C is 3980 kg m^{-3}, calculate the density at 1000 °C. Mean coefficient of linear expansion over the temperature range 20–1000 °C is 8.0 MK^{-1}. [Answer: 1.0078 cm; 3886 kg m^{-3}]

10. Assuming that a glaze will 'craze' if the tensile strain is greater than 0.1%, find the critical relationship for the onset of crazing between the mean expansion coefficients of a glaze and of a thick section ceramic body to which it is applied. The transformation temperature of the glaze can be assumed to be 725 °C and room temperature 25 °C [Answer: $\alpha_g - \alpha_b > 1.43\,\mathrm{MK^{-1}}$].

11. (i) Calculate the ratio of the partial pressures of CO and CO_2 at 1400 °C that would provide an oxygen partial pressure of 10^{-15} bar given that

$$2CO + O_2 = 2CO_2$$
$$\Delta G^{\ominus}_{1673} = -260\ \mathrm{kJ}$$

(ii) Estimate the oxygen partial pressure in wet hydrogen (dew point, 25 °C) at 1 atm and 1700 °C given that

$$H_2 + \tfrac{1}{2}O_2 = H_2O$$
$$\Delta G^{\ominus}_{1973} = -138\ \mathrm{kJ}$$

(vapour pressure over water at 25 °C = 3.17 kPa). [Answers: 2.76×10^3; 4.9×10^{-11} bar]

12. Estimate the oxygen partial pressure below which alumina will be reduced to the metal at 1700 °C, given that

$$2Al + \tfrac{3}{2}O_2 = Al_2O_3$$
$$\Delta G^{\ominus}_{1973} = -1053\ \mathrm{kJ}$$

[Answer: $\sim 2.5 \times 10^{-19}$ bar].

13. What precautions could be taken to prevent the take up of impurities when reducing the particle size of zirconia from 10 μm to 1 μm? Describe the beneficial effects that addition of a deflocculating agent might be expected to have on the ball-milling process.

14. A product is to be made by cold pressing. Describe the attributes of a good binder and the benefits expected from its use.

15. What shapes are suited to fabrication by extrusion, cold pressing or slip-casting? What are the advantages and disadvantages of isostatic-pressing?

16. Discuss the control of grain-growth during sintering.

17. Because of the volatility of PbO at sintering temperatures it is not easy to control the composition of ceramics containing it. What steps might be taken to alleviate the problem?

18. A slip for band-casting contains, on a weight basis, 100 parts alumina powder, 9 parts non-volatile organics and 35 parts toluene. The thickness of the dried film

which adheres to the band on which it is cast is 0.6 of the cast thickness. What is the porosity of the tape (i) after drying and (ii) after the binder has been removed? On firing the shrinkage is a further 5% in thickness and the final total porosity is 5%. Estimate the shrinkage in the plane of the film. [Answers: ≈ 17%; ≈ 38%; ≈ 17%]

19. A few grammes of 100 μm diameter polythene spheres are warm-pressed into a solid block. Describe the expected electrical characteristics of the block. In a separate experiment each sphere is first coated with a nanometre thick layer of aluminium prior to compaction. Describe how the electrical properties of the block would be modified and estimate the quantity of aluminium in the 'composite' on the bases of volume and weight. Comment on the relevance of your finding to electroceramics research and development 30 p.p.m.; 90 p.p.m.
[Answers: Insulator; conductor; about 30 p.p.m on volume basis.]

Bibliography

1. Reed, J.S. (1995) *Principles of Ceramic Processing*, 2nd edn, J. Wiley and Sons Inc., New York.
2. Rahaman, M.N. (1995) *Ceramic Processing and Sintering*, Marcel Dekker Inc., New York.
3. Cahn, R.W. *et al.* (eds) (1996) *Materials Science and Technology*, Volumes **17A** and **17B**, R.J. Brook (ed.), 'Processing of ceramics', Parts 1 and 2, Verlagsgesellscaft mbH, Weinheim.
4. Shaw, D.J. (1980) *Introduction to Colloid and Surface Chemistry*, 3rd edn, Butterworths, London.
5. Segal, D. (1996) Chemical preparation of powders. In: Materials Science and Technology, R.J. Brook (ed.), 'Processing of ceramics', Part 1, Verlagsgesellschaft mbH, Weinheim. **17A**, pp. 69–98.
6. Lewis, J.A. (2000) Colloidal processing of ceramics. *J. Am. Ceram. Soc.* **83**, (10), 2341–59.
7. Hellebrand, H. (1996) Tape casting. In: Materials Science and Technology, R.J. Brook (ed.), 'Processing of ceramics', Part 1, Verlagsgesellschaft mbH, Weinheim. **17A**, pp. 189–265.
8. Su, B., Pearce, D.H. and Button, T.W. (2001) Routes to net shape electroceramic devices and thick films. *J. Eur. Ceram. Soc.* **21**, 2005–9.
9. Evans, J.R.G. (1996) Injection moulding. In: Materials Science and Technology, R.J. Brook (ed.), 'Processing of ceramics', Part 1, Verlagsgesellschaft mbH, Weinheim. **17A**, pp. 268–311.
10. Bowen, L.J., *et al.* (1993) Injection moulded fine-scale piezoelectric composite transducer. *1993 IEEE Ultrasonics Symposium*, pp. 499–503.
11. Will, J. *et al.* (2000) Fabrication of thin electrolytes for second-generation solid oxide fuel cells. *Solid State Ionics* **131**, 79–96.
12. McMillan, P.W. (1979) *Glass-ceramics*, 2nd edn, Academic Press, New York.
13. Lewis, M.H. (ed.) (1989) *Glasses and Glass-ceramics*, Chapman & Hall, London.

14. Gentile, A.L. and Ainger, F.W. (1996) Single crystals. In: Materials Science and Technology, R.J. Brook (ed.), 'Processing of ceramics', Part 1, Verlagsgesellschaft mbH, Weinheim. **17A**, pp. 313–42.
15. Remeika, J.P. (1954) A method for growing barium titanate single crystals. *J. Am. Chem. Soc.* **76**, 940.

4

CERAMIC CONDUCTORS

Chapter 2 provides the elementary background physics essential to an understanding of the following discussion of conductive ceramics. These include conductors capable of sustaining their mechanical integrity at high (>1500 °C) temperatures, ohmic resistors with properties little affected by temperature and voltage, resistors with voltage-dependent resistivities, resistors with large negative temperature coefficients, resistors with large positive temperature coefficients, fast-ion conductors, sensors and superconductors.

The discussion may be supplemented by reference to the text edited by R.C. Buchanan [1].

4.1 High-temperature Heating Elements and Electrodes

The principal application for high-temperature heating elements and electrodes is as furnace elements for temperatures at which most materials either oxidize or melt. Strictly speaking, for this application the term 'resistor' is more apt than 'conductor' since resistance is necessary for Joule heating. It is advantageous for such components to be ohmic in their behaviour, but it is not essential as suitably controlled power supplies can be devised to compensate for both non-linear voltage–current characteristics and high temperature coefficients of resistance. Resistivities of 0.01–1 Ω m are convenient as this allows the impedance of rods up to 1 m long and 0.5–2 cm in diameter to match into readily available power supplies; it is possible but impracticable to form ceramics into replicas of the wire windings that utilize metals with their much lower resistivities (~ 10^{-7} Ω m). The nearest to this is to form a ceramic tube and then to cut a spiral slot through the wall with a pitch of about 1 cm. Suitable transformers for feeding very low resistances are available if necessary. Highly resistive elements necessitate high

Electroceramics: Materials, Properties, Applications. 2nd Edition. Edited by A. J. Moulson and J. M. Herbert.
 ISBN 0 471 49747 9 (hardback) 0 471 49748 7 (paperback)

voltage power supplies which lead to difficulties in furnace design since refractories all become conductive when hot so that it is difficult to avoid current leakage and often an accompanying risk of thermal breakdown (see Section 5.2.2).

In cases where reducing atmospheres can be tolerated, graphite or refractory metals such as molybdenum and tungsten can be used, while platinum and its alloys can be used safely in air up to 1500 °C. Ceramics allow the use of air atmospheres at relatively low cost.

4.1.1 Silicon carbide

Silicon carbide ceramic is noted for its hardness and for its high strength maintained to high temperatures. A passivating silica layer protects the material from corrosion in oxidizing atmospheres, the rate of reaction between SiO_2 and SiC limiting the practical operating temperature in air to about 1650 °C.

It is widely used as an abrasive, as a refractory, for high temperature furnace elements and for voltage-sensitive resistors ('varistors') (see Section 4.3). For most purposes it is made by the method devised by Acheson in 1891. He accidentally formed SiC, initially taking it to be a mixture of carbon and corundum (Al_2O_3), or so it is believed, which explains the trade name 'Carborundum'. The Carborundum Company was formed in 1891 to produce SiC for grinding wheels. The Acheson process for producing SiC has not changed significantly since 1891 and in its essentials involves passing a large electric current through a graphite resistor packed around with a mixture of sand and coke. Temperatures of approximately 2500 °C are reached in the reaction zone producing SiC and CO.

Silicon carbide is covalently bonded with a structure similar to that of diamond. There are two basic structures. One is a cubic form, β-SiC which transforms irreversibly at about 2000 °C to one of a large number of hexagonal polytypes, and the other is a rhombohedral form also with many polytypes. Both the hexagonal and rhombohedral forms are commonly referred to as α-SiC.

The properties that have led to the established commercial importance of SiC as a ceramic include its hardness (approximately 9 on the Mohs scale), its hot strength and its electrical conductivity. The combination of high strength, particularly at high temperatures, high thermal conductivity and low thermal expansivity which combine to impart good thermal shock resistance, have placed the ceramic in the fore as a candidate for advanced high temperature engineering applications. In the present discussion its electrical properties are the prime consideration.

Pure cubic SiC is a semiconductor with a band gap of approximately 2.3 eV and, as expected, is transparent with a pale yellow appearance in transmitted light. The hexagonal form has a band gap of $\sim$3 eV. The wide band gap of SiC

Fig. 4.1 Heating elements
(i) SiC lump as produced by the Acheson process (approx. 150 mm across).
(ii) Sintered (recrystallized) element manufactured in 3 sections (700 mm long).
(iii) Reaction-bonded, spiralled element (150 mm long).
(iv) Illustrating-range of manufactured SiC elements.
(v) Molybdenum disilicide ($MoSi_2$) elements.
(Courtesy of Kanthal AB Hallstahammer, Sweden.)

offered the realization of blue light emitting diodes (LEDs) and there was volume production of the devices over the period 1990–5. However they have been totally replaced by the far more efficient GaN variety.

There has been impressive progress in the technology of growing single crystal SiC for semiconductor devices. Silicon and gallium arsenide melt and therefore the Czochralski melt-growth (see Section 3.11) approach to growing single crystals can be exploited. Because silicon carbide does not melt the same approach cannot be exploited, but good quality wafers can be produced via vapour phase epitaxy. Polycrystalline SiC is heated to about 2400 °C when it sublimes, the vapour condensing on to a cooler SiC 'seed' crystal. At present SiC single crystal wafers of diameter larger than 5 cm are commercially available. This breakthrough, coupled with the wide band gap and excellent thermo-mechanical properties of silicon carbide, has made available material having potential for various devices. These include high-power transistors for switched mode power supplies, improved microwave devices for radar and communications, radiation-hardened devices, UV sensors, and high temperature sensors for management of automotive and aircraft engines.

As produced commercially as a mass of small crystals (see Fig 4.1) the colour of SiC ranges from black, through shades of grey, to blue, green and pale yellow. The colours arise from a variety of impurities including boron, aluminium, nitrogen and phosphorus, and are useful for selecting grades for particular applications. The purer grades contain at least 99.5 wt% SiC, but with sufficient impurities to render them conductive at room temperature.

There are three principal methods of manufacturing the types of SiC heating element shown in Fig. 4.1:

1. pressureless sintering (also described as 'recrystallization'),
2. reaction-bonding; and
3. *in situ* formation of SiC from carbon and SiO.

Method 1, which accounts for approximately 95% of element manufacture, is the straightforward sintering ('recrystallization') of α-SiC grit which is formed into a rod or tube and sintered in a carbon furnace at approximately 2500 °C. To give the component the necessary low resistance cold-end terminations the pore volume of a predetermined length of the end sections is infiltrated with silicon or a silicon alloy.

The above outlines the manufacturing route for one-piece elements. In fact most rod elements are of three-piece construction in which the low-resistance cold end sections (silicon or silicon alloy infiltrated) are manufactured separately from the high-resistance hot centre section. The three sections are then joined by a reaction-bonding process. This is the most economic approach to manufacturing elements and, importantly, offers flexibility in quickly responding to customer demand.

Method 2 produces what is usually referred to as self-bonded SiC by a process essentially the same as that developed for nuclear applications [2] and known as 'Refel*' silicon carbide. Refel silicon carbide contains a finely dispersed residual silicon phase accounting for between 5 and 10% of the total volume, the precise amount depending upon the porosity of the original preform. In the case of self-bonded silicon carbide elements the residual silicon is volatilized away by a final heat-treatment at over 2500 °C when recrystallization occurs resulting in grain growth and further bonding between primary and secondary carbide phases.

The element fabrication route starts with a mix of α-SiC grains, carbon powder and plasticizers being formed into the required shape by standard ceramic processing, usually extrusion or injection-moulding. The 'green' form is then brought into contact with molten silicon which, by capillarity, permeates the pore space, siliciding the carbon and forming a secondary phase of SiC, which bonds to the primary SiC phase.

Although the melting point of silicon is about 1410 °C, because of the need to remove the 'native' oxide layer coating the SiC so that the silicon wets the grains, the infiltration is usually carried out at temperatures $>$2000 °C when α-SiC forms. The removal of the native silica layer depends upon the ambient and in neutral atmospheres reaction-bonding can be effected at lower temperatures forming β-SiC.

The material is fired again in a recrystallization process which involves heat-treating the rods at approximately 2500 °C in an inert atmosphere. This process transforms any β-SiC to the α-modification and partially removes residual silicon, permitting control over electrical resistivity. The resulting ceramic has a low porosity and consequently a long service life.

The resistance of the hot section of a tubular element may be adjusted to the required value by spiralling, which is readily accomplished in the 'green' state, the resistance of the heating section being varied by adjusting the pitch of the cut. Alternatively, for rods, low resistance 'cold ends' may be joined to high resistance hot-zone lengths by reaction-bonding.

In method 3, which currently is followed only to an insignificant extent, a carbon tube is heated at a temperature of about 1900 °C in a bed of sand (SiO_2) and coke (carbon). The tube may be directly resistance-heated or heated indirectly by a sacrificial carbon tube of smaller diameter. Silicon monoxide (SiO) from the SiO_2-C reaction infiltrates the carbon tube transforming it to β-SiC. After the siliciding stage the SiC tube is readily removed from the bed and any residual carbon burnt out. In this form the tubes have a uniform electrical resistance along their length and the necessary higher resistance heating section is made by diamond-sawing a spiral through the tube wall. The tube has a porosity

* 'Reactor Fuel Element Laboratories'.

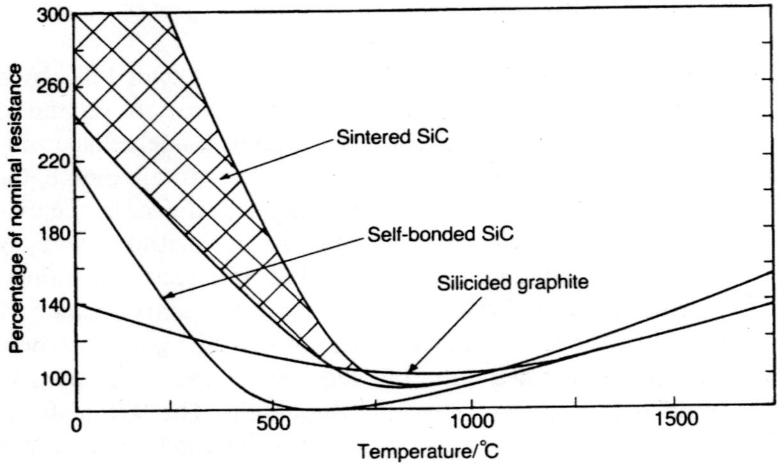

Fig. 4.2 Resistance–temperature characteristics of SiC elements prepared by various methods.

of about 30 vol.% and a large internal surface area. To reduce the rate of oxidation during use the tubes usually are protected by a silicate glaze fired to about 1450 °C.

In most cases the ends of the elements are flame-sprayed with aluminium to make good electrical contacts with the power leads.

SiC heating elements, depending upon the grade, are routinely used at temperatures up to about 1500 °C in air, and even up to about 1650 °C for short periods. The unglazed variety rely on a thin native passivating silica film for their protection against oxidation. Service life therefore depends strongly on the atmosphere in which they operate, which affects the stability of the film, and on temperature which, of course, affects reaction rates.

The general form of the resistance–temperature characteristics for SiC elements is shown in Fig. 4.2. Below approximately 800 °C the resistance varies from batch to batch; above this temperature the variability is, practically speaking, absent. Although the variety of SiC elements available precludes the possibility of quoting a single electrical resistance characteristic it is useful to assume a resistivity value at 1000 °C of about $10^{-3}\,\Omega\,\mathrm{m}$.

The variability in the resistance–temperature characteristics at temperatures below about 800 °C can be attributed to impurities, the materials behaving as very complex, doped semiconductors (see Section 2.6.2 and Problem 2.10). The positive temperature coefficient of resistance observed at the higher temperatures suggests the effect of decreasing charge-carrier mobility with increasing temperature (see Section 2.6.2), but because of the complex nature of the materials both from the chemical and microstructural standpoints, this has to be regarded as speculation.

As far as applications are concerned SiC heating elements are very widely used and likely to be found wherever high-temperature technology is involved – in the laboratory and particularly in the ceramics, glass and metallurgical industries. The intrinsic hot strength of silicon carbide is an important factor in the use of heating elements permitting horizontal installation without supports.

4.1.2 Molybdenum disilicide

Many metals form conductive silicides, which, like SiC, are resistant to oxidation through the formation of stable 'passivating' layers of silicates or silica on their surfaces at high temperatures. Molybdenum disilicide ($MoSi_2$) has been developed as a heating element for use in air at temperatures above 1500 °C. Its resistivity behaves as is expected for a metal, increasing from about $2.5 \times 10^{-7}\ \Omega\,m$ at room temperature to about $4 \times 10^{-6}\ \Omega\,m$ at 1800 °C.

A commercial $MoSi_2$ heating element known as 'Kanthal Super' is a cermet consisting of a mixture of particles bonded together with an aluminosilicate glass phase which forms about 20% of the total volume. Common forms of element are shown in Fig. 4.1. To fabricate the elements a mixture of fine $MoSi_2$ powder with a carefully chosen clay is extruded into rods of suitable diameters for the terminal sections and heating zones. The rods are dried, sintered and cut to various lengths. The heating zones are bent to the required shapes at high temperature and then welded to the larger-diameter terminal sections. The best grade of $MoSi_2$ element is capable of operating up to 1800 °C.

Because of the nature of the phase composition a $MoSi_2$ heating element will soften at high temperatures and therefore is usually mounted vertically; if the furnace design calls for horizontal mounting then some refractory support has to be supplied, as illustrated in the figure. This contrasts with silicon carbide elements which maintain their rigidity at normal maximum operating temperatures.

4.1.3 Lanthanum chromite

Lanthanum chromite ($LaCrO_3$) was developed during the 1960s for electrodes in magnetohydrodynamic (MHD) generators where hot electrically conducting gas is passed through a duct across which there is a strong magnetic field. The e.m.f. induced at right angles to both the gas flow and the magnetic field develops a voltage difference between electrodes located on opposite sides of the duct. The gas temperature must be close to 2000 °C and the gas seeded with potassium to render it conducting. The electrode material is required to be electronically conducting and resistant to corrosion by potassium and, allowing for some cooling, it must be able to withstand a temperature of 1500 °C for at least

10 000 h. $LaCrO_3$ was considered a prime candidate for this application since it combines a melting point of 2500 °C with high electronic conductivity (about 100 S m^{-1} at 1400 °C) and resistance to corrosion. There is now little interest in MHD systems but $LaCrO_3$ is established as a specialized heating element.

$LaCrO_3$ is one of the family of lanthanide perovskites RTO_3, where R is a lanthanide and T is a period 4 transition element. In the cubic unit cell R occupies the cube corners, T the cube centre and O the face-centre positions. The coordination numbers of T and R are 6 and 8 respectively. $LaCrO_3$ loses chromium at high temperatures, leaving an excess of O^{2-} ions. The excess charge is neutralized by the formation of Cr^{4+} which results in p-type semiconductivity with 'hole hopping' via the localized 3d states of the Cr^{3+} and Cr^{4+} ions. The concentration of Cr^{4+} can be enhanced by the substitution of strontium for lanthanum. A 1 mol.% addition of SrO causes the conductivity to increase by a factor of approximately 10 (see Section 2.6.2).

Heating elements are formed by normal ceramic processing. Strontium acts as a sintering aid as well as promoting conductivity, and cobalt can be added to limit grain growth. Sintering takes place in a reducing atmosphere ($p_{O_2} \sim 10^{-12}$ atm or 10^{-7} Pa) at temperatures close to 1700 °C and is followed by an anneal in oxygen that establishes the high conductivity.

Satisfactory conductivity is maintained up to 1800 °C in air but falls off at low oxygen pressures so that the upper temperature limit is reduced to 1400 °C when the pressure is reduced to 0.1 Pa. A further limitation arises from the volatility of Cr_2O_3 which may contaminate the furnace charge. The combination of high melting point, high electronic conductivity and resistance to corrosion has led to the adoption of lanthanum chromite for the interconnect in high temperature solid oxide fuel cells (see Section 4.5.3).

4.1.4 Tin oxide

Tin oxide (SnO_4) has found applications in high-temperature conductors, ohmic resistors, transparent thin-film electrodes and gas sensors.

It crystallizes in the tetragonal rutile structure (see Fig. 5.27) with cell dimensions $a = 474$ pm and $c = 319$ pm; in the single-crystal form it is known by its mineralogical name, cassiterite. It is a wide band gap semiconductor, with the full valence band derived from the O 2p level and the empty conduction band from the Sn 5s level. The band gap at 0 K is approximately 3.7 eV, and therefore pure stoichiometric SnO_2 is a good insulator at room temperature when its resistivity is probably of the order of 10^6 Ω m.

In practice both natural and synthetic crystals are oxygen deficient, leading to donor levels approximately 0.1 eV below the bottom of the conduction band and consequently to n-type semiconductivity. Doping the crystal with group V elements also induces n-type semiconductivity; the usual dopant is antimony. The ground state electronic configuration of the Sb atom is $5s^2p^3$, and when it replaces Sn^{4+} ($r_6 = 69$ pm) in the SnO_2 lattice it does so as Sb^{5+} ($r_6 = 61$ pm). The n-type conductivity is due to electrons ionized from the $5s^1$ state.

Antimony-doped SnO_2 is a complex system and is far from completely understood. The interpretation of data is complicated by the fact that both the stoichiometry of the host lattice and the oxidation state of the dopant are dependent upon ambient oxygen partial pressure and temperature. The successful exploitation of semiconducting SnO_2 has been achieved largely through development work guided by a general understanding of underlying principles, as indeed is the case for much of electroceramics technology.

An important application of SnO_2 in ceramic form is in conducting electrodes for melting special glasses, such as those used for optical components and lead 'crystal' tableware. The ideal glass-melting electrode should have a high electrical conductivity at room temperature and, of course, at glass-melting temperatures and a high resistance to corrosion by the glass. In addition, it must not discolour the glass. SnO_2 is the only material, apart from platinum, which fulfils these requirements for glasses containing lead oxide.

SnO_2 itself does not readily sinter to a dense ceramic and so sintering aids such as ZnO ($r_6(Zn^{2+}) = 75$ pm) and CuO ($r_6(Cu^{2+}) = 73$ pm) are added, together with group V elements such as antimony and arsenic to induce semiconductivity. Some of the sintering aids enter the lattice, tending to negate the effect of the dopants added to induce semiconductivity, and successful compositions, typically containing more than 98 wt% SnO_2, have been arrived at by trial and error.

The oxide powder, together with binders, is commonly isostatically pressed into the forms shown in Fig. 4.3 and fired in oxidizing conditions at approximately 1400 °C. Under these conditions sintered densities in the range 6500–6700 kg m^{-3} (theoretical density for SnO_2 is 7002 kg m^{-3}) are achieved. The largest electrodes made in this way are in the form of cylinders about 600 mm long and 150 mm in diameter weighing approximately 60 kg. Cooling from the sintering temperature is carried out, in part, in a nitrogen atmosphere with the object of creating electron-compensated oxygen vacancies and so enhancing room temperature conductivity which is typically of the order of 10^3 S m^{-1}. The high conductivity minimizes Joule heating in the electrode region outside the molten glass.

The form of the glass-melting furnace is shown in Fig. 4.4 although in practice the electrodes may well be a substantial part of the furnace walls. The charge ('cullet') is preheated, using gas or oil, to about 1000 °C when it has sufficient conductivity to be directly heated to the required temperature (1300–1600 °C) by power dissipated internally. By supplying the heat from within the body of the glass melt rather than from the outside, the free surface temperature is kept relatively low and thus the excessive loss of volatile elements, particularly lead, is avoided. The process is economic since the heat is generated in the glass, where it is required. The elements are resistant to attack by glass and so have long service lives, typically 2 years.

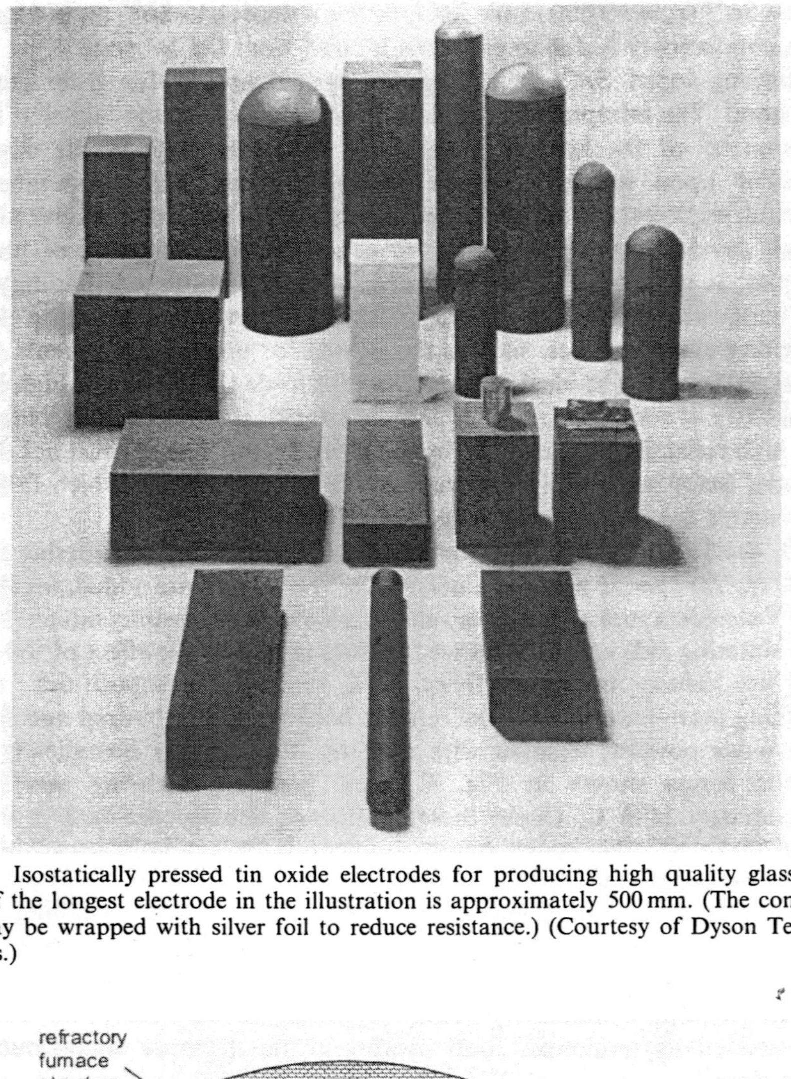

Fig. 4.3 Isostatically pressed tin oxide electrodes for producing high quality glasses; the length of the longest electrode in the illustration is approximately 500 mm. (The contacting stubs may be wrapped with silver foil to reduce resistance.) (Courtesy of Dyson Technical Ceramics.)

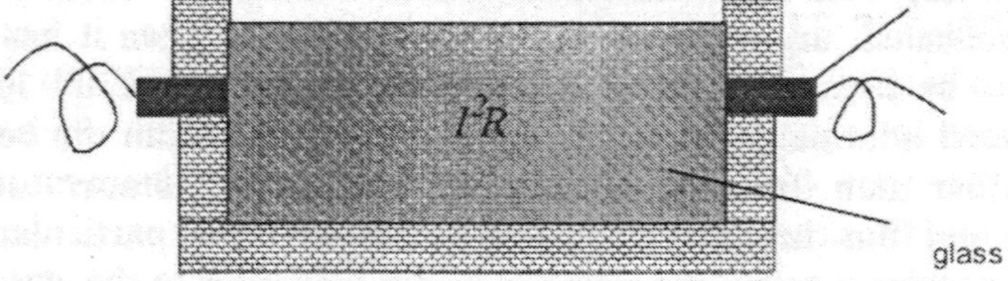

Fig. 4.4 Heating a glass melt: the power (I^2R) is introduced directly into the melt via tin oxide electrodes.

4.1.5 Zirconia

The 'Nernst filament', once used as a light source, consisted of zirconia (ZrO_2) doped with thoria and ceria. Preheating the filament to about 600 °C using an external source reduces its resistance to a value which permits direct Joule heating to be effective when temperatures of up to 1800 °C can be attained. More recently the development of stabilized ZrO_2 (see Section 4.5.3) has made it possible to use ZrO_2-based ceramics as furnace elements, although they also always require preheating to reduce the resistance to a level at which Joule heating is effective. Its high negative temperature coefficient of resistance makes it necessary to have a resistor, preferably with a high positive temperature coefficient, in series with it to limit current.

Once the element has reached temperatures exceeding about 700 °C it can also be used as a susceptor and heated by eddy currents generated by an induction coil and a high frequency (0.1–10 MHz) power source.

ZrO_2 was one of the earliest ceramics in which conduction by oxygen ions was observed and now this characteristic is exploited in fuel cell and sensor technologies (see Sections 4.5.3 and 4.6.1).

4.2 Ohmic Resistors

Most resistors for electrical and electronic applications are required to be ohmic and to have small temperature coefficients of resistance. The major requirement in electronics is for resistors in the range 10^3–10^8 Ω, while materials with suitable electrical properties usually have resistivities less than 10^{-6} Ω m. Fabrication of a 10^5 Ω resistor of length 110 mm from a material with a resistivity of 10^{-6} Ω m requires a cross-sectional area of 10^{-12} m^2, i.e. a strip 1 μm thick and 1 μm wide, for example. This is not a technical impossibility, but other more economic routes to resistor manufacture have been established based on the following two principles.

1. Very thin conductive layers are deposited on an insulating substrate and large length-to-width ratios are obtained by etching a suitable pattern.
2. The conductive material is diluted with an insulating phase.

These methods are often combined.

4.2.1 Thin films

Thin films of thickness typically 10 nm are readily formed in a vacuum chamber by evaporation, 'sputtering', or chemical vapour deposition (CVD). Many metals and metal alloys, e.g. aluminium, silver, gold and Ni–Cr, can be evaporated from the molten state and condensed onto suitable substrates. Ni–Cr alloys with resistivity values of about $10^{-6}\,\Omega\,m$ are deposited in thin-film form and provide a basis for the manufacturer of high-value resistors.

Thin-film oxides are usually formed by 'sputtering'. The chamber is filled with argon at a pressure of typically about 1 Pa to which a small amount of oxygen is added. The target is usually a disc of the oxide to be sputtered fixed to a metal plate. The substrate on which the film is to be deposited rests on a metal plate. A high-frequency (about 1 MHz) high-voltage (about 5 kV) is applied between the two plates and a plasma is developed. Gaseous ions bombard the source and detach clusters of ions or molecules from its surface which pass through the plasma and are deposited on the substrate. SnO_2, In_2O_3 and mixtures of these oxides, e.g. $90In_2O_3$–$10SnO_2$, are deposited as transparent conductive films by this method. They are essential to the functioning of many electro-optical devices.

Indium tin oxide (ITO) films are also commonly deposited by CVD in the manufacture of film resistors. Glass, or sometimes steatite, rods are heated to about 700 °C in air of controlled humidity and a mixture of tin tetrachloride and antimony pentachloride is then introduced for a few seconds. Reaction with water occurs on the surface of the rods, resulting in the formation of a thin film of mixed oxides which is firmly attached to the substrate. The antimony lowers the resistivity of the SnO_2 and also enables the temperature coefficient of resistance to be controlled near to zero; the SnO_2 alone gives a high negative coefficient. A small fraction of the O^{2-} ions in the deposit are replaced by Cl^- ions, and these also lower the resistivity because the lower negative charge of Cl^- compared with O^{2-} is compensated by an electron in the conduction band. The resulting units require a protective layer of impervious lacquer to inhibit the effects of moisture but are very stable and reliable in use; because of the strong adhesion of the film open circuits do not occur through detachment of small pieces, which occasionally happens with sputtered carbon films. The presence of moisture and soluble ionic material on the film can set up electrolytic action which results in a portion of the film being converted to metallic tin with a resultant diminution in resistance, but care in production makes this a very rare occurrence.

Enhanced length-to-width ratios can be achieved by cutting a spiral groove through the resistive film and into the cylindrical rod substrate. Resistance values can be kept within close tolerances by stopping the groove cutting automatically when a preset value is reached. Patterned deposits on flat surfaces can be formed by a number of processes, e.g. by an air jet carrying abrasive particles or by volatilization by an infrared laser beam or an electron beam. Frequently photolithography is used when a layer of photosensitive material, called a 'resist', is formed on the deposited film. The resist is exposed to a pattern of light that renders it insoluble and the unexposed regions are then removed with a solvent. The deposit is then immersed in a reagent that dissolves the sputtered material and not the resist. Finally, the remaining resist is removed. Examples of thin-film resistors are shown in Fig. 4.5.

4.2.2 Thick films

Rather thicker films with thicknesses typically in the range 10–15 μm are made by what is termed the 'thick-film' or 'silk-screen' technique. Silk screening is a well-established method for printing artwork. The screens were formerly silk but are now meshes of either nylon or stainless steel. Patterns are formed by coating them with a resist, exposing them to a pattern of light and washing out the unexposed parts. The screen is held taut in a frame that is fixed 1–3 mm above the surface to be printed. A paint of a stiff creamy consistency is swept across the screen by a hard rubber squeegee with sufficient pressure to force the screen, now loaded with paint, into contact with the underlying surface. The consistency of the paint is such that, as the screen rises from the surface, it flows over the spaces left by the threads or wires of the screen so that there is only a small variation in the thickness of the deposit.

Resistors are formed by taking a conductive powder and mixing it with powdered glazes and an organic vehicle that imparts the necessary rheological properties. Mixing is carried out on a set of rollers running at different speeds as described under calendering (see Section 3.6.7), and this ensures a very high degree of dispersion of the components. A required resistor pattern is silk screened onto a substrate, usually a 96% alumina. A carefully controlled firing schedule ensures that the organic solvents evaporate (100–150 °C), the remaining organic compounds evaporate or are burned out (200–450 °C) and, finally, the inorganic glass and active resistive components mature and bond to the substrate (850 °C for about 10 min). The complete firing takes about 1 h. Metal conductors and contact pads can be made using the same technique. Examples of thick-film devices and circuits are shown in Fig. 4.5.

Although it is simple to prepare pastes and use them to make resistors, the structure of the resulting components is complex and the conduction mechanisms remain a matter for debate.

The active components are usually highly conductive oxides (10^5–10^6 S m^{-1}) such as PdO, RuO_2, $Bi_2Ru_2O_7$ and $Bi_2Ir_2O_7$. Electrically they behave as metals and have low positive temperature coefficients of resistivity (TCR). The glaze is usually a lead borosilicate, of composition typically (in wt%) $52PbO$–$35SiO_2$–$10B_2O_3$–$3Al_2O_3$. The resistors normally give high resistivities and negative temperature coefficients for low concentrations of the conductive component, and low resistivities and positive temperature coefficients for high concentrations, as shown in Fig. 4.6†. This behaviour differs markedly from that to be expected from a random distribution of conductive particles in an insulating matrix. The conductivity of dispersions of metal particles of mean size around

† In characterizing the resistance of the printed films used in microelectronics it is the practice to define a 'surface resistivity', $\rho_s = \rho/h$, in which h is the film thickness. It is also common practice to express ρ_s in 'units' of 'ohms per square' or simply $\Omega/\square$; ρ_s is equivalent to the resistance measured between opposite edges of a printed square.

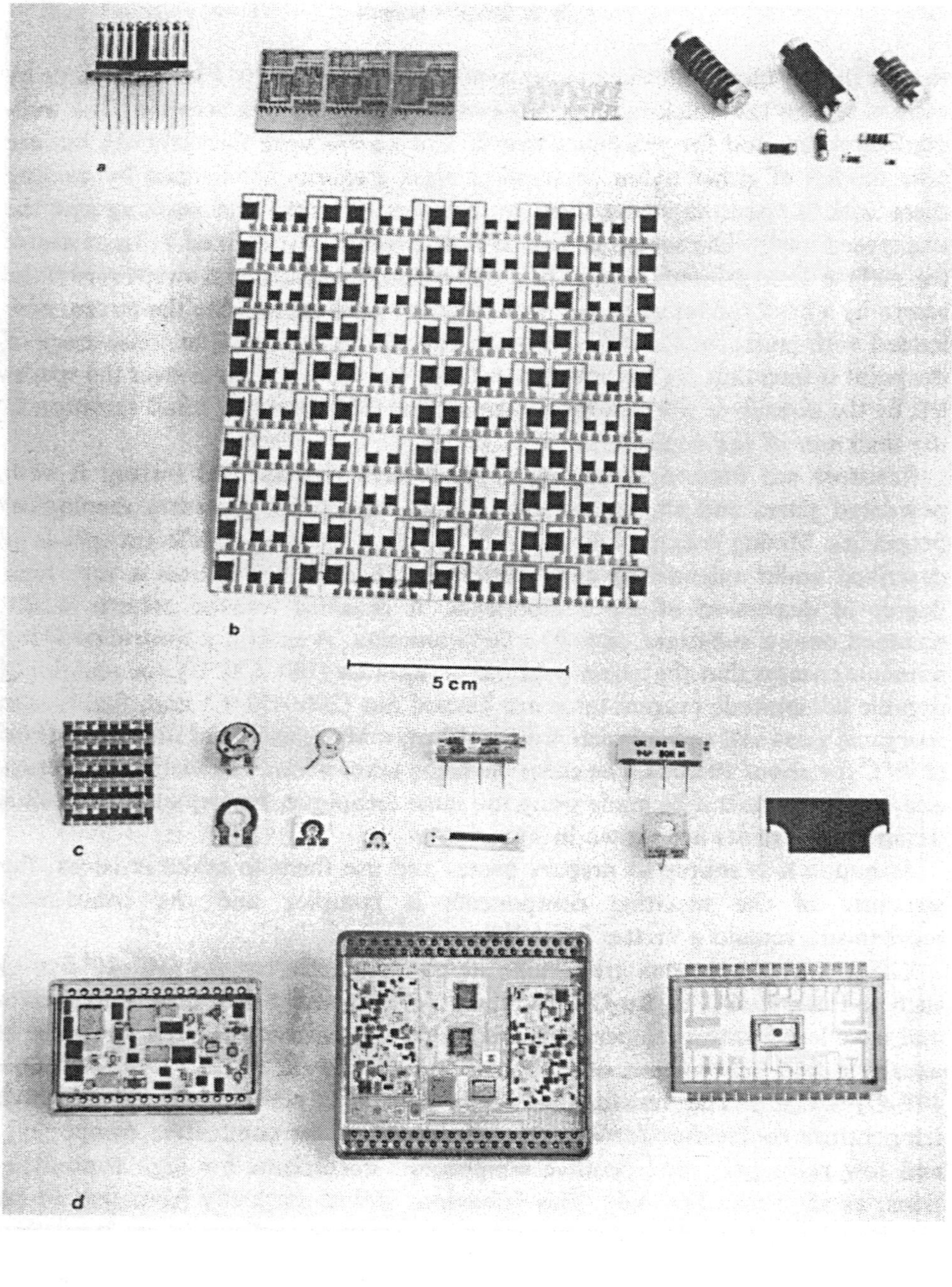

Fig. 4.5 Film devices and circuits: (a) thin-film resistors on glass and steatite substrates; (b) thick-film resistor networks on 'snapstrate' alumina substrate; (c) various thick-film resistors; (d) hybrid microcircuits. (Components kindly supplied by General Hybrid, C-MAC and Welwyn Components Ltd.)

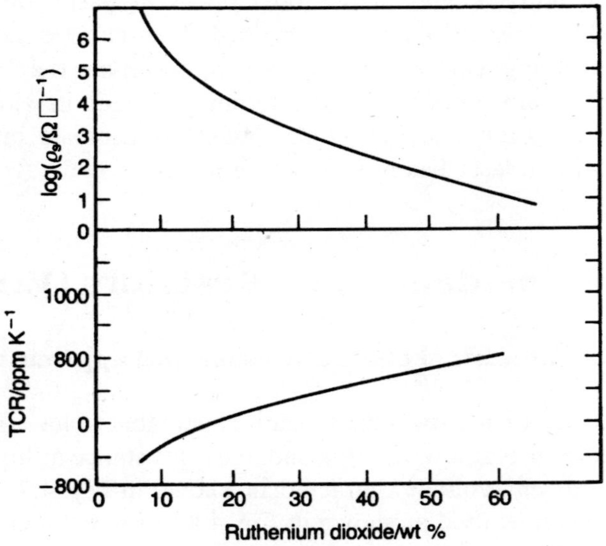

Fig. 4.6 Electrical characteristics of RuO_2 thick-film resistors.

100 μm in an insulating matrix is very low when the metal concentration is small and then increases by several orders of magnitude over a concentration range of a few per cent around 10 vol.% (see Fig. 2.49), with only a small increase for higher concentrations. This is understandable as a consequence of the establishment of continuous contact between particles from electrode to electrode at a well-defined concentration.

Transmission electron micrographs show that the dispersed conductive particles form convoluted chains with the particles either close together or in contact. According to this model there will be chains of conductive particles joining the electrodes even at high dilutions. The resistance is provided by two components, that due to the interior of the conductive particles and that due to the areas of near contact between them. The former will provide a positive temperature coefficient of resistance whilst the latter, if it depends on some form of thermally activated semiconductive behaviour, will provide a negative coefficient. So far it has not been possible to determine the properties of the material between the surfaces of the particles. The solubility of the conductive oxides in the glaze has been shown to be very small but, as the intergranular layer is very thin, it may be enough to endow the glaze with a sufficient level of conductivity.

The glaze component is resistant to moisture so that units are stable in normal ambients. Their mode of fabrication allows whole patterns of resistors to be made at once, although strict control of production conditions is necessary to ensure reproducibility. Generally several batches of paste are made and the values they yield on processing are determined. The required value can then be obtained by blending two batches whose values span that required, using Lichtenecker's relation (Eq. (2.131)) to determine the ratio in which the two end-members must be combined. Values can finally be trimmed into tolerance by volatilization using a laser beam or by jet abrasion.

4.3 Voltage-dependent Resistors (Varistors)

4.3.1 Electrical characteristics and applications

There are a number of situations in which it is valuable to have a resistor which offers a high resistance at low voltages and a low resistance at high voltages as is the case in the current–voltage characteristic shown in Fig. 4.7.

Such a device can be used as shown in Fig. 4.8 to protect a circuit from high-voltage transients by providing a path across the power supply that takes only a small current under normal conditions but takes a large current if the voltage rises abnormally, thus preventing high-voltage pulses from reaching the circuit. High-voltage transients are of sufficiently frequent occurrence in most power supplies to make the protection of sensitive circuits essential. Transistors and integrated circuits are particularly susceptible to damage by transients.

Ceramics based on SiC and ZnO are two materials in everyday use that have the characteristic shown in Fig. 4.7. In both cases it has been established that the resistance is controlled by the region in which the ceramic grains contact one another.

The microstructure of a ceramic for a voltage-dependent resistor (VDR) can be visualized as shown in Fig. 4.9(a) with IGLs of varying thickness between grains that differ in size by about a factor of 10. An idealized structure which is

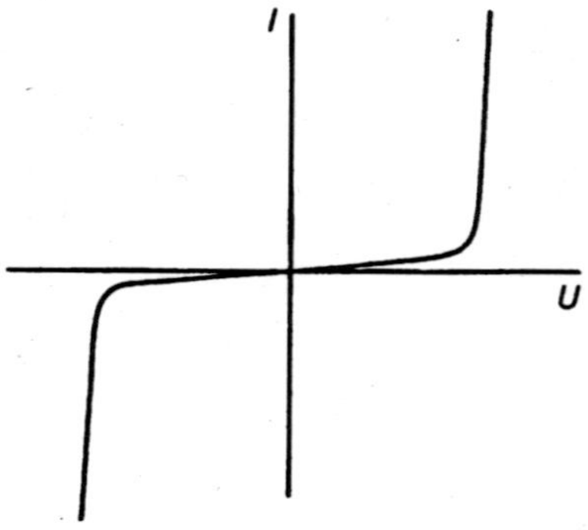

Fig. 4.7 Typical current–voltage relation for a voltage-dependent resistor.

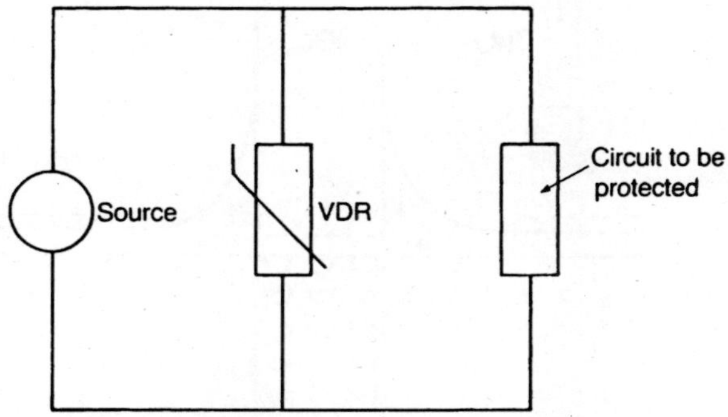

Fig. 4.8 Use of a voltage-dependent resistor (VDR) to protect a circuit against transients.

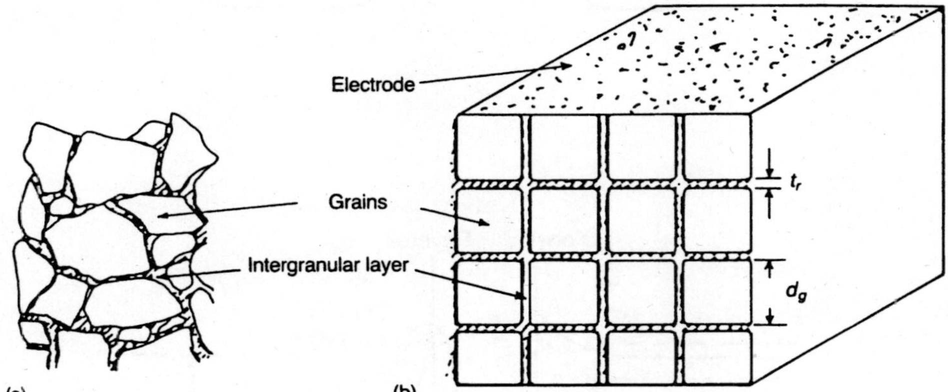

Fig. 4.9 Illustrations of (a) actual and (b) idealized microstructure of a varistor.

useful for calculating average properties is shown in Fig. 4.9(b), but it must be appreciated that in practice the high-current paths are likely to be via those particles separated only by the thinnest IGLs, so that regions of high current density will form tortuous routes between the electrodes.

The VDR behaviour in ZnO varistors is governed by electron states that are formed on the surfaces of crystals as a consequence of the discontinuity. These surface states act as acceptors for electrons from the n-type semiconductor. Electrons will be withdrawn from the region near the surface and replaced by a positive space charge. In this way oppositely oriented Schottky barriers will be created at the surfaces of neighbouring crystals so that a high resistance will be offered to electron flow in either direction (Fig. 4.10(a)). The situation with an applied field is shown in Fig. 4.10(b). With low applied fields small thermally

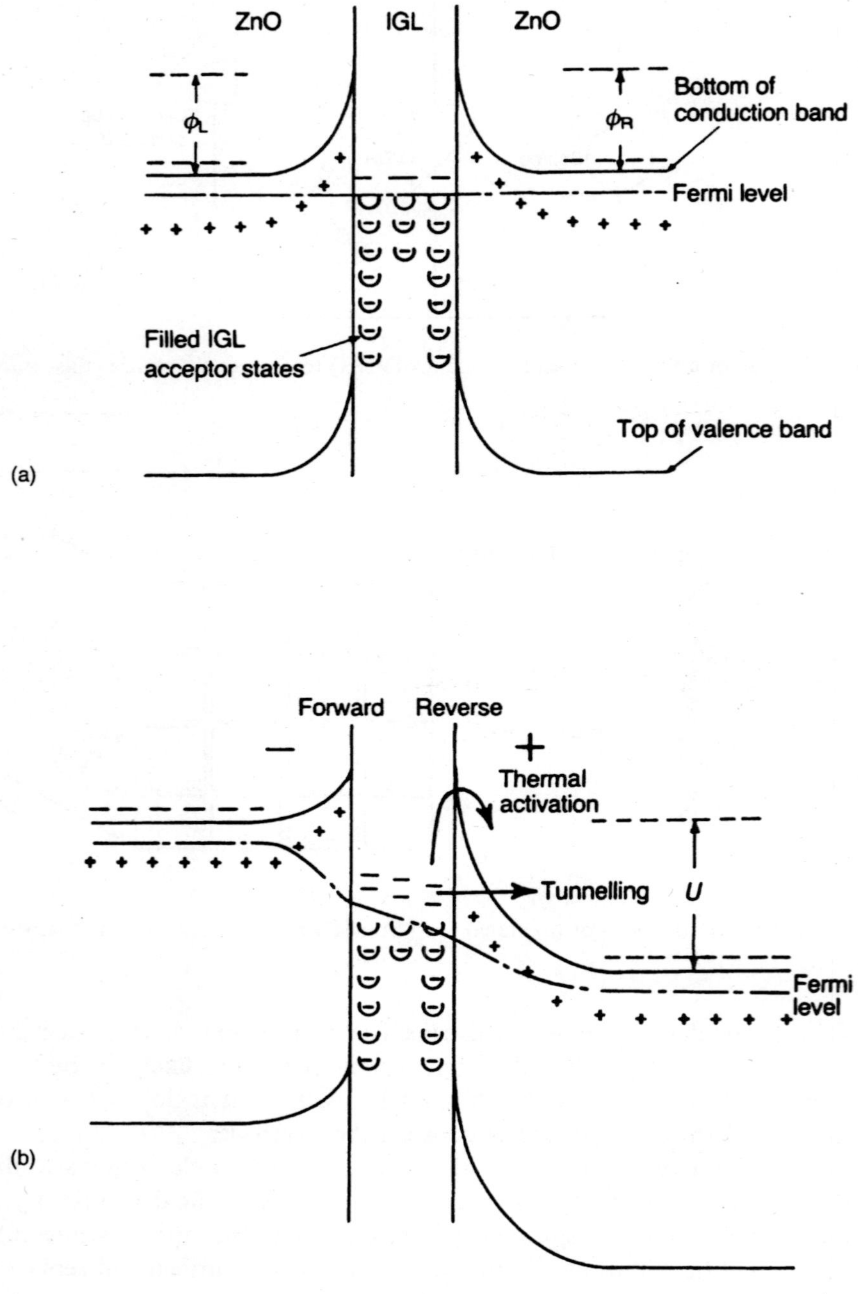

Fig. 4.10 Proposed electronic structure at a junction between semiconducting ZnO grains: (a) no voltage applied; (b) with applied voltage.

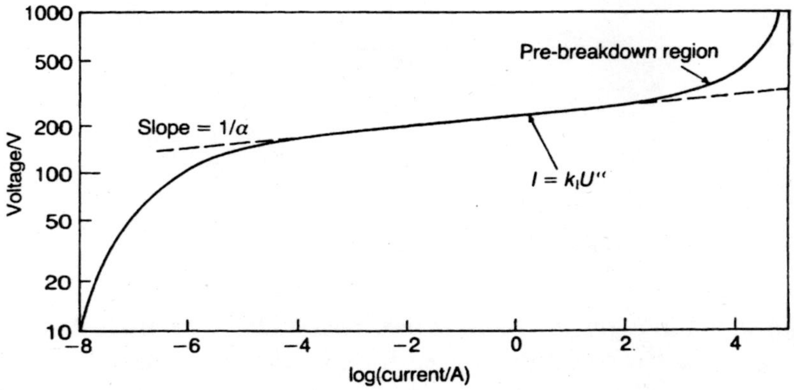

Fig. 4.11 Typical varistor voltage–current characteristic.

activated currents pass over the reverse biased junction, and at high fields tunnelling through the junction will occur, accounting for the low resistance. The behaviour is similar in some respects to that of Zener diodes, and VDR characteristics are similar to those of two Zener diodes connected back to back.

A typical varistor voltage–current characteristic is shown in Fig. 4.11. The linear part can be represented by the relation

$$I = k_I U^{\alpha} \tag{4.1}$$

where k_I is a constant and α falls off at low voltages. If I_1 and I_2 are the currents at voltages that differ by a factor of 10,

$$\alpha = \log_{10}\left(\frac{I_1}{I_2}\right) \qquad I_1 > I_2$$

Alternatively

$$U = k_V I^{\beta} \tag{4.2}$$

where $\beta = 1/\alpha$ and $k_V = k_I^{-1/\alpha}$. The resistance at a given voltage is

$$R = k_V I^{\beta-1} = \frac{1}{k_I} U^{-(\alpha-1)} \tag{4.3}$$

and the power dissipated is

$$P = IU = k_I U^{\alpha+1} \tag{4.4}$$

Eq. (4.4) indicates that the voltage permanently applied to a VDR must be carefully limited. For instance with $\alpha = 25$ a 10% increase in voltage would increase the power dissipation by a factor of about 2.5. Since VDRs usually have negative temperature coefficients of resistivity, it can be seen that a runaway condition can easily be precipitated.

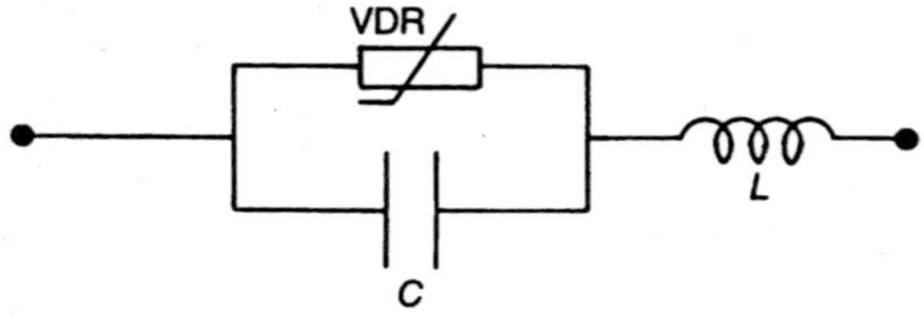

Fig. 4.12 Equivalent circuit of a VDR.

Certain limitations on the performance of a VDR arise from its secondary properties. It has a capacitance derived from its electrodes and its permittivity, and an inductance due to the length of its leads and the area of its electrodes. Its equivalent circuit is shown in Fig. 4.12.

The effectiveness with which a VDR absorbs the very rapid transients that may occur accidentally in a power supply depends on its speed of response. The response time of the voltage-sensitive material is generally adequate, being typically 0.5 ns. The capacitance in parallel with the VDR will partially absorb the transient and delay the rise in voltage, but this will not affect the protective action of the device. The inductance may have an important effect because the steep rise time of a transient results in the presence of very high frequency voltage components so that the impedance due to even a small inductance in series with a VDR becomes significant. The leads to the VDR must therefore be as short as possible.

Another use for VDRs is the suppression of sparks in switches and relay contacts in highly inductive circuits, and it is instructive to outline the principle of operation.

If the inductance L in the simple circuit shown in Fig. 4.13 has a d.c. resistance of 100 Ω, the current through it with the switch closed is 0.24 A. Opening the switch sets the charge in the LC loop oscillating, and the peak instantaneous current is 0.24 A. Because the maximum energy stored in the capacitor ($\frac{1}{2}CU^2$) must be equal to that stored in the inductor ($\frac{1}{2}LI^2$), it follows that

$$U = I\left(\frac{L}{C}\right)^{1/2} \tag{4.5}$$

Thus, if $L = 0.05\,\text{H}$ and $C = 100\,\text{pF}$, the maximum instantaneous voltage developed is 5366 V, which is sufficient to cause a spark to be struck across the switch contacts with consequent damage.

Figure 4.13 shows the introduction of a VDR with $\alpha = 5$ and a characteristic such that the current through it with the switch closed is 0.024 A so that $k_1 = 0.024/24^5$. When the switch is opened the instantaneous current through the VDR is approximately 0.24 A since the impedance of the capacitor and

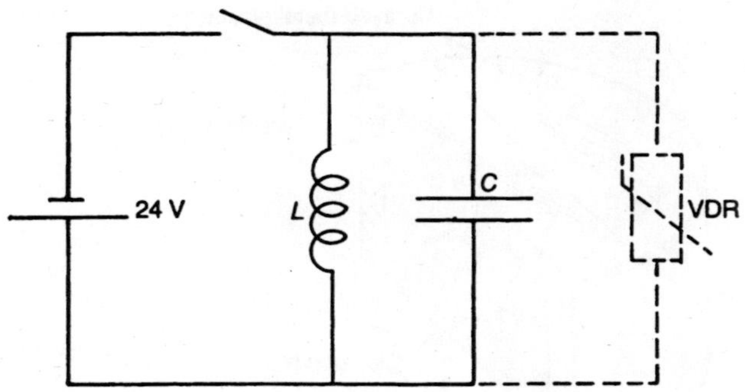

Fig. 4.13 Use of a VDR to suppress sparking in a highly inductive circuit.

inductor at the resonant frequency is very large (about 22 kΩ). The resulting voltage across the VDR is

$$U = \left(\frac{I}{k_{\mathrm{I}}}\right)^{1/\alpha} = 38\,\mathrm{V} \tag{4.6}$$

The total voltage across the switch will now be 38 V + 24 V = 62V which is insufficient to cause sparking.

The earlier pre-1970 VDRs were based on SiC and had α values up to about 5. These are still used to some extent where large amounts of energy are to be dissipated, as in lightning arresters. In general VDRs are based on ZnO and have α values up to about 70, but more commonly in the range 25–45. They are widely used to protect circuits against transients.

4.3.2 Silicon carbide

SiC VDRs are made from selected grades of the material produced primarily as an abrasive. The first basis of selection is colour: the blacker material is a p-type semiconductor and is preferred for VCRs. Particle size is usually in the range 50–150 μm. The SiC powder is mixed with clay and other siliceous components, shaped by pressing and fired at about 1300 °C. The final body consists of SiC particles closely bonded by a siliceous glassy matrix. The final properties depend on empirically determined details of the processing. Additions such as graphite may be made and the firing atmosphere may be air, nitrogen or hydrogen.

Metal electrodes are usually applied by flame-spraying, and the unit is then subjected to a series of high current pulses (about 10 MA m^{-2} for 10 μs) with alternating polarity – a treatment that stabilizes the properties. Units are usually

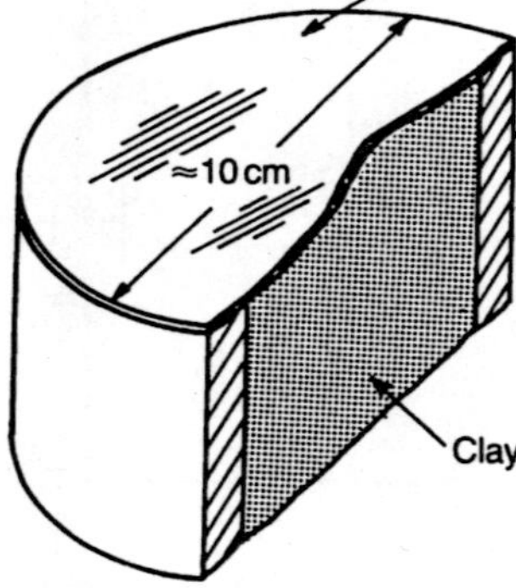

Fig. 4.14 Section through a commercial SiC surge arrester.

sealed into porcelain shells with an epoxy resin to prevent the ingress of moisture (Fig. 4.14).

While there is no doubt that the VDR behaviour of SiC units is derived from the contact areas between the particles, which very probably contain a thin layer of silica, the precise mechanism is uncertain although the tunnelling of carriers through potential barriers is likely to be involved.

A range of SiC varistors is shown in Fig. 4.15.

4.3.3 Zinc oxide

Zinc oxide (ZnO) can exhibit a wide range of electrical properties depending on minor constituents and sintering conditions. It is also remarkable for the ease with which high-density ceramics can be achieved over a wide range of sintering temperatures and for its excellent resistance to thermal shock.

Zinc oxide has the wurtzite (ZnS) structure comprising hexagonal close-packed oxygen ions with half the tetrahedral interstices containing Zn^{2+} ions. On heating to high temperatures in air it loses oxygen, thus creating oxygen vacancies, and early models attributed the resulting non-stoichiometry to the formation of zinc atoms on interstitial sites. It is now accepted that the defect is a neutral oxygen vacancy and that thermal energy at room temperature is sufficient to ionize the vacancy, thus promoting electrons into the conduction band. The energy gap at room temperature for a crystal free from donors is approximately 3.2 eV. Conductivities exceeding $10^2\,\mathrm{S\,m^{-1}}$ can be obtained in ceramics that have been cooled sufficiently rapidly after sintering to preserve a high concentration of oxygen vacancies.

Varistor compositions based on ZnO contain a number of dopants at a level of approximately 1 mol.%; a typical formulation is (in mol.%) 96.5ZnO–0.5Bi_2O_3–

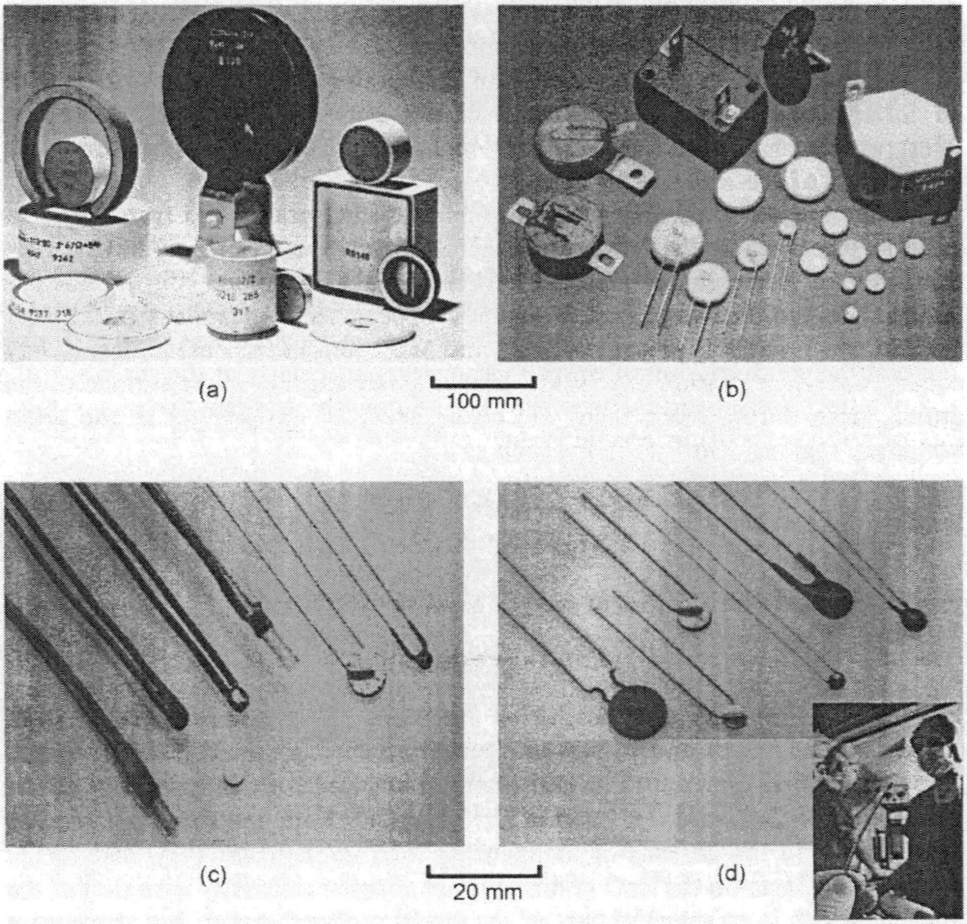

Fig. 4.15 (a) Silicon carbide varistors, and (b) zinc oxide varistors. (Courtesy of P.D. Devices Ltd. UK.) (c) PTC thermistors for motor protection and (d) NTC thermistors: inset, catheter fitted with biomedical NTC thermistor for measuring blood temperature. (Courtesy of Thermometrics.)

1.0CoO–0.5MnO–1.0Sb_2O_3–0.5Cr_2O_3. The standard ceramic powder processing procedure is followed, although calcination at about 800 °C is sometimes omitted. Sintering takes place at about 1250 °C in air. Reducing conditions must be avoided because zinc boils at 913 °C and is therefore rapidly lost if it is formed at high temperatures. The rate of cooling after sintering affects the non-linearity: quenching to room temperature from 900 °C reduces the α value.

Electrodes usually consist of fired-on silver paint with a small glaze content. It may be necessary to remove a high-resistivity surface layer from the ceramic before silvering. The contacts formed in this way are unlikely to be ohmic, but their non-ohmic behaviour is not perceptible compared with the extremely non-linear behaviour of the bulk of the ceramic.

The structure of the ceramic consists of ZnO grains of diameter 10–50 μm with an intergranular phase varying in thickness between 1 nm and 1 μm. The intergranular phase, which usually has a high bismuth content, is insulating with a resistivity of the order of $10^6\,\Omega\,\mathrm{m}$.

Conduction seems to take place through the ZnO grains and from grain to grain in places where the grains are either in direct contact or only have a very thin layer of intergranular material between them. The minor constituents are divided between the two phases. Manganese appears to be the most significant of the additives, probably present as Mn^{2+} and Mn^{3+} ions in the ZnO lattice. It may control the concentration of oxygen vacancies, particularly at the surface of the grains, since during firing there will be a supply of oxygen gas in the grain boundary regions. Thus reactions such as

$$2Mn^{2+} \rightarrow 2Mn^{3+} + 2e' \tag{4.7}$$

$$\tfrac{1}{2}O_2(g) + 2e' \rightarrow O^{2-} \tag{4.8}$$

may compensate for the loss of oxygen from the ZnO lattice:

$$O^{2-} \rightarrow \tfrac{1}{2}O_2(g) + V_O \tag{4.9}$$

Mn^{2+} and Mn^{3+} may also behave as acceptors for electrons liberated from oxygen vacancy traps and so prevent them from entering the conduction band. Cobalt and chromium may also contribute to reactions of this type through the coexistence of Co^{2+} and Co^{3+} and of Cr^{3+} and Cr^{4+} ions, but they appear to be less effective in the absence of manganese. Such mechanisms may lead to the presence of a layer on the ZnO grains that has a higher resistivity than that of the interior, which is an essential part of the model outlined earlier, but the precise mechanism has yet to be determined.

Bi_2O_3, Sb_2O_3 and other constituents provide a liquid phase that may assist sintering and control the grain size. It may also lead to a distribution of intergranular contacts that enhances the VDR effect. It has been found that subjecting ZnO VDRs to a.c. or d.c. fields of order 10–100 V mm^{-1} results in a significant change in their characteristics. Since the change results in an increase in current and in power dissipated as heat, so that a catastrophic runaway condition becomes possible, the process is termed 'degradation'. The units can be stabilized by annealing for 2 h in air at 700–800 °C. The treatment results in a fall in resistivity at low voltages but the altered characteristic is stable.

The degradation has been attributed to the diffusion of interstitial zinc ions Zn_i^+ in the depletion region near the surface of the ZnO grains. These ions migrate to the grain surface under the influence of a field and neutralize part of the negative charge at the interface with the formation of neutral Zn_i interstitials. This lowers the effective height of the interface barrier and so lowers the resistivity. The annealing treatment allows the diffusion of ions to lower energy states which are not affected by the subsequent application of a field.

While ZnO is outstanding for the high α values that can be attained, other systems that contain barrier layers, for instance the positive temperature coefficient resistors based on $BaTiO_3$, also show the effect, but alternatives to ZnO have not been developed commercially.

4.4 Temperature-sensitive Resistors

There are numerous uses for resistors with high values of the temperature coefficient of resistance (TCR) and they may be negative (NTC) or positive (PTC). An obvious application is in temperature indicators that use negligible power to monitor resistance changes. Compensation for the variation of the properties of other components with temperature may sometimes be possible; in this case the applied power may be appreciable and the resulting effect on the temperature-sensitive resistor (TSR) must be taken into account.

There will be a time interval between the application of a voltage to a TSR and the establishment of its equilibrium temperature and resistance. Thus NTC resistors can be used to delay the establishment of a final current and power level, while PTC units can be used to give an initially high current that falls back to a required level. PTC units can be used to maintain a comparatively constant current from a source of variable voltage since the increase in resistance resulting from power increase due to a voltage increase may be sufficient to inhibit any current increase.

Both NTC and PTC units can be used to indicate changes in ambient conditions since the power they draw from a source depends on the heat that they can dissipate into their surroundings. PTC units have the advantage that they are unlikely to overheat since an increase in temperature cuts down the power that they need to dissipate. Precautions must be taken with NTC units to ensure that runaway conditions cannot arise, because an increase in their temperature increases the power that they can draw from a constant voltage source (see Section 5.2.2).

In a ceramic a large temperature coefficient of resistivity can arise from three causes:

1. the intrinsic characteristic of a semiconductor which leads to an exponential fall in resistivity over a wide temperature range;
2. a structural transition which is accompanied by a change in the conduction mechanism from semiconducting to metallic (this usually results in a large fall in resistivity over a small temperature range);

3. a rapid change in dielectric properties in certain ceramics which affects the electronic properties in the intergranular region to give rise to a large increase in resistivity with temperature over small temperature ranges.

The choice of material exhibiting mechanisms 2 and 3 is limited to those in which the effect occurs at a practically useful temperature, usually somewhat above the normal ambient. The second mechanism has not led to significant applications and is discussed only briefly below. Mechanism 3 has led to important PTC devices, and familiarity with the ground covered in the sections concerned with $BaTiO_3$ is a prerequisite to an understanding of the PTC effect (see Sections 2.6.2 and 2.7.3).

4.4.1 Negative temperature coefficient resistors (NTC Thermistors)

The TCR of a semiconductor is expected to be negative (see Section 2.6) whether the conducting electrons move in a conduction band, as for example in SiC, or 'hop' between localized sites as is believed to occur in lithium-doped NiO (see Section 2.6.2). In each case the resistivity ρ depends on temperature according to

$$\rho(T) = \rho_\infty \exp\left(\frac{B}{T}\right) \tag{4.10}$$

where ρ_∞ is approximately independent of temperature and B is a constant related to the energy required to activate the electrons to conduct. Differentiating Eq. (4.10) leads to the TCR value α_R:

$$\alpha_R = \frac{1}{\rho}\frac{d\rho}{dT} = -\frac{B}{T^2} \tag{4.11}$$

For example, if $B = 2700\,K$, $\alpha_R = -3\%\ K^{-1}$ at 300 K; alternatively α_R can be expressed as 30 000 p.p.m. $°C^{-1}$ or 30 000 MK^{-1}. It follows that the larger is the value of B, the greater is the temperature coefficient; in contrast, the resistivity will be high at low temperatures because of the paucity of carriers, which is a disadvantage in many practical applications. There is a large choice of NTC materials, but those most used in practice are based on solid solutions of oxides with the spinel structure, e.g. Fe_3O_4–$ZnCr_2O_4$ and Fe_3O_4–$MgCr_2O_4$. A series that gives favourable combinations of low resistivity and high coefficients is based on Mn_3O_4 with a partial replacement of the Mn by Ni, Co and Cu, as shown in Table 4.1.

Mn_3O_4 is a normal spinel with Mn^{2+} on tetrahedral sites and Mn^{3+} on octahedral sites. Therefore it is not very conductive since it does not contain ions of the same element with differing charges on similar sites, as required for electron hopping. The substitution of Ni for Mn increases the conductivity. Ni^{2+}

Table 4.1 Properties of thermistor compositions based on Mn_3O_4 at 25 °C

Composition (cat.%)				$\rho_{25}/\Omega\,m$	B/K	$\alpha_R/\%K^{-1}$
Mn	Co	Ni	Cu			
56	8	16	20	10^{-1}	2580	−2.9
65	9	19	7	1	2000	−2.2
70	10	20		10	3600	−4.0
85		15		10^2	4250	−4.7
94		6		10^3	4600	−5.1

ρ_{25} is the value at 25 °C.

occupies octahedral sites and the charge balance is maintained by the conversion of Mn^{3+} to Mn^{4+}, thus providing a basis for hopping. Co^{2+} and Cu^{+} and Cu^{2+} have similar effects, but the site occupancies and ion charges are difficult to predict or determine in the more complex mixtures.

Components in the form of rods and discs are produced by the normal ceramic route, but the firing schedule and atmosphere have to be precisely controlled in order to obtain the required oxidation states. For instance, a prolonged delay when cooling in air through the range 700–500 °C could promote the formation of an excess of Mn^{4+}. Also, the operational temperature range for the device during use is specified as −50 to +250 °C. At higher temperatures a slow interchange of oxygen with the atmosphere takes place, causing a permanent change in the resistance value and its temperature characteristic.

Mixtures of the rare earth oxides, e.g. 70 cat.% Sm and 30 cat.% Tb, can be used at temperatures up to 1000 °C as they have no tendency to lose or gain oxygen. Since their room temperature resistivity is of the order of $10^7\,\Omega\,m$, they are not convenient for room temperature use. At 600 °C the resistivity is about $25\,\Omega\,m$ with $B = 6500\,K$, giving $\alpha_R = -0.9\%\,K^{-1}$.

Electrodes are commonly silver fired on with an admixture of glaze, but electroless nickel and vacuum-deposited metals can also be used. A final coating of a glaze over the whole body improves the long-term stability.

Miniature 'bead' thermistors, which are particularly valuable as temperature sensors, can be made by arranging two sets of fine platinum wires at right angles with a separation of a fraction of a millimetre between the sets. The intersections of the wires are then enclosed by small beads of paste containing the thermistor material in powder form. The beads are dried out and sintered in the same way as bulk units or fused individually with a laser beam or an oxidizing flame and annealed. The beads are then separated with two platinum lead wires which can be attached to a probe. Their small mass enables them to reach a rapid thermal equilibrium with their surroundings.

The behaviour of an NTC resistor on load is complicated by its increase in internal temperature when it passes a current. As a rough approximation it can be assumed that the difference between the temperature T of the resistor and the temperature T_0 of its surroundings is proportional to the power being dissipated, i.e.

$$T - T_0 = k_{th} UI = k_{th} P \tag{4.12}$$

where P is the power, U is the voltage drop across the thermistor, I is the current and k_{th} is a constant depending on the mounting, shape and surface finish of the thermistor as well as the ambient conditions. Because the power dissipated is proportional to $1/k_{th}$, this can be regarded as a 'dissipation factor'. Evidently k_{th} can vary between wide limits, but the following example from experiment gives the order of magnitude involved. Small rods 5–50 mm long and 3–10 mm in diameter with copper leads about 10 mm long and 0.5–2 mm in diameter mounted in air have a heat dissipation per degree Celsius excess over that of the ambient given by

$$1/k_{th} = g_1 A + g_2 d^2 \tag{4.13}$$

where A is the surface area and d is the lead diameter. If A is in square millimetres and d is in millimetres, $g_1 \approx 0.04$ and $g_2 \approx 10$. As indicated by Eq. (4.13), when the thermistors are very small the heat dissipation is governed by conduction down the leads rather than by convection currents in the surrounding air, which have the dominating influence in the case of larger units.

Substituting Eq. (4.12) into Eq. (4.10) and referring to the resistance of the device rather than to the resistivity of the material gives

$$R(T) = R_\infty \exp\left(\frac{B}{T_0 + k_{th} P}\right) \tag{4.14}$$

$R(T)$ can be calculated as a function of P from Eq. (4.14); U and I can then be derived from the relations

$$\begin{aligned} U^2 &= PR(T) \\ I^2 &= P/R(T) \end{aligned} \tag{4.15}$$

P and I are shown as functions of U in Fig. 4.16 and R is shown as a function of U in Fig. 4.17. The curves represent equilibrium conditions, and it is evident that no equilibrium can exist above a certain *maintained* maximum voltage. If a higher maintained voltage is applied, the current will go on rising indefinitely until the accompanying high temperature destroys the unit. In practice there must always be a temperature-insensitive resistor in series with a thermistor if sufficient power to raise its temperature appreciably is to be applied.

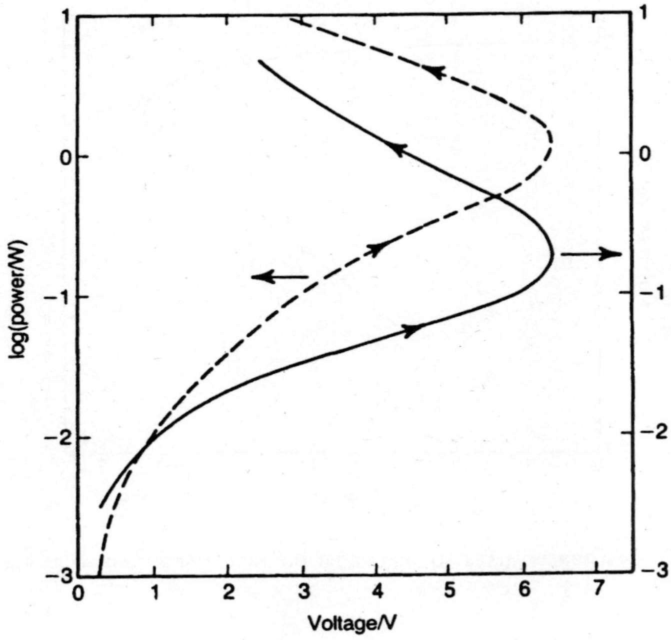

Fig. 4.16 Power and current versus voltage for an NTC thermistor with $R_\infty = 4.54 \times 10^{-3}\ \Omega$, $B = 3000\ \text{K}$, $T_0 = 300\ \text{K}$ and $T - T_0 = 30P\ \text{K}$.

The maximum voltage that can be applied to a thermistor, without a runaway condition being established, can be calculated from the voltage–power relation obtained by substituting Eq. (4.14) into Eq. (4.15):

$$U = P^{1/2}\left\{R_\infty \exp\left(\frac{B}{T_0 + k_{\text{th}}P}\right)\right\}^{1/2} \tag{4.16}$$

The power $P(U_{\max})$ corresponding to the maximum voltage $U_{\max}$ that can be maintained across the thermistor without runaway occurring can be found by differentiating Eq. (4.16) and putting $\mathrm{d}U/\mathrm{d}P = 0$. Under the condition that $4T_0^2/(B - 2T_0)^2 \ll 1$,

$$P(U_{\max}) = \frac{1}{k_{\text{th}}}\frac{T_0^2}{B - 2T_0} \tag{4.17}$$

The corresponding expression for $U_{\max}$ can be found by substituting $P(U_{\max})$ from Eq. (4.17) into Eq. (4.16) to give

$$U_{\max} = \left(\frac{1}{k_{\text{th}}}\right)^{1/2}\frac{T_0}{(B - 2T_0)^{1/2}}R_\infty^{1/2}\exp\left\{\frac{B(B - 2T_0)}{2T_0(B - T_0)}\right\} \tag{4.18}$$

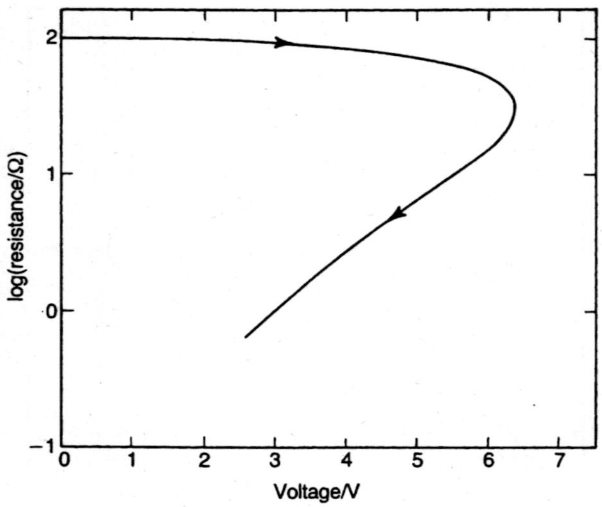

Fig. 4.17 Resistance versus voltage under the same conditions as in Fig. 4.16.

It is apparent that U_{max} increases with the ability of the thermistor to dissipate heat (proportional to $1/k_{th}$). Also, because the exponential term dominates, U_{max} decreases as T_0 increases, as expected.

Some of the properties and uses of thermistors, other than temperature sensing, can be appreciated from the simple circuit shown in Fig. 4.18. A fixed voltage U is applied to an NTC thermistor of resistance $R(T)$ in series with a load resistance R_L which is invariant with temperature. In this case there is the complication that, as the thermistor warms up and falls in resistance, the voltage across it also falls. The situation is analysed as follows:

$$I = \frac{U}{R(T) + R_L} \tag{4.19}$$

and the voltage U_{th} across the thermistor is

$$U_{th} = R(T)I = \frac{U\,R(T)}{R(T) + R_L} \tag{4.20}$$

The power P dissipated in the thermistor is

$$P = IU_{th} = \frac{U^2 R(T)}{\{R(T) + R_L\}^2} \tag{4.21}$$

Eliminating P between Eqs (4.14) and (4.21) and rearranging yields

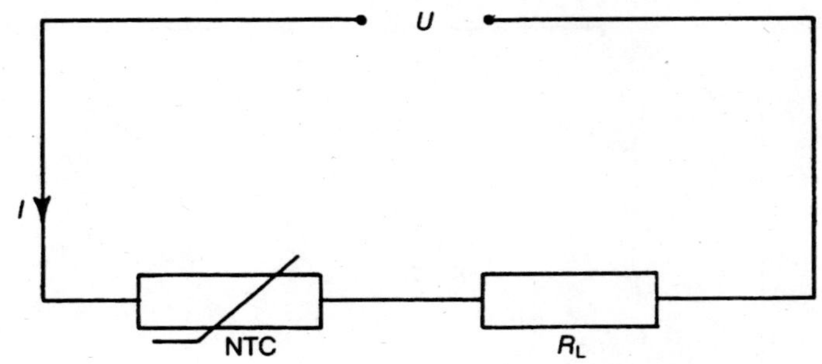

Fig. 4.18 NTC thermistor with a series load.

$$U^2 = \frac{\{R(T) + R_L\}^2}{k_{th} R(T)} \left[\frac{B}{\ln\{R(T)/R_\infty\}} - T_0 \right] \qquad (4.22)$$

Eq. (4.22) can be used to plot the relationship between U and $R(T)$ for given values of the other parameters. The relationship for $B = 3000$ K, $T_0 = 300$ K, $R_\infty = 4.54 \times 10^{-3}\ \Omega$, $R_L = 50\ \Omega$ and $k_{th} = 30\ \mathrm{K\,W^{-1}}$ is shown in Fig. 4.19 as an example. The current I (Eq. (4.19)) and the temperature of the thermistor calculated from Eqs (4.21) and (4.12) are also shown. It can be seen from the figure that, for $U = 50$ V, $R(T) \approx 5\ \Omega$ and $I \approx 0.9$ A and, when temperature equilibrium has been established, $T - T_0 = 127$ K ($T = 154\ °$C).

The room temperature resistance $R(300)$ of the thermistor, calculated from the equivalent to Eq. (4.10) and using $R_\infty = 4.54 \times 10^{-3}\ \Omega$, is $100\ \Omega$ so that immediately the switch is closed and before the temperature of the thermistor has changed the current is 0.33 A and the power dissipated in the load is 5.4 W. After the thermistor has reached temperature equilibrium with its surroundings, the current rises to 0.9 A and power dissipation in the load rises to 40 W. Thus a thermistor can be used to delay the development of full power in a load during the period it takes to reach its final temperature. The penalty is the power required to maintain the thermistor at temperature, in this case about 4 W. Using a thermistor with a higher TCR, i.e. a higher B value, would not greatly affect the ratio of the power in the load when first switched on to that at equilibrium, but it would decrease the power wasted in the thermistor, provided that $R(T_0)$ is the same in both instances.

A further consideration in the use of NTC resistors arises from the distribution of temperature within them. Clearly the inner part of the resistor must be hotter than the surface region in order to maintain an outward heat flow, but, in addition, the negative coefficient of resistivity will result in the inner part having a lower resistance and a higher current passing through it, so that the temperature gradient between the inner and outer parts will be enhanced. If the thermistor is grossly overloaded, the results can be catastrophic; for instance,

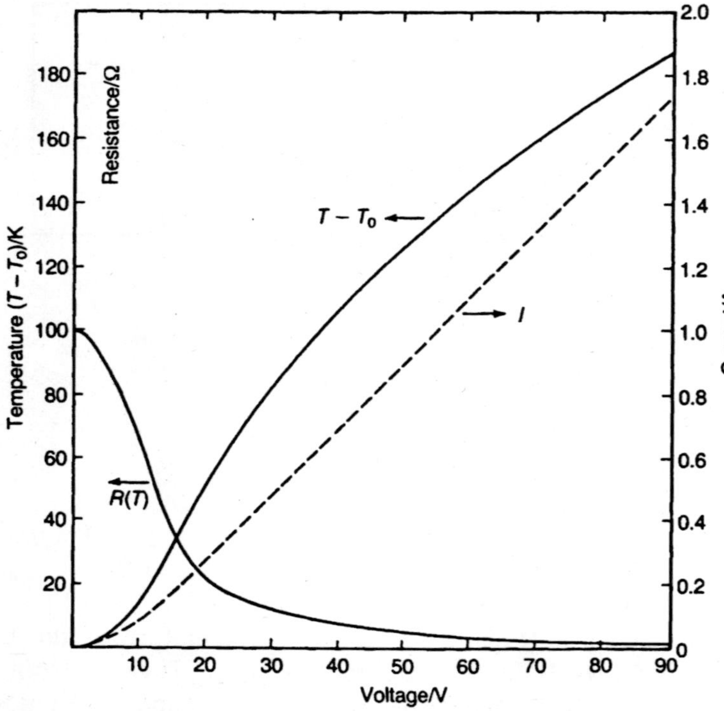

Fig. 4.19 Resistance, current and temperature versus voltage for an NTC thermistor with series load: $P = (T - T_0)/30\,\mathrm{W}$.

the central core of a rod-shaped NTC resistor may melt. In practical cases a severe mechanical stress can arise from the temperature gradient and may result in fracture or the development of internal cracks. Thermal stress due to differences in thermal expansion coefficients may also cause the metallization to separate from the ceramic. The probability of thermal stress failure depends on the same thermal and mechanical properties as thermal shock failure, which is discussed in Section 5.3.

At low power levels NTC thermistors are widely used wherever temperature needs to be controlled or accurately measured, for example in automotive engines, air-conditioning units, hair dryers, and in the medical field for heart catheters, fever thermometers, etc. They are also used to maintain picture stability in television receivers by compensating for increases in the resistance of the beam-focusing coils as temperature rises in the cabinet. The dependence of resistance on the rate of heat dissipation enables them to be used as indicators that fuel tanks are filled to a prescribed level and also in instruments for the measurement of the velocity of fluids.

Examples of NTC thermistors are shown in Fig. 4.15.

4.4.2 Positive temperature coefficient resistors (PTC thermistors)

PTC resistors could be classified as critical temperature resistors because, in the case of the most widely used type, the positive coefficient is associated with the ferroelectric Curie point.

A typical PTC characteristic is shown in Fig. 4.20. In the instance illustrated the material has the negative resistivity–temperature characteristic associated with normal semiconductors up to about 100 °C (AB) and above about 200 °C (CD), while between these temperatures (BC) there is an increase of several orders of magnitude in resistivity. The underlying physics of the effect as outlined below draws on the discussions in Sections 2.6.2 and 2.7.3.

The PTC effect is exhibited by specially doped and processed $BaTiO_3$. Because the effect is not observed in the single-crystal form of the material its cause must be assumed to lie in processes associated with grain boundaries. Attention here is focused on lanthanum-doped $BaTiO_3$ (BLT), although other donor dopants would be satisfactory, e.g. yttrium (A site) or niobium, tantalum or antimony (B site).

Electron acceptor states in the grain boundary together with nearby ionized donor states give rise to an electrical double layer, as shown in Fig. 4.21. In consequence, conduction band electrons moving up to a grain boundary from the interior of a grain are confronted by a potential barrier of height ϕ. To obtain an expression for ϕ the simplifying assumption is made that the positive charge density is constant out to a distance d from the grain boundary, where it falls to zero, as shown in Fig. 4.22. It is also assumed that the potential varies

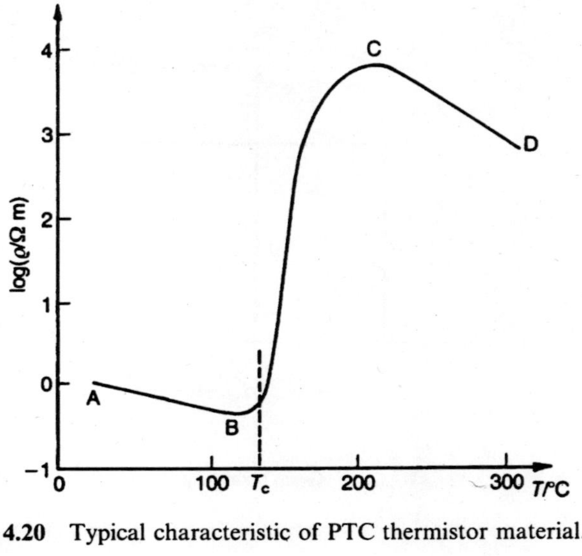

Fig. 4.20 Typical characteristic of PTC thermistor material.

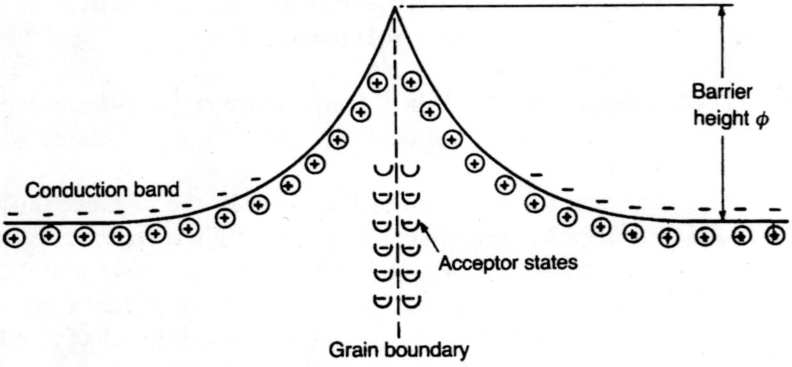

Fig. 4.21 Electrical double layer at a grain boundary.

only with the distance x from the grain boundary. The barrier height is found by first integrating the one-dimensional form of Poisson's equation

$$\frac{d^2V}{dx^2} = -\frac{\rho}{\varepsilon} \tag{4.23}$$

in which V and ρ are respectively the electrostatic potential and the charge density at x and ε is the permittivity. If the boundary condition $E = -dV/dx = 0$ at $x = d$ is assumed and $V = 0$ is arbitrarily fixed at $x = 0$, the potential V_d at $x = d$ is given by

$$V_d = -\rho\frac{d^2}{2\varepsilon} \tag{4.24}$$

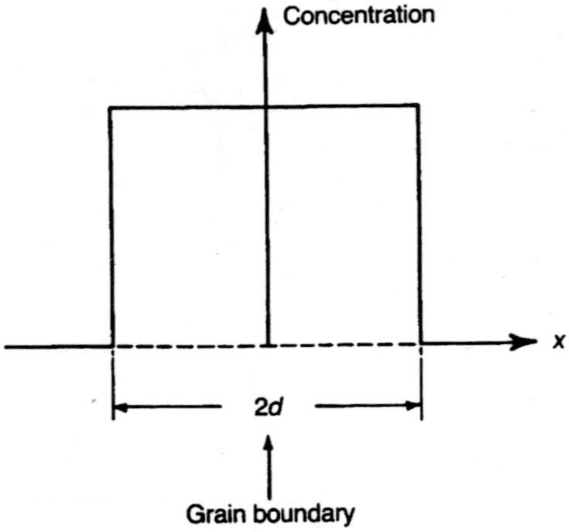

Fig. 4.22 Assumed positive charge density distribution in the vicinity of a grain boundary.

From Fig. 4.22 $\rho d = \frac{1}{2} N_s |e|$ where N_s is the surface density of acceptor states near the grain boundary and $|e|$ is the magnitude of the electronic charge. Therefore the height ϕ of the barrier to an electron becomes

$$\phi = -eV_d = \frac{e^2 N_s^2}{8\varepsilon n} \tag{4.25}$$

where $n = \rho/|e|$ is the volume density of donor states in the grain.

The probability that electrons are able to surmount the barrier is measured by the Boltzmann factor $\exp(-\phi/kT)$, leading to the following proportionality for the resistance R_{gb} of a grain boundary:

$$R_{gb} \propto \exp\left(\frac{\phi}{kT}\right) \tag{4.26}$$

Because BLT is ferroelectric, above its Curie temperature $\varepsilon = C/(T - \theta)$ (see Eq. (2.90)), where C is the Curie constant and θ is the Curie–Weiss temperature. Therefore

$$R_{gb} \propto \exp\left\{\frac{e^2 N_s^2}{8nkC}\left(1 - \frac{\theta}{T}\right)\right\} \quad T > \theta \tag{4.27}$$

The PTC effect is seen to have its origins in the resistance of the grain boundary region which increases exponentially with temperature above the ferroelectric–paraelectric transition temperature. It therefore depends on the number of grain boundaries per unit volume of ceramic, i.e. on the microstructure, and of course on the acceptor and donor state densities N_s and n.

It is evident that the double layers at the grain boundaries constitute Schottky barriers which are similar in some respects to those formed in VDR resistors. In accord with this it is found that the resistivity–temperature relation of PTC material is voltage sensitive. The low-temperature resistivity may be reduced by a factor of 4 by an increase in applied field from 1 to 80 kV m^{-1}, and the ratio of maximum to minimum resistivities, above and below T_c, may be reduced from five to three orders of magnitude.

Because the PTC effect is partly a consequence of the steeply falling permittivity just above the ferroelectric transition temperature, why does the resistance not similarly increase with a strongly decreasing permittivity as temperature falls below the Curie temperature? Below T_c the material is ferroelectric with each grain comprising domains terminating on grain boundaries. A discontinuity in polarization at the grain boundary necessitates a polarization charge at the grain boundary whose sign depends upon the nature of the discontinuity. This surface charge partially cancels the double-layer effect and removes, or at least reduces, the barrier in places. The barriers throughout a polycrystalline ceramic may be short circuited by such a process so that the material as a whole has low resistivity. The effect will depend upon details of the domain configuration, the magnitude of the polarization and the density of surface acceptor states. The lack of precise information about these precludes the possibility of predicting electrical properties in this temperature region.

In its essentials the model serves as a sound basis for understanding the PTC effect, but little is known regarding the nature of the acceptor states. It has been proposed that they arise because the ceramic does not achieve thermodynamic equilibrium during processing. On this basis a somewhat speculative mechanism for the formation of acceptors is outlined below.

At the sintering temperature the excess charge of $La_{Ba}^{\cdot}$ is largely compensated by the promotion of electrons into the conduction band (see Eq. (2.57)). On cooling, some of the electrons are replaced by V_{Ti}'''' (see Eq. (2.59)):

$$4TiO_2 + O_2(g) + 4e' \rightleftharpoons 3Ti_{Ti} + V_{Ti}'''' + 8O_O + \text{`}TiO_2\text{'} \quad (4.28)$$

It can be seen from Eq. (4.28) that the interchange of electrons and vacant cation sites requires the presence of oxygen gas and the formation of TiO_2 in a phase separate from that of $BaTiO_3$. Oxygen is available at the surface of the grains through diffusion along the grain boundaries. The extension of the reaction to the interior of a grain requires the diffusion of oxygen ions through the crystal lattice, which is a much slower process than grain boundary diffusion, especially as the concentration of oxygen vacancies is minimal in the presence of donor ions (Fig. 2.13). Therefore it is possible that, on rapid cooling to room temperature, V_{Ti}'''' will be at a higher concentration in the surface layers of the grains where they will act as the acceptors postulated in Fig. 4.21. The 'TiO_2' may form part of an intergranular phase based on $Ba_6Ti_{17}O_{40}$ (see Fig. 5.41) which would require the diffusion of V_{Ti}'''' into the bulk of the grains, but it is unlikely that such a highly charged entity would be particularly mobile, so providing a further reason for a higher V_{Ti}'''' concentration at the grain surface. It has been found empirically that PTC properties are improved when the acceptor ion Mn_{Ti}' is present (at about the 0.05 cat.% level) in the intergranular region. The overall increase in resistivity during the transition is made larger and the resistivity at lower temperatures is reduced.

The PTC effect is distinguished from the majority of other critical temperature effects in the ease with which the critical temperature can be shifted by altering the composition. The replacement of barium in $BaTiO_3$ by strontium lowers the critical temperature by 4 °C per percentage atomic replacement, whilst replacement by lead raises the critical temperature by 4.3 °C per percentage atomic replacement (see Fig. 2.47). Since the critical temperature for $BaTiO_3$ is 120–130 °C, it is a simple matter to prepare ceramics with PTC regions anywhere between −100 °C and +250 °C, although the highest temperature coefficients are found in barium titanate compositions without major quantities of substituents.

The fabrication route for PTC thermistors is typical of that employed for modern electroceramics except in so far as special attention is given to maintaining high purity and to the firing schedule.

The basic composition is usually derived from oxides or carbonates, e.g. $BaCO_3$, $SrCO_3$, TiO_2, La_2O_3 etc., which are mixed in a polyethylene-lined ball-mill using mullite or agate balls and deionized water. Alternatively, and if economically viable, the mix can be synthesized from organometallic compounds, usually in conjunction with soluble inorganic salts.

The mix is dried and calcined ($\approx 1000\,°C$) when the semiconducting ceramic is formed. The calcine is ball-milled, in a similar mill to that used before, to a size of about 1 μm. At this stage other dopants and binders can be added, e.g. $MnSO_4$ and polyvinyl alcohol, or the mix might be blended with $PbTiO_3$ if the device is to have a high switching temperature; the special treatment given to the lead compound is necessary because of the high volatility of PbO. The slurry is then granulated, usually by spray drying, when it is ready for pressing into discs.

Sintering at about 1350 °C in air and the subsequent cooling stages have to be carefully controlled since this is when the barrier-layer characteristics are established. The conditions must be such as to allow a barrier layer of optimum thickness (0.1–1 μm) to form, while the grains grow, ideally uniformly, to the optimum size, normally about 50 μm.

After the sintering stage, electrodes are applied, usually either by electroless nickel plating or by painting or screening on specially adapted silver paint. Leads are then soldered to the electrodes when, for many applications, the device is complete; in other cases it may be encapsulated in epoxy or silicone resins. Examples are illustrated in Fig. 4.15.

A typical current–voltage relation for a PTC body in thermal equilibrium is shown in Fig. 4.23. At low voltages (EF) the relation is approximately ohmic (Fig. 4.20, AB); then, as the temperature of the thermistor reaches the regime of steeply rising resistance (Fig. 4.20, BC), its temperature rises only slowly with increasing voltage and the current falls to give a correspondingly slow increase in power dissipation. If the increase in voltage is sufficient to bring the temperature above the region of rising resistance (Fig. 4.20, CD), the temperature coefficient becomes negative and a rapid increase in current and temperature results.

If the rate of heat dissipation from the thermistor is changed, the location of FG in Fig. 4.23 will shift. The temperature of the element changes only slightly, but the power changes to a level corresponding to the new rate of heat dissipation. If the voltage is kept constant the current becomes a measure of the rate of heat dissipation, and this relation is used in a number of devices to sense a change in environment. There is a marked change in heat dissipation when a probe at thermal equilibrium in air is plunged into a liquid at the same temperature as the air. Devices for indicating the level of liquids in tanks are based on this change.

The relative constancy in the temperature of a PTC device, despite changes in both the voltage supply and ambient conditions, when it is maintained on the steeply rising limb of its resistance–temperature characteristic has led to its use as

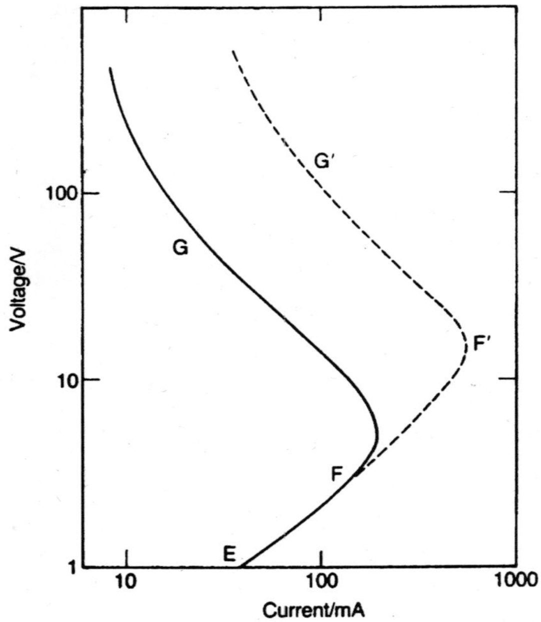

Fig. 4.23 Current–voltage characteristic for a PTC thermistor in thermal equilibrium.

a heating element in miniature ovens for quartz crystals acting as constant-frequency sources. It has also been made use of in hair-driers and space-heaters. Air is blown through the perforations in a block of ceramic (Fig. 4.24) that is designed to heat the emergent air to a suitable temperature. Even when the fan is switched off the heater temperature only rises by about 20 °C so that no damage results.

PTC elements make useful temperature indicators because the sharp rise in resistivity above the Curie point is very easily detected or used to operate a control mechanism. Small tubular elements a millimetre in diameter and a few millimetres long can be inserted in motor or transformer windings and used to detect overheating or to control it directly by increasing the overall resistance of the winding when it becomes too hot.

The non-equilibrium properties of PTC elements can also be utilized. If the room temperature resistance is low there will be a high current surge when a voltage is applied and this will fall to 1% or less of its initial value when the element heats up above its Curie point. Such devices are used in demagnetizing ('degaussing') coils in T.V. monitors and in place of capacitors in motor starters.

Just as in the case of NTC thermistors (see Section 4.4.1), the temperature distribution within a PTC thermistor can be far from uniform in certain non-equilibrium applications. For example, thin discs in motor-starter units have

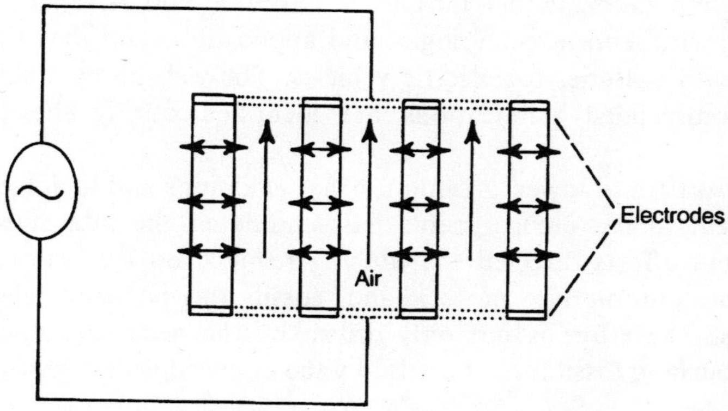

Fig. 4.24 Section through a PTC ceramic hair-drier heating element: horizontal arrows indicate heat flow out of the ceramic.

been known to split along a plane parallel to their electrodes. Highly homogeneous flaw-free ceramics are needed to survive under such severe conditions.

It can be seen that PTC elements give rise to a greater diversity of applications than NTC units, but their production in high yield to meet close specifications demands great care and attention.

4.5 Fuel Cells and Batteries

4.5.1 The stimulus for developing fuel cells and batteries

> Increasing awareness of environmental factors and limited energy resources have led to a profound evolution in the way we view the generation and supply of energy. Although fossil and nuclear sources will remain the most important energy providers for many more years, flexible technological solutions which involve alternative means of energy supply and storage need to be developed urgently.
>
> The search for cleaner, cheaper, smaller and more efficient energy technologies has been driven by recent developments in materials science and engineering. The aim of this collection of reviews is therefore to focus on what materials-based solutions can offer and to show how the rational design and improvement of chemical and physical properties of these materials can lead to energy alternatives that can compete with existing technologies [3].

The treatment in the present overview places emphasis on the roles ceramics and their processing are playing in two of the various alternative energy technologies.

There is an extensive literature dealing with fuel cell and battery technologies and the interested reader has much choice. The monograph by N.Q. Minh and T. Takahashi [4] deals in depth with the science and technology of ceramic fuel cells, especially with the electroceramics aspects. The comprehensive texts by K.

Kordesch and G. Simander [5] and J. Larminie and A. Dicks [6] deal more generally with fuel cell technologies and applications and that by D.A.J. Rand *et al.* [7] with batteries for electric vehicles. The web offers much information; other recommended publications are identified during the course of the discussion.

The imperative to lower 'greenhouse' gas emissions and to achieve cleaner city and, indeed, global environments has stimulated the intensive research and development efforts directed to reducing reliance on the internal combustion engine for automotive power, and fossil fuel-powered electrical power generators. The effort is not only driven by the need to reduce the harmful effects of burning fossil fuels, but also by the knowledge that supply of these fuels is limited.

An important stimulus to battery development was the 1989 Los Angeles legislation which has as its objective an increasing proportion of new cars sold in California to be 'Zero Emission Vehicles' which means, in effect, that cars must be powered by hydrogen fuel cells or batteries. The legislation requires that at least 10% of all new vehicles marketed in California must be ZEVs by 2003. The California Air Resources Board (CARB) initiative was followed by similar legislation in New York and Massachusetts. Many hundreds of thousands of electric vehicles are expected to be on the road in California by 2010.

Students are reminded of the upper thermodynamic limit set on the efficiency of a heat engine, for example the internal combustion and gas-turbine engines. The ideal and totally unrealistic engine would operate on the so-called Carnot cycle where the 'working substance' (e.g. the gas) is taken in at the high temperature (T_h) and pressure and after doing external work is exhausted at the lower temperature (T_c) and lower pressure. The Carnot efficiency, η, is given by

$$\eta = 1 - T_c/T_h \tag{4.29}$$

As an example, if the hot temperature is 1273 K (1000 °C) and the cold temperature 373 K (100 °C) then the efficiency is approximately 70%. In practice the operation of a real engine does not follow the Carnot cycle and the efficiency is considerably lower. For a medium sized motor car with an internal combustion engine the fuel efficiency is about 12%, much of the wasted 88% demanding water cooling. There are continuous improvements made in petrol and diesel engine technologies and in the fuels and projections suggest that thermal efficiencies a little over 50% will eventually be achieved.

For a modern gas-fired power station in the UK the electrical efficiency is about 45% and for a coal-fired station about 37%. There are combined cycle plants in which the waste heat from the gas turbines is used to produce steam that in turn drives a steam turbine generator. Such combinations, which are expensive, can deliver overall (electrical+useful heat) efficiencies of up to 55%. Another approach to increasing overall thermal efficiencies is via combined heat and power installations where the waste heat is used typically for district heating or for an industrial process. For such CHP plants efficiencies of 80% are achieved.

It is important to recognize that all the potential improvements to increase the thermal efficiencies of heat engines do nothing to reduce emission of the 'greenhouse' gases CO_2 and NO_x. The achievement of this goal probably lies with the fuel cell and rechargeable battery. But then there is a range of competing fuel cell and battery types and choices will depend on many complex factors, not least the impact new technologies will have on global economies, for instance the oil industries, the infrastructures geared to the present day automotive internal combustion engine and, not least, political considerations.

The subject is too complex to attempt an intelligent summary here; over the next decade or two the scenario will unfold with the various technologies most probably first finding their own particular niche markets and eventually dominating the scene.

Although the research and development effort is focused on a range of fuel cell and electrical storage battery types the discussion here is confined to those whose function depends on conducting ceramics. To appreciate the technological challenges involved in developing them to a commercially viable reality it is necessary to have an understanding of the relevant basic science.

Electrochemical cells are devices that convert chemical energy directly into electrical energy thus circumventing the fundamental efficiency limit set by the Carnot cycle. This is the case whether the device is the familiar battery or the less familiar fuel cell.

A fuel cell is a device in which the reactants, for example hydrogen and oxygen, are each continuously supplied to opposite sides of a suitable membrane and converted to electrical energy.

A 'secondary' electrochemical cell is simply one that can be recharged as in the case of the Na/S cell discussed below (in contrast a 'primary' cell, such as the common torch battery, is exhausted after use and cannot be recharged). During charging the chemical reaction is driven in reverse by applying an e.m.f. in the sense to oppose the forward direction e.m.f.

The voltages developed by individual electrochemical cells are small, typically in the range 1–2 V and for them to constitute useful power sources they must be connected in series so that the voltage is multiplied by the number of cells, and in parallel to increase current availability.

The fuel cells and batteries we are concerned with here (and the sensors to be discussed later) rely on electroceramics having high electrical conductivities and, ideally, charge-carriers that are either ions or electrons but not a mix of the two. They are discussed later.

4.5.2 Basics of fuel cells and batteries

Fuel cells

Overview The English chemist Humphrey Davy wrote in 1812 'If a piece of zinc and a piece of copper be brought in contact with each other, they will form a weak electrical combination, of which the zinc will be positive, and the copper negative . . .' so initiating the history of the electrochemical cell. But it was Michael Faraday who, in 1834, laid the foundations of quantitative electrochemistry by relating the quantity of a substance electrolysed to the amount of electrical charge involved.

It is from these foundations that 'electrochemistry' has evolved and which now provides the scientific basis to the technology of electrochemical cells.

Fuel cells are not new, the concept dating back to 1839 and the observations of a Welsh physicist, William Grove. One of the first practical cells was developed at Cambridge University in the UK by an engineer, 'Tom' Bacon, a descendant of the 17th century philosopher, Francis Bacon. Bacon constructed a stack of 40 alkaline fuel cells (AFCs) which could develop 5 kW of electrical power. Interest continued at a low level and in 1958 General Electric developed a polymer electrolyte membrane (PEM) cell which was used in the Gemini space programme; in the 1960s Pratt and Whitney developed Bacon's AFC technology which was used in the Apollo space programme. During the 1990s interest in fuel cell technology grew at a rapid pace, particularly in the automotive industry, and the growth in interest is unabated.

There are many obstacles to the replacement of the internal combustion engine for automotive power by fuel cell technology, a major one being the existing infrastructure designed to service the petrol- and diesel-driven vehicles. As far as reducing greenhouse gases is concerned the favoured fuel is hydrogen since water is the only reaction product. However, because of the existing infrastructure (e.g. petrol service stations and garages) there would be short-term economic advantages to be gained through using cells fuelled by methanol or petrol; unfortunately one of the reaction products is the greenhouse gas CO_2, although the higher efficiency of the fuel cell would mean a pro rata reduction in these.

The almost limitless abundance of hydrogen on the planet in the form of water has stimulated intensive research effort, especially in Japan and the US, directed towards developing the so-called 'hydrogen economy'. This demands economically viable routes to producing adequate supplies of hydrogen, for example via the electrolysis of water. There are various 'renewable' ways of powering the electrolysis of water to obtain hydrogen and oxygen, including for example, wind power, tidal and wave power. There is also photoelectrolysis which harnesses the sun's energy via the photovoltaic effect to electrolyse water directly. The challenges are real but the potential benefits immeasurable. To achieve the 'hydrogen economy' would be an unquestionable 'giant step for mankind'.

The situation is an extremely complex one and future evolution of the large-scale exploitation of fuel cell technology impossible to predict. But there are significant developments. For example, DaimlerChrysler announced in 2000 its plans to build a fleet of hydrogen-fuelled buses, the hydrogen being stored in tanks under pressure. This is an initial but very significant step. There are alternative and far more convenient ways of storing hydrogen 'on board', for example the metal hydrides can store hydrogen within their crystal structure and release it on heating [3].

One of the important operational considerations is the range the car can travel on a single fuelling (or, in the case of the battery, on a single charging) and the power availability. It is clear that if fuel cell and battery technologies are to be acceptable to the motorist then they must compete favourably with the internal combustion engine in these respects. The vehicle has, of course, to transport its own 'engine' and fuel as well as passengers. The comparisons are made with the help of the power-to-weight and energy-to-weight ratios, the former determining the acceleration and hill-climbing capabilities of the vehicle and the latter its range. On these bases Fig. 4.25 shows fuel cells' characteristics to be very competitive even though they are unlikely to match internal combustion engines when it comes to 'high performance' characteristics.

The various types of fuel cell currently under development and described by Larminie and Dicks [6] are listed below.

1. The polymer electrolyte membrane or proton exchange membrane (PEM) cell.
2. The alkaline electrolyte cell (AFC).
3. The phosphoric acid cell (PAFC).
4. The molten carbonate cell (MCFC).
5. The solid oxide ceramic cell (SOFC).

The SOFC will be the major concern here although it will be helpful to consider first the elements of the PEM cell since not only is it a very strong contender for large-scale use, but its basic science is simple.

Elementary fuel cell science The essential of a fuel cell is the electrolyte, a material which conducts electricity by the transport of an ionic species. On one side of the electrolyte membrane there is a source of the ion species at a particular chemical potential and on the other side a 'sink' for the ions at a relatively lower chemical potential.

The PEM cell is the cleanest fuel cell since the fuel is hydrogen, the oxidant oxygen and the product water. Although it clearly falls outside the scope of a text focused on electroceramics there are good reasons for prefacing the present discussion with a brief outline of those elements of the science and technology basic to it and common to the ceramics-based fuel cells. Also, for an intelligent

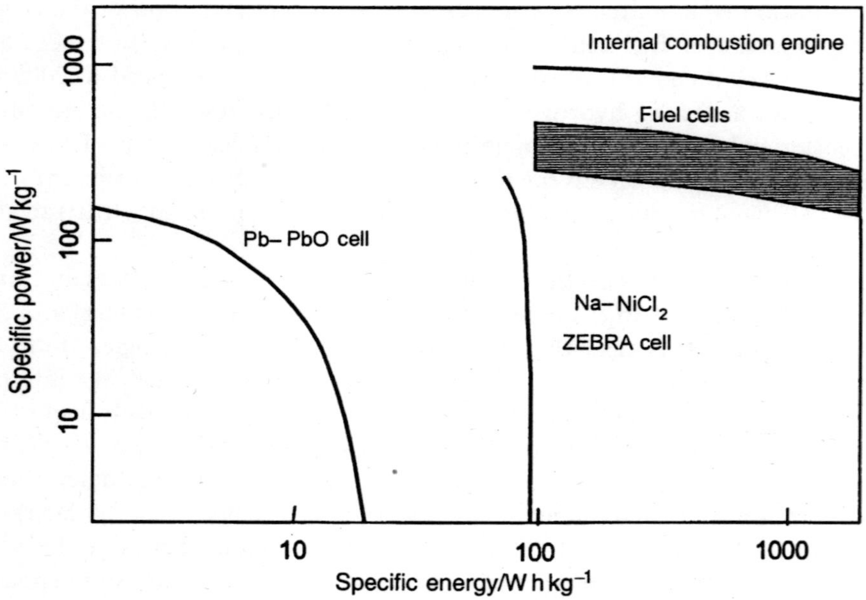

Fig. 4.25 Ragone plot comparing the internal combustion engine with the ZEBRA battery and fuel cells (very approximate). [The plot was introduced for comparing batteries (standardized to a weight of 300 kg); to include the performances of engines and fuel cells in a meaningful way their masses, together with the fuel carried, should be standardized to 300 kg.].

appreciation of the status of ceramics-based cells, the 'solid oxide fuel cells' (or SOFCs), they must be viewed in the context of fuel cell technology as a whole which includes the PEM cell as an important part.

The elements of the PEM cell are shown in Fig. 4.26 together with the reactions occurring at the anode and cathode.

The student will recall that the anode is the electrode to which the anions (the negatively charged species) in the electrolyte migrate and the cathode the electrode to which the cations (the positively charged species) migrate.

$$\textit{Anode oxidation reaction:} \quad H_2 \rightarrow 2H^{\cdot} + 2e' \tag{4.30}$$

$$\textit{Cathode reduction reaction:} \quad 2H^{\cdot} + \tfrac{1}{2}O_2 + 2e' \rightarrow H_2O \tag{4.31}$$

$$\textit{Overall cell reaction:} \quad H_2 + \tfrac{1}{2}O_2 \rightarrow H_2O \tag{4.32}$$

The theoretical cell electromotive force, the e.m.f, can be calculated on the basis that the free energy released due to the chemical reaction is available to do external work, in this case electrical work. The change in the Gibbs free energy ΔG represents the total electrical energy available. From the definition of the volt, the electrical energy involved in transferring Q coulombs through E volts is

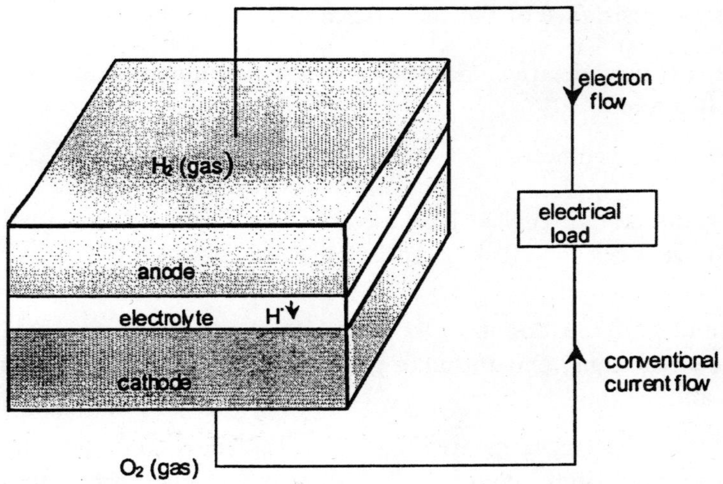

Fig. 4.26 The elements of the hydrogen fuel cell. *Note*: (i) The student is reminded that a chemical species which loses electrons is oxidized; one that gains electrons, reduced. (ii) The proton H^+ is transported through the polymer electrolyte attached to water molecules, $(H_2O)_nH^+$ which causes a 'water management' problem. Current research is aimed at developing proton conductors able to operate in the region of 200 °C when the proton migrates unattached to water molecules.

QE joules. If the transferred charge is carried by 1 mole of electrons, then $Q = F$, F being the Faraday constant (approx. 9.648×10^4 C mol^{-1}). If each chemical species carries a charge of $|z|e$, z being the oxidation number (plus or minus depending on whether the species carries a positive or negative charge), then the electrical work done in transferring 1 mole of the species is $|z|FE$. Equating the chemical energy made available to the electrical work done leads to the equation

$$\Delta G = -|z|FE \tag{4.33}$$

From tables the free energy change for reaction (4.32) taking place at 298 K (25 °C) is −237 kJ so that from Eq. (4.33), with $|z| = 2$, $E = 1.23$ V.

In practice the PEM cell is likely to operate at approximately 80 °C (353 K) when the theoretical e.m.f. is 1.18 V.

It is important to appreciate that the e.m.f values relate to open circuit conditions, that is when the cell is delivering no current. When current is drawn from the cell power is dissipated internally (internal resistive effects), resulting in a terminal p.d. of approximately 0.8 V; this is the effective voltage for doing work on an external electrical load.

A further factor that determines deliverable power (VI) is the current (I) that can be drawn from the cell. Current will be limited by the following factors:

1. the rate of the chemical reaction at either anode or cathode, the slowest being rate-determining, and

2. the internal resistance of the cell determined in part by

   (a) the intrinsic electrical properties of the cell components, especially the electrolyte and

   (b) electrolyte geometry.

[The student should recall G (conductance) $= \sigma A/l$, (see Section 2.6.1)]
The optimised cell therefore comprises the following:

1. anode and cathode designed to maximize the rates of the oxidation and reduction reactions, and to make good electronic contact with the external circuit, and
2. an electrolyte having large area and small thickness consisting of a material having a high ionic and zero electronic conductance; any electronic contribution to conductance will tend to internally short-circuit the cell, lowering the terminal p.d. and contributing to power wastage.

To maximize the rate of a reaction occurring at a gas/solid interface requires the total reactive surface be maximized, and be catalytic. Therefore the anode and cathode must have a high specific surface area (total internal area per unit mass or volume) with the proviso that the flow of reactant molecules to the reaction sites must not be impeded. As stated above, both anode and cathode should be good electronic conductors.

It is inappropriate to pursue here optimization of the electrolyte design for a PEM cell. The essentials of the cell are a thin polymer membrane coated on each surface with carbon mixed with platinum particles acting as the catalyst.

Having optimized the design of a single cell there arises the challenge of interconnecting many such cells in series to form a stack to increase total voltage (or in parallel to increase current flow and to remove the risk of total failure should a single cell fail). This challenge is met by introducing a bipolar plate (also termed an 'interconnect').

The interconnect normally links the anode of one cell to the cathode of the next. It must, of course, be an electronic conductor and also a gas barrier preventing the direct meeting of fuel and oxidant gases. Fig. 4.27 illustrates how the interconnection is achieved in the case of the so-called 'planar' fuel cell stack. In the later discussion of the ceramics-based cells a tubular configuration is described, but the principles are the same.

In the case of the high temperature SOFC discussed below the principles outlined above equally apply. The technical differences are that the cell runs typically on hydrocarbon fuels (e.g. 'natural' or coal-gas) and that the electrolyte is an oxygen ion conductor rather than a proton conductor. The complex fuel molecules, in the presence of the water molecule and at the high operating

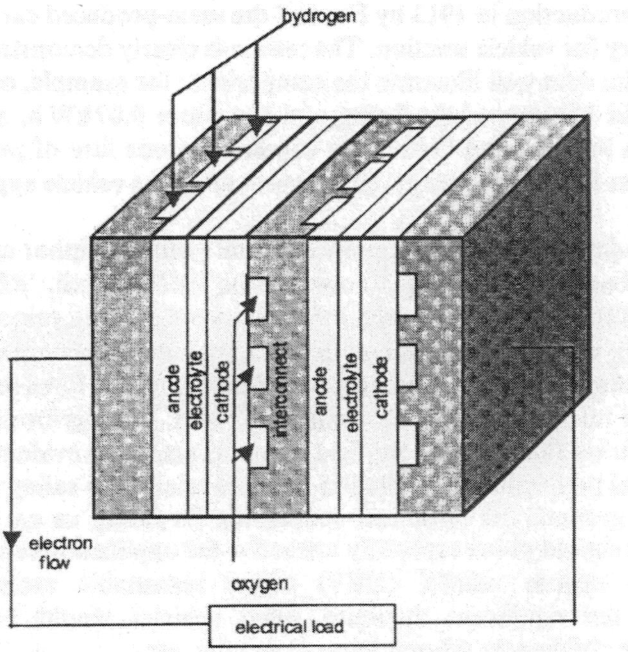

Fig. 4.27 Illustrating the connecting of two fuel cells via bipolar plates ('interconnects').

temperatures (600–1000 °C), are 'reformed' into H_2 and CO. Oxygen supplied at the cathode reacts with electrons flowing in the external circuit to form oxygen ions (O^{2-}) which diffuse through the electrolyte to the anode where they react with the fuel to form H_2O and CO_2. The open circuit voltage (e.m.f.) for a SOFC depends, of course, on the nature of the fuel and temperature but is typically approximately 1.1 V.

Storage batteries exploiting β''-alumina

Overview The battery is simply an electrochemical cell just as the fuel cell. The difference is that in the case of the fuel cell fuel is continuously supplied and electrical power extracted, whereas in the case of the storage battery, as the name implies, chemical energy is stored and electrical energy extracted on demand.

As for fuel cell technology there is an extensive 'battery' literature and various competing types, including lead/acid, Ni/metal hydride, Na/S, Na/NiCl and Li-ion, and each has advantages and disadvantages. The subject is dealt with comprehensively in [7].

The familiar lead/acid battery dates back to 1860, and by the early part of the 20th century there were many electric vehicles in operation throughout the world, mainly as taxicabs, personal cars, delivery vans, trams and buses.

However, the introduction in 1913 by Ford of the mass-produced car completely ousted the battery for vehicle traction. The reason is clearly demonstrated in Fig. 4.25. More precise data well illustrate the comparison; for example, one litre of a modern lead/acid battery weighs 2.4 kg and can store 0.07 kW h, sufficient to drive a vehicle a few hundred metres. By comparison one litre of petrol weighs 0.85 kg and stores 11 kW h of energy, sufficient to drive the vehicle approximately 10 km.

Strong contenders for automotive power are the sodium/sulphur and sodium/nickel chloride batteries, the latter known as the ZEBRA cell. 'ZEBRA' was originally (*c.* 1979) an acronym devised for commercial security reasons but now it stands for the very apt 'Zero Emissions Batteries Research Activity'. Several European car manufacturers including BMW and Mercedes have incorporated the ZEBRA cell into prototype cars, vans and buses. The performance of the battery far outstrips that of the lead/acid counterpart, as is evident from Fig. 4.25. Its technical performance coupled to features relating to safety and overall economics give grounds for optimism concerning its future as an automotive power source. It should prove especially attractive for applications such as in the 'neighbourhood electric vehicle' (NEV) where reasonable range and not acceleration is the significant attribute. Such vehicles would be used for shopping, taking children to school, leisure pursuits, etc.

A further possible and important application for the Na/S and ZEBRA cells is in 'load levelling'. By this is meant the storage of electrical energy from power supply generators when demand is relatively low and its release into the supply network (the 'grid') at peak demand times. In this way the generators can be run continuously at their optimal efficiency speeds.

The Na/S and ZEBRA batteries, which incorporate ceramic electrolytes, will be discussed in detail below but first the elementary basic science is summarized.

Elementary Na/S battery science The basic science of the Na/S cell is identical to that of the fuel cell. In both cases the essential is an electrolyte. On one side of the electrolyte membrane there is a source of sodium ions at a particular chemical potential, and on the other side a 'sink' for the ions at a relatively lower chemical potential. The elements of the process are illustrated in Fig. 4.28.

The ideal electrolyte has a high conductivity with Na ions being the only charge-carrier, that is, the transport number for Na is unity. (The 'transport number' for a particular mobile charged species is defined as the conductivity contributed by the species expressed as a fraction of the total conductivity.) Consider the discharge of the cell. At the anode two Na ions enter the electrolyte having given up an electron each; that is the sodium is chemically oxidized. At the cathode a polysulphide unit gains two electrons and is chemically reduced.

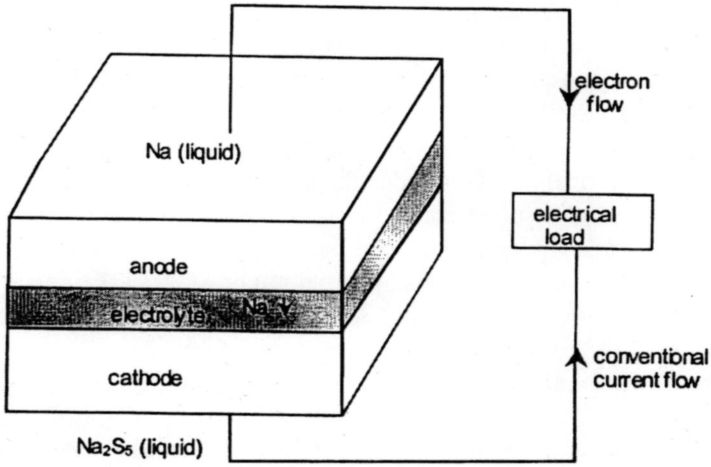

Fig. 4.28 Elements of the Na/S battery.

This process leads to the development of a voltage difference between anode and cathode and when the external circuit is 'open', this potential difference is the 'electromotive force' (the e.m.f.) of the cell. When there is an electrical load in the external circuit, for example a motor, then a current will flow and electrochemical energy is converted into mechanical energy.

Just as in the case of the fuel cell, current also has to flow through the electrolyte doing work against its electrical resistance, generating heat and leading to a terminal p.d. value lower than the e.m.f.

It follows from this brief outline that if maximum useful power is to be extracted from the cell then the electrolyte resistance must be kept to a minimum. This is dependent not only on the material but also on the geometry of the membrane which must be as thin as is practicable. As in the case of the fuel cell, the e.m.f can be calculated from thermodynamic principles. In fact the chemical reactions occurring when the sodium ions react with the sulphur are rather complex and the sodium:sulphur ratios change as the battery discharges. The first product as the cell discharges is the compound Na_2S_5.

At the anode the half cell reaction may be written:

$$2Na \rightarrow 2Na^{\cdot} + 2e' \qquad (4.34)$$

and at the cathode

$$5S + 2e' + 2Na^{\cdot} \rightarrow Na_2S_5 \qquad (4.35)$$

The overall cell reaction is, therefore

$$2Na + 5S \rightarrow Na_2S_5 \qquad (4.36)$$

When all the sulphur has been converted to Na_2S_5 this compound is gradually converted to Na_2S_3 as the discharge proceeds. The sodium/sulphur cell operates at a temperature of approximately 350 °C (the melting points of both Na_2S_5 and Na_2S_3 are close to 240 °C).

There has been a reduction in free energy of the chemical system which is manifested as electrical energy. Under the assumption that the chemical reaction occurs reversibly, that is no energy is 'lost' but only exchanged between chemical and electrical forms, the cell e.m.f. can be calculated. The change in free energy (ΔG) for reaction (4.36) at 350 °C is approximately −400 kJ. For each mole of the sulphide formed two moles of Na^+ ions are transported across the cell. The calculation (Eq. (4.33)) leads to a cell e.m.f. of 2.08 V.

As with the fuel cell this small voltage must be multiplied up by connecting many cells in series, and in parallel to increase current availability and battery reliability. This is a relatively straightforward procedure with the separately packaged individual cells being connected with metallic conductors to form a battery.

Elementary 'ZEBRA' battery science There is no new principle involved. The basic reaction is:

$$2Na + NiCl_2 \rightarrow Ni + 2NaCl \tag{4.37}$$

The open circuit e.m.f., calculated from the free energy change for the reaction at 250 °C is approximately 2.59 V.

4.5.3 Electroceramics for fuel cells and batteries

Solid fast-ion conductors

Fast-ion conductors: general comments The essential element of ceramics-based fuel cells and batteries is the electrolyte, a solid, fast-ion conductor.

Ionic transport under an applied field takes place to some extent in all ionic solids but generally involves the movement of ions into vacant sites over large energy barriers resulting in very low mobilities, even at high temperatures. Certain materials known as 'fast-ion conductors' or 'superionic conductors', exhibit higher ionic mobilities and opened up the possibility of practical applications, particularly in fuel-cell, battery and sensor technologies. In liquid ionic conductors both positive (cations) and negative (anions) ions act as current-carriers, although one or the other may dominate with a higher transport number. In solids there is usually only one species of mobile ion, and it may be a cation or anion.

The precise structural conditions for fast-ion conduction have yet to be established (especially when the effects of microstructure are involved), but there clearly must be a high concentration of sites into which the mobile ion can move (see Section 2.6.3). These may be lattice vacancies or they may be features of the crystal structure such as tunnels or under populated planes within which specific ions can move by a mechanism with a low activation energy.

The size and charge of mobile ions and the size of the spaces within a lattice through which they may move are important factors. Small singly charged cations such as Li^+ and Na^+ are mobile in a number of environments, in contrast to the more highly charged and smaller cations. The reason for this is the polarization of the surrounding anions by the smaller more highly charged cations. This interaction constitutes a high-energy barrier to the movement of the cations. However, there are factors other than geometry and size that have important effects. For example, on a rigid sphere model, the interstices between the anions (radius r_A) in a face-centred cubic lattice would only allow penetration of a cation of radius $0.15\,r_A$. However, in the case of I^- of radius 220 pm, where the interstices offer an aperture of only 34 pm radius, the transport number of Li^+ ($r_3 \sim 55$ pm) is significant, and the room temperature ionic conductivity of LiI is $5\times10^{-5}\,S\,m^{-1}$. This can be attributed to the following factors:

- ions are not rigid spheres,
- simple concepts relating to the interaction of massive bodies are not applicable in detail to the interaction of ions; and
- thermal energy will lead to random enlargements of the interstices and to the creation of anion and cation vacancies facilitating the movement of ions.

With respect to the last point, it is significant that room temperature ($\sim$300 K) is approximately 0.4 the melting point of LiI ($\sim$723 K) and so the concentration of thermally induced lattice vacancies will be high. LiI, which is used as a solid electrolyte in cardiac pacemakers, can be contrasted with $RbAg_4I_5$, a fast-ion conductor that, in polycrystalline form, has an ionic conductivity due to Ag^+ ions of $21\,S\,m^{-1}$ at 25 °C. Even though silver has a large radius ($r_6 = 115$ pm), the $RbAg_4I_5$ structure is such as to offer apertures approaching this diameter for the Ag^+ ions.

The fast-ion ceramic conductors of interest here are cubic stabilized zirconia (CSZ), an oxygen ion conductor, and a sodium aluminate (β''-alumina), a sodium ion conductor. Both are discussed in detail below. The overview by T.A. Ramanarayanan *et al.* [8] covering CSZ and its applications is recommended.

Cubic stabilized zirconia (CSZ) Pure zirconia (ZrO_2) is either chemically extracted and purified from the mineral zircon ($ZrSiO_4$) or purified from baddeleyite. It occurs as three crystalline polymorphs with monoclinic, tetragonal and cubic structures. The monoclinic form is stable up to 1170 °C

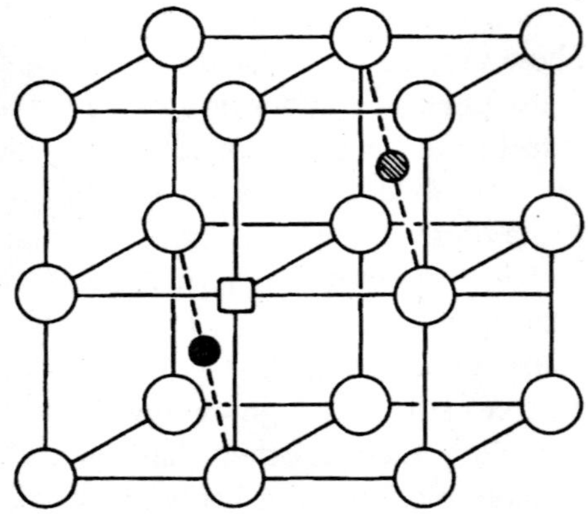

Fig. 4.29 The ideal fluorite structure showing half a unit cell including a dopant cation and a charge-compensating oxygen vacancy.

when it transforms to the tetragonal modification, which remains stable up to 2370 °C; from 2370 °C to the melting point (2680 °C) the cubic form is stable. The monoclinic-tetragonal conversion is accompanied by a contraction in volume of approximately 5% which can cause mechanical failure in ceramic pieces. This difficulty is overcome in CSZ by stabilizing the cubic form down to room temperature by substituting lower-valence cations for some of the zirconium.

Cubic ZrO_2 has the fluorite structure with the O^{2-} ions arranged in simple cubic packing and half the interstices in this lattice occupied by Zr^{4+} ions (Fig. 4.29); the substitution of lower-valence cations leads to O^{2-} ion vacancies as indicated. The vacancies which stabilize the structure also lead to high mobility in the oxygen sub-lattice and to behaviour as a fast-ion conductor.

The elements that stabilize the cubic fluorite structure in zirconia include all the lanthanides, scandium, yttrium, magnesium, calcium, manganese and indium. The main qualification would appear to be an ionic radius close to that of Zr^{4+} ($r_8 = 84$ pm). Ce^{3+} with $r_8 = 114$ pm is one of the largest ions fulfilling this function. Ca^{2+} ($r_8 = 112$ pm) is the most commonly used substituent at about the 15 mol.% level. Yttrium ($r_8 = 101$ pm) stabilizes the cubic phase when present in the 13–68 mol.% range, but maximum conductivity is obtained with 7–8 mol.% which results in some admixture of tetragonal and monoclinic phases with the cubic. Calcium is also sometimes used at the 7–8 mol.% level, resulting in a partially stabilized zirconia. Scandium has been found to give a material with a higher conductivity, which is particularly valuable at lower temperatures. From the fact that the conductivity reaches its maximum at an intermediate concentration of lower-valence ions, it appears that only a fraction of the vacant O^{2-} sites contribute to O^{2-} mobility when the concentration of vacant sites is high. Ions of variable valence need to be excluded if the electronic conductivity is to be minimized. Thorium is added to yttria-containing bodies as a grain growth inhibitor.

For refractory applications, when the 'mixed oxide' route is followed, temperatures as high as 1900 °C are necessary to sinter to high density. The high interest in SOFCs has stimulated efforts to develop novel fabrication routes to zirconia ceramics and yttria-stabilized sub-micron sized powders have been prepared which sinter to 95% theoretical density at a temperature as low as 1150 °C [4].

Chemically prepared, highly reactive and uniform zirconia powders are commercially available. One production route is the co-precipitation of a mixed solution of $ZrOCl_2$ (derived from zircon) and YCl_3 (derived from yttria) solution followed by calcinations, milling and spray-drying. The best commercially available, chemically prepared, powders of 8% Y_2O_3/ZrO_2 (8YSZ) can be sintered to full density at 1400 °C.

Careful attention has to be given to the purity of the precursors to avoid detrimental effects on conductivity. In a polycrystalline ceramic the conductivities of grain boundaries and bulk contribute to overall conductivity. In the case of polycrystalline YSZ, because of its unusually high intrinsic (bulk) conductivity the grain boundaries are far less conductive than the crystal, typically by a factor of ~100. The effect the grain boundaries have on overall conductivity will depend on grain size and, of course, on impurity content (e.g. silica), since impurities tend to concentrate there. It is the effort to understand more of the various contributors to overall conductivity which has led to the application of impedance spectroscopy (see Section 2.7.5).

β″-alumina The β-aluminas are a family of non-stoichiometric aluminates of which the most important have the approximate formulae equivalent to $Na_2O11Al_2O_3$ (β-alumina), and $Na_2O5Al_2O_3$ (β″-alumina). They have a layer structure (Fig. 4.30 (a)) with layers approximately 1 nm thick consisting of blocks of close-packed O^{2-} ions in which the Al^{3+} ions occupy octahedral and tetrahedral interstices in the same arrangement as the Mg^{2+} and the Al^{3+} in the spinel structure. The spinel layers are separated by mirror planes containing Na^+ and O^{2-} ions. The Na^+ ions can move quite freely within this plane with the sort of concerted motion indicated in Fig 4.30(b); as a result the conductivity is high in these planes but negligible in the perpendicular direction.

It is instructive to write the compositions of the β- and β″-aluminas with the oxygen lattice content the same in both cases, that is $Na_3Al_{33}O_{51}$ and

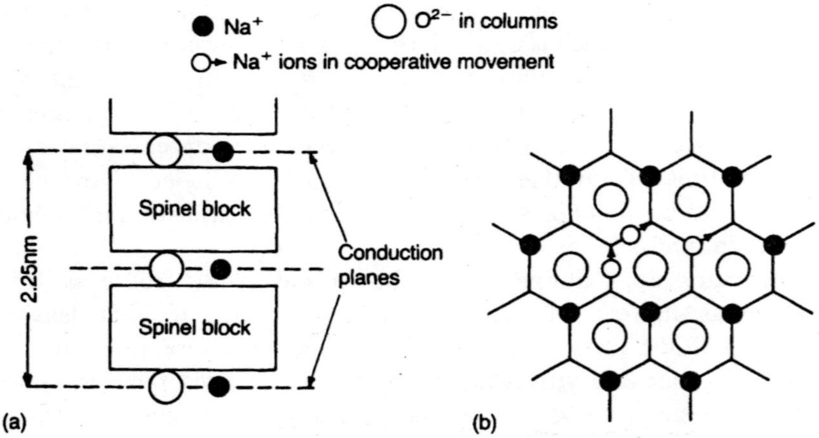

Fig. 4.30 Structure of β''-alumina: (a) alternating spinel blocks and conduction planes; (b) migration pathway of Na^+ ions indicating paths of concerted motion.

$Na_6Al_{32}V_{Al}O_{51}$, respectively. This suggests that compared to β-alumina the β'' modification contains an aluminium vacancy compensated by three extra sodium ions in the conduction plane. The consequent higher conductivity of the β'' modification makes it favoured for battery electrolytes. The conductivity of polycrystalline β-alumina at 350 °C (the temperature appropriate to battery operation) is about $5\,S\,m^{-1}$ and for polycrystalline β''-alumina about $50\,S\,m^{-1}$.

The preparation of 'pure' β''-alumina is not easy and dopants (e.g. Mg and Li) are added to stabilize the modification. The dopants have the added advantage of reducing interstitial oxygen in the conduction plane and therefore facilitating the movement of the sodium ions.

A route described by J.L. Sudworth *et al.* [9] for the commercial production of an almost pure β''-alumina tube starts with the mineral boehmite {AlO(OH)} which is calcined at ~800 °C to produce a microcrystalline mixture of the γ-, δ- and θ-forms of alumina. The calcined powder is mixed with sodium carbonate and lithium hydroxide and fired at ~1200 °C when complete conversion to β''-alumina occurs having a composition close to (wt%) $90Al_2O_3$, $9.2Na_2O$ and $0.8Li_2O$. The β''-alumina is wet-ground to a particle size of approximately 1–2 μm after which the slurry is spray-dried to produce a free-flowing powder suited to forming the tubes by isostatic pressing.

Sintering is a critical step with schedules extending over typically 24 h and a peak temperature of approximately 1600 °C. At this temperature there is an appreciable sodium vapour pressure over beta aluminas and special precautions are taken to maintain the required composition. The component may be enclosed in a sealed spinel ($MgAl_2O_4$) saggar which can be re-used.

β''-Alumina tends to suffer from discontinuous grain-growth during sintering with the result that an undesirable duplex (excessively large grains embedded in a small grain-sized matrix) develops. Since mechanical strength is reduced by the presence of large grains the sintering schedule must be adjusted so that the rate of densification is very much greater than the rate of grain-growth. In this way a uniformly small grain-sized ceramic can be made and the necessary optimal technical strength (~200 MPa) achieved.

Other fabrication routes have been tried including forming by electrophoretic deposition and densification by 'fast-firing'. In 'fast-firing' (or 'zone-sintering') the tubes may be fed sequentially through an inductively-heated furnace with the peak temperature approximately 1700 °C. The sintering takes only a few minutes and so Na-loss is negligible, and the tube is rotated as it passes through the furnace to prevent bending.

The solid oxide fuel cell (SOFC)

Details of the ceramics and approaches to fabrication involved in SOFC technology are many and complex and well beyond the scope of the present text. The following summary provides the framework to which the details can be added. The articles by N.Q. Minh [10], B.C.H. Steele [11,12], B.C.H. Steele and A. Heinzel [3, pp. 345–52], and S.C. Singhal [13] serve as good entries to the more specialized literature.

Electroceramics exploited The four major functions electroceramics perform in the present generation of SOFCs are as the electrolyte, the anode, the cathode and the interconnect. The favoured ceramics are discussed below.

Electrolyte–cubic stabilized zironia Almost without exception cubic stabilized zirconia is the chosen ceramic for the electrolyte in SOFCs. This is because of its adequate conductivity and almost total absence of electronic conductivity, and because it is stable against the wide range of oxygen partial pressures (~1 atm. to ~10^{-20} atm.) encountered in a fuel cell. Also, because of a combination of availability and cost the favoured compound is yttria-stabilized zirconia, ZrO_2+8–10 mol.% Y_2O_3 (YSZ).

The ceramic is required in the form of dense, polycrystalline layers, as thin as practicable to minimise cell resistance, consistent with the need to ensure mechanical integrity. Adopted layer thickness is, of course, also determined by the operating temperature of the cell. Self-supporting plates 150 μm thick can be used at 900–1000 °C and supported thick films (down to ~15 μm) to as low as ~700 °C.

Fig. 4.31 shows Arrhenius plots for YSZ together with other candidate electrolyte ceramics illustrating how the combination of temperature and ceramic

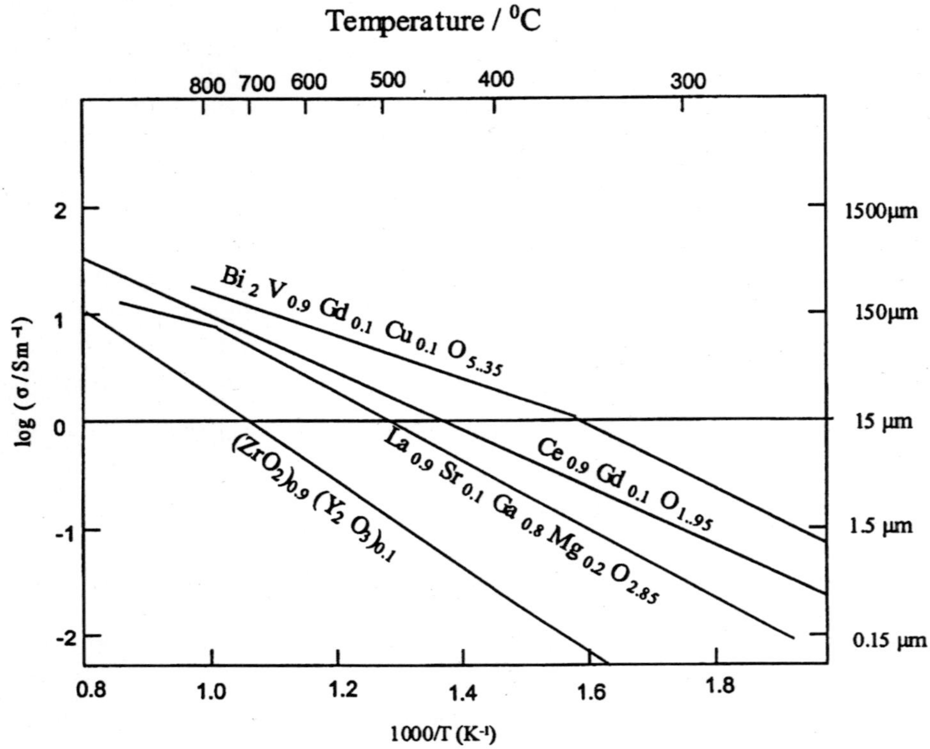

Fig. 4.31 Ionic conductivities for candidate electrolyte ceramics. The arbitrary assumption that for a planar cell format a resistance of $<15\,\mu\Omega\,m^{-2}$ is required places an upper limit on the permitted thickness of the electrolyte; lower values of conductivity demand thinner membranes whilst higher values permit correspondingly thicker membranes. Electrolyte thicknesses greater than approximately 150 μm are considered mechanically self-supporting. After B.C.H. Steele [11].

type influences the electrolyte thickness for a given areal resistance value, arbitrarily but realistically set at $15\,\mu\Omega\,m^{-2}$.

Anode The anode, or fuel electrode, must be stable against reduction, be electronically conducting, have a high specific surface area and a connected porosity allowing free passage of the fuel gases to the anode/electrolyte interface; it must also facilitate the counter flow of the oxidation products away from the interface. Although the temperature and reducing atmosphere conditions permit the use of partially sintered metallic nickel as an anode, prolonged use would lead to further sintering and undesirable microstructural changes; there is also the need to match thermal expansivities. Coating the partially sintered nickel with YSZ gives a better thermal expansion match between anode and electrolyte, reducing thermal fatigue during operation and improving adhesion between anode and electrolyte. A further advantage is that the microstructure offers an increase in reactive area at the three-phase boundary where oxidation of the fuel occurs.

Cathode In contrast to the anode the cathode operates in an oxidizing environment but, like the anode, it must have high electronic conductivity and a pore structure enabling the gaseous oxidant to reach the cathode/electrolyte interface.

The favoured material is modified lanthanum manganite (e.g. $La_{0.8}Sr_{0.2}MnO_{3+x}$) which has the perovskite structure. It is a p-type semiconductor the electron transport occurring by electron-hopping (see Section 2.6.2) between the +3 and +4 states of the Mn ion. The strontium-doping enhances the conductivity.

Interconnect The property requirements for the interconnect ceramic are the most demanding. The conductivity should be as high as possible and, ideally, 100% electronic. This is necessary to reduce internal resistance and to avoid severe problems which would arise were ionic space-charge polarization (see Section 2.7.1) to develop.

Because the interconnect is exposed to high temperatures and oxidizing environments it must be stable against both. It must also serve as an effective barrier to the oxidant and fuel gases so that direct reaction is avoided, and it is required to be chemically inert with respect to the cathode, anode and the inter-cell connector material, for example nickel (see Fig. 4.32(a)).

These requirements are satisfied by a doped lanthanum chromite (see Section 4.1.3) e.g. La (Ca, Mg, Sr, etc.) CrO_3. However for SOFCs operating in the temperature range 500–750 °C a stainless steel interconnect plate can be used.

Geometries and construction The two basic geometries for the SOFC are planar and tubular.

Planar geometry Although there are variations in detail, the design of the planar or 'flat plate' cell, and how cells are interconnected are essentially as illustrated in Fig. 4.27. The construction may involve a self-supporting, coated electrolyte or, if the operating temperature permits, the anode–electrolyte–cathode structure may be supported on a stainless steel bipolar plate.

Brief summaries of the various processing technologies used in fabricating cell components are given in Section 3.6.9.They include tape-casting, calendering, screen-printing, electrophoretic deposition, electrochemical vapour deposition, chemical vapour deposition, spray-pyrolysis and sol–gel.

As an example the electrolyte may be tape-cast or calendered. Flat and dense plates of YSZ having thickness down to 150 μm are readily fabricated. The anode and cathode materials can then be applied (by a slurry) to the faces of the electrolyte and fired on. Alternatively the anode, electrolyte and cathode layers may be formed together by tape-casting and stacking, or by calendering, and the stack co-fired. Similarly, the interconnect ceramic may be tape-cast or hot-pressed. Whichever the approach, the elements of the assembled stack must make good electrical contact one with the other.

The most challenging technology is concerned with effecting, with the aid of metals, glasses and glass-ceramics, gas-tight seals at the edges of the stack. Whichever the sealing technology, serious problems arise because of the inevitable thermal expansion mismatches and the resulting mechanical stresses developed during cooling. Apart from immediate failure such stresses can lead to 'fatigue' effects (slow crack-growth) and therefore to in-service unreliability over required working lifetimes of order 10^4 hours. During this lifetime the cell may be temperature-cycled (from an operating temperature of $\sim 700\,°C$ to room temperature) making further severe demands on reliability, and failure induced through chemical and electrochemical reactions between the sealants and the stack ceramics are an additional hazard.

Reliability could be improved by a reduction in working temperature, maintaining adequate cell conductance. This demands decreasing the electrolyte membrane thickness, consistent with impermeability to gas molecules and mechanical soundness, or by using more conductive electrolytes such as $Ce_{0.9}Gd_{0.1}O_{1.95}$ (see Fig. 4.31). Unfortunately ceria exhibits some electronic conductivity under the low oxygen potentials encountered in a fuel cell, the resulting internal partial short-circuiting leading to a small reduction in cell voltage.

Tubular geometry The following discussion draws specifically on the Siemens–Westinghouse [13] programme. The arrangement of the cell components is illustrated in Fig. 4.32(a).

The fabrication of the long (~ 1.5 m) tubular structure starts with the cathode which is extruded (~ 20 mm dia. and ~ 2 mm wall thickness) with additions to the lanthanum manganate powder of organics (e.g. starch and cellulose: see Section 3.9) in order to develop the necessary porosity. The tube is sintered in air at $\sim 1500\,°C$. The structure must be chemically stable with respect to the subsequent processing of the electrolyte, anode and interconnect layers and have compatible thermal expansivity. It must, of course, have adequate strength to be self-supporting and also to support the electrolyte and cathode.

The electrolyte (YSZ) layer ($\sim 40\,\mu m$ thick) is applied to the cathode by EVD (see Section 3.6.9). Because this is an expensive route efforts are made to use more cost-effective methods, for example plasma-spraying and electrophoretic deposition followed by sintering.

The anode (fuel electrode) consists of an approximately 120 μm thick Ni/YSZ cermet layer. This is achieved simply by applying a slurry of nickel and YSZ powders to the electrolyte and sintering. A cross-section of the 3-layer microstructure is shown in Fig. 4.33.

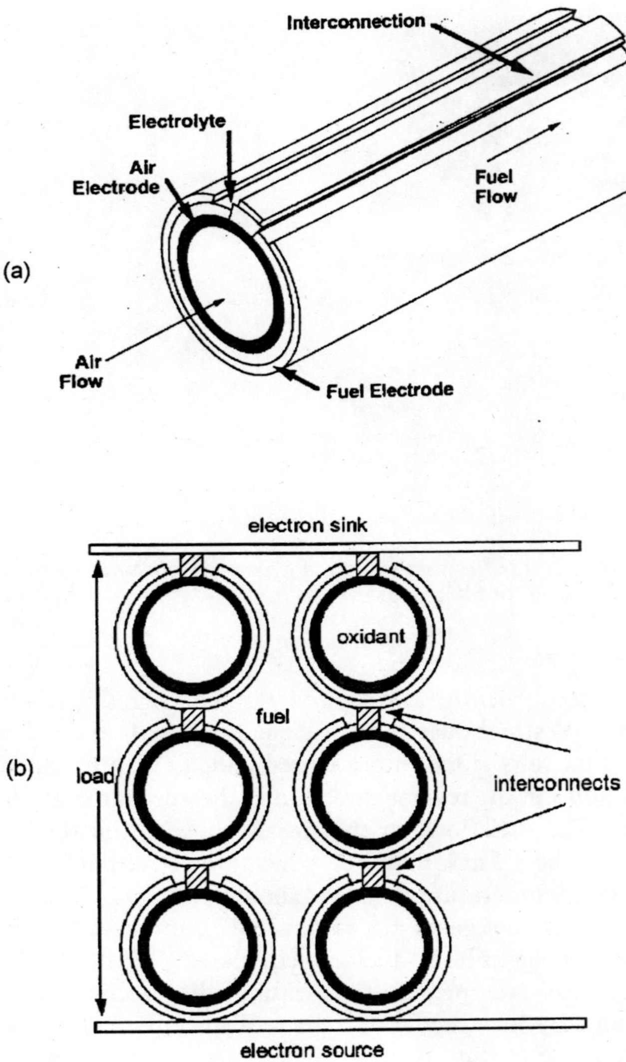

Fig. 4.32 (a) The arrangement of the electrodes in a tubular SOFC [13]. (b) Schematic of a stack of series/parallel fuel cells. The diagram should be viewed in conjunction with (a).

The lanthanum chromite interconnect strip is applied along the length of the tube by plasma-spraying. Since this is required to pass through the electrolyte and anode it is necessary to use appropriate masking during their deposition. The connection between cells is made via a soft nickel felt so that no dangerous mechanical stresses are placed on the tubes.

The manner in which the separate cells are connected, along their lengths, in a series/parallel array is shown schematically in Fig. 4.32(b).

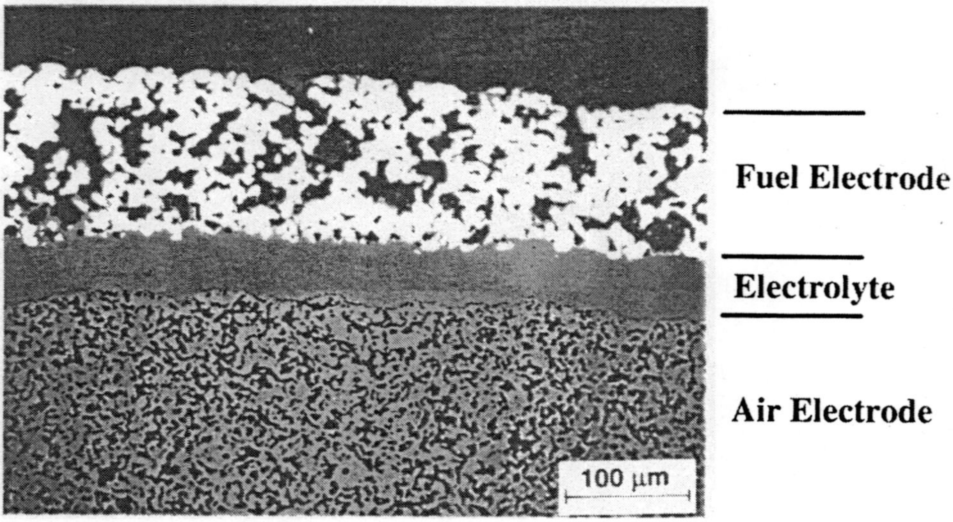

Fig. 4.33 Microstructure of a cross-section of a Siemens–Westinghouse SOFC. (Reproduced with the kind permission of S.C. Singhal.)

Operational considerations and forecasts The significant advantage offered by the Siemens–Westinghouse tubular design is the absence of critical gas-tight seals. The oxidant (air) is fed into a closed-ended cell tube via a long feed tube after which it flows in the reverse direction in the annular space between the feed and cell tubes. The fuel flows in the spaces between the stacked cells towards their open ends. The 1.5 m long tubes, which are active along almost their entire lengths, run at a temperature of approximately 900 °C. Some of the emergent unused oxidant and fuel gases are combusted to pre-heat the incoming air and fuel, and a part of the unburnt fuel is recirculated. 'Waste' heat can be used for other purposes, for example, district heating, the precise operating conditions being determined by the specific objectives. Typically the cell voltage is ~0.7 V, depending upon the current drawn. Such systems have been built and operate successfully delivering about 100 kW of electrical power with negligible degradation problems.

Recently a 200 kW high pressure (~3 atm.) system has been constructed by the same collaborators and tubular cells have been tested at pressures as high as 15 atm. Operating the system at higher pressures increases the efficiency because the free energy change, ΔG, for the oxidation reaction is increased and therefore so too is the cell voltage (see Eq. 4.33).

The additional advantage is that it allows the hot emergent gases to drive a turbine and the 200 kW, 3 atm. system is coupled to a small gas turbine generating 50 kW. The predictions are that such combined cell and gas turbine systems can achieve overall efficiencies as high as 70% and it seems that the future of the tubular design lies with systems having outputs greater than 1 MW.

A major disadvantage associated with the tubular design is the low power density ($\sim 0.3\,\mathrm{MW\,m^{-3}}$). There are two main reasons for this. One is the inherent low stacking density of the individual fuel cells; the other is the high internal resistance of the cell arising because of the long circumferential paths the current has to take in the cell tubes to reach the interconnects. A further disadvantage is cost arising from the rather sophisticated fabrication technologies (EVD) involved.

The planar format presents many fabrication challenges, as mentioned above, but the bonus is an estimated power density of approximately $2\,\mathrm{MW\,m^{-3}}$ and flexibility for constructing power sources to a required rating. A present objective is to build a relatively small power (up to ~ 10 kW) module which could be mass-produced and then integrated with similar modules to meet power requirements ranging from on-board transport to static, large scale distribution. All the technologies for fabricating the cell are cheap and suited to mass production. The electrolyte and interconnect can be tape-cast, calendered or pressed and the porous electrodes formed by slurry-coating. The major challenge is to achieve a reliable high temperature edge-sealing technology. However, as is usually the case, once the problem has been clearly defined a solution will be found, and even if this is an expensive solution widespread adoption soon lowers the price (the initially very costly, fully dense alumina tube, later universally exploited in high pressure public lighting, is a classic example). If it proves possible to successfully exploit the more highly conducting Gd-doped ceria for the electrolyte then the lower operating temperatures ($\sim 500\,^{\circ}\mathrm{C}$) will very significantly ease the sealing challenge.

In conclusion, fuel cell technology has to be seen 'in the round'. The PEM cell is certainly favoured for automotive power. Its serious drawback has nothing to do with the cell technology per se, rather with the fact that it runs on hydrogen and the infrastructure to supply the gas does not exist, at least not to any significant extent. It can be combined with a fuel reforming plant to produce hydrogen from a hydrocarbon but then, because the system is 'cold', the incorporation of such a reformer carries with it its own penalties regarding efficiency. The cell is regarded as a power source for far more than the motor car; for example it is seen as the candidate for powering lap-top computers, torches and for any equipment presently powered by batteries. Some projections envisage the domestic home powered totally from PEM cells.

The intermediate temperature SOFC offers an advantage over the PEM cell of a predicted higher efficiency, 45–50% compared to 30%. It can also be integrated with a reformer which, utilizing some of the 'waste' heat, produces useable fuels, (H_2 and CO) from a hydrocarbon fuel.

The tubular technology as described above is suited only to static power-generation. In contrast the planar format is far more versatile and, especially if it could be coupled with a lower operating temperature of approximately 500 °C, would be a strong candidate for auxiliary power supplies for motor vehicles, for example to provide air-conditioning when the internal combustion engine is not running, and for the supply of domestic electricity and heating. For such applications the stack would be required to reliably withstand many cycles from room temperature to the operating temperature.

Na/S and Na/NiCl batteries

Na/S cell There are various cell designs and details can be found elsewhere [7]. The principle of operation is illustrated in Fig. 4.28 and a schematic of one design of cell shown in Fig. 4.34. The cell contains just the amount of sodium for it all to be combined with the sulphur for complete discharge. The small space between the stainless steel liner and the electrolyte contains a fine stainless steel mesh which, by capillarity, can raise the height of the liquid sodium which enters the gap through the small hole at the bottom of the liner. The sodium can be pulled to a height of at least 200 mm at 300 °C.

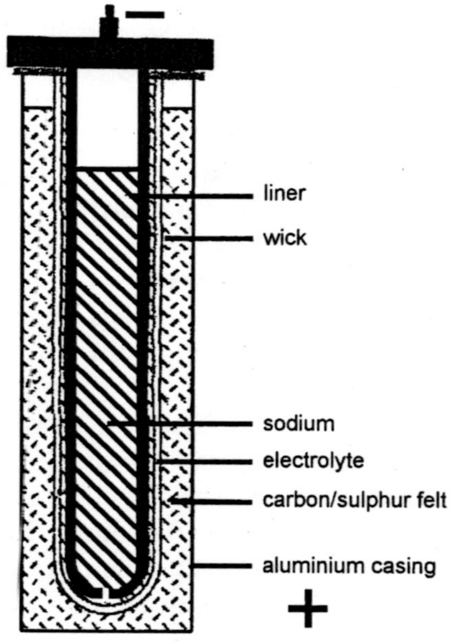

Fig. 4.34 Schematic of Na/S cell.

Na/NiCl$_2$ ('ZEBRA' cell) To illustrate the principle a schematic of one design of cell is shown in Fig. 4.35. Cell design and technology have evolved very considerably and the reader is referred for details to [7] and to the paper by R.C. Galloway and S. Haslam [14].

One innovative aspect of the ZEBRA technology is that the cell is constructed with the reaction 'products', metallic nickel and sodium chloride in situ; that is the cell is in the discharged state, greatly facilitating fabrication. The liquid sodium and nickel chloride are formed by charging the cell.

The $NaAlCl_4$ is a liquid electrolyte at the normal operating temperature of $\sim 270\,°C$ (maximum of $350\,°C$) of the cell, interfacing both with the porous Ni/NaCl electrode and the β''-alumina. After assembly the cell is charged, liquid sodium (melting point $\sim 98\,°C$) being formed. Excess NaCl in the liquid electrolyte ensures that Na ions are not removed from the solid electrolyte which would otherwise cause an increase in internal resistance.

In order to increase the active area of the electrolyte and so increase drawn current capability, the circular section tube has evolved to a convoluted form in which the cross section has four lobes.

The future for batteries based upon β''-alumina As far as the 'zero emission vehicle' is concerned the foreseeable future depends upon the battery, and there are competing technologies. In its 1998 report the United States Advanced Battery Consortium (USABC) identified the following four battery technologies – the nickel/metal hydride(NiMH), the lithium ion (Li-Ion) and the high temperature sodium ion batteries (Na/S and the Na/NiCl$_2$-ZEBRA) as most promising as far as meeting the post 2000 performance objectives. Electric

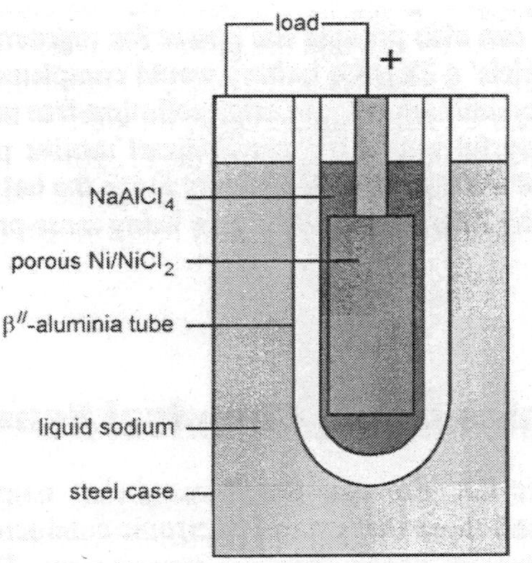

Fig. 4.35 Schematic of the Na/NiCl$_2$ 'ZEBRA' cell (charged state).

vehicles powered by all technologies are running but, of the two based on β''-alumina, the ZEBRA battery is the most promising.

There are many technical matters to be taken into account in considering the viability of a particular technology. The specific energy and specific power which, on a single charging determine respectively travelling range and acceleration and hill-climbing capabilities, are critical. Another important technical characteristic is energy density on a volume basis since this, of course, determines the space occupied by the battery pack. Most importantly for a motive power source there are 'safety' considerations and the sodium-based batteries are posing no problems in this respect.

Thermal management is also a critical matter. At first sight a comparison between the high temperature (~300 °C) sodium cells and the ambient temperature NiMH and Li-Ion cells would seem to favour the latter. The ZEBRA cell must be maintained at the operating temperature even when the battery is resting. If the battery is out of use for such a time that it cools to ambient temperature, then recharging has to be a carefully controlled process, taking a day or more as the battery is raised to its operating temperature. However there are very significant advantages. The sodium battery is unaffected by extremes in ambient whereas, in contrast, the near room temperature cells require elaborate thermal management. As an additional bonus the stored heat is immediately available for car interior heating or window defrosting.

The battery poses no 'safety' problems and is ideally suited to the totally electric car, van and bus. Batteries can be manufactured with voltages as high as the 600 V which is standard for many electric buses and trains, and therefore motors and drive systems are already available. The battery and fuel cell can work together, the fuel cell being ideally suited to providing a steady power output and the battery providing the boost necessary for climbing hills and acceleration. The battery can also provide the power for regenerative braking.

In the 'series hybrid vehicle' a ZEBRA battery would complement the internal combustion engine. This combination could offer pollution-free motoring within cities, with the more powerful but 'dirty' petrol/diesel motive power used for longer journeys. In the 100 kW h to 10 MW h energy range the batteries would be suited to load-levelling. The ZEBRA battery is now being mass-produced (MES-DEA, Stabio, Italy).

4.6 Ceramics-based Chemical Sensors

Ceramic chemical sensors fall into two broad categories, namely those that exploit solid electrolytes and those that exploit electronic conductors. In all cases the sensors respond to changes in the chemical environment. The operational principles and typical applications are described below.

4.6.1 Sensors based on solid electrolytes

Sensing oxygen with cubic stabilized zirconia

Potentiometric mode There is no essentially different principle involved from that on which the fuel cell is based. The distinction is that in the case of the fuel cell the required output is power whereas with the sensor it is either a small voltage or small current that monitors some chemical characteristic of the ambient.

A common type of oxygen sensor takes the form of an yttria stabilized zirconia (YSZ, see earlier) tube electroded on the inner and outer surfaces with a porous catalytic platinum electrode. The electrode allows rapid equilibrium to be established between the ambient, the electrode and the tube. Such a system is shown schematically in Fig. 4.36.

It is usual to write the electrochemical cell in the form

$$p'_{O_2}\text{ ('test')},\quad \text{Pt}|\text{YSZ}|\text{ Pt},\ p''_{O_2}\text{ (reference)}$$

in which p'_{O_2} and p''_{O_2} are partial pressures.

Platinum catalyses the dissociation and recombination of the oxygen molecules so that O^{2-} ions can be formed at one electrode and converted to O_2 molecules at the other. The e.m.f. developed by such a cell is given by the Nernst equation (see Eq. (4.33)). In the present case $z = 4$ because an oxygen molecule consists of two atoms each acquiring two electrons on being ionized. Hence, the equation becomes

$$E = RT/4F\,\{\ln(p''_{O_2}/p'_{O_2})\} \tag{4.38}$$

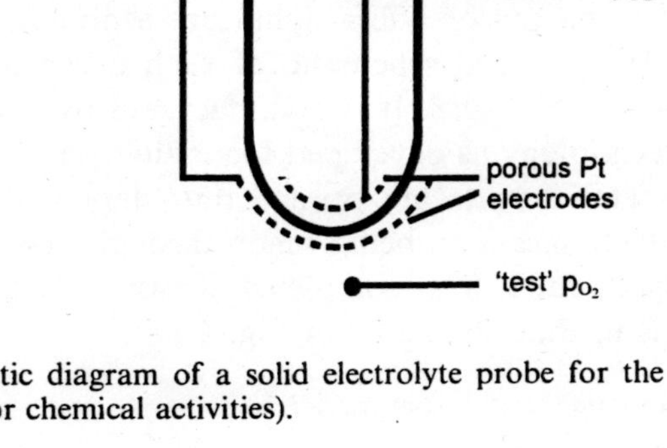

Fig. 4.36 Schematic diagram of a solid electrolyte probe for the measurement of oxygen partial pressures (or chemical activities).

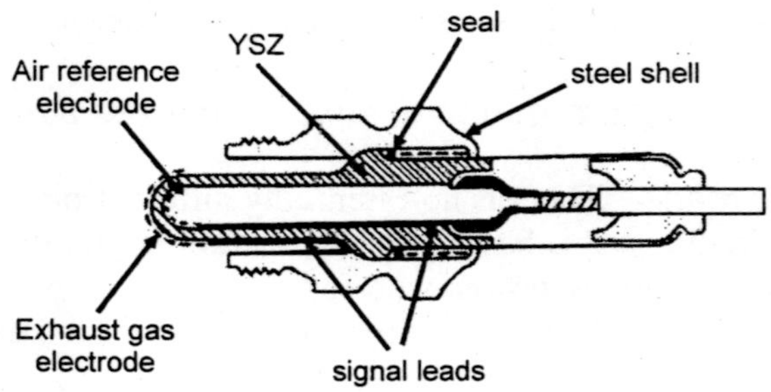

Fig. 4.37 Form of a 'lambda' automobile exhaust sensor (the forms may vary from model to model but the essentials are the same).

or, more correctly

$$E = RT/4F\,\{\ln(a''_{O_2}/a'_{O_2})\} \tag{4.39}$$

where a''_{O_2} and a'_{O_2} are the chemical 'activities'. (In writing the partial pressures the assumption is being made that the gases are behaving as ideal.)

YSZ oxygen sensors are used to monitor automobile exhaust gas emissions so that the fuel-to-air ratio can be kept at the optimum from the point of view of both pollution and engine efficiency. Pollutant gases generated by the internal combustion engine include CO, hydrocarbons and oxides of nitrogen (NO_x). Catalytic converters in the engine exhaust line are able to oxidize the CO and hydrocarbons to CO_2 and H_2O, and reduce the NO_x gases to N_2 provided the catalytic unit is operated close to the λ* (lambda) = 1 point, defined as the condition when precisely the correct amount of air (the 'stoichiometric' amount) is involved in the combustion to completely oxidize the fuel.

The general form of a 'lambda' sensor is illustrated in Fig. 4.37. The closed end projects into the hot exhaust gas stream heating the sensor to a temperature at which it is sufficiently conductive for the e.m.f. to be measured by a high impedance meter. The input impedance of the meter must be 100 times that of the cell for 1% precision. Corrections for changes in cell temperature (Eq. (4.38)) can be made by monitoring the cell resistance.

A planar format lambda sensor offers significant economic advantages over the tubular design. An exploded schematic of such a sensor for monitoring exhaust gases is shown in Fig. 4.38. It is manufactured by essentially the same ceramic fabrication technology as developed for multilayer circuits (see Section 5.4.3). The various metallic parts are printed onto tape-cast 'green' ceramic-loaded sheets, the interconnections being made through vias. The assembly is laminated and finally co-fired. The completed sensor is housed in a manner similar in its essentials to that illustrated in Fig. 4.37.

* λ = the 'operating' air : fuel ratio/the 'stoichiometric' air : fuel ratio.

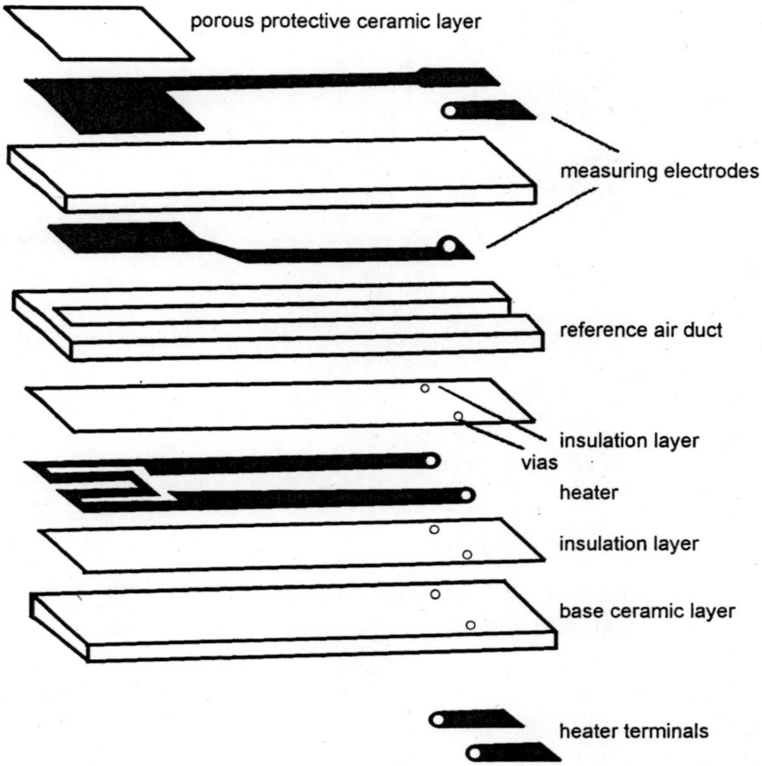

Fig. 4.38 Planar sensor format (adapted from the *Automotive Electronics Handbook*, 1994).

In the case of the lambda sensor the YSZ needs to have adequate mechanical strength to withstand vibrations etc. Partially stabilised zirconia (PSZ) ceramic is a mix of tetragonal and monoclinic crystals and exhibits exceptionally high fracture-toughness. For this reason a compromise is reached at approximately 4 mol.% yttria addition which produces the high toughness PSZ at the same time offering adequate ionic conductivity.

Another important application is in the refining of steels when the oxygen content must be controlled at the parts per million level and monitored continuously 'on line' and many oxygen sensors are currently used in the steel industry for this purpose. The principle of operation is as described for the lambda sensor and one form is shown schematically in Fig. 4.39. In this case the reference activity is established by a chromium metal/chromium oxide mix rather than being defined by air.

Essentially the sensor comprises an YSZ element one side of which is in contact with the steel melt and the other with the metal–metal oxide reference mixture. The steel serves as one electrical contact to the zirconia and the

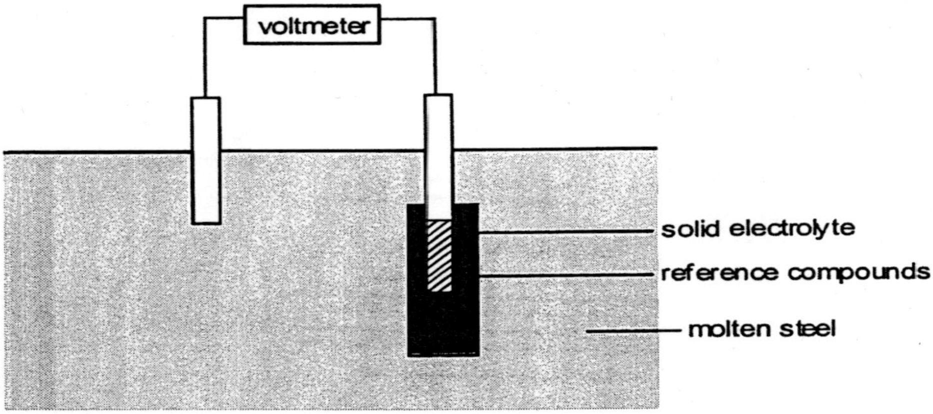

Fig. 4.39 Schematic of an oxygen meter for monitoring the oxygen content of steel during refining (adapted from [15]).

chromium metal to the other. The oxygen activity in the Cr/Cr_2O_3 mix is known from thermochemical data relating to the equilibrium reaction:

$$Cr_2O_3 \longleftrightarrow 2Cr + 3/2O_2 \tag{4.40}$$

A common form of disposable sensor is a long cardboard tube with the sensor elements built into one end. The tube is plunged into the molten steel and the measurements made within a few seconds.

Amperometric mode An advantage of the lambda sensor as described above is its sensitivity close to the $\lambda = 1$ point where the oxygen activity in the exhaust changes rapidly with burn conditions and consequently so too does the cell e.m.f. The disadvantage is that under 'lean-burn' (oxygen-rich) conditions the log dependence on oxygen activity renders the cell insufficiently sensitive to provide effective control. This shortcoming has led to the introduction of sensors used in the amperometric mode discussed below and illustrated schematically in Fig. 4.40.

The requirement is to measure the concentration (of order 1 vol.%) of oxygen in a gas. The oxygen diffuses down the capillary and by Fick's Law the flux, J, is

$$J = -DA\delta c/l \tag{4.41}$$

in which D is the oxygen gas diffusion coefficient, l the capillary length and A its cross-section; δc is the concentration difference between the capillary ends.

The oxygen ions are driven into the electrolyte by an applied e.m.f. and the current (I), limited by the supply of oxygen ions, measured. It follows that

$I = 4eJ$ (there are four electrons carried by each oxygen molecule)

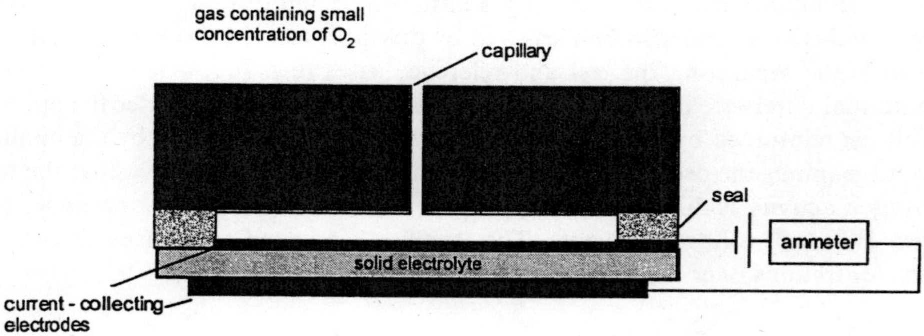

Fig. 4.40 Schematic illustrating the principle of the amperometric mode.

or

$$I = 4e\,DA\delta c/l \tag{4.42}$$

Assuming that the oxygen concentration at the surface of the electrolyte is zero, and with D, A and l all constant, then

$$C_{(\text{oxygen})} \propto I \tag{4.43}$$

where $C_{(\text{oxygen})}$ is the oxygen concentration in the gas.

The capillary tube can be replaced by a porous ceramic appropriately limiting gas flow. Because of the constant urge to reduce component size and cost multilayer technology is exploited in the fabrication of sensors.

The sensor illustrated in Fig. 4.41 for measuring NO in oxygen well illustrates the complexities in design which multilayer technology can readily cope with. In the first stage the gas mixture enters the sensor where the oxygen is pumped away. The potentiometric sensor monitors the oxygen level and controls the pumping. The NO diffuses through the second barrier and dissociates at the amperometric sensor electrode, the oxygen ions being driven through the sensor into the air reference compartment.

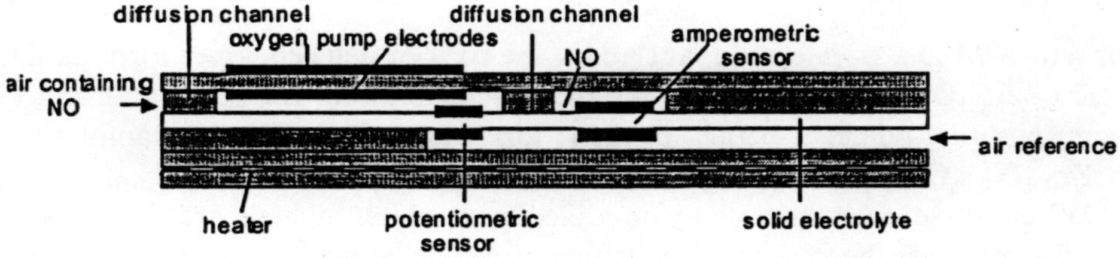

Fig. 4.41 A planar form of NO sensor the entire fabrication exploiting multilayer technology (adapted from [16]).

As an alternative to the limited gas diffusion technique described above, high oxygen activities can also be measured by driving a current though the wall of a membrane separating the test and reference gases (e.g. Fig. 4.36). The driving potential difference is reversed so that the current is measured under the applied voltage reinforced by the cell e.m.f. for one half of the cycle and by the applied voltage minus the cell e.m.f. for the next half cycle. It can be shown that the test oxygen activity is directly related to the ratio of the currents flowing under the two different voltage gradients. The technique is suited to measuring oxygen concentrations over a wide range [17].

Sensors for chemical species other than oxygen

The following examples serve to illustrate the versatility of the experimental approach provided suitable electrolytes can be found. The technology is reviewed by D.J. Fray [15,18].

Sensing hydrogen in molten aluminium Because of the detrimental effect dissolved hydrogen can have on the mechanical properties of many metals it is monitored in metallurgical processing. For this the requirement is for a proton conductor and suitable ceramics have been identified.

It has been known for many years [19] that 'water' can be incorporated into silica in the form of hydroxyl groups which have a characteristic absorption in the infrared. This was a problem encountered in the early days of fibre optics communications' technology and solved by excluding water from the processing atmospheres. This 'water' can be readily introduced by high temperature annealing in water vapour and removed by heating in a water-free atmosphere.

More recently, H. Iwahara *et al.* [20] reported that some compounds having the perovskite structure (see Section 2.7.3) become proton conductors if hydrogen is introduced into the crystal, and the solubility of 'water' and proton mobility in 'perovskites' are now actively researched topics [21]. The perovskites which can be tailored to exhibit high protonic conductivity have compositions of the type

$$AB_{1-x}M_xO_{3-x}$$

in which M is a trivalent dopant and x the oxygen deficiency per formula unit. $CaZr_{0.9}In_{0.1}O_{3-x}$ is an example, the indium substituting for zirconium and the compensating defect being oxygen ion vacancies. Other examples are $SrZr_{0.95}Y_{0.05}O_{3-x}$, $BaCe_{0.8}Y_{0.2}O_{3-x}$, $BaCe_{0.9}Nd_{0.1}O_{3-x}$ and $SrCe_{0.95}Yb_{0.05}O_{3-x}$.

When these compositionally modified perovskites, which are p-type semiconductors, are heated to high temperatures in a water vapour atmosphere protons are incorporated into the structure, most probably is the form of hydroxyl groups, with the proton able to 'hop' from oxygen to oxygen. The proton mobility will depend upon the strength of the hydrogen bond which will, in turn, depend critically on structural details. The proton transport number can be as high as 0.99. Arrhenius plots of the conductivities of the perovskites listed above are shown in Fig. 4.42.

The exploitation of proton conductors is discussed by H. Iwahara [22]. Figure 4.43 is a schematic of a commercial sensor for measuring the hydrogen content of molten aluminium [23]. Silicon nitride and carbon are used in the construction because both are corrosion-resistant to molten aluminium. The electrolyte tube is $CaZr_{0.9}In_{0.1}O_{3-x}$ ceramic. The reference hydrogen activity is fixed by a mixture of 1 vol.% hydrogen in argon fed to the platinised outer sensor surface via the spring-loaded steel tube. The hydrogen gas collected in the inverted sensor tube is in equilibrium with the hydrogen dissolved in the molten aluminium. The measured e.m.f. is given by the Nernst equation

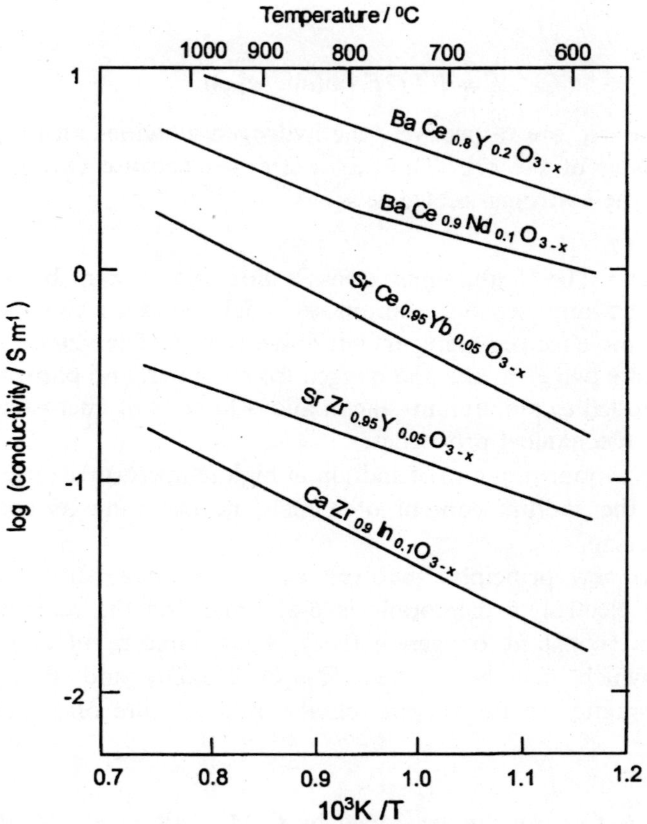

Fig. 4.42 Electrical conductivities of modified 'perovskite' proton conductors (adapted from [22]).

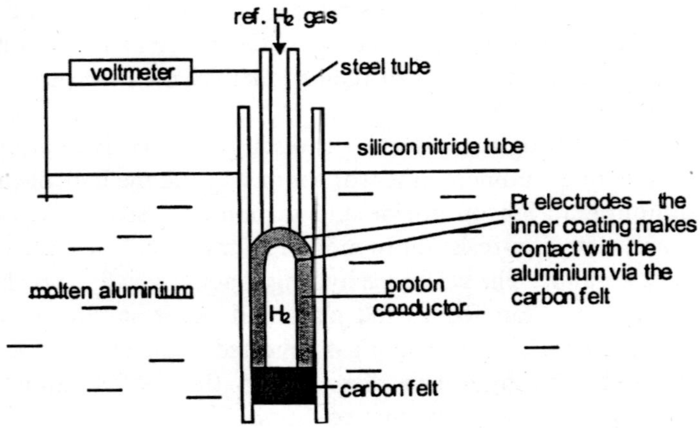

Fig. 4.43 A protonic electrolyte used for measuring the hydrogen content of molten aluminium (adapted from TYK Corporation (Japan) literature; see [23]).

$$E = RT/2F\ \{\ln(a''_{H_2}/a'_{H_2})\} \tag{4.44}$$

in which a''_{H_2} and a'_{H_2} are respectively the hydrogen activities on the measuring and reference sides of the cell. The factor 2 arises because two electrons are associated with the hydrogen molecule.

Sensing sodium The β-aluminas (see Section 4.5.3) can be successfully exploited as a sodium sensor. Sodium is widely used in the metallurgical industry, for example for removing arsenic and antimony from zinc and lead, for removing the same two elements and oxygen from copper, and phosphorus from iron. It is also added to aluminium–silicon alloys to control microstructure and, in consequence, mechanical properties.

The very high vapour pressure of sodium at high temperatures poses problems for controlling the sodium content of a melt, necessitating its constant and accurate monitoring.

There are no new principles involved as far as an electrolytic sensor is concerned. The electrolyte membrane is β-alumina and the reference sodium activity fixed, at constant oxygen activity, by a mixture of sodium ferrite ($Na_{10}Fe_{16}O_{29}$, which can be written $5Na_2O\ 8Fe_2O_3$) and Fe_2O_3, just as it is common practice to fix oxygen activity by a mixture of a metal and its oxide.

Sensing CO_2 A CO_2 sensor developed by G.M. Kale *et al.* [24] illustrates a principle permitting the sensing of species other than that involving the mobile ion in a solid electrolyte. For example sensing CO_2 employing a Na-ion

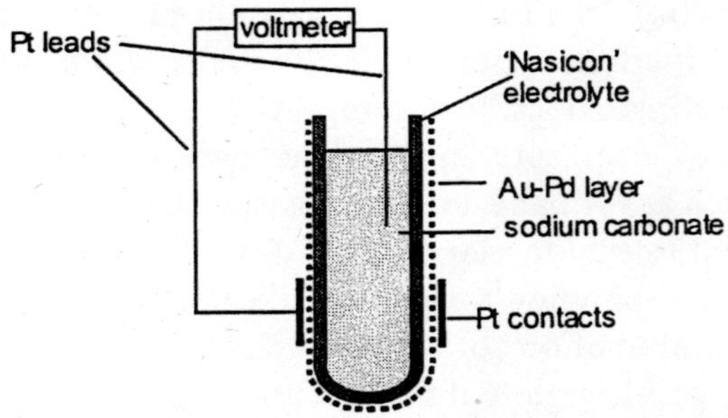

Fig. 4.44 Schematic of a CO_2 sensor incorporating sodium carbonate as a 'linking' compound. N.B. Because of the small size of the sensor (the outside diameter of the tube is only approx. 4 mm) equilibration between the atmosphere and sodium carbonate is rapid at 400 °C.

conductor, in this case 'Nasicon'*. The experimental arrangement is shown in Fig. 4.44. The Nasicon tube is packed with Na_2CO_3 which serves as an 'activity-linking' compound. Because of the equilibrium reaction

$$Na_2CO_3 \longleftrightarrow Na_2O + CO_2 \tag{4.45}$$

any change in CO_2 activity in the gas phase in contact with the solid Na_2CO_3 causes a change in Na_2O activity. At the surface of the electrolyte the activity of Na_2O in the Nasicon changes to the same value. This change in activity can be sensed through the corresponding change in cell e.m.f.

At the platinum contact with the solid Na_2CO_3 the electrochemical reaction is:

$$Na_2CO_3 \longleftrightarrow 2Na^{\cdot} + 2e' + CO_2 + \tfrac{1}{2}O_2 \tag{4.46}$$

And at the Nasicon/gas interface,

$$2Na^{\cdot} + 2e' + \tfrac{1}{2}O_2 \longleftrightarrow Na_2O \tag{4.47}$$

The oxygen partial pressures are the same on both sides of the cell so that the overall cell reaction is given by Eq. (4.45).

4.6.2 Gas-sensors based on electronically conducting ceramics

The science of gas-sensors based on changes in electronic conductivity of a semiconducting ceramic is a complex matter depending upon a combination of surface chemistry and electron transport in mainly transition metal oxides. As is usual in exploiting electroceramics, the applications' technology runs ahead of

* Nasicon is a generic term for sodium ion conducting oxides containing elements such as Zr, Si, and P. The values of their conductivities are similar to that of β-alumina.

the basic understanding. The following discussion provides the base for further study for which the literature sources [25–29] are recommended.

Many semiconducting oxides, for example doped SnO_2, ZnO, and TiO_2, show changes in electrical resistivity in the presence of small concentrations of flammable gases such as propane, and toxic gases such as carbon monoxide. This phenomenon has led to the development of a wide variety of gas-sensors which are finding a rapidly expanding range of applications.

Because of a combination of unique physical properties, namely high transparency in the visible region of the spectrum, semiconducting characteristics and good thermal, mechanical and chemical stabilities, tin oxide is exploited for resistors (see Section 4.2.1), furnace electrodes (see Section 4.1.4) and in optoelectronic devices (see Chapter 8) as well as dominating in gas-sensor technology. Unless otherwise stated the discussion is based upon n-type semiconducting SnO_2 and the reader is referred to Section 4.1.4 for introductory remarks relating to its solid state science.

The sensors detect gases because of a change in electrical resistance which accompanies the reaction between adsorbed species such as O_2^-, O^- and O^{2-} and the ambient gases being sensed, the 'target' gases. Figure 4.45 illustrates how the electron distribution in a semiconductor is modified as oxygen interacts with the surface leading to the adsorption of the most likely species, O^-.

(Note. The Debye length (L_D), although not introduced into the present simplified discussion, is a parameter frequently referred to in the gas-sensor literature. It was originally introduced into ionic solution theory and later applied to semiconductor theory where it is especially applicable to semiconductor/metal and semiconductor/semiconductor junctions. It is a measure of the distance beyond which the disturbance at the junction has effectively no influence on the electron distribution and therefore closely related to 'd' (see Eq. (4.49)). It is a material parameter given by $L_D = (\varepsilon kT/e^2 c_o)^{1/2}$ where c_o is the undisturbed electron concentration, essentially the extrinsic electron concentration in the case of doped n-type tin oxide, and the other symbols have their usual meaning.)

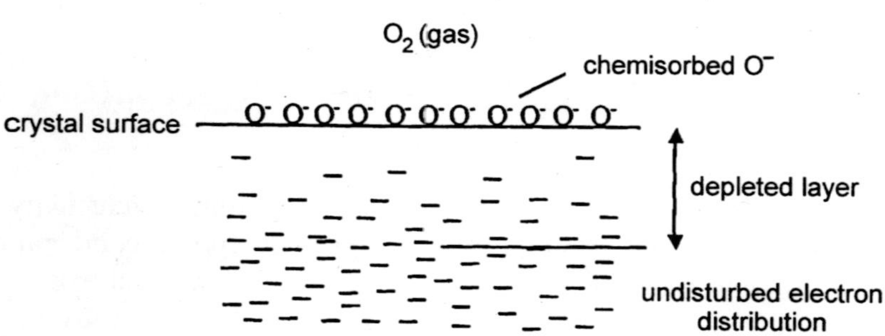

Fig. 4.45 Schematic illustrating the electron-depleted layer at a surface of SnO_2 carrying chemisorbed oxygen.

With the chemisorption of, for example, oxygen a surface density of electron acceptor states leads to the establishment of a Schottky barrier. The process is essentially the same as that which occurs in the case of the PTC thermistor (see Figs 4.21 and 4.10). The electron potential barrier height (φ) is given by

$$\varphi = e^2 N_s^2 / 2\varepsilon N_d \tag{4.48}$$

where N_s and N_d are respectively the surface density of 'acceptors' and the volume density of 'donors'.

Under the same simplifying assumptions as for the PTC thermistor (where the surface states are shared between two grains) the depleted layer thickness d is given by

$$d = (2\varepsilon\varphi / e^2 N_d)^{1/2} = N_s / N_d \tag{4.49}$$

Summarizing the basics of the gas-sensing mechanism, oxygen molecules from the surroundings become chemisorbed (that is chemically bonded) to the surface atoms, the process being described by the reaction:

$$O_2(g) + 2e' = 2O^-(ad) \tag{4.50}$$

As the adsorption proceeds a positive space charge develops in the surface layer of the oxide as electrons transfer from donor defects to the adsorbed oxygen. The electrostatic field generated in the surface regions of the crystal tends to oppose the charge transfer process and eventually it stops. The thickness of the charge-depleted layer is determined by the availability of electrons able to transfer to the surface states and by the surface concentration of adsorbed oxygen. Measurements indicate the thickness of depletion layers to lie in the range of a few nanometres to approximately 500 nm. Reaction of reducing gases at the surface removes some of the adsorbed oxygen and releases electrons which are then available for conduction.

The above description is on the basis of the semiconductor being n-type. In the case of a p-type semiconductor (e.g. $Cr_{2-x}Ti_xO_3$; $0.01 < x < 0.45$ [29]) then a decrease in resistance would accompany an increase in the amount of adsorbed oxygen.

The most common form of gas sensor is based on a porous, sintered ceramic as shown schematically in Fig. 4.46. As adsorption occurs over the surface of the semiconductor grains, the barriers to charge transport develop, especially at the grain boundaries and at particle contact areas (the 'neck' regions).

The ideal sensor would have high 'sensitivity' and be 'selective'. That is, it would show a large resistance change for a small change in the target gas concentration, and have the ability to discriminate between different gases. It would also show a reproducible response over the required life-time and be economically viable.

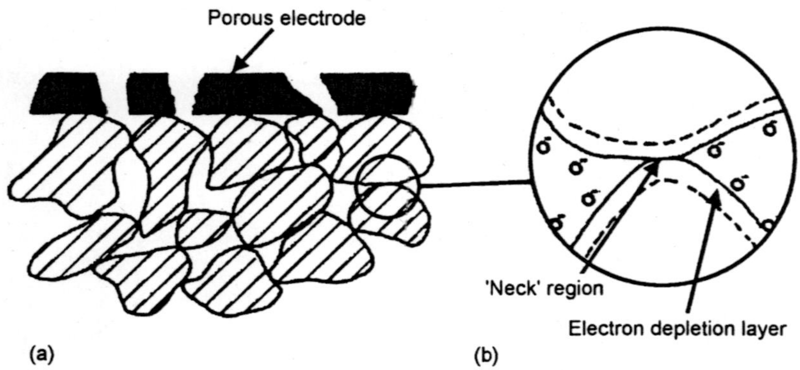

Fig. 4.46 (a) Schematic diagram of a section through a porous SnO_2 compact with a porous electrode; (b) detail of the 'neck' region between the grains showing the effect of adsorbed oxygen.

The sensitivity of a polycrystalline gas-sensing ceramic depends upon sensor composition, ceramic particle size (since this determines reactive surface area, and particularly its relationship to L_D), sensor geometry, temperature (which strongly affects adsorption processes and oxide conductivity) and accessibility of the sensor bulk material (i.e. permeability) to the gas or gases.

It would be inappropriate here to consider the effects of these variables in detail since optimising sensor performance is still essentially a matter of 'technical development' involving not only them but also the sensor housing and associated circuitry. Optimisation must, of course, be achieved within the inevitable cost constraints.

The sensor compositions usually include noble metals, particularly palladium and platinum, since they are found to increase sensitivity through catalysing the reaction between the oxygen adsorbate and reducing gases. In this role the additive is said to be acting as a 'chemical sensitizer'.

The noble metal might also act in a quite different way, as an 'electron sensitizer'. Because of its very fine particle size ($<10\,nm$) relative to the typical SnO_2 grain size there are many metal/semiconductor contact points. Also, because of the high specific surface area ($\sim 50\,m^2\,g^{-1}$) of the particulate metal it, too, presents a large adsorbent surface available to oxygen species, the adsorption process removing electrons from the metal. Because the work function of the metal (5.12 eV) is larger than the electron affinity of the semiconductor (4.45 eV) electrons will flow out of the semiconductor into the metal, increasing the depletion layer thickness. This is very likely an over-simplification of what is a complex charge transfer process, and the role of Pt in influencing the sensitivity to CO of Sb-doped SnO_2 is addressed in detail by L. Morris and D.E. Williams [29].

'Selectivity' is mainly achieved by 'trial and error'. It can be engineered through specific dopants to the tin oxide, for example Fe_2O_3 to sensitize for NO_2 or V_2O_5 and Pd to sensitize for CO and other combustible gases, as illustrated below. It can also be engineered by interposing selective diffusion barriers between the sensor and the ambient.

'Sensitivity' is a strongly microstructure-dependent property, the depletion layer thickness relative to the particle size of the sensing ceramic being the particularly important parameter.

It seems likely from Fig. 4.46 that the greatest sensitivity would be achieved when the particle size is close to twice the depletion layer thickness. Sensor ceramics therefore have a particle size typically well into the sub-micron range, and often in the range 10–100 nm. Figure 4.47 demonstrates the dependence of sensitivity on grain size, the onset of the steep rise at around 10 nm being associated with the grain size approximating to the depletion layer thickness.

The now classic 'Taguchi' sensor design, dating from the 1960s, is shown in Fig. 4.48(a) and a present day format in Fig. 4.48(b).

The planar format sensor is most likely to be fabricated by screen-printing, the sensing layer being typically 20–30 μm thick. The doped tin oxide can be prepared chemically, for example by precipitating $Sn(OH)_4$, together with hydroxides of the dopants, by passing NH_3 through an aqueous solution of $SnCl_4$ together with dopant chlorides. The grain size will depend on the subsequent calcining temperature and time, typically 800 °C for 1 h.

The screen-printing paste is made up in the manner described in Section 4.2.2 and the patterns printed. The heater element and its leads would typically be platinum and the interdigitated electrodes gold. As illustrated below (Fig. 4.49) the heater might also be a composition based on ruthenium oxide (see Section 4.2.2). The contacting tabs are also screen-printed gold.

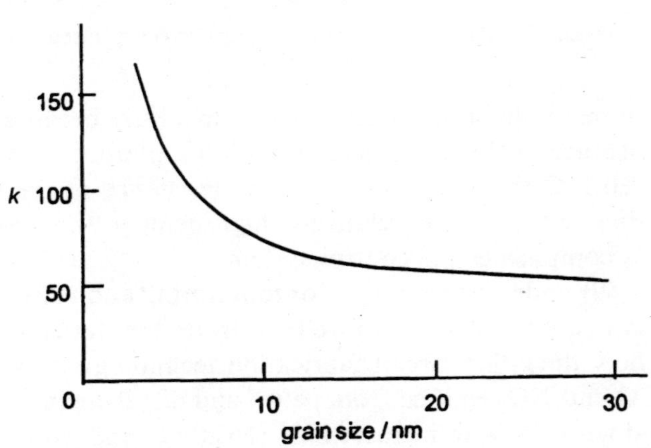

Fig. 4.47 Dependence on grain size of the sensitivity (k) of a tin oxide sensor to 100 p.p.m. hydrogen in air. Adapted from [28]. (k is defined as the sensor resistance measured in air to that measured in the hydrogen-containing air.)

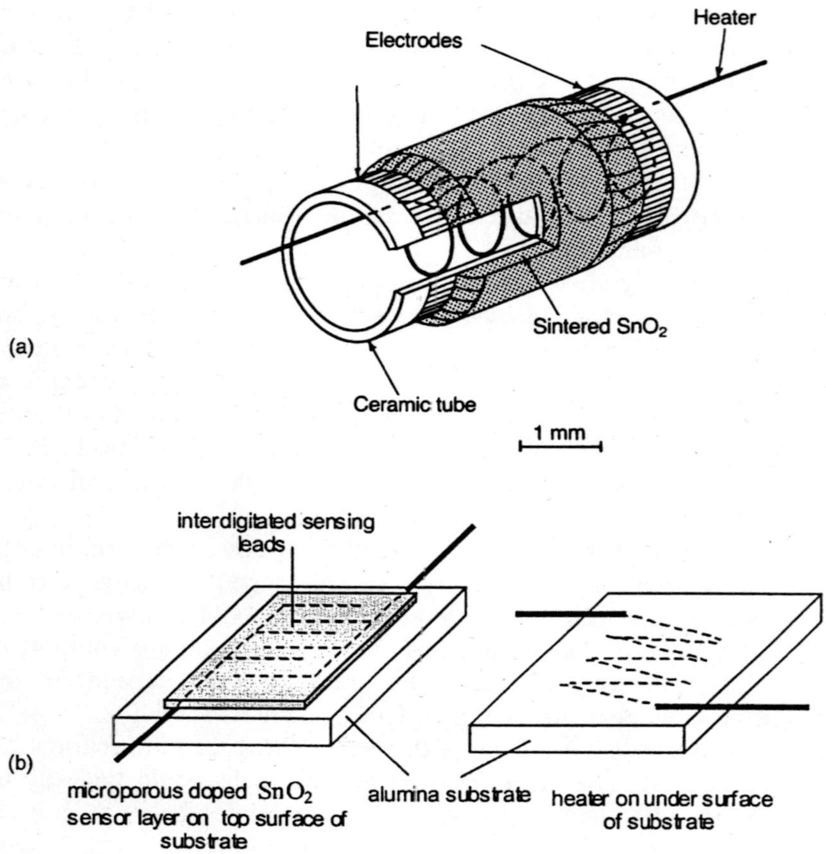

Fig. 4.48 (a) The classic 'Taguchi sensor'; (b) the typical screen-printed planar format.

It is necessary to maintain close control over temperature because of its effect upon the base resistance of the semiconductor. With a platinum heater this can be effected to $\sim \pm 0.1\,°C$ through monitoring the resistance of the heater itself using a Wheatstone bridge. The relatively high temperature coefficient of resistivity of RuO_2 complicates the control system.

A sensor (Fig. 4.49) under development for monitoring and activating control over the air quality in automobile cabins well illustrates the status of present-day technology, and how thick-film circuit fabrication technology is exploited. The target gases are CO and NO_2 emitted from petrol and diesel engines respectively. The SnO_2 is doped with Fe_2O_3 to increase NO_2 sensitivity and with V_2O_5 and Pd to enhance sensitivity to CO. The heater pad is RuO_2. The two sensing areas have a covering of porous silica to maintain cleanliness. The measured sensitivity to NO_2 is 0.5 p.p.m. and to CO 30 p.p.m., and the long-term stability is good.

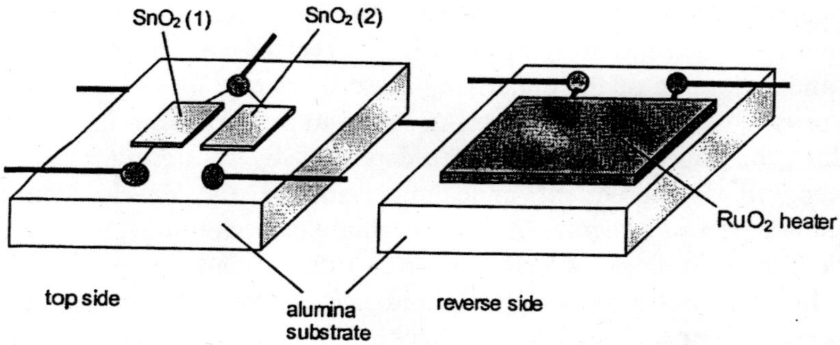

Fig. 4.49 Twin tin oxide-based sensors for monitoring NO_2 and CO in automobile passenger compartments. NB: alumina substrate size approximately 2×2.6×0.254 mm. (Adapted from Oto, K. *et al.* (2001), Sensors and Actuators, **B77**, 525–8.)

There is a wide variety of chemical sensor systems developed for specificity to certain gases and vapours. For example WO_3-based sensors for O_3 and NH_3, In_2O_3/SnO_3 for NO and SnO_2/Al_2O_3 for H_2S. A comprehensive list is given in [28].

Apart from chemical factors, the detailed physical form of the sensor can have an important influence on selectivity. It is apparent from Fig. 4.48(b) that the target gas has to diffuse through the microporous sensor material towards the sensing electrodes. The gas flow will depend upon the mean free path of the gas molecules in relation to the diameter of the channels in the solid. The mean free path of gas molecules usually present in a functioning sensor will be of the same size order as the channel sizes making the gas interdiffusion process a difficult one to model mathematically. There will certainly be gas concentration gradients across the sensor thickness, and different gradients for different gases. A graded microporosity, readily accomplished via screen-printing is a route to influencing selectivity.

The effective thickness of sensing material overlying the electrodes will depend upon the electrode spacing, the electric field between the electrodes penetrating more deeply into the sensor layer the more widely spaced the electrodes. Electrode spacing is, therefore, an important design parameter which can influence selectivity. Indeed, varying the electrode spacing can be used as a research tool to probe the processes ongoing through the depth of the sensor layer.

Water vapour is an inadvertent and very variable 'contaminant' in most sensing situations. The operating temperature of the sensor eliminates the risk of the involvement of water molecules at the semiconductor surface but not that of hydroxyl groups. In the case of the SnO_2-based sensor it seems that OH^- behaves in much the same way as O^- so that the effect of the presence of water vapour can be large. The effects in the case of the Cr_2O_3/TiO_2 (p-type) sensor system are far less marked, for reasons yet to be elucidated [30].

It is apparent that with so many variables the approach to designing effective sensors has to be essentially empirical but, nevertheless, one founded on a working understanding of the underlying basic sciences.

At the research level there is interest in thin film sensors. Thin films, typically 100 nm thick, of doped tin oxide can be deposited by sputtering, evaporation, sol–gel, etc. when subsequent annealing leads to the development of a nanocrystalline porous structure. Although 'thin' film technology, in comparison with thick film technology, would appear to offer the advantage of relatively lower production costs because of the avoidance of powder and screen-printing paste processing steps, there are drawbacks. Thin films offer a much smaller reactive area than sintered thick film ceramics and so the sensitivity of sensors fabricated from them is correspondingly lower; for the same reason they are more susceptible to surface contamination.

The established and potential applications of ceramic gas sensors are many and varied. Principal among them is CO sensing because of the highly toxic nature of the gas. It is also the gas given off in the early stages of a fire and so its detection can give warning of an impending fire in, for example, computer areas and television sets, and of smouldering cables in electrical equipment generally.

The installation of gas sensors to warn of gas leaks and the attendant risk of explosion is increasing in the developed world, especially in situations where bottled gas is stored, for example in caravans, trailers and boats. In the UK it is becoming commonplace to install them in homes along with smoke alarms.

By responding to fumes in cooking areas, car parks, laboratories and similar places, gas sensors can be used to control ventilation fans. In industrial situations the sensor can monitor concentrations of carbon monoxide, ammonia, solvent vapours, hydrocarbon gases, ozone etc., and because of the growing consciousness of environmental pollution and the safety aspects of industrial processes the applications will multiply.

Gas sensors are being employed to characterise the 'noses' of perfumes, fruits, wines and other foodstuffs and to measure the effectiveness of deodorants. The range of sensor types and applications is expanding at a remarkable rate with the term 'electronic noses' introduced to describe the technology.

Semiconducting lambda-sensors based on titania have been developed and used for car engine management. However for a variety of reasons, one being sensitivity to surface contamination, the electrolytic sensor dominates.

A variety of commercial gas sensors is illustrated in Fig. 4.50.

4.6.3 Humidity sensors

Almost all ceramics that contain interconnected porosity exhibit a fall in resistivity when exposed to humid atmospheres at normal ambient temperatures. Water is adsorbed from low-humidity atmospheres by most oxide surfaces when it

Fig. 4.50 A selection of commercial gas sensors. The alumina tile carrying the sensor is approximately 3×3 mm^2. An activated charcoal filter may be mounted inside the top of the housing. (Courtesy of City Technology Ltd.)

dissociates, with OH^- attaching itself to a surface cation and the hydroxonium ion H_3O^+ attaching itself to a surface O^{2-} ion. Movement of the ions to alternative sites under the influence of a field gives an initial reduction in resistivity. With increased humidity further layers of water are adsorbed and the high electrostatic fields resulting from the surface states lead to a far greater degree of dissociation than in liquid water and therefore to a further reduction in resistivity. At high humidities liquid water may condense in the pores as a result of the curvature of the liquid surface induced in the confined space by surface tension.

The formation of liquid depends on pore diameter, surface tension and the contact angle between the solid, liquid and gas phases. The last factor is difficult to estimate in general but is likely to be close to zero between water, humid air and a surface consisting of adsorbed water. It can be deduced that, when the contact angle is zero, the vapour pressure p in equilibrium with a concave liquid meniscus relative to the vapour pressure p_0 in equilibrium with a flat surface is given by

$$\ln\left(\frac{p_0}{p}\right) = \frac{2\gamma V_m}{R_0 T r} \tag{4.51}$$

where γ is the surface tension, V_m is the molar volume of the liquid, R_0 is the gas constant, T is the temperature and r is the radius of curvature of the meniscus. Since the humidity h is $100p/p_0$ and the diameter d of a pore in which water will condense is given by $d = 2r$, the relation between pore diameter and humidity leading to condensation is

$$d = \frac{4\gamma V_m}{R_0 T \ln(100/h)} \tag{4.52}$$

If $\gamma = 0.073\,\mathrm{N\,m^{-1}}$, $V_m = 0.018 \times 10^{-3}\,\mathrm{m^3}$, $T = 293\,\mathrm{K}$ and $R = 8.315\,\mathrm{J\,K^{-1}\,mol^{-1}}$, d is approximately 3 nm when $h = 50\%$ and approximately 42 nm when $h = 95\%$. It is therefore evident that condensation will occur only at the narrowest points in the microstructure, i.e. in the 'neck' regions.

The speed of response of a humidity element will depend on the rate of diffusion of the ambient atmosphere into the pores of the sensing element. The element must therefore be thin in section and may contain a system of larger pores to reduce the length of the finer pore pathways into which gas must penetrate. Electrodes covering major surfaces must be porous.

The response of the surfaces of the element to moisture must remain constant, and in practice this is one of the most difficult conditions to meet. Over a period of time the adsorbed molecules are thermally activated to the lowest-energy sites and the response to a change in humidity may then be either sluggish or permanently altered. Furthermore, molecules other than water may be adsorbed increasingly with the passage of time, modifying the response to moisture. One of the most effective solutions to this problem is periodically to heat the sensor to a temperature of $\sim 500\,°\mathrm{C}$ so that the surface is restored to a reproducible state. One of the advantages of ceramic over organic elements is that they can be repeatedly reconditioned in this way.

A sensor developed to detect the moisture content of the atmosphere in a microwave oven and so indicate the onset of the cooking process serves to illustrate the principles. The sensing element consists of a solid solution of TiO_2 in $MgCr_2O_4$ with a minor phase of $MgTi_2O_5$. It is formed by mixing the constituent oxides, pressing them into compacts 0.25 mm thick at the relatively low pressure of about 70 MPa (about 10^4 p.s.i.) and firing for several hours at about 1350 °C to yield a sintered body with, typically, 35% porosity. The electrodes are RuO_2 with the minimum amount of added glaze for adhesion so that the layer is porous.

$MgCr_2O_4$ is a normal spinel with Mg^{2+} ions on tetrahedral sites and Cr^{3+} ions on octahedral sites. 30 mol.% TiO_2 can be incorporated without losing the spinel structure. It is thought that Ti^{4+} and Mg^{2+} ions occupy neighbouring octahedral sites, and the excess charge is compensated by the formation of Cr^{2+} ions which occupy tetrahedral sites. It is a p-type semiconductor, although this does not appear to be relevant to its behaviour as a humidity sensor, in contrast with its behaviour as a gas sensor discussed above. Typical resistance–humidity characteristics are shown in Fig. 4.51. It is believed that the high charge on the Cr^{3+} ions exposed in the pore surfaces promotes the adsorption and dissociation of water molecules according to

$$2H_2O \rightleftharpoons H_3O^+ + OH^- \tag{4.53}$$

leading to the conduction mechanism outlined above. The resistance increases irreversibly when the element is exposed to high humidities for prolonged

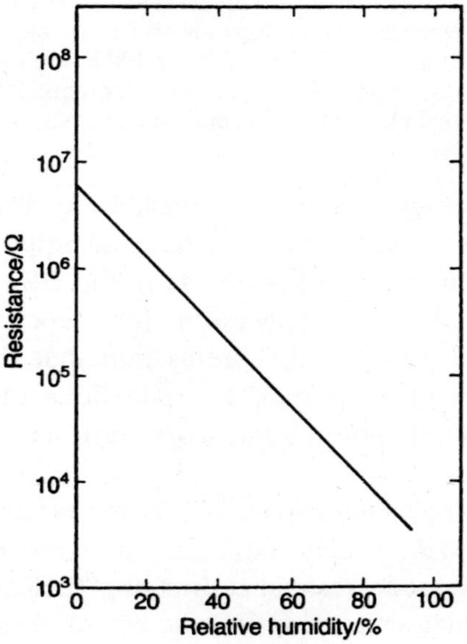

Fig. 4.51 Resistance versus humidity for $MgCr_2O_4 + 30\,mol.\%\ TiO_2$.

periods. This may be due to reduced dissociation in the adsorbed layer as the ions reach lower energy sites.

In practice the sensor is a plate 4-mm square mounted within a heater coil, and its characteristic is periodically restored to the standard state by heating to 450 °C for a few seconds.

4.7 High Transition Temperature Superconductors

4.7.1 Overview

The following extract from a paper [31] published in 1968 makes interesting reading in the light of subsequent developments.

> Industrial interest in a physical phenomenon lies in a 'quality' or 'outstanding feature' of that phenomenon. And in fact, the absence of electrical resistance unlocks a wealth of new possibilities. Although the highest transition temperature so far found is still as low as 20.05 K, the possibility of using underground cooled superconducting cables for electricity supply in certain densely populated areas such as London is already being seriously examined to see what advantages it might offer. There is also interest in the use of superconducting elements for logic circuits ('cryotrons'). Some very interesting possible uses include superconducting magnet coils which can be used to give very high magnetic fields, sometimes with flux densities greater than 10 Wb/m^2 (100 000 gauss) over volumes large enough to

> be of interest in technical applications. One wonders how many other even more impressive changes in electrical technology are to be expected if superconducting materials with a transition temperature of 50 or 100 K are ever produced . . .
>
> Nevertheless, it is to be expected that, even if superconductors with transition temperatures well above 20 K are never found, superconductivity will still find its applications in technology.

Since then there has been the totally unexpected breakthrough with the superconducting ceramics operating comfortably at liquid nitrogen temperature (77 K), and stimulating intensive R&D efforts in the US, Japan and Europe.

Probably the most obvious application for superconductors is in the transmission of electrical power and it seems that that this is close to being a reality. There are many other potential applications in the power electrical engineering field, for example power generators, motors, transformers, leads and fault current-limiters.

An important and established application is for electromagnets for nuclear magnetic resonance (NMR) widely exploited as a research tool, and for its medical counterpart, magnetic resonance imaging (MRI), a powerful medical diagnostic tool. Strong magnetic fields are also essential for high-energy particle accelerators for nuclear research.

Until superconducting magnets became available the generation of magnetic fields of $10\,\mathrm{MA\,m^{-1}}$ (approximately 12 T in air) involved power dissipation of order 1 MW in the windings. Not only did such a magnet require a large power source (a large power station generator delivers about 600 MW) but most is dissipated as heat which had to be removed by coolants. Superconducting magnets exploiting Nb–Ti and other intermetallic compounds have been used for many years, and it is now commonplace to produce fields of 12 T in quite compact and easy to run electromagnets.

Exploiting the magnetic fields generated by superconducting coils, and also those resulting from trapped magnetic induction within a superconducting bulk material, is a very active R&D field. Frictionless bearings and magnetic levitation for transport systems are among the many objectives.

For 'superconducting electronics' interest is focussed on materials in thin film form, the major objectives being to achieve faster computers and improved communications systems. Signal transit times are reduced by using superconducting interconnects, and the Josephson junction is the basis of a bistable logic device. In communications technology superconducting filters and antennae offer significantly improved selectivity and sensitivity.

4.7.2 The phenomenon of superconductivity

The following summary of the essentials of the phenomenon and theory of superconductivity serves to introduce terminology and permits the 'high transition temperatures (T_c) superconductors' (HTSs) to be seen in a historical

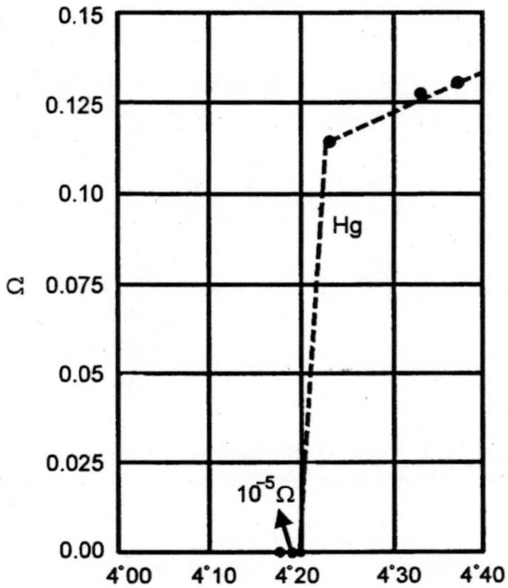

Fig. 4.52 Resistance of mercury versus temperature. (After H. Kammerlingh Onnes (1911) *Akad. van Wetenschappen, Amsterdam*, **14**, 113, 818. Taken from *Introduction to Solid State Physics*, C. Kittel, John Wiley, Chichester, 1986, p. 318.)

perspective. The texts by G. Vidali [32] and V.Z. Kresin and S.A. Wolf [33] are recommended for supplementary reading.

The history of superconductivity has its origins in the experiments of H. Kamerlingh Onnes who in 1908 at the University of Leiden succeeded in liquefying helium and reaching temperatures down to 1 K. In 1911, while pursuing the study of the electrical resistivity of metals at these very low temperatures, he discovered superconductivity in mercury. As the temperature was reduced the resistivity fell sharply at about 4 K and at the T_c, fell to zero as shown in Fig. 4.52. A current induced in a superconducting loop circuit has been observed to persist without decay for over 2 years, setting an upper limit on the resistivity of $10^{-28}\,\Omega$ m; for practical purposes this is zero resistivity.

Since 1911 superconductivity has been observed in many metallic elements and intermetallics, principal among them being tin (3.7 K), tantalum (4.5 K), lead (7.2 K), niobium (9.3 K), Nb_3Sn (18 K) and Nb_3Ge (20.9 K).

There are two types of superconductor, Type I and Type II, characterized by the way in which they respond to an applied magnetic field.

If a Type I superconductor such as lead is placed in a small magnetic field (e.g. a few mT) and cooled, then at T_c the magnetic field is expelled from the interior of the specimen. This is the Meissner effect, which is fundamental to the superconducting state; it is not simply characteristic of a material which happens to be a (fictitious) perfect conductor. The total absence of an electric field in a

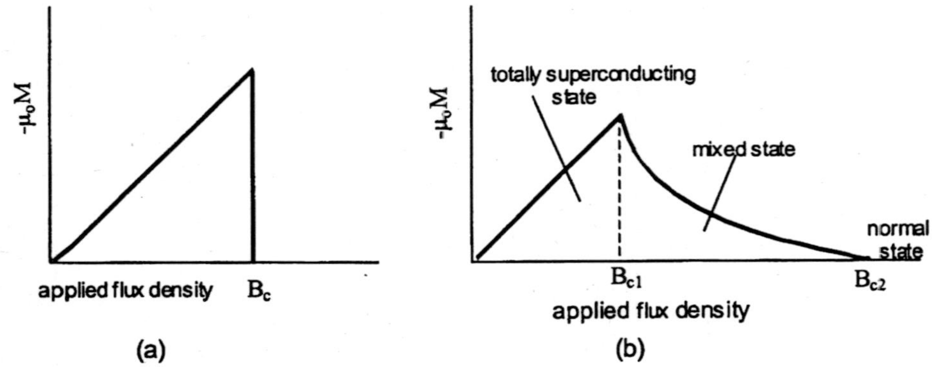

Fig. 4.53 The induced magnetic moment ($\mu_0 M$) as a function of applied field for (a) a Type I superconductor and (b) a Type II superconductor.

material having zero resistivity only demands that the magnetic induction (B) is constant, not necessarily zero.

Because in the presence of an applied magnetic field the interior of the superconductor has *zero* magnetic induction it can formally be described as a perfect diamagnet (see Section 9.1.5). The 'diamagnetic effect' is of course not due to a distribution of diamagnetism throughout the volume, but arises because of 'screening currents' which flow in the very near surface region of the superconductor. The externally applied magnetic field penetrates the superconductor to a small depth λ, characteristic of the particular superconductor.

Below T_c the applied magnetic field can be increased up to a critical value B_c, when the superconductor changes to the normal state. This behaviour is characteristic of a Type I superconductor.

In the case of a Type II superconductor below its T_c, the applied magnetic induction is totally excluded until a critical value B_{c1} is reached when some flux penetration occurs through filaments known as *flux vortices*. The filaments are of normal, that is non-superconducting material, and carry a magnetic flux. The flux is quantized, the single quantum, a *fluxon*, being of magnitude $h/2e = 2.0678 \times 10^{-15}\,\mathrm{T\,m^2}$. The filament itself is surrounded by screening currents as a magnetic field along the axis of a solenoid is surrounded by circular currents (see Section 9.1.2). The material is therefore in a 'mixed', superconducting/normal state. As the applied flux density is increased so the penetration increases until at an applied flux density B_{c2} the material as a whole becomes normal.

The different behaviours are illustrated in Fig. 4.53(a) and (b). Some of the alloys, for example Nb_3Sn (18 K), and all the high temperature superconductors are Type II.

Because useful currents carried by a Type I superconductor generate magnetic fields it follows that there are also critical current densities J_c, corresponding to the critical applied magnetic fields. In a Type II superconductor the situation is complicated by the extent of flux 'pinning' discussed in Section 4.7.4 below.

These are the essential phenomenological characteristics of superconductivity. The first significant step in a theoretical interpretation was taken in 1950 when H. Fröhlich and J. Bardeen showed that electron–phonon interactions are capable of coupling two electrons together as if there is a direct attractive force between them.

In 1956 L.N. Cooper published a theoretical analysis describing the addition of a single electron pair to a normal metal at 0 K. This showed that if there is an attraction between the two electrons, however weak, their combined energy is always less than that of normal conduction electrons in their highest energy state, the Fermi energy. In 1957 J. Bardeen, L.N. Cooper and J.R. Schrieffer extended Cooper's treatment to a large population of interacting electrons and showed that they would form an assembly of so-called 'Cooper pairs'. This has become known as the 'BCS theory'.

Cooper pairs move in the 'classic' superconductors in such a way that at equilibrium their combined momenta is unchanged; whatever momentum change one of the pair suffers due to interaction with a phonon, the change produced in the partner is equal and opposite. Therefore phonon scattering has no net effect on the forward momentum of the electron population accelerating in an electric field. Because electron–phonon interactions are also responsible for electrical resistance at normal temperatures, it is not surprising that relatively stable Cooper pairs are found in poorly conducting metals with corresponding relatively high values of T_c. The energy required to split up a Cooper pair at 0 K is of the order of kT_c (actually $1.8\,kT_c$).

Fröhlich also predicted the so-called 'isotope effect' which associates a higher T_c with light isotopes of a metal than with heavier isotopes. This seems reasonable since the lighter the atom the more easily it is moved and therefore able to couple to the electron motions. The BCS theory explains the isotope effect in terms of phonon frequencies.

Although the BCS theory offers an explanation for the behaviour of superconductors known prior to 1986 it is generally not applicable to the high temperature ceramic superconductors. In fact the BCS theory predicts a maximum T_c of approximately 30 K and this has now been comfortably passed. It is surprising that, despite intensive efforts over more than 15 years, there is still no generally accepted theory to explain the mechanism of high-temperature superconductivity. However, it is believed that whatever theory is finally accepted it will involve paired electrons, although the pairing mechanism may well differ from that invoked in the BCS theory.

W.L. Ginzburg and L. Landau developed a phenomenological theory, the 'GL theory', which makes no assumptions concerning the basic mechanism responsible for superconductivity. In this respect the approach is similar to the Devonshire phenomenological theory as applied to ferroelectrics (see Section

2.7.1). The GL theory produces useful rationalizations concerning superconductors in general. One of these concerns the 'coherence length' ξ, a characteristic of a superconductor and a measure of the distance superconductor properties extend into a normal material. The coherence length is related to T_c, the smaller ξ the higher T_c. As examples, for Pb ($T_c = 7.2\,K$) $\xi = 87\,nm$ and for the high temperature superconductor, $Bi_2Sr_2CaCu_2O_x$ ($T_c = 89\,K$) $\xi = 1.8\,nm$. The very small ξ-values for the ceramic superconductors have significant implications as far as their processing is concerned.

The relative magnitudes of λ and ξ determine whether a superconductor is Type I or II. If $\lambda > \xi$ the superconductor is Type II.

Understanding and controlling the behaviour of the 'vortices' which develop as an applied magnetic flux penetrates Type II superconductors is critical to their successful exploitation. If the superconductor is carrying a current, then the vortices experience the expected electromagnetic Lorentz interaction force and move. In fact, under an applied magnetic field, vortices are continually entering the material, moving through it and leaving. This motion of a magnetic vortex induces an e.m.f. which leads to an effective electrical resistance and energy dissipation; in the specialist literature this resistive effect is referred to as a 'viscous drag' opposing the vortex movement. If the process continued and the temperature exceeded T_c the material would revert to the normal state. Therefore basic research is focused on learning more concerning how the vortex population is arranged and how it moves. In particular, materials' engineers seek means of preventing its movement, that is by 'pinning' at irradiation-induced defects, dislocations and microstructural defects, including grain boundaries, second phases, etc.

Preventing vortex motion by 'pinning' is also critical to the development of the HTS magnets discussed below.

4.7.3 Ceramic high-T_c superconductors (HTSs)

Access to the *Handbook of Superconducting Materials* edited by D.A. Cardwell and D.S. Ginley [34] and the *Handbook of Superconductivity* edited by C.P. Poole Jr [35] is recommended. The review by R.J. Cava [36], focused on the structures of the high-T_c superconductors, should also be consulted.

A significant step in the history of the HTSs was the discovery in 1966 of superconductivity in the oxygen-deficient perovskite $SrTiO_{3-\delta}$, containing some barium or calcium substituted for strontium. Although the T_c value was very low (0.55 K), in retrospect it can be seen as the first superconducting ceramic. In 1979 a T_c of approximately 13 K was discovered for $BaPb_{0.75}Bi_{0.25}O_3$, which also has the perovskite structure.

The breakthrough came in 1986 when J.G. Bednorz and K.A. Müller [37] reported superconductivity with $T_c = 35\,K$ in the semiconducting ceramic,

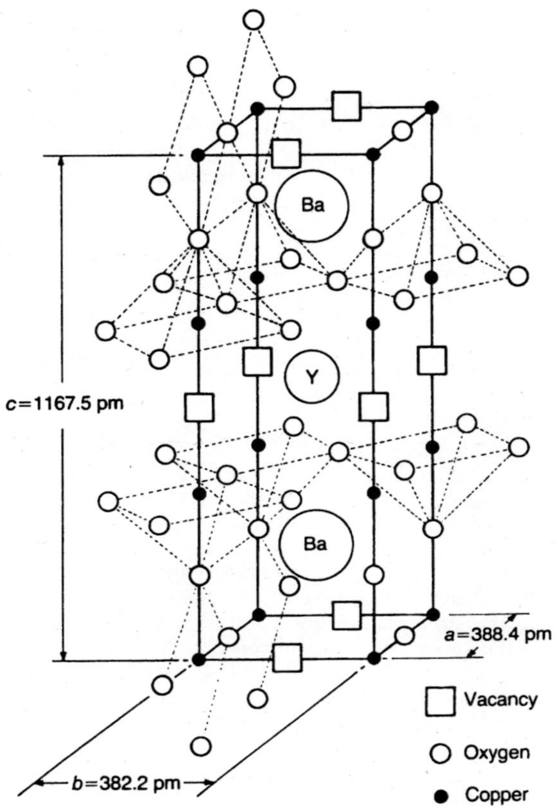

Fig. 4.54 The structure of $YBa_2Cu_3O_{7-\delta}$.

$(La,Ba)_2CuO_4$, which has the layer perovskite K_2NiF_4 structure. The discovery won them a Nobel prize.

This discovery stimulated what can only be described as frenzied research in many laboratories throughout the world which quickly resulted in the discovery of $YBa_2Cu_3O_{7-\delta}$, the so-called '123' phase, or YBCO, with a transition at a temperature as high as 93 K. There are also bismuth-based high-T_c superconductors, which are described below.

The structure of YBCO is closely related to that of perovskite as evident in Fig. 4.54, which shows the unit cell consisting of three perovskite-type cubes with Cu ions at the corners and O ions at the centres of the cube edges. The top and bottom cubes contain Ba and the centre cube Y.

The practical significance of the discovery, apart from the fact that it suggested the possibility of a room temperature superconductor, is that the T_c is above the boiling point of liquid nitrogen (~77 K), a cheap refrigerant.

The remarkable progress stimulated by the work of Bednorz and Müller is illustrated in Fig 4.55.

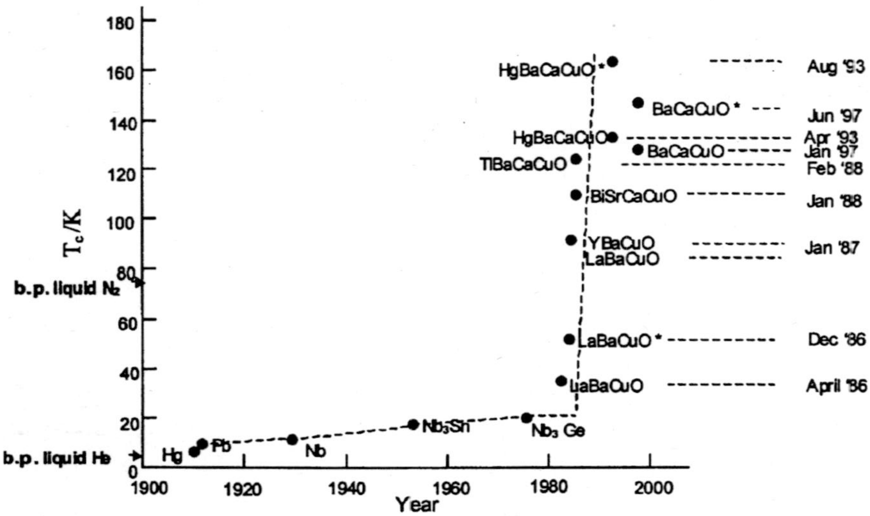

Fig. 4.55 T_c against time illustrating the remarkable development following the discovery in 1986 of the high temperature superconductors. (*under pressure) (Adapted from Chu, C.W. (2000) 'High temperature superconducting materials: present status, future challenges, and one recent example – the superconducting ferromagnet', *Physica C*, **341–348**, 25–30.)

The presence of Cu–O planes is a critically important structural feature of the HTSs. Two such planes are evident in Fig. 4.54 separated by a plane containing Y ions and oxygen vacancies. It is accepted that the Cu–O planes are a requirement for high temperature superconductivity in this class of ceramic. The structure of the prototype HTS, $Bi_2Sr_2CaCu_2O_x$ (designated 'Bi-2212') has two such adjacent planes and a T_c of 80 K, and a similar compound, $Bi_2Sr_2Ca_2Cu_3O_x$ (Bi-2223) has three sets and a T_c of 110 K. There are also thallium-based analogues e.g. $Tl_2Ca_2Ba_2Cu_2O_x$ with a T_c of 125 K, but thallium is toxic.

A compound which held out great promise has the precise and surprising composition, $Ca_{0.84}Sr_{0.16}CuO_2$. It consists entirely of Cu–O planes separated by a layer of the alkaline earth elements, randomly arranged but in a fixed ratio. In the light of the experience outlined above, the fact that the compound consists entirely of Cu–O planes pointed to exceptionally high T_c values, if it could be suitably compositionally modified, and attempts to achieve this were undertaken. The task proved difficult until synthesis at high pressure and high temperature (typically 50 kbar; 1000 °C) was tried. A modified composition was prepared and, although not of great significance in itself, the work stimulated a strong interest in high pressure synthesis, an interest which has continued.

The record for the highest T_c is held by pressure-formed $HgBa_2Ca_2Cu_3O_9$ with 134 K measured at normal pressure and 166 K under high pressure.

The perovskite structure is, of course, of special significance in the electroceramics context since the ferroelectric perovskites are dominant in the ceramic capacitor, PTC thermistor and electromechanical transducer industries. The structure favours the existence of soft modes (low frequency phonons) as evidenced by its tendency to instability, for example the ferroelectric–paraelectric transition. Instability is evident in the case of the '123' compound which exhibits a tetragonal-orthorhombic transition in the region of 700 °C (the exact temperature depends on the oxygen content). Extensive twinning, very reminiscent of ferroelectric domain structures, is observed.

4.7.4 The properties, processing and applications of HTSs

Reference to the texts by F.C. Moon [38] and that edited by H. Maeda and K. Togano [39] is recommended. The Proceedings issue [40] covers the present status of the materials, processing and applications and the article by D. Larbalestier *et al.* [3, pp. 368–77] reviews HTSs for power applications.

Important characteristics

Efforts directed to develop HTSs to commercial products are confined to the RE–123 compounds, where RE = rare earth element, Y, Nd, Sm, etc., and the Bi-based materials. As far as the former type is concerned the discussion is confined to YBCO, the most prominent of the RE–123 family. YBCO has the advantage of a higher J_c value at 77 K than the Bi-based materials ($\sim 10^{10}$ A m^{-2} compared to $\sim 10^9$ A m^{-2}) although the comparison is not an easy one to make since the values are so dependent on specimen form, doping, applied field, etc. More importantly, the Bi-based powders have a flat, clay-like morphology and this is very significant when it comes to fabrication. In contrast YBCO particles tend to be equiaxed.

For the reasons summarized below, the coherence length ξ, and irreversibility field H_{irr} are two parameters of special significance in the processing of HTSs.

The coherence length (ξ) To attain one of the principal goals, namely to achieve maximum values of J_c and B_{c2}, the spacing of features which are not intrinsically superconducting, such as second phases and grain boundaries should be $\gg \xi$.

The grain boundaries constitute barriers to the supercurrent flow and their width should, ideally, be less than ξ. Values of ξ for the high-T_c superconductors lie in the range 1–100 nm (depending upon applied field strength) and so in their processing close attention has to be given to optimising microstructure and texture.

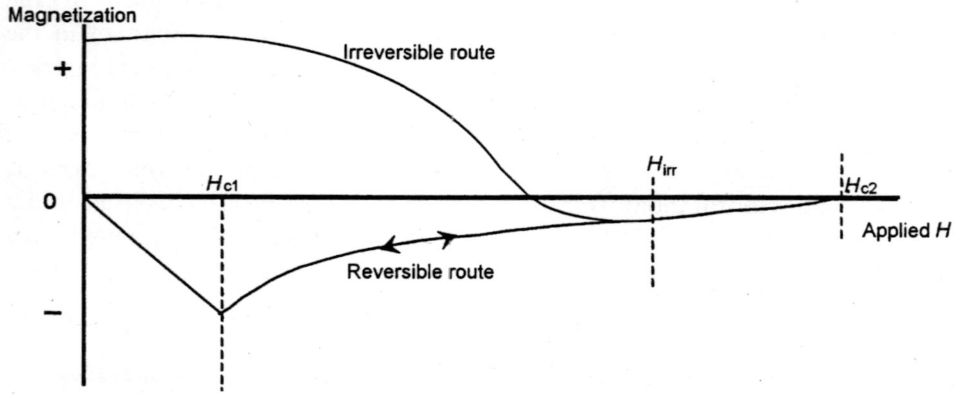

Fig. 4.56 The reversible and irreversible routes taken by the magnetization of a Type II superconductor as the applied magnetic field is reduced from a value larger than H_{c2}.

The irreversibility field (H_{irr}) Figure 4.56 illustrates the events as a magnetic field is applied to and then removed from a Type II superconductor below its critical temperature. At applied fields up to H_{c1} the response of the material is as expected for a Type I material. That is the field is excluded (the Meissner effect) and the material behaves as a perfect diamagnetic. At applied fields above H_{c1} there is some flux penetration and this increases until the material becomes normal at H_{c2}.

As the applied field is removed the route the M–H curve takes depends upon the extent to which the vortices are 'pinned'. With no pinning the magnetization remains negative and fully reversible, that is in the opposite sense to the applied field.

If the vortex population is pinned so that it is not in equilibrium with the applied field, then the M–H curve takes the irreversible route as illustrated. The pinned vortices contribute an M in the same sense as the applied H since they *are* the penetrated field, and as the applied field is reduced the +M manifests itself just as it does for an ordinary permanent magnet. The pinned field is aptly described as the 'trapped field'.

The irreversibility field (H_{irr}), which has to be measured experimentally, is an important parameter since it determines the size of the final M developed. As expected H_{irr} depends upon temperature, and irreversibility field v. temperature curves are shown in Fig. 4.57. It is apparent that at liquid nitrogen temperature YBCO is suited to permanent magnet applications whereas Bi-2223 is not. Because of the energy dissipation accompanying vortex motion and referred to earlier, vortex pinning also raises J_c.

In summary, under the usual operating conditions the ceramic HTSs are characterized by the presence of flux vortices and materials engineers are

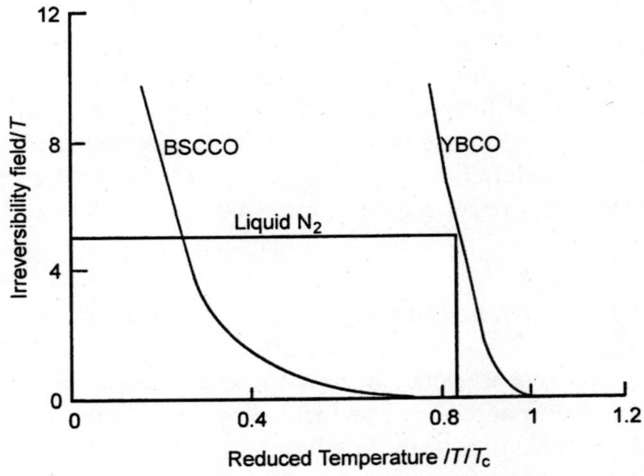

Fig. 4.57 Irreversibility curves for candidate HTS ceramics: at liquid N_2 temperature T/T_c (YBCO) $\sim$0.83 and for Bi-2223 $\sim$0.79. (After D.A. Cardwell [41].)

learning how to 'pin' them. This is necessary because in current-carrying applications their movement is accompanied by energy dissipation, and in the production of quasi permanent magnets they constitute the 'trapped' magnetic flux, so determining the strength of the magnet.

Processing high-T_c superconductors

Intensive efforts are being made to exploit bulk and tape HTSs for a variety of applications including magnets for magnetic levitation ('maglev') and for energy storage, current leads and fault current-limiters and cables for power generation, control and transmission.

As discussed in Chapter 3, the fabrication of bulk oxide ceramics usually starts with a powder and this can be prepared via solid state reaction (also loosely referred to as the 'mixed oxide' route), or by 'wet chemical' routes.

As an example, the preparation of Bi-2223 compound starts with Bi_2O_3, PbO, $SrCO_3$, $CaCO_3$ and CuO powders which are well mixed, in a typical case in acetone in a mill using zirconia media. The mix is calcined at 800–900 °C and the calcine ground and granulated. To avoid the occurrence of carbon residues in the grain boundaries, which are said to have a deleterious effect upon J_c and T_c, certain of the processing steps are carried out in pure oxygen or *in vacuo* [42].

There are a variety of routes described as 'wet chemical'. For example an acid (HNO_3) solution of the metal nitrates is prepared and the oxalates precipitated by the addition of oxalic acid and ammonia, to control pH. Because of mixing on an atomic scale the reaction to form the oxides can be carried out at a relatively low temperature of about 700 °C.

'Spray-drying/calcination' and 'spray pyrolysis' would typically start with a solution of nitrates of the metals. In the former case, the solution is sprayed into a warm (~250 °C) enclosure where the granules are collected and subsequently calcined and, in the latter, into an enclosure at a temperature sufficiently high so as to simultaneously effect the decomposition and reaction to form the oxides.

Leads and 'fault current-limiters' Current leads and fault current-limiters are the first large scale commercial applications of Bi-based HTSs. Current leads significantly improve efficiency of liquid helium-cooled magnets.

In principle, components can be formed by any standard ceramic fabrication route. The densification step involves a partial melting and a controlled recrystallization, and attention has to be given to sintering/annealing atmospheres to achieve the required stoichiometry. Leads have also been formed by melt-casting.

Because the J_c in the *a–b* plane of Bi-HTSs is several orders of magnitude larger than that in the c-direction, it is necessary to have the latter normal to the lead length. The following tape-casting route is one which was developed to achieve optimum texture.

The organic-based tape-casting slurry consisted of the usual solvents, plasticizers and binders (see Section 3.6.6) and Bi-2212 powder (approximately 3 μm particle size). The slurry also contained MgO fibres of diameter and length in the range 5 μm and 100 μm respectively. These fibres formed the substrate onto which the platelets grew epitaxially during the partial melting–recrystallization stage. The slurry was tape-cast into sheets, a process which aligned the MgO fibres in the plane of the sheet and in the casting direction. The tape-cast sheets were cut into strips, stacked and consolidated by warm-pressing. Short lengths of silver foil were incorporated into the ends of the stack prior to laminating to provide the contacts. After the organics were burnt off the lead was fired in oxygen when partial melting and recrystallization occurs. The final step was an anneal in argon with 2 vol.% oxygen to adjust the stoichiometry. The result was a well textured structure as shown in Fig. 4.58(a). Complete leads are shown in Fig. 4.58(b).

An alternative manufacturing route exploits the powder-in-tube technology described below, except that a metal with poorer thermal conductivity replaces the silver matrix.

The function of a current lead is to improve the efficiency of liquid helium-cooled superconducting electromagnets. The current is fed to the coils in three stages. The first comprises cooled copper leads (the resistivity of a metal falls with falling temperature) which are joined to the HTS leads, connected in turn to the low temperature superconductor (e.g. Nb_3Sn). The heat management is such

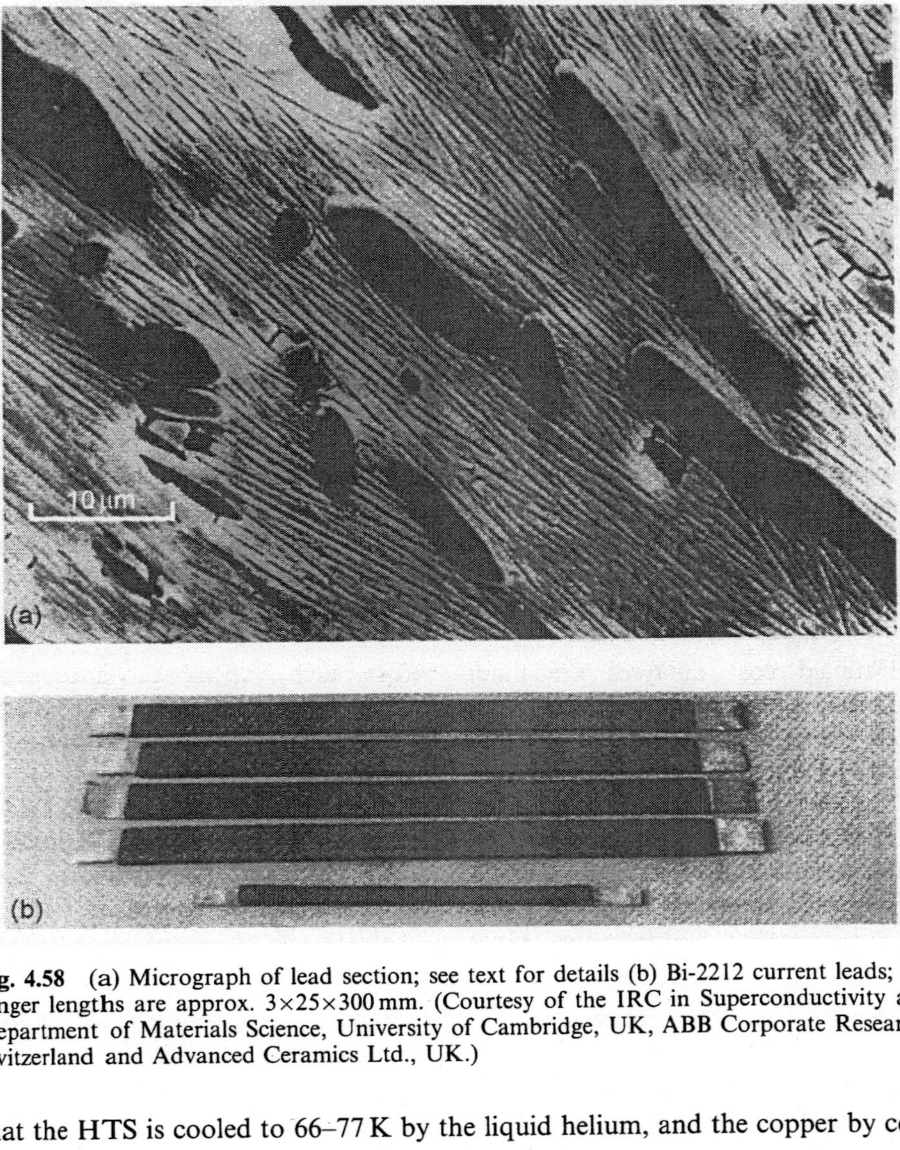

Fig. 4.58 (a) Micrograph of lead section; see text for details (b) Bi-2212 current leads; the longer lengths are approx. 3×25×300 mm. (Courtesy of the IRC in Superconductivity and Department of Materials Science, University of Cambridge, UK, ABB Corporate Research, Switzerland and Advanced Ceramics Ltd., UK.)

that the HTS is cooled to 66–77 K by the liquid helium, and the copper by cold He vapour.

A fault current-limiter is a component which protects power transmission and distribution systems from surges caused by, for example, a lightning strike, fulfilling a function similar to that of a varistor (see Section 4.3.1). The limiter should be capable of reducing the fault current to a fraction of its peak value in less than a cycle. Because for this application the requirement is for low J_c, fault current limiters are already a commercial product. In the case of the lead shown in Fig. 4.58(b) the fault current is limited to a safe value within 5 ms of the arrival of the current 'spike'.

Cables via the 'powder-in-tube' (PIT) method The approach is especially suited to processing Bi-2223 into leads and cables for power applications. A silver or silver alloy tube, filled with the partially reacted precursor powders formulated to yield Bi-2223, is drawn down to a wire 1–2 mm diameter. The wire is rolled into a tape if that is the required form, usually with a width-to-thickness ratio of approximately 10:1. The composite is then heated to 800–900 °C when the powder partially melts. The recrystallization process is controlled and the pure Bi-2223 phase develops with large grains oriented so that the Cu–O planes lie parallel to the silver surface to optimize J_c.

Cables for handling power can be made up from the tapes which are wound in a manner similar to standard cabling practice with a surrounding corrugated (to facilitate bending) steel duct for carrying the liquid nitrogen refrigerant. Lengths totalling in excess of 150 km have been manufactured for electromagnet applications. A commercially important objective arising because of the demand for increased power, is the replacement of existing electric power transmission cables with high-T_c superconducting cables. This is economically attractive in large and highly developed and congested cities such as Tokyo, London and New York, where space in the existing underground ducts is running out and enlarging them would be unacceptably disruptive to day-to-day city life.

Detailed cost analyses are made before such expensive ventures are undertaken, but there seems little doubt that the next decade or so will see HTS cables steadily replacing metallic conductors for power distribution in locations where the economic benefits are justified.

The strong inverse dependence of critical current density on magnetic field at 77 K is a limitation associated with Bi-2223 tape. This necessitates operation at lower temperatures for certain applications (e.g. electromagnets) when the cable is subjected to strong magnetic fields. This limitation does not arise with the (RE)BCO system and intensive efforts are being made to manufacture suitable tapes.

Engineering the required texture in a (RE)BCO tape is a much more complex matter than in the case of the Bi-2223 tape. One approach is to prepare a nickel tape with a carefully controlled crystalline surface texture. This forms the substrate for the development of a correspondingly textured thermally grown oxide layer which, in turn, forms the substrate for the epitaxial growth of MgO by electron beam evaporation. The textured MgO serves as the substrate for the epitaxial growth of the $\sim 1\,\mu m$ thick (RE)BCO via pulsed laser deposition.

High field gradient 'magnetic separation' is an old technology being increasingly exploited for the purification of liquids. Applications include the removal of heavy metal ions from laboratory waste water, purification of china clay for the paper industry, recycling waste water used in the steel industry and recycling of grinding sludge in glass polishing industries, for example in the manufacture of cathode ray tubes. Bi-2223-based electromagnets are being introduced leading to significant increases in efficiency. It is anticipated that the Bi-2223 cables will eventually be replaced with those incorporating (RE)BCO.

Tapes by dip-coating The process is essentially as described in Section 3.6.6. A silver or silver alloy tape is passed through a slurry of the partially reacted Bi-2223 powder dispersed in an organic vehicle. The coated tape is heated at ~800 °C in air to remove organics and then at nearly 900 °C when partial melting and controlled recrystallization takes place.

Bulk magnets Because of the many potential applications, including magnetic levitation for transport, magnetic separation, bearings and energy storage, the development of 'quasi-permanent magnets' is an important branch of HTS technology. The practical maximum field developed by the best rare earth permanent magnets is ~1.0 T. From Fig. 4.57 the H_{irr} field for YBCO at 77 K is approximately 5 T, and trapped fields in excess of 3 T have been reported at this temperature.

The various common processing routes to producing bulk (RE)BCO can be exploited. Although there seems to be no obstacle to obtaining high T_c values with sintered material, there is with regard to obtaining high critical current densities J_c. The basic reason for this is the presence of 'defects' of various kinds, particularly grain-boundaries, which interfere with current flow. This has led to a focus on the growth of single crystal forms of (RE)BCO by 'melt-processing'. By this approach, crystals possessing values of H_{irr} and J_c attractive for permanent magnets are being grown. The technology is reviewed by D.A. Cardwell [41] (see Fig. 4.59) and by M. Murakami [43].

Melt-processed crystals contain many defects, including dislocations, twin planes and non-superconducting particles, which act as the essential pinning centres. The deliberate engineering of a fine dispersion of precipitates of the non-superconducting '211' phase, Y_2BaCuO_5, has been a major breakthrough permitting trapping of the strongest fields, for example 10 T at 42 K. A penalty is the occurrence of microcracking around the inclusions due to differential thermal expansivities between inclusion and matrix.

During the magnetizing process, and in the final fully magnetized state, complex mechanical stress fields which can be large enough to cause fracture are developed. The stresses originate from the Lorentz forces (see Section 9.1.2) on the fluxoids which, through flux-pinning, are transferred to the material. In the final fully magnetized state, when the magnetizing field is removed, the screening currents associated with the *non-uniform* fluxoid distribution result in the critical current density J_c flowing throughout the volume. The stresses are the result, then, of the interaction between J_c and the penetrated induction B_p. They are tensile throughout the bulk and any defect of 'critical' dimensions (or larger) will lead to immediate fracture or, in the case of smaller defects, possible eventual

Fig. 4.59 Large grain of 'melt-processed' $YBa_2Cu_3O_{7-\delta}$ (YBCO) fabricated in the IRC in Superconductivity, University of Cambridge, UK. (Courtesy of of D.A. Cardwell.)

fracture following sub-critical crack-growth. T.H. Johansen [44,45] discusses the rather complex underlying physics.

Considering a specific example (see Question 20), if a cylindrical, circular section superconductor is magnetized along its length the critical current density J_c, flows throughout the volume in circular paths in planes perpendicular to the axis of the cylinder. The interaction of J_c with the penetrated induction B_p, leads to tensile stresses of magnitude $\approx B_p^2/2\mu_0$.

The cross-breaking strength of YBCO is very variable with typical values lying in the range 25–50 MPa. The strengths are significantly increased by the inclusion of stress-relieving silver particles, by resin-impregnation and by 'steel-banding'; with 20 wt% Ag addition strengths as high as 100 MPa have been achieved.

Permanent magnets are exploited in magnetic bearings for a variety of applications ranging from machine tools to energy storage. In an experimental energy storage application flywheels, having a construction based on carbon fibre-reinforced plastic and mounted on YBCO magnetic bearings, are spun up to very high speeds (~50 000 r.p.m.) during low power demand periods and their kinetic energy drawn off at peak demand times. Japanese research is aimed at developing flywheels having a stored energy of 10 MW h, a quantity appropriate to load-levelling at the power distribution sub-station stage (see Section 4.5.2).

4.7.5 Superconducting electronics – thin films

The Josephson junction

In 1962 a postgraduate student, Brian Josephson, working in the University of Cambridge, and later to win a Nobel Prize, predicted that Cooper pairs should be able to tunnel through a thin (approximately 1 nm) insulating barrier from one superconductor to another with no electrical resistance [46]. This 'quantum tunnelling' was confirmed by experiment and is known as the 'Josephson effect'. The superconducting electronic devices exploit Josephson junctions.

The Josephson effect can be explained in terms of coupling across the barrier of the wave functions describing the Cooper pair populations in the superconductors on either side of the barrier. The wave functions maintain their meaning for a distance approximating to the coherence length ξ mentioned above. Fig. 4.60(a) shows a barrier separating two superconductors and Fig. 4.60(b) the current flowing at zero applied voltage. If the tunnelling current reaches a critical value (I_c) (typically <1 mA), then a voltage appears across the junction, and for larger currents there is an oscillatory component with a frequency dependent upon the applied voltage. When a voltage is applied across the junction it has no effect on the tunnelling current until a critical value (U_c) is reached beyond which the current/voltage relationship is ohmic. The critical voltage is related to the energy required to split the Cooper pair and a result of the BCS theory is that this energy is given very nearly by $3.53\,kT_c$ Critical voltages lie in the range 1–100 μV.

An application of Josephson junctions is in superconducting quantum interference devices (SQUIDS). Figure 4.61 shows two Josephson junctions passing a current with a magnetic induction B, through the ring of area A.

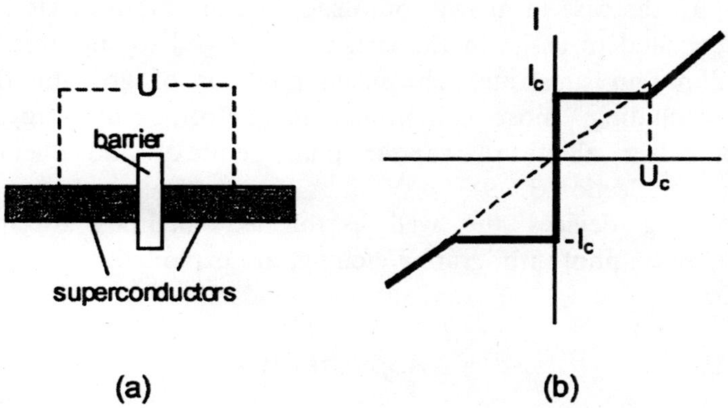

Fig. 4.60 (a) A Josephson junction. (b) The current–voltage characteristic of a Josephson junction.

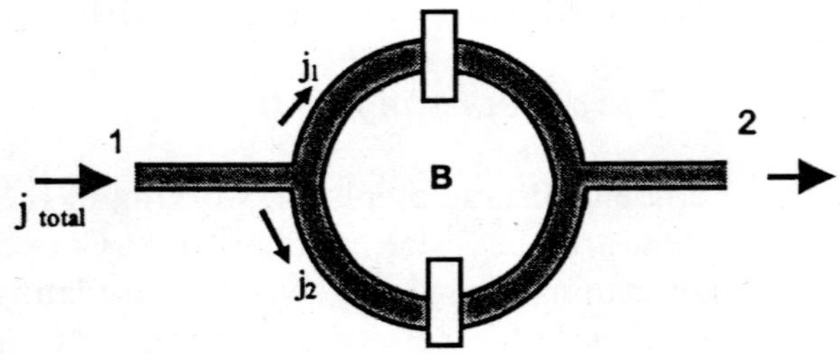

Fig. 4.61 Two Josephson junctions comprising a superconducting quantum interference device (SQUID).

The currents passing through the junctions are arranged to be slightly larger than the critical currents so that a voltage appears across them. A magnetic induction through the ring results in an interference effect arising because the wave functions describing the Cooper pairs have a different wavelength depending upon whether the currents pass in a clockwise or anticlockwise direction with respect to the direction of B. The result is that the voltage appearing across the junctions varies periodically with the strength of B. The SQUID is the basis of single quantum flux (SQF) devices, capable of measuring extremely small magnetic fields of order $10^{-6}\Phi_o$ where $\Phi_o = h/2e$, the quantum of magnetic flux, or 'fluxon' (the concept has already been encountered above in connection with the magnetic induction penetrating a Type II superconductor).

Growing films and forming devices

High-T_c superconductor electronics is based on thin (typically ~100 nm) films of YBCO. The films are grown epitaxially on to, typically, MgO or sapphire substrates. In the case of growth onto sapphire an intermediate buffer layer of ceria is deposited to assist in the lattice match and to suppress interdiffusion between film and substrate. The usual methods of growing thin films are exploited including, most commonly, magnetron sputtering, pulsed laser deposition (laser ablation), vapour phase epitaxy and chemical solution deposition (see Section 3.6.9).

For forming devices the well established methods for patterning in microelectronics, photolithography/etching, are exploited.

Applications

The switching between states, a characteristic of Josephson junctions, is applicable to binary-based logic circuits; there is no voltage across the junction for currents below the critical current I_c, but one appears for currents greater than I_c. The SFQ device can also be exploited as a binary logic element, the presence or absence of a fluxon in the SQUID ring corresponding to a '1' or '0'. The attraction of such superconductor devices lies in the very fast switching times, of the order 10^{-12} s. However there is much development work to be done before HTSs make a significant impact on a technology dominated by silicon and gallium arsenide.

The SQUID is exploited for the detection and measurement of extremely small magnetic fields and finds medical, geophysical and military applications. A medical application is in magnetoencephalography (MEG), the recording of the very small magnetic effects caused by neural currents in the brain. The measurements must, of course, be made in an environment free from extraneous magnetic field pollution. Even the earth's magnetic field is approximately 10^9 times stronger than those being measured and so magnetic screening is of paramount importance. Very effective screening enabling the MEG equipment to be used in the vicinity of strong sources of magnetic interference (lift motors and transformers) is achieved by surrounding the patient and SQUID sensor array by a large (1.65 m×0.65 m) hollow Ni cylinder, plasma spray-coated with a 1 mm thick Bi-2223 layer and cooled by He vapour.

Thin film YBCO devices are achieving commercial success in microwave communications technology. The explosive growth in mobile phone usage is forcing the operating frequency upwards from the current 800–900 MHz band to the 2 GHz band. Because the propagation losses are larger at the higher frequency the base station receiver sensitivities need to be optimized. Also, because of the increased number of subscribers signal selectivity is critical. YBCO-based thin and thick film antennae and filters are under development for these applications. The surface resistivity of YBCO at 10 GHz and 77 K is nearly two orders of magnitude lower than that of copper at the same frequency and temperature. This results in very high Q-values (high selectivity) and an YBCO cavity resonator (see Section 5.6.5) with a Q-value of 400 000 at 10 GHz is in service.

4.7.6 The future for HTSs

The silver matrix Bi-2223 cables are a commercial reality but they suffer from a very steep dependence of the critical current on magnetic field at 77 K and so for some applications use is restricted to the lower temperatures. The critical current for (RE)BCO depends far less strongly on magnetic field and so offers a wider range of possible applications. However, the full potential of the (RE)BCO family will be realized only when the technology for fabricating cables is perfected so as to minimize the 'weak link' grain boundary effects. When this technology is fully developed it will be possible to generate large volume ($>10\,m^3$) fields of strength >5 T and the applications field opens up, especially in conjunction with trapped field magnets for high speed magnetically levitated transport systems.

With regard to 'trapped field magnets' the technology has been steadily improving with both the development of Nd-BCO and Sm-BCO varieties which are potentially capable of trapping stronger fields than YBCO at 77 K. The mechanical strength of the (RE)-BCO ceramics is a limiting factor and they are prone to fracture under the stresses developed as a result of the strong magnetic fields to which they can be subjected. Also the ingress of water and its subsequent freezing leads to fracture, a problem which has been alleviated by resin-impregnation.

Understanding of vortex pinning mechanisms will steadily develop and with it the technology of optimising compositions and structures of materials for trapped field magnets, particularly the melt-textured, large grain varieties.

The introduction of high-T_c superconductor technology into the mobile communications arena will continue.

Computer circuits based on the very fast switching effects associated with Josephson junctions are very much at the experimental stage and their development will pose many problems. However, because circuit line widths in superconducting logic circuits are not as demanding as in conventional semiconducting circuits, devices can be made from YBCO using very conventional photolithographic equipment and techniques. If inexpensive superconducting microcircuits become a commercial reality then they will make a very significant impact upon the technology as a whole.

In summary, the past decade has seen impressive developments in HTS technology embracing cables, bulk materials and electronics. The progress seems set to continue especially with regard to materials design and processing technologies. Penetration of components into the commercial market will inevitably grow.

Problems

1. A thin-film nichrome resistor in the form of a rectangle of dimensions 1 mm×5 mm is deposited on to a glass substrate. Estimate the thickness of the film for the resistance measured between electrodes contacting the 1 mm edges to be 1.0 kΩ. The resistivity of 'nichrome' may be taken to be $1.07 \times 10^{-6}\ \Omega$ m.

   Calculate the surface resistivity of the nichrome film. [Answers: 5.35 nm; 200 Ω/□]

2. An electroded disc of thickness 1 mm is connected to a 2 V supply. Using the idealized model given in Fig. 4.9(b) estimate the electric field across the IGL given that $t_r = 1$ nm and $d_g = 20\ \mu$m. Assume the resistance of the IGLs to be dominant. [Answer: $\sim 10^7\ \text{V m}^{-1}$]

3. The α value for a ZnO varistor is 15 and $k_I = 9.3 \times 10^{-38}$, the potential difference and current being measured in volts and amps respectively. Calculate the p.d. across the varistor for currents of 0.1, 1, 10, 100 and 1000 A. If a device needs protection against transient voltages in excess of 300 V, show how a varistor having the characteristic defined above can be connected to afford protection. [Answers: 252, 294, 343, 400, 466 V]

4. An NTC thermistor has a B value of 3000 K. What is its temperature coefficient of resistance at 27 °C? [Answer: $-3.3\%\ \text{K}^{-1}$]

5. The resistance (R) of an NTC thermistor varies with temperature according to $R = R_\infty \exp(B/T)$. Calculate the critical maintained voltage drop across the thermistor above which thermal runaway would occur for an ambient temperature of (i) 300 K and (ii) 360 K. The thermistor characteristics are as follows: $R_\infty = 4.5 \times 10^{-3}\ \Omega$, $B = 3000\ K$; $1/k_{th} = 3.3 \times 10^{-2}\ \text{W K}^{-1}$.

   Calculate the new critical voltages if the surroundings of the thermistor change so that the heat dissipation factor ($1/k_{th}$) is modified to $5.0 \times 10^{-2}\ \text{W K}^{-1}$. [Answers: 6.35 V, 3.36 V; 7.82 V, 4.13 V]

6. Suggest a basic composition for a PTC thermistor that will 'switch' at around 30 °C.

7. The tracer diffusion coefficient for oxygen ions in a particular cubic stabilized zirconia is measured and found to fit the relationship

$$D = 1.8 \times 10^{-6} \exp\left(-\frac{Q}{kT}\right) \text{m}^2\ \text{s}^{-1} \quad \text{where } Q = 1.35\ \text{eV}.$$

   Assuming a transport number for oxygen ions of unity, estimate the electrical conductivity at 1000 °C. Assume a unit cell of side 280 pm. [Answer: $\approx 2.0\ \text{S m}^{-1}$]

8. The CSZ material described in Question 7 is used to form an oxygen meter which is measuring an oxygen pressure of 10^{-10} atm relative to a reference oxygen pressure of 1 atm. Under the operating conditions it is believed that oxygen permeates across the wall of the cell at a rate limited by the transport of positive 'holes'. Estimate the molar oxygen flux given that the membrane is 1 mm thick and that the transport number for 'holes' at 1000 °C is 10^{-3}. [Answer: $3.3 \times 10^{-6}\ \text{mol m}^{-2}\ \text{s}^{-1}$ of O_2]

9. A disc-shaped piece (1 cm^2 area × 1 mm thick) of the CSZ referred to in Questions 7 and 8, carrying porous electrodes across the two faces, serves as a membrane separating an oxygen atmosphere from an enclosure of volume $10^{-4}\ \text{m}^3$. If the membrane is maintained at 100 °C and the enclosure at 300 °C, estimate the rise in oxygen pressure in the enclosure if oxygen is pumped into it for 1 h by a potential difference of 1 V maintained between the disc faces. [Answer: ~90 kPa (0.9 atm)]

10. One side of an oxygen meter is exposed to oxygen in equilibrium with a mixture of Ni and NiO at 1300 K, and the other side is exposed to air at 101 kPa. If the cell e.m.f. is

0.596 V, calculate the standard free energy change at 1300 K for the reaction $2Ni+O_2 = 2NiO$. [Answer: −247 kJ]

11. SnO_2 crystallizes in the rutile structure with cell dimensions $a = 473.7$ pm; $c = 318.6$ pm, and there are two formula units per cell. Calculate its theoretical crystal density. Relative atomic mass of Sn = 118.7. [Answer: 7.002 Mg m^{-3}]

12. What volume of oxygen gas at STP is equivalent to a monolayer of oxygen ions chemisorbed on 1 g of a lightly sintered compact of 200 nm diameter tin oxide particles? [Answer: ~1 ml]

13. A lightly sintered compact of monosized 20 nm diameter, Sb_2O_5-doped (1 wt%) tin oxide particles is conditioned in an enclosure so that the surface is atomically clean. With the compact held at 300 °C oxygen gas is steadily admitted to the enclosure and the electrical resistance of the compact monitored. At a stage in the process the resistance shows a marked increase. Estimate the fraction of surface covered by the chemically adsorbed species at this stage. [Answer: ~2%. Hint: consider Eq. 4.49]

14. Describe the construction of a tin oxide-based CO gas-sensor paying particular attention to essential design features.

15. (i) The sun's energy is derived from a nuclear reaction which, overall, results in the fusing of the nuclei of 4 hydrogen atoms to form the nuclei of 2 helium atoms, together with 2 positrons. In the process approximately 0.7% of the mass of the hydrogen is converted into radiation energy. Estimate the quantity of this energy relative to that derived from combusting hydrogen according to the reaction

$$H_2 + \tfrac{1}{2}O_2 \rightarrow H_2O : \Delta H^0_{298} = -286\,\text{kJ}$$

How does the energy released according to the above chemical reaction compare with that available from a practical hydrogen fuel cell, given that $\Delta G^0_{298} = -228$ kJ? Explain why there is a difference.
(ii) Estimate the time a standard cylinder of hydrogen would last if the gas were used to fuel a PEM cell stack supplying the *typical* UK family home with electricity for non-heating purposes during winter, stating and justifying assumptions made. What advantages, if any, would be derived from using a low temperature (~500 °C) SOFC for the same purpose? A standard cylinder contains about 7 m^3 of gas at STP (cylinder pressure 125 atm. (12.6 MPa)). [Answers: (i) 1 kg of hydrogen converted to helium releases an amount of energy equivalent to that released by burning at least 4.4 kt of hydrogen at constant pressure. (ii) A reasonable estimate would be ~1 to 2 days]

16. Give a quantitative estimate of the relative annual energy requirements for maintaining living comfort (23 °C) in two locations where the outside temperatures are averaging, respectively, 13 °C and 33 °C. The effects of relative humidity on comfort may be ignored.

17. Discuss the relative merits of the tubular and planar SOFC designs. With special reference to the electroceramics involved and to their processing, sketch a feasible design for a planar SOFC stack running at 900 °C.

Such a stack, running at 900 °C, is made up of 100 individual cells connected in series. The YSZ electrolyte in each cell is a layer having dimensions $100 \times 100\,mm \times 200\,\mu m$. From the data in Fig. 4.31 estimate the power deliverable into a resistive load of optimum value.

Suggest and quantitatively justify a strategy for reducing the dissipated power density in the stack, whilst maintaining the same power output in a resistive load. [Answer: ~6.25 kW]

18. At a stage in the discharge of a Na/S battery operating at 350 °C all the sulphur has been converted to the polysulphide Na_2S_5. Further discharge leads to the formation of Na_2S_3 according to the reaction

$$4Na(l) + 3Na_2S_5(l) \rightarrow 5Na_2S_3(l)$$

when the cell e.m.f. is 1.78 V. Given that the free energy of formation of $Na_2S_5(l)$ at 350 °C, is $-400\,kJ\,mol^{-1}$, estimate the free energy of formation of $Na_2S_3(l)$ at the same temperature. [Answer: $-377\,kJ\,mol^{-1}$]

19. Under normal operation at 250 °C a ZEBRA type cell discharges according to the overall reaction:

$$2Na(l) + FeCl_2(s) \rightarrow Fe(s) + 2NaCl(s)$$

for which the cell e.m.f. is 2.35 V. Given that the standard free energy of formation of NaCl at 250 °C is $-367\,kJ\,mol^{-1}$ calculate the standard free energy of formation of $FeCl_2$ at the same temperature. What are the operational disadvantages associated with the Fe cell when compared with the Ni ZEBRA cell? Ref. [7] discusses the matter. [Answer $-280\,kJ\,mol^{-1}$]

20. A HTS in the form of a circular cylinder, length 6 cm and diameter 3 cm, is magnetized along its length. The critical current density J_c is flowing throughout the volume and the strength of the trapped magnetic induction measured at the centre of the flat end-faces and parallel to the axis (B_o), is 5 T.

(i) On a diagram show the directions of the currents flowing and the principal stresses, and

(ii) estimate the magnitude of the stress and, given that $K_{Ic} = 1\,MPa\,m^{1/2}$, the size of the strength-limiting defect. Repeat the calculations for a similarly measured trapped induction of 10 T.

(iii) Comment on the risk of failure during the magnetizing process.

Consulting [44] and [45] will be necessary. [Answers: (i) 40 MPa and ~600 μm; (ii) 160 MPa and ~40 μm]

Bibliography

1. Buchanan, R.C. (ed.) (1991) *Ceramic Materials for Electronics*, 2nd edn, Marcel Dekker Inc., New York.
2. Forrest, C.W., Kennedy, P. and Shennan, J.V. (1972) The fabrication and properties of self-bonded silicon carbide bodies, in Special Ceramics 5, The British Ceramic Research Association, Stoke-on Trent, UK.
3. Various authors (2001) Materials for clean energy, *Nature*, **414**, 331–77.
4. Minh, N.Q. and Takahashi, T. (1995) *Science and Technology of Ceramic Fuel Cells*, Elsevier, Amsterdam.
5. Kordesch, K. and Simander, G. (1996) *Fuel Cells and their Applications*, VCH, Weinheim, Germany.
6. Larminie, J. and Dicks, A. (2000) *Fuel Cell Systems Explained*, John Wiley and Sons Ltd., Chichester.
7. Rand, D.A.J., Woods, R. and Dell, R.M. (1998) *Batteries for Electric Vehicles*, John Wiley and Sons Ltd., Chichester.
8. Ramanarayanan, T.A., Singhal, S.C. and Wachsman, E.D. (2001) *High Temperature Ion Conducting Ceramics*, The Electrochemical Society: *Interface*, **10**, Summer, 22–7.
9. Sudworth, J.L. *et al.* (2000) Toward commercialisation of the beta-alumina family of ionic conductors, *MRS Bull.* **25**, 22–6.
10. Minh, N.Q. (1993) Ceramic fuel cells, *J. Am. Ceram. Soc.* **76**, 563–88.
11. Steele B.C.H. (1996) Materials for high-temperature fuel cells, *Phil. Trans. Roy. Soc. Lond. A* **354**, 1695–710.
12. Steele, B.C.H. (2001) Material science and engineering: the enabling technology for the commercialisation of fuel cell systems, *J. Mater. Sci.* **36(5)**, 1053–68.
13. Singhal, S.C. (2000) Advances in solid oxide fuel cell technology, *Solid State Ionics* **135**, 305–13.
14. Galloway, R.C. and Haslam, S. (1999) The ZEBRA electric vehicle battery: power and energy improvements, *J. Power Sources* **80**, 164–70.
15. Fray, D.J. (1996) The use of solid electrolytes as sensors for applications in molten metals, *Solid State Ionics* **86/8**, 1045–54.
16. Göpel, W., Reinhardt, G. and Rösch, M. (2000) Trends in the development of solid state amperometric and potentiometric high temperature sensors, *Solid State Ionics*, **136/7**, 519–31.
17. Gibson, R.W., Kumar, R.V. and Fray, D.J. (1999) Novel sensors for monitoring high oxygen concentrations, *Solid State Ionics*, **121**, 43–50.
18. Fray, D.J. (2000) Potentiometric gas sensors for use at high temperatures, *Mater. Sci. Technol.* **16**, 237–42.
19. Moulson, A.J. and Roberts, J.P. (1958) Water in silica glass, *Trans. Br. Ceram. Soc.* **59**, 388–99.
20. Iwahara, H., Esaka, T., Uchida, H. and Maeda, N. (1981) Proton conduction in sintered oxides and its application to steam electrolysis for hydrogen production, *Solid State Ionics* **3/4**, 359–63.

21. Yamaguchi, S., *et al.* (2000) Basicity and hydroxyl capacity of proton-conducting perovskites, *Solid State Ionics* **136/7** 191–5.
22. Iwahara, H. (1995) Technological challenges in the application of proton conducting ceramics, *Solid State Ionics* **77**, 289–98.
23. Schwandt, C. and Fray, D.J. (2000) Hydrogen sensing in molten aluminium using a commercial electrochemical sensor, *Ionics* **6**, 222–9.
24. Kale, G.M., Davidson, A.J. and Fray, D.J. (1996) Investigation into an improved design of CO_2 sensor, *Solid State Ionics* **86/8**, 1107–10.
25. Moseley, P.T., Stoneham, A.M. and Williams, D.E. (1991) Oxide semiconductors: patterns of gas response behaviour according to material type. In: *Technology and Mechanisms in Gas Sensing*, Moseley, P.T., Norris, J.O.W. and D.E. Williams (eds), Adam Hilger, Bristol.
26. Lantto, V. (1992) Semiconductor gas sensors based on SnO_2 thick films. In: *Gas Sensors*, G. Sberveglieri (ed.), Kluwer Academic Publishers, Dordrecht, pp. 117–67.
27. Göpel, W. and Schierbaum, K.D. (1995) SnO_2 sensors: current status and future prospects, *Sensors and Actuators* **B26–27**, 1–12.
28. Shimizu, Y. and Egashira, M. (1999) Basic aspects and challenges of semiconductor gas sensors. *MRS Bull.*, 18–24.
29. Morris, L. and Williams, D.E. (2001) Pt(II) as an electronically active site in the room temperature CO response of Pt-modified gas sensitive resistors, *J. Chem. Phys. B.* **105**, 7272–9.
30. Williams, D.E. (1999) Semiconducting oxides as gas-sensitive resistors, *Sensors And Actuators B-Chemical*, **B57**, pp. 1–16.
31. Volger, J. (1968) Superconductivity, *Philips Tech. Rev.* **29**, 1–16.
32. Vidali, G. (1993) *Superconductivity: the Next Revolution?*, Cambridge University Press, Cambridge.
33. Kresin, V.Z. and Wolf, S.A. (1990) *Fundamentals of Superconductivity*, Plenum Press, New York.
34. Cardwell, D.A. and Ginley, D.S. (eds) (2002) *Handbook of Superconducting Materials*, Institute of Physics, London.
35. Poole, C.P. Jr. (ed.) (2000) *Handbook of Superconductivity*, Academic Press, London.
36. Cava, R.J. (2000) Oxide Superconductors, *J. Am. Ceram. Soc.* **83**, 5–28.
37. Bednorz, J.G. and Müller, K.A. (1986) Possible high-T_c superconductivity in the Ba-La-Cu-O system, *Z. Phys. B-Condensed Matter*, **64**, 189.
38. Moon, F.C. (1994) *Superconducting Levitation*, J. Wiley and Sons, Inc., New York.
39. Maeda, H. and Togano, K. (eds) (1996) *Bismuth-based High-Temperature Superconductors*, Marcel Dekker, Inc., New York.
40. Proceedings of the 13th International Symposium on Superconductivity – 'Advances in Superconductivity XIII' Physica C, (2001) **357–360**.
41. Cardwell, D.A. (1998) Processing and properties of large grain (RE)BCO, *Mater. Sci. Eng. B* **53**, 1–10.
42. Yamada, Y. *et al.* (1991) Microstructural and transport properties of high J_c Bi (2223)/Ag tapes, *Physica C* **185–189**, 2483–4.
43. Murakami, M. (ed.) (1992) *Melt-processed High-temperature Superconductors*, World Scientific, London.

44. **Johansen, T.H.** (2000) Flux-pinning-induced stress and magnetostriction in bulk superconductors, *Supercond. Sci. Technol.* **13**, R.121–37.
45. **Johansen, T.H.** (2000) Pinning-induced stress during activation of bulk HTSs as trapped-field magnets, *Supercond. Sci. Technol.* **13**, 830–5.
46. **Josephson, B.D.** (1962) Possible new effects in superconductive tunnelling, *Phys. Lett.* **1**, 251.

5

DIELECTRICS AND INSULATORS

'Ceramic dialectrics and insulators' is a wide-ranging and complex topic embracing many types of ceramic, physical and chemical processes and applications. Broadly speaking over the past century there have been clearly discernible trends; for example, in the case of electric power transmission to higher line voltages for greater transmission efficiency. Radio broadcasting and radar technologies made their appearance and as they developed to operate at steadily increasing frequencies and power levels, so too did the demand for ceramics tailor-made for a variety of essential components. But it is the 'computer' and 'communications' technologies which have had the most significant impact on ceramic insulators and dielectrics. Here, because of the development of silicon and GaAs semiconductor integrated circuits, the trend has been to lower operating voltages and to frequencies in the 1–10 GHz range. The need for 'miniaturization' has accompanied the demand for increased computing power and the advent of satellite and mobile communications.

With the increase in complexity of equipment, for example a mobile phone of the next generation and suited to 'Bluetooth' technology may well contain a few hundred capacitors, about half that number of resistors and perhaps a few tens of inductors. This dramatic increase in passive component packing density is placing a strain on surface mount technology, and on reliability because of the large number of solder joins. The problem is being addressed through efforts to develop fully integrated three dimensional ceramic packages with the active, semiconductor circuits and components which cannot be integrated because of their size, mounted on the surface. This technology is based on low temperature co-fired ceramics (LTCC).

To structure a discussion around all these developments is not an easy task and we have attempted to assist in intelligibility by presenting it in two parts.

Electroceramics: Materials, Properties, Applications. 2nd Edition. Edited by A. J. Moulson and J. M. Herbert.
 ISBN 0 471 49747 9 (hardback) 0 471 49748 7 (paperback)

Part I is focused on capacitors, providing the oportunity to introduce the many important ideas relating to their performance and, indeed, to the wider application of dielectrics and insulators. Part II is concerned with the important ceramic types and their applications.

The monographs [1] and [2] are recommended for supplementing the discussions.

PART I CAPACITATIVE APPLICATIONS

5.1 Background

Dielectrics and insulators can be defined as materials with high electrical resistivities. Dielectrics fulfil circuit functions for which their permittivities ε and dissipation factors $\tan\delta$ are also of primary importance. Insulators are used principally to hold conductive elements in position and to prevent them from coming in contact with one another. A good dielectric is, of course, necessarily a good insulator, but the converse is by no means true.

Just as there is no perfect insulator, neither is there a perfect dielectric, and in the various usage contexts some parameters are more suitable than others for describing the departure from perfection. For example, a power engineer would focus attention on the loss factor $\varepsilon'' = \varepsilon' \tan\delta$ (Eq. (2.102)) because it is dissipation of energy and the attendant heating and wastage which are the chief concerns. An electronics engineer would be more concerned with $\tan\delta$, and certainly so when dealing with oscillator and filter circuits which exploit the electrical resonance phenomenon. The sharpness of tuning at resonance depends on the 'quality factor' or Q of the circuit: the higher is Q the lower is the damping and the sharper is the resonance. Q is determined by the components comprising the circuit and, in particular, by the dielectric properties of the materials used; $\tan\delta$ is a material property and its reciprocal is Q. In dealing with dielectric heating a significant parameter is the dielectric conductivity $\sigma_{\mathrm{a.c.}} = \omega\varepsilon'' = \omega\varepsilon'\tan\delta$ (Eq. (2.98)). These various parameters are a measure of essentially the same thing, i.e. the dissipation of energy in the dielectric in an alternating electric field. In some applications of ceramics in electronics their insulating rather than dielectric properties are significant. Examples of such applications are substrates on which circuits are constructed, parts of variable air capacitors and coil formers. These are essential constructional parts whose function is simply to support circuits, capacitor vanes or coils respectively, and the ideal material would have $\varepsilon_r = 1$ and $\tan\delta = 0$ $(Q = \infty)$ across the desired frequency and temperature ranges – in fact a solid vacuum would be ideal!

The uses to which insulating ceramics are put are many and varied; although a very large number of types have been developed to meet particular demands it is possible to discern certain trends, and it is around these that the following discussion is developed. One such trend has been meeting the demands set by the increase in line voltages as power transmission networks have developed; another is the move towards higher frequencies as telecommunications systems have advanced. With regard to the latter trend, it is shown in Section 2.7.2, that the power dissipated in an insulator or a dielectric is proportional to frequency. This explains the demand for low-loss dielectrics for high-frequency applications. High-loss-factor dielectrics for high frequencies cannot be tolerated, partly because excessive power dissipation can lead to unacceptable rises in temperature, and even breakdown in high-power devices, and partly, and more importantly, because the resonances in tuned circuits become less sharp so that the precise selection of well-defined frequency bands is not possible.

Before embarking on a detailed consideration of the application of dielectrics and insulators, it is opportune to focus attention briefly on 'dielectric strength' and 'thermal shock resistance'. Both properties demand careful consideration in certain applications of dielectrics and insulators. They are by no means simple to define and, generally speaking, it is necessary only to develop some appreciation of how component and operational parameters determine them.

5.2 Dielectric Strength

Dielectric strength is defined as the electric field just sufficient to initiate breakdown of the dielectric. The failure of dielectrics under electrical stress is a complex phenomenon of very considerable practical importance. Theories have been developed to explain what is termed the *intrinsic strength* of a material. Single crystals have usually been used to measure this, with the specimen geometry and electrode arrangement carefully designed and the ambient conditions closely controlled. Under such conditions reasonably reproducible data and satisfactory agreement with theory are obtained. Unfortunately electric strength depends markedly on material homogeneity, specimen geometry, electrode shape and disposition, stress mode (d.c., a.c. or pulsed) and ambient conditions, and in practice inadequate control is often exercised over these variables.

In the industrial situation *thermal breakdown* is the most significant mode of failure and is avoided through experience rather than by application of theory. Nevertheless it is important to appreciate the mechanisms leading to thermal breakdown. A third mode of failure, referred to as *discharge breakdown*, is of importance in ceramics because it has its origins in porosity.

5.2.1 Test conditions

Electric strength data are meaningful only if the test conditions are adequately defined: for example, if d.c. loading is employed the rate of voltage increase should be specified, if pulsed voltages are used the rise time should be specified and if a.c. loading is adopted the frequency and waveform should be specified.

The importance of sample geometry can be appreciated from Fig. 5.1. With the geometry (a), although a large volume of the specimen is stressed, failure is likely to be initiated from the electrode edges where the average electrical stress is magnified by a significant but uncertain factor. With the geometry (b) failure would probably occur at the centre where the stress is a maximum and known. Therefore more meaningful data are likely to be obtained using geometry (b).

When testing electric strength there is the risk of 'flash-over' across the specimen surface between the electrodes. This is avoided by making the measurement with the specimen immersed in an insulating liquid, such as transformer oil, which displaces air from the ceramic surface.

5.2.2 Breakdown mechanisms

Intrinsic breakdown

When a homogeneous specimen is subjected to a steadily increasing voltage under well-controlled laboratory conditions, a small current begins to flow which increases to a saturation value. As the voltage is further increased a stage is reached when the current suddenly rises steeply from the saturation value in a time as short as 10^{-8} s and breakdown occurs. The very short rise time suggests that the breakdown process is electronic in character.

Intrinsic breakdown is explained as follows. When the field is applied the small number of electrons in thermal equilibrium in the conduction band gain kinetic energy. This energy may be sufficient to ionize constituent ions, thus increasing the number of electrons participating in the process. The result may be an electron avalanche and complete failure. Intrinsic breakdown strengths are typically of the order of $100\,\mathrm{MV\,m^{-1}}$.

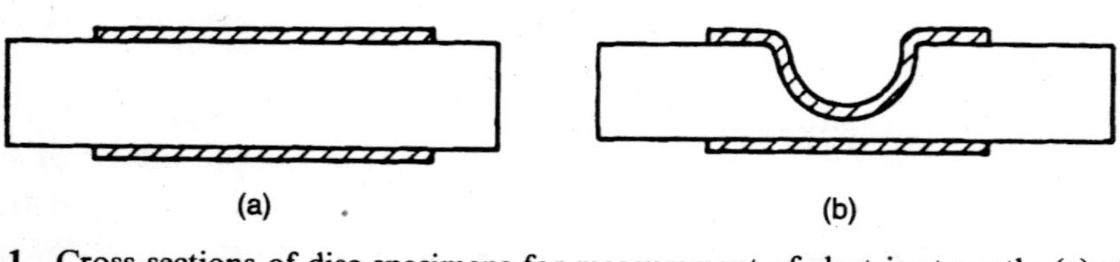

Fig. 5.1 Cross-sections of disc specimens for measurement of electric strength: (a) standard flat disc; (b) recessed disc.

Thermal breakdown

The term 'thermal breakdown' is restricted to those cases in which the breakdown process can be described in terms of the thermal properties of the dielectric.

The finite d.c. conductivity of a good dielectric results in Joule heating; under a.c. fields there is additional energy dissipation. If heat is generated in the dielectric faster than it can be dissipated to its surroundings, the resulting rise in temperature leads to an increase in conductivity and to dielectric loss. These are the conditions leading eventually to 'runaway', culminating in breakdown. The situation is identical with that already discussed in detail in Section 4.4.1, in the context of the negative temperature coefficient (NTC) thermistor.

Attempts have been made to develop a comprehensive theory of thermal breakdown, but solutions to the governing differential equation can be found only for the simplest of geometries. Another serious obstacle to achieving a realistic theoretical description is that the functional relationships connecting charge movement with field and temperature, and thermal diffusivity with temperature, are invariably very poorly defined. The expression

$$U_{\mathrm{b}} \propto \left(\frac{\lambda}{\omega\alpha_{\mathrm{d}}\varepsilon'\tan\delta}\right)^{1/2}\varphi \tag{5.1}$$

for the breakdown voltage U_{b} of a flat disc with a.c. applied across its thickness identifies the important parameters which include the temperature coefficient of loss factor α_{d} and φ which is a function of specimen thickness and heat transfer to the environment. It is evident that the variables determining thermal breakdown fall into one of two categories. One is concerned with material properties, i.e. thermal conductivity λ and loss factor, and the other with operational conditions, i.e. frequency, specimen dimensions and heat transfer to the surroundings. In practice, lack of sufficiently precise information regarding the various parameters precludes the possibility of developing an equation corresponding to (5.1).

The effect of temperature on electrical breakdown for an aluminous porcelain shown in Fig. 5.2 illustrates the expected marked decrease in strength arising through thermal breakdown of the glassy phase initiated above a critical ambient temperature (see Section 2.6.3).

Discharge breakdown

A ceramic is rarely homogeneous, a common inhomogeneity being porosity. There is strong evidence to show that breakdown can be initiated at pores and that the occurrence of gas discharges within pores is an important factor. There

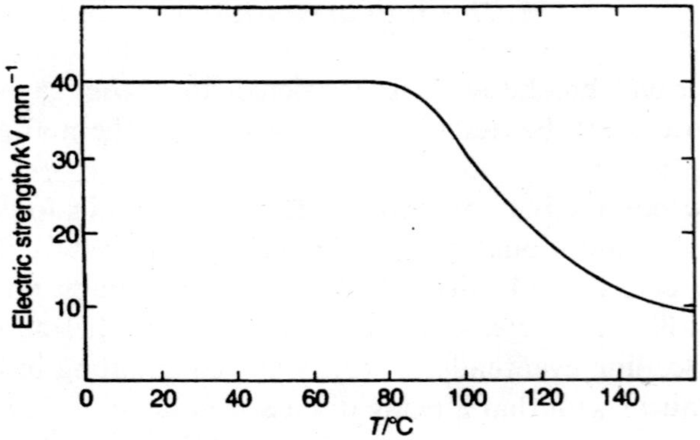

Fig. 5.2 Dependence of electric strength on ambient temperature for an aluminous porcelain.

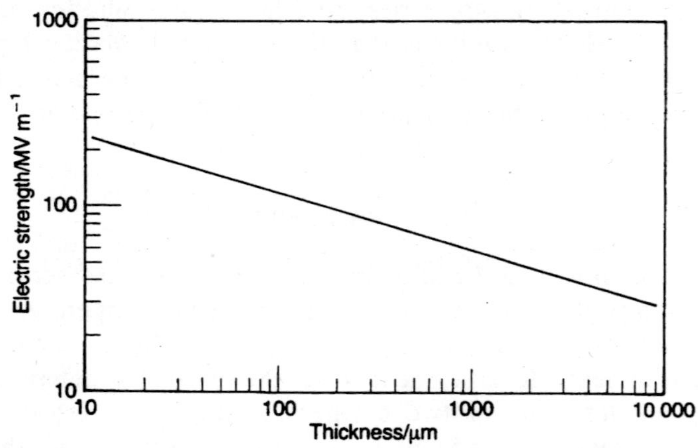

Fig. 5.3 Dependence of the electric strength of alumina ceramic on specimen thickness.

is a close apparent analogy with the mechanical strength of ceramics, where probability of failure depends upon the occurrence of defects, one of which is critical under the imposed conditions. Just as with the mechanical analogue, electric strength depends upon specimen size, because a decrease in size reduces the probability of occurrence of a critical defect at the given stress. A typical plot of electric strength against specimen thickness is shown in Fig. 5.3.

The mechanism of breakdown originating from discharges in a pore is by no means self-evident, but the starting point is that the electric field in a pore is greater than that in the surrounding dielectric. For a disc-shaped cavity with its plane normal to the applied field E, the field E_c within the cavity is given by

$$E_c = \frac{\varepsilon_r}{\varepsilon_{rc}} E \tag{5.2}$$

where ε_{rc} is the relative permittivity of the cavity gas and has a value close to unity, and ε_r is the relative permittivity of the dielectric. For a spherical pore

$$E_c = \tfrac{3}{2}E \tag{5.3}$$

provided that $\varepsilon_r \gg \varepsilon_{rc}$.

When the voltage applied to a porous dielectric is steadily increased a value is reached when a discharge occurs in a particular pore. Under d.c. conditions the discharge is quickly extinguished as the field falls in the cavity. Then, following charge leakage through the material, the sites will recharge until discharge occurs again. The interval between discharges will depend upon the charge leakage time $\rho\varepsilon$ (see Eq. (5.10)), which can have minimum values of approximately 10^2 s for ceramics at room temperature.

How discharges within pores lead to total breakdown of the dielectric is a matter for debate. It may result from the propagation of a discharge 'streamer' through the ceramic, possibly progressing from pore to pore and encouraged by the increased electric stress to which the material between discharging pores is subjected and perhaps also by enhanced stress in the material close to a discharging pore. Alternatively, failure might be a direct consequence of the generation of heat by the discharge. Estimates of the rise in temperature in the material immediately adjacent to a discharging pore range from a few degrees Celsius for a low-permittivity ceramic such as alumina to about 10^3°C for high-permittivity ferroelectric ceramics. Clearly the larger the pore is the more likely it is to lead to breakdown.

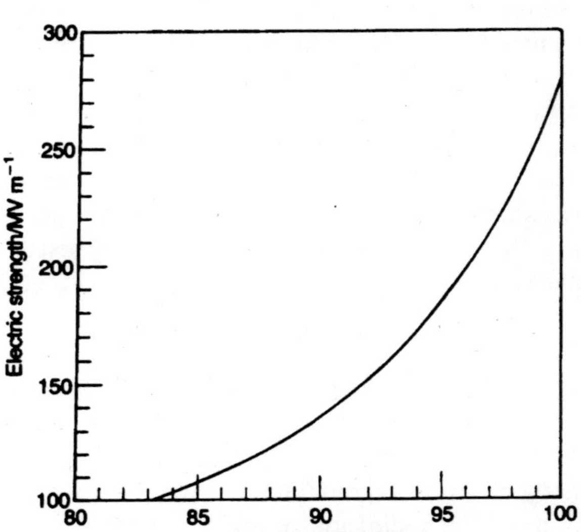

Fig. 5.4 Dependence of electric strength on specimen density (high-purity alumina).

Because discharges occur every half-cycle under a.c. conditions, breakdown is more likely than in the case of applied d.c., when the leakage time determines the discharge rate. Accordingly, a.c. breakdown voltages are lower than those for d.c.

Although there may be doubt concerning the mechanisms by which discharges lead to failure there is none regarding the fact that high electric strength is favoured by low porosity. A typical plot of density against electric breakdown strength is shown in Fig. 5.4.

Long-term effects

In some materials the prolonged application of electric stress at a level well below that causing breakdown in the normal rapid tests results in a deterioration in resistivity that may lead to breakdown. There are a large number of possible causes, and the effects may not show up in the brief tests carried out during development and production operations.

Among the possibilities are the effect of the weather and atmospheric pollution on the properties of the exposed surfaces of components. They will become roughened and will adsorb increasing amounts of moisture and conductive impurities. Surface discharges may occur which will result in further deterioration due to local high temperatures and the sputtering of metallic impurities from attached conductors.

Electrochemical action takes place under d.c. stress both on the surface and in the bulk of materials. It may cause silver to migrate over surfaces and along grain boundaries, thus lowering resistance. Sodium ions in glassy phases migrate in a d.c. field and vacant oxygen sites migrate in crystalline phases. Such changes result in local variations in electrical stress that contribute to breakdown.

The effects of differing geometries, materials and operating conditions provide a large number of possibilities for eventual failure which can only be mitigated by good design based on experience. Prolonged tests under practical running conditions are necessary before a new material becomes generally acceptable. Gradual processes that may lead to failure are discussed in the sections concerned with particular materials (see Sections 5.4.1 and 5.6.2).

5.3 Thermal Shock Resistance

Thermal shock resistance (TSR) is of importance in both the fabrication and applications of many electronic ceramics.

Precise evaluation of TSR is difficult for many reasons, one being that a failure criterion which depends on the application has to be adopted. For example, a refractory lining to a furnace may repeatedly crack during thermal cycling but, because its structure retains its overall integrity, it has not 'failed'. However, a glass beaker is usually regarded as having failed when it cracks. A criterion for failure therefore has to be defined for every usage situation, and the same applies to TSR testing. A small crack developed in a porous specimen need not constitute 'failure', but in a dense ceramic intended for advanced engineering applications it probably would. TSR is very dependent upon component shape and this, together with the precise mode of testing, is an important variable. Despite these real difficulties a useful guide to TSR is the quantity $\lambda\sigma_c/\alpha_L Y$, where λ is the thermal conductivity, σ_c is the strength, Y is the Young modulus and α_L is the coefficient of linear expansion. Therefore materials with high λ and σ_c and low Y and α_L are favoured for high-thermal-shock applications. Because σ_c/Y is approximately constant (about 10^{-3}) for many ceramics, the significant parameters determining TSR are λ and α_L. However, it must be emphasized that this is only a rough guide and the matter of TSR, like electric strength, is extremely complex.

5.4 Capacitors

Capacitors can fulfil various functions in electrical circuits including blocking, coupling and decoupling, a.c.–d.c. separation, filtering and energy storage. They block direct currents but pass alternating currents, and therefore can couple alternating currents from one part of a circuit to another while decoupling d.c. voltages. A capacitor can separate direct from alternating currents when the capacitance value is chosen such that the reactance $1/\omega C$ is low at the frequency of interest; similarly it can discriminate between different frequencies. The charge storage capability is exploited in, for example, the photoflash unit of a camera.

The operational capabilities of the various capacitor types are compared with the help of the characteristics described below.

5.4.1 Capacitor characteristics

Volumetric efficiency

Volumetric efficiency, as the name implies, is a measure of the capacitance that can be accommodated in a given size of capacitor. In the case of a parallel-plate capacitor of area A and electrode separation h,

$$C = \varepsilon_r \varepsilon_0 \frac{A}{h} \tag{5.4}$$

and the volumetric efficiency C/V is given by

$$\frac{C}{V} = \frac{\varepsilon_r \varepsilon_0}{h^2} \tag{5.5}$$

that is it is directly proportional to the relative permittivity and inversely proportional to the square of the dielectric thickness. Volumetric efficiency is an important parameter, particularly in the context of multilayer ceramic capacitors (see Section 5.4.3) where its maximization depends upon the development of novel ceramic compositions and the technology of fabricating extremely thin and electrically sound dielectric layers.

Of course, for a given working voltage U_w, the dielectric thickness cannot be reduced indefinitely because the electric field $E = U_w/h$ would eventually attain the breakdown value E_b. In fact, to achieve acceptable reliability the working voltage must be less than the normal breakdown voltage $E_b h$ by a factor η so that

$$\eta U_w = E_b h \tag{5.6}$$

From Eqs (5.5) and (5.6) the maximum permissible energy density is given by

$$\frac{CU_w^2}{2V} = \frac{\varepsilon_r \varepsilon_0 E_b^2}{2\eta^2} \tag{5.7}$$

Equation (5.7) indicates that $\varepsilon_r E_b^2$ is a figure of merit for dielectrics that are to be used at high fields. The various types of capacitor are compared on the bases of volumetric efficiencies and typical working energy densities in Table 5.1.

d.c. resistance

Ideally, the d.c. resistance of a capacitor is infinite, but in practice it will have a finite value R_L. In the case of the parallel-plate capacitor, and confining attention to the dielectric,

$$R_L = \rho \frac{h}{A} \tag{5.8}$$

where ρ is the resistivity of the dielectric. A charged capacitor will discharge through its own resistance according to

$$Q(t) = Q_0 \exp\left(-\frac{t}{\tau}\right) \tag{5.9}$$

in which $Q(t)$ is the charge remaining at time t, Q_0 is the original charge and $\tau = R_L C$ is the time constant of the capacitor. τ depends only on the dielectric material, as is evident from substituting from Eqs (5.4) and (5.8):

$$R_L C = \frac{h}{A} \rho \frac{\varepsilon_r \varepsilon_0 A}{h} = \varepsilon_r \varepsilon_0 \rho \tag{5.10}$$

Although the above analysis is applicable to the great majority of standard capacitors, there are exceptions. For example, where very high working voltages (>1 kV) are involved, the d.c. resistance may be determined by that of

Table 5.1 Typical values of volumetric efficiency and energy density for the various types of capacitor

Capacitor type	*Volumetric efficiency* (μF cm^{-3})	*Typical working energy density** (mJ cm^{-3})
Electrolytics		
Aluminium	10	500
Tantalum chip	600	30
Double layer carbon ('Supercapacitors')	$\sim 3 \times 10^6$	~2500
*Polymer film***		
Wound	0.02–2	1–10
Multilayer polymer	~ 5	~ 1
Mica	5×10^{-3}	not relevant***
Single layer ceramic		
NP0/C0G	$\sim 10^{-4}$	0.25
Z5U/Y5V	8×10^{-2}	~ 0.1
Ceramic multilayer		
NP0/C0G	0.5	1.5
X7R	30	35
Z5U/Y5V	300	40

The values, derived from manufacturers' data sheets, are only intended to permit general comparisons to be made; capacitor coatings and cases are not allowed for in estimating volumes.
NP0/C0G, X7R, Z5U/Y5V are defined in Table 5.10 and Fig. 5.40.
*The values are calculated based on stated maximum 'working' voltages.
**The quoted values are an estimate from data for polystyrene, polypropene, polyester and polycarbonate capacitors.
***Mica capacitors are used for stability in RF circuits where the energy density is irrelevant.

interelectrode surfaces. To avoid this the outside surfaces of such capacitors need to be kept scrupulously clean and dry.

Equivalent parallel and series resistance

When an a.c. voltage is applied to a perfect capacitor, no energy is dissipated. However, a real capacitor dissipates energy because of lead and electrode resistances, d.c. leakage resistance and, most importantly, dielectric losses. These account for the capacitor's 'dissipation factor' or 'loss tangent' $\tan\delta$. It is sometimes convenient to regard the 'lossy' capacitor as an ideal capacitor shunted by a resistance R_p or in series with a resistance r_s, as shown in Fig. 5.5.

It is readily shown that

$$R_p = \frac{1}{\omega C \tan\delta} \qquad \text{and} \qquad r_s = \frac{\tan\delta}{\omega C} \tag{5.11}$$

According to Eq. (5.11) R_p and r_s are explicitly inversely proportional to ω; however, it must be remembered that, in general, C and $\tan\delta$ depend on ω and temperature. Therefore the matter is more complex than appears at first sight.

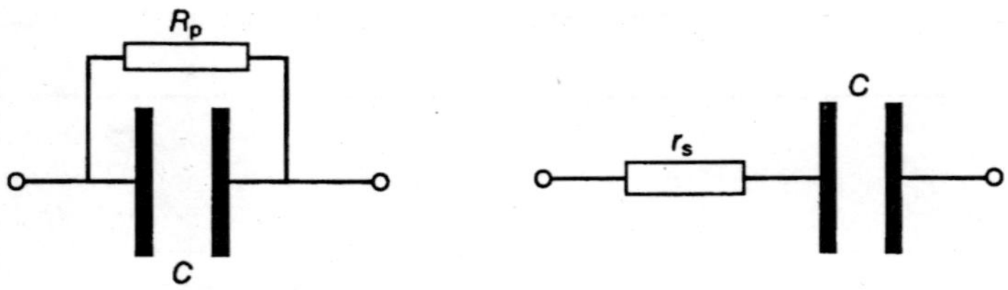

Fig. 5.5 Parallel and series resistance models of a lossy capacitor.

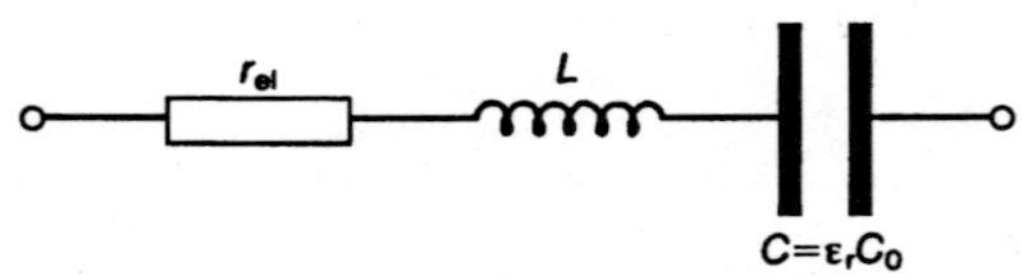

Fig. 5.6 Capacitor with series inductance and resistance.

In considering the practical application of capacitors, the equivalent series resistance (e.s.r.) is more significant since it carries the total current passed by the capacitor.

Resonance frequency

The equivalent circuit for a capacitor shown in Fig. 5.6 is more representative at high frequencies than those shown in Fig. 5.5, as it takes separate account of the resistance r_{el} and the inductance L of the leads and electrodes and of losses in the capacitor dielectric (for the meaning of $C^* = \varepsilon_r^* C_0$, see Section 2.7.2).

It is instructive to examine how the equivalent circuit for a typical capacitor responds to frequency. As an example we choose a mica capacitor with the following characteristics, which are assumed to be independent of frequency: $C = 100\,\text{pF}$, $L = 100\,\text{nH}$, $\tan\delta = 2\times 10^{-4}$ and $r_{el} = 1\,\Omega$. Now

$$Z = r_{el} + \mathrm{j}\omega L - \frac{\mathrm{j}}{\omega C^*} \tag{5.12}$$

$$= r_{el} + \mathrm{j}\left\{\omega L - \frac{1}{\omega C_0(\varepsilon_r' - \mathrm{j}\varepsilon_r'')}\right\}$$

$$= r_{el} + \mathrm{j}\left(\omega L - \frac{1+\mathrm{j}\tan\delta}{\omega C'(1+\tan^2\delta)}\right) \tag{5.13}$$

Because $\tan^2\delta \ll 1$, it follows that

$$Z \approx r_{el} + \frac{\tan\delta}{\omega C'} + \mathrm{j}\left(\omega L - \frac{1}{\omega C'}\right) \tag{5.14}$$

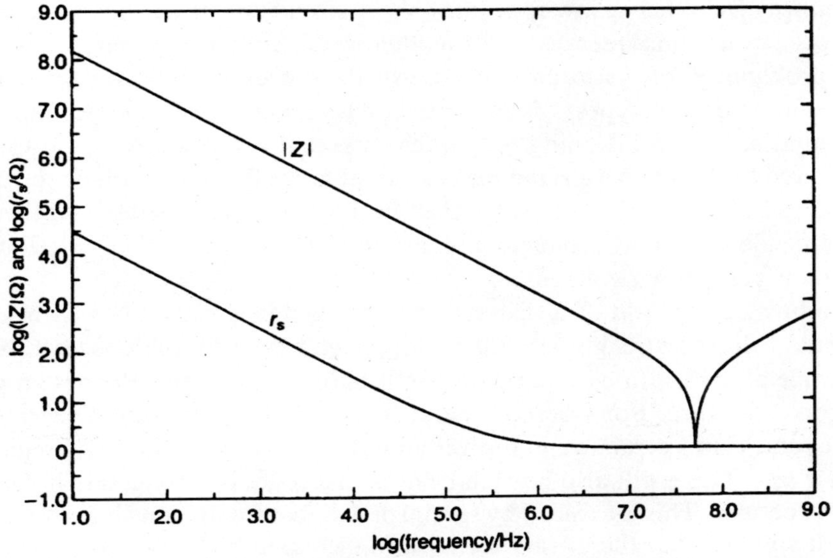

Fig. 5.7 Frequency response of the equivalent series resistance r_s and the impedance $|Z|$ for a typical mica capacitor ($C = 100\,\mathrm{pF}$, $L = 100\,\mathrm{nH}$, $R_{el}=1\,\Omega$, $\tan\delta = 0.0002$).

in which the sum of the first two terms is the equivalent series resistance r_s. Evidently, as ω increases, r_s decreases and approximates to r_{el}. Figure 5.7 shows r_s and $|Z|$ plotted as functions of ω, where

$$r_s = r_{el} + \frac{\tan\delta}{\omega C'}$$

and

$$|Z| = \left\{ r_s^2 + \left(\omega L - \frac{1}{\omega C'} \right)^2 \right\}^{1/2} \tag{5.15}$$

Clearly the resonance effect places an upper limit on the frequency at which the capacitor can normally be used. Above resonance the reactance of a capacitor is inductive (ωL), and this is significant for some applications. For example, ceramic multilayer capacitors, which are discussed later, are commonly used to decouple high-speed computer circuits, and a prime function is to eliminate noise which has frequency components above the resonance frequency. Therefore the inductive reactance must be kept to a minimum.

Breakdown and degradation

Breakdown and degradation have been discussed in general terms in Section 5.2.2. Breakdown in a capacitor results in the replacement of a reactive insulating component by either a low-resistance short circuit or an open circuit, usually with disastrous consequences as far as the overall circuit function is concerned. The probability of its occurrence must therefore be kept to an absolute minimum. Certain types of capacitor, e.g. metallized polyester film and electrolytics, have self-healing properties such that a breakdown causes a brief short circuit followed by a rapid restoration of normal reactive behaviour. So far no method of making a ceramic unit self-healing has been devised, and therefore recommended working voltages are usually about a factor of 5 below the minimum breakdown voltage.

The breakdown field of a dielectric in a capacitor is generally well below its intrinsic dielectric strength since it is largely governed by the concentration of flaws (metallic inclusions, pores, surface roughness, etc.) and the design of the electrode structure. For instance, metal from the electrode may penetrate into surface cracks in a ceramic and form conductive spikes which will have high fields at their tips. There will also be high fields at the edges of conductors in parallel-plate structures. This is avoided by special design features in units intended for use at high voltages, but these features result in increased bulk and cost.

The effects of moisture must be limited by ensuring that the residual porosity in ceramics consists of isolated pores so that water does not penetrate into the bulk. Units must be protected from humidity either by encapsulation in polymers with a low permeability to water vapour or by enclosure in sealed metal cans.

Tests for long-term deterioration can be speeded up in some cases by performing them at higher temperatures and using an Arrhenius relation to estimate the probability of failure under normal working conditions. However, it is essential to check the validity of the relation between failure rate and temperature since there may be effects that are stronger at low temperatures than at high. For instance, liquid water will disappear above 100 °C and so any electrochemical effects occurring at lower temperatures will be eliminated.

Degradation must be distinguished from the process known as 'ageing' (cf. Section 2.7.3) which occurs in ferroelectric materials. Ageing is due to the presence of ferroelectric domains and to a change in the mobility of the walls between them as time passes. The changes that occur in the properties do not lead to an increase in the probability of breakdown, only to a limited fall in permittivity and small changes in other properties. Since the dissipation factor diminishes, ageing has some beneficial aspects.

5.4.2 Non-ceramic capacitors

It is desirable to have, as a background to a discussion of ceramic capacitors, some knowledge of alternative dielectrics and the capacitor structures used with them. The principal types are described below and their capacitance ranges indicated in Fig. 5.8.

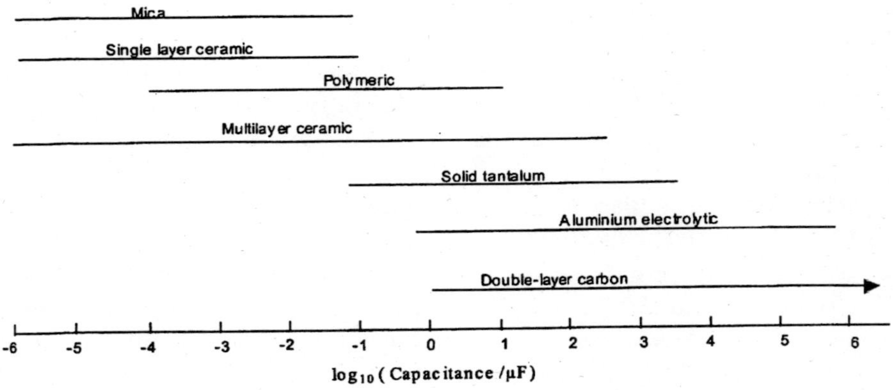

Fig. 5.8 Capacitance ranges covered by commercial capacitor types.

Polymer-film capacitors

Essentially polymer-film capacitors comprise dielectric films (polymer or paper or both together) interleaved with aluminium electrodes, either as aluminium foil or, more commonly, in the form of a layer evaporated directly on the dielectric, and rolled together. They are sealed in an aluminium can or in epoxy resin. Because the dielectric films and evaporated electrodes have thicknesses of only a few microns and about 0.025 μm respectively, volumetric efficiencies can be high. The dielectric films are polystyrene, polypropylene, polyester, polycarbonate or paper; paper dielectrics are always impregnated with an insulating liquid.

The use of metallized electrodes on certain polymers provides the capacitor with self-healing properties during manufacture and throughout its working life. If breakdown should occur at a fault in the dielectric layer, the local heating causes the deposited aluminium in the vicinity to evaporate, leaving an area bare of metal around the hole produced by the breakdown.

Polystyrene capacitors have exceptionally low $\tan\delta$ values ($<10^{-3}$), making them well suited for frequency-selective circuits in telecommunications equipment. Polymer capacitors are widely used for power-factor correction in fluorescent lighting units, and in start/run circuitry for medium-type electric motors used in washing machines, tumble-dryers and copying machines for example. They are also used in filter circuits to suppress radio frequencies transmitted along main leads. Such interference 'noise' may originate from mechanical switches, furnace controllers and switch mode power supplies; it not only spoils radio and television reception but can also cause serious faults in data-processing and computer equipment.

Aluminium electrolytic capacitors

The aluminium electrolytic capacitor comprises two high-purity (99.99% Al) aluminium foil electrodes, approximately 50 μm thick, interleaved with porous paper and wound into the form of a cylinder. One electrode – the anode – carries an anodically formed alumina layer approximately 0.1 μm thick. The completed winding is placed in an aluminium can where the porous paper is vacuum impregnated with one of a number of possible electrolytes (e.g. adipic acid or ammonium pentaborate). The 'formed' aluminium foil is the anode, the Al_2O_3 layer is the dielectric and the electrolyte together with the 'unformed' aluminium foil is the cathode. After the capacitors have been sealed in the can, they are 're-formed' by subjecting them to a d.c. potential sufficient to heal any possible damage to the oxide layer caused during manufacture. Provided that the anode is maintained at a positive potential during use, minor breakdowns in the Al_2O_3 dielectric will be healed by electrolytic action. Both aluminium foils are etched to increase their effective areas (by a factor of over 20 for low-voltage capacitors) and this, together with the very thin dielectric, leads to high volumetric efficiencies.

Aluminium electrolytic capacitors are exploited in a range of applications and their relatively low cost makes them attractive for printed circuits for car radios, stereo equipment, pocket calculators, digital clocks, etc. Also, the very high value capacitors are used in large photo-flash equipment and for voltage smoothing.

A particularly important application for aluminium electrolytic capacitors is in switch mode power supplies (SMPSs) which are now extensively used, especially in computer systems. In this application the capacitor is used essentially to smooth a rectified voltage, but it inevitably passes a ripple current I which, because of the capacitor's e.s.r., r_s, leads to power losses I^2r_s. The switching frequency determines the size of an SMPS, and frequencies have increased from about 50 kHz to about 300 kHz over the past decade. This has led to the multilayer ceramic capacitor's challenging the aluminium electrolytic in this important application, and the signs are that it will continue to do so.

Tantalum electrolytic capacitors

There are two types of tantalum electrolytic capacitor: 'wet' and 'solid'. Both varieties consist of a porous anode made by sintering tantalum powder at 1800 °C in vacuum. In the wet type the porous structure is impregnated with sulphuric acid, anodized to form a thin layer of Ta_2O_5 and encapsulated in a tantalum container that also serves as the cathode. The use of sulphuric acid gives a lower e.s.r. than that of the aluminium electrolytic and increases the temperature range within which the unit can be run. In the solid type the liquid electrolyte is replaced by MnO_2 and the cathode can be graphite paint overlaid with silver. The units may be encapsulated in a polymer.

Because tantalum capacitors are very stable with respect to temperature and time and have high reliability, they are widely used, particularly in large main-frame computers, military systems and telecommunications.

Double-layer carbon capacitors

When two electrodes are immersed in an electrolyte and a small potential difference is applied between them, anions (negatively charged) move close to the anode and cations (positively charged) move close to the cathode. Provided the applied voltage is below the dissociation voltage for the electrolyte, electric double layers are established close to the electrodes. The very small separation (0.5–1.0 nm) of the opposite charges is mainly resonsible for the high electrical capacitance associated with the double layers. The principle has been known for more than a hundred years, but it is only within the last decade that interest in double-layer carbon capacitors (DLCCs) has intensified.

In the DLCC the electrodes are carbon as this has an exceptional combination of chemical enertness, electrical conductivity and low cost along with availability as a material with high surface area to volume ratio. A variety of electrolytes is used. If the electrolyte is aqueous then the applied voltage must be limited to below approximately 1.2 V to avoid dissociation of the water molecule. Some non-aqueous electrolytes are used that raise the operating voltage to close to 3 V.

The capacitance which can be packed into a given volume depends, of course, upon the surface area (typically approximately $2500\,m^2\,g^{-1}$) of the carbon electrodes and this is commonly a woven carbon fibre cloth. A non-conducting porous membrane separates anode from cathode. Single cells offering a capacitance greater than 2000 F at 3 V are available commercially. Cells are connected in series to raise the operating voltage.

The applications for the capacitors are many and varied but fall into three main categories. One is to provide back-up power when batteries are changed or the main power fails in, for example, video-recorders and cameras, or for TV channel and car radio settings. In another type of application the capacitor is used as the main power source, for example for toys and watches. It is also used for the illumination of public information signs, the primary source being solar power.

Other possible, and more significant, applications are in connection with road transport, for example, 'load-levelling' and 'regenerative braking'. In load-levelling large capacitor power packs would be used in tandem with the primary power source which may be a battery, a fuel cell or an internal combustion engine (see Section 4.5.1). In regenerative braking the capacitor has the capability to store energy extracted during braking, making it available for subsequent acceleration.

A drawback inherent in the carbon capacitor is the low working voltage of the basic cell. The high working voltages, for example 500 V, required for some 'transport' applications, necessitates connecting many capacitors in series, at the cost of volumetric efficiency.

The capacitor is discussed in detail by R. Kötz and M. Carlen [3].

Mica capacitors

Mica for use as a dielectric is obtained from the mineral muscovite ($KAl_2(Si_3Al)O_{10}(OH)_2$) which can be cleaved into single-crystal plates between 0.25 and 50 μm thick.

One common construction consists of mica plates carrying fired-on silver electrodes stacked and clamped together to form a set of capacitors connected in parallel. The assembly is encapsulated in a thermosetting resin to provide protection against the ingress of moisture which would increase both the capacitance and tan δ.

Special features of mica capacitors are long-term stability (for example, $\Delta C/C \approx 0.03\%$ over three years), a low temperature coefficient of capacitance (TCC) (+10–80 MK^{-1}) and a low tan δ.

5.4.3 Ceramic capacitors

Of the various geometrical forms of ceramic capacitor, discs, tubes and multilayer, world-wide production is dominated by the multilayer variety.

Since the first multilayer ceramic capacitor (MLCC) was introduced in the early part of World War II there have been two principal development trends. One is towards smaller sizes and higher capacitance values, that is towards maximizing volumetric efficiency, and the other is cost reduction. These developments have had to be accompanied by improved reliability, which assumes increasing relevance as the number of capacitors in a given piece of equipment, for example a PC or mobile phone, steadily increases.

These developments have only been possible because of a continuing improvement in understanding of the basic solid state science of the ceramic dielectrics, of the electrode metals and of the interaction between the two, and particularly of the technologies of ceramic powder production and the MLCC fabrication processes.

Classes of dielectric

Ceramic dielectrics and insulators cover a wide range of properties, from steatite with a relative permittivity of 6 to complex ferroelectric compositions with relative permittivities exceeding 20 000. For the purposes of this discussion insulators will be classed with low permittivity dielectrics, although their dielectric loss may be too high for use in capacitors. Reference should be made to Table 5.10 and Fig. 5.40.

Class I dielectrics usually include low- and medium-permittivity ceramics with dissipation factors less than 0.003. Medium-permittivity covers an ε_r range of 15–500 with stable temperature coefficients of permittivity that lie between +100 and −2000 MK^{-1}.

Class II/III dielectrics consist of high-permittivity ceramics based on ferroelectrics. They have ε_r values between 2000 and 20 000 and properties that vary more with temperature, field strength and frequency than Class I dielectrics. Their dissipation factors are generally below 0.03 but may exceed this level in some temperature ranges and in many cases become much higher when high a.c. fields are applied. Their main value lies in their high volumetric efficiency (see Table 5.1).

Class IV dielectrics contain a conductive phase that effectively reduces the thickness of dielectric in capacitors by at least an order of magnitude. Very simple structures such as small discs and tubes with two parallel electrodes can give capacitances of over 1 μF. Disadvantages are low working voltages, mostly between 2 and 25 V and high losses. 'Barrier layer' capacitors are now very little used and regarded as, practically speaking, obsolete.

Form and fabrication

Ceramic capacitors must either be low in cost in order to compete with polymer-film units or must possess special qualities that assure them of a market. The structure and method of manufacture of a capacitor are governed by a combination of these requirements.

Discs and tubes Discs can be formed by dry pressing calcined and milled powders containing some 5–10 vol.% of an organic binder. Alternatively, they can be cut from extruded ribbon or band-cast tape. The pieces are fired in small stacks and, when suitably formulated, do not adhere strongly to one another during sintering. The furnace atmosphere must be fully oxidizing for compositions (other than Class IV) containing TiO_2, both at the peak temperature and during cooling. After firing, silver paint is applied to the major surfaces and the discs are briefly refired in a single layer at 600–800 °C.

The next stages are usually fully automated. The discs are fed between clips formed from the tinned copper wire that will eventually form the leads. They are first warmed over a heated solder bath which is then briefly raised so as to

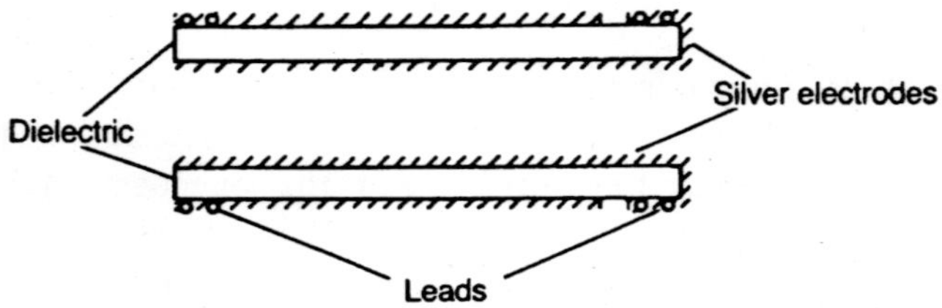

Fig. 5.9 Longitudinal section showing silvering of a tubular capacitor.

immerse the discs and coat them with solder. The preliminary warming lessens the thermal shock (see Section 5.3) of immersion in molten solder. The time of immersion must be brief since silver dissolves rapidly in the molten metal. The units are next coated with a polymer, usually an epoxy resin, baked to cure the resin and tested for capacitance value and voltage breakdown.

Tubes are formed by extrusion. They have the advantage of being less fragile than flat pieces and are more suitable in some types of circuit assembly. After sintering, the tubes are completely coated with silver. Automatic machines then grind the silver from one end and remove a ring of silver from the outer surface near the other end (Fig. 5.9). The leads are looped round each end and soldered in place by immersion in solder.

As an alternative to solder, nickel can be deposited from a solution containing nickel salts and strong reducing agents such as sodium hypophosphite. This process is termed 'electroless plating'. The surface of the ceramic must first be activated by successive immersion in dilute stannous chloride and palladium chloride solutions. The adhesion of the nickel is inferior to that of fired-on silver but adequate to secure a wrapped-round wire on a tube. Nickel has the advantage of insolubility in solder but acquires a tough layer of oxide if exposed to air for a prolonged period and cannot then be wetted by solder.

Lead and electrode inductance can be somewhat less in discs than in tubes so that discs have some advantage at higher frequencies. The two shapes are similar in volumetric efficiency since their bulk largely consists of the encapsulating resin. The tubular geometry is suited to the manufacture of 'feed-through' capacitors. A schematic diagram of the cross-section of such a capacitor is shown in Fig. 5.10. These are used as bypass capacitors in television and FM tuners.

Disc and tubular shapes are used for all classes of dielectric since they are lowest in cost. Using Class I dielectrics they cover the 0.1–1000 pF range, with Class II/III they cover 1000–100 000 pF and with Class IV they cover 0.1–2 μF. Except for Class IV, the safe working voltages are usually at least 100 V although in electronic circuits they are not likely to encounter applied voltages of more than 10 V. Dielectric thicknesses lie in the 50 μm to 2 mm range with the thicker units able to withstand the normal mains supply. Discs range in diameter (or side) from 2 to 30 mm whilst tubes may be 5–60 mm long × 1–10 mm in diameter; these are bare dimensions without an encapsulant.

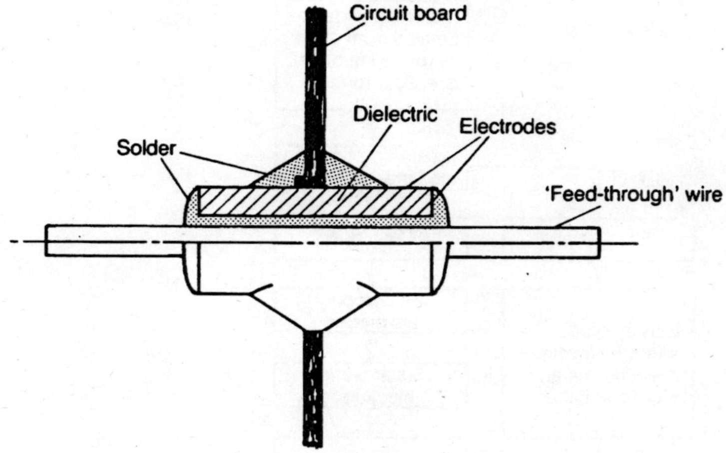

Fig. 5.10 Cross-section of 'feed-through' tubular capacitor.

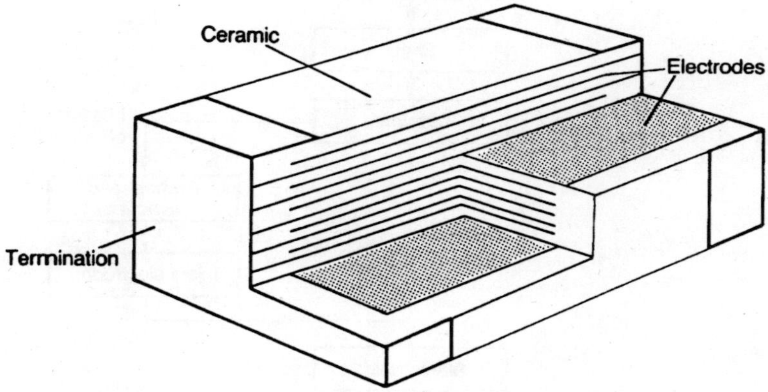

Fig. 5.11 Schematic diagram of a multilayer ceramic capacitor construction.

Multilayer capacitors A critical step in the manufacture of multilayer capacitors is, of course, the barium titanate-based starting powders, and the various routes for producing these are described in Section 3.4. The multilayer capacitor structure (Fig. 5.11) enables the maximum capacitance available from a thin dielectric to be packed into the minimum space in a mechanically robust form.

Broadly speaking there are two significant manufacturing processes, the 'dry' and the 'wet'; there is a third, the 'fugitive electrode' process, developed to replace the expensive palladium-containing electrodes with a cheap metal, but it has ceased to be of any commercial significance. The steps in the various processes are summarized in Fig. 5.12.

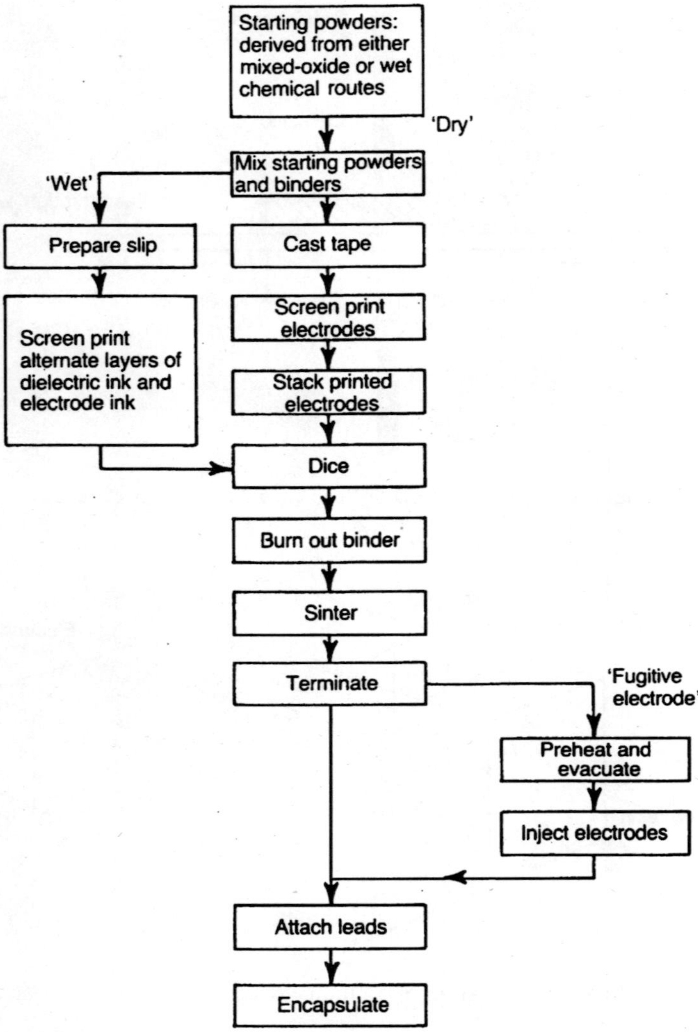

Fig. 5.12 Outline of the fabrication process for multilayer ceramic capacitors.

The thickness of the dielectric layers produced in the 'dry' process is typically in the range 5–8 μm with 3 μm being achieved and $\sim 2\,\mu$m a target for some manufacturers, the objective being to maximize volumetric efficiency. The dielectric layers manufactured by the 'wet' process are thicker, typically $\sim 20\,\mu$m. The overall capacitor dimensions range from approximately $1.0 \times 0.5 \times 0.5$ mm to $6 \times 6 \times 2.25$ mm, and the corresonding capacitance values from 1 pF to 300 μF. Generally speaking the dielectrics employed are Class I or Class II/III.

In the 'dry' process the dielectric powder is formed into either an organic- or water-based slip and tape-cast (see Section 3.6.6) to form a continuous strip. The

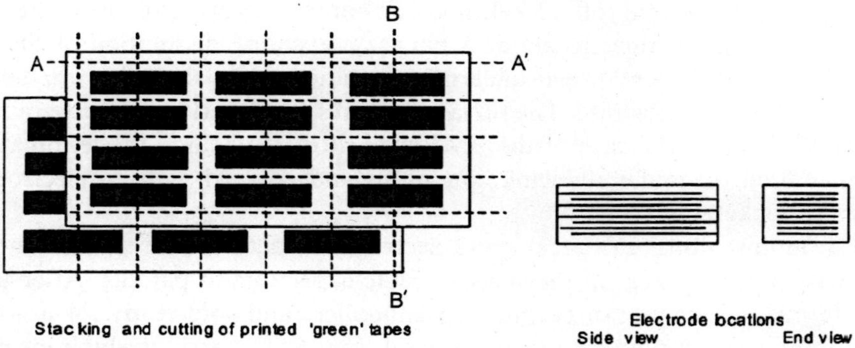

Fig. 5.13 Stacking arrangement and cutting of laminated printed sheets for forming multilayer ceramic capacitor.

cast tape is cut into sheets onto which the electrode pattern is printed as shown in Fig. 5.13. The electroded sheets are stacked and consolidated under pressure at about 70 °C. The consolidated stack is diced by cutting along lines such as AA′ and BB′, to expose electrodes of successive layers at opposite end-faces. The polymer binder, which may consist of up to 35 vol.% of the green body, is next carefully removed by heating in air. This is a critical step which must be accomplished without any disruption of the structure. Following removal of the organic material, the 'chips' are fired to the ceramic sintering temperature during which process the electrodes must remain solid and in place.

In the 'wet' process a slip carrying the ceramic powder is laid down, by screen-printing for example, onto a temporary carrier such as a glass tile. The process can be repeated to build up the required thickness of the dielectric onto which the electrodes are screen-printed. The next dielectric layer is then laid down and the process repeated. The mutlilayer structure is diced as described above, and the individual chips removed from the tile for the subsequent stages, as for the 'dry' process.

Extreme care has to be exercised to avoid contaminating the tape-cast sheets with 'dust' which would seriously reduce yield and impair reliability, and a high level of cleanliness, including 'clean room' standards, are required for processing the thinner layers.

The inks for screen-printing the electrodes contain the sub-micron metal powder, either a Ag–Pd alloy or a base metal, usually nickel (melting point, ~1455 °C) but sometimes copper (melting point ~1084 °C). Palladium (melting point, ~1554 °C) and silver (melting point, ~961 °C) form solid solutions with melting points approximately proportional to the content of the end members.

The sintered chips are next 'terminated' which involves coating the ends with a paint, in the case of Ag–Pd electroding consisting of a mixture of powdered silver and glass frit, and a suitable organic vehicle, which is fired on at about 800 °C.

Thin (~2 μm) layers of nickel and tin are electroplated over the silver termination, the former to act as a barrier preventing dissolution of the silver during the subsequent 'wave-soldering' operation when the chips are 'surface-mounted' onto a substrate. The tin layer ensures good wetting of the termination by the solder. In the case of the BME capacitor (see below) the termination is copper fired on under reducing conditions and covered with an electroplated layer of nickel.

The terminations must make good electrical contact with the exposed alternate electrodes, connecting up the stack of plate capacitors in parallel. After testing the terminated chips can be put into bandoliers and sold ready for automatic mounting onto printed circuits (Chapter 4, Fig. 4.5). Alternatively, leads can be soldered to the terminations and the chip encapsulated in a suitable polymer and then bandoliered.

The high cost of palladium has been a stimulus to reducing its content in the Ag/Pd alloy or of eliminating it entirely. The former strategy necessitates developing ceramic dielectrics which sinter at correspondingly lower temperatures (see Section 5.7.2) or, in the case of base metal electroded (BME) MLCCs, are stable against reduction firing (see Section 5.7.3).

BME technology presently (2002) accounts for at least 50% of total MLCC production and the trend seems set to continue for as long as the price of palladium dictates. The thermodynamic and chemical factors involved in the production of a reliable BME capacitor are discussed in Section 5.7.3.

As in the case of the BME capacitor, the 'fugitive electrode' process was developed to avoid the high expense involved in using palladium. In this process the 'electrodes' are printed using an ink consisting of an organic vehicle carrying coarse (~ 10 μm) uncalcined powders ($BaCO_3$, TiO_2, ZrO_2). Following sintering, porous 'electrode' cavities 2–5 μm thick are retained in the ceramic. After the chip is terminated with porous silver it is evacuated and then, under pressure, the electrode cavities are impregnated with a low melting metal or alloy, for example lead. The technology is not now exploited to a significant extent, if at all.

High-voltage and high-power capacitors From the point of view of sustaining high voltages, the structures so far discussed have the disadvantage that the electrodes remain parallel up to their edges. For example, in a parallel-plate capacitor the field is normal to the electrodes in central regions remote from their edges but extends into the dielectric beyond the electrode edges with the result that there is an electric field concentration near the edges as indicated in Fig. 5.14(a). As a result the field strength close to the electrode edge may be as much as twice the average value. This is of little consequence when the average field is a small fraction of the breakdown strength but becomes extremely important in high-voltage units.

The field at the edges can be reduced by shaping the dielectric into the form of two cups as shown in Fig. 5.14(b). This shape ensures that the electrode edges are

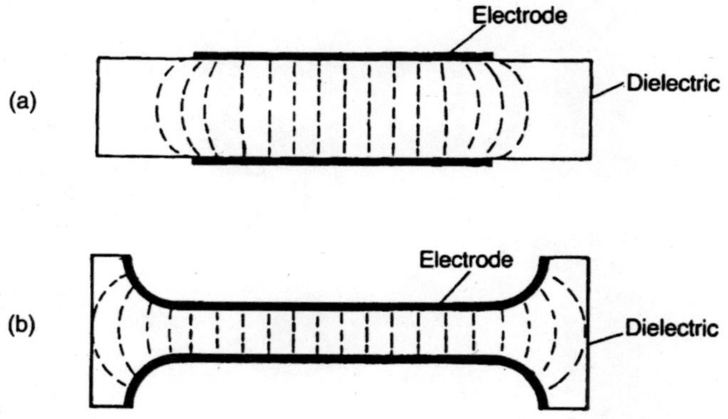

Fig. 5.14 (a) Converging lines of force at electrode edges; (b) dielectric shaping for high voltage capacitors.

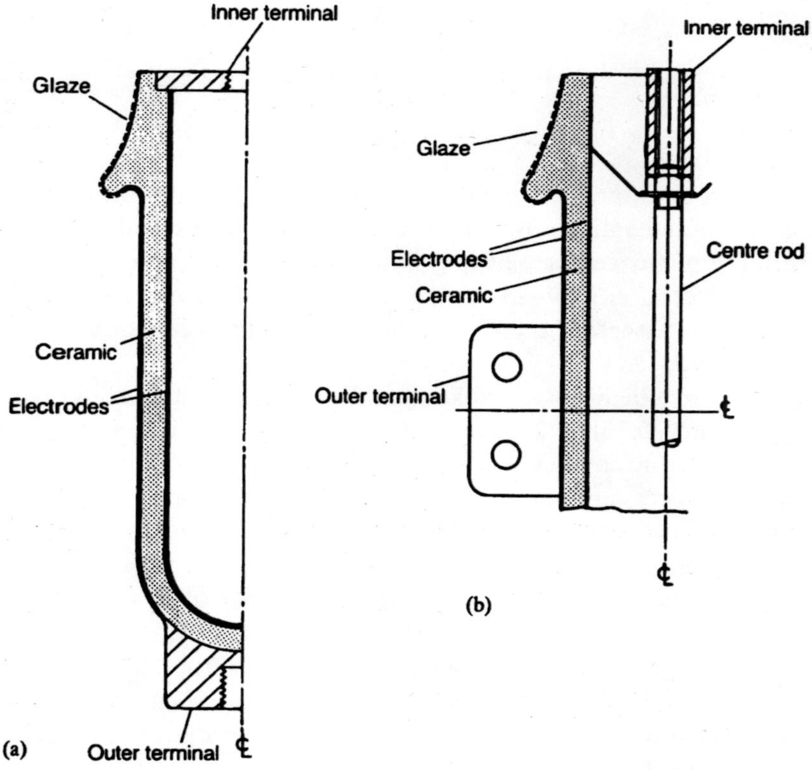

Fig. 5.15 Sections through high-voltage high-power capacitors illustrating electrode edge configurations to avoid field concentration; (a) pot, and (b) cylinder.

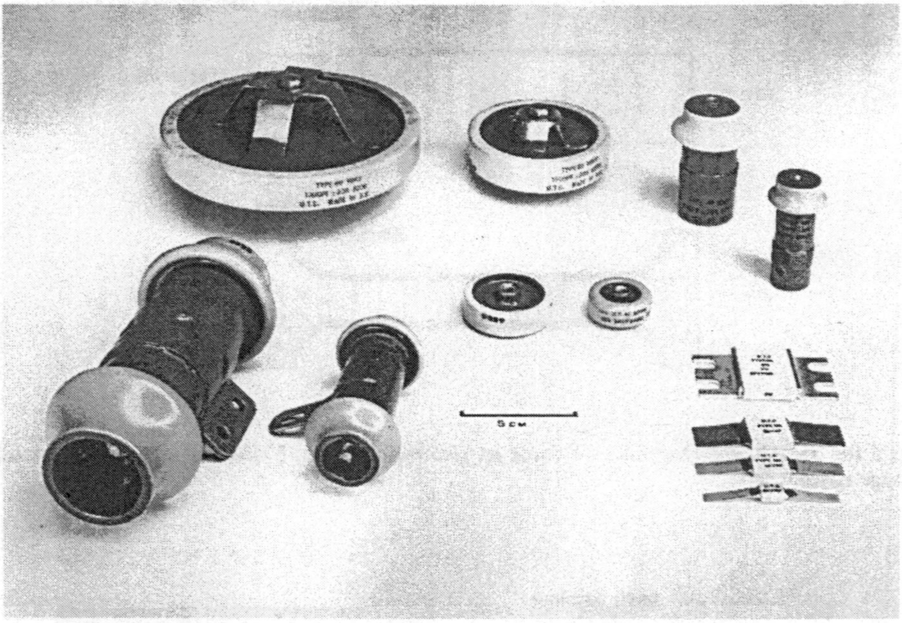

Fig. 5.16 A range of high-power transmitter capacitors (components supplied by Morgan Electro Ceramics).

separated by several times the dielectric thickness in the centre of the unit. It also increases the path length across the surface between the electrodes and so lessens the possibility of surface breakdown. Surface breakdown occurs at lower fields when the dielectric is exposed to the atmosphere. Therefore in high-voltage applications it is essential to encapsulate materials in closely adherent glaze or polymeric substances.

The ceramic in a high-power capacitor is disc shaped, cylindrical or pot shaped (see Figs 5.15 and 5.16) with thicknesses of up to 10 mm. It is essential that the dielectric is free from cavities and has a uniform density in order to avoid plasma discharges in pores and local high-field regions. Pressing is therefore best undertaken hydrostatically with the final shaping of the edges carried out on the unfired piece on a lathe.

Metallization usually comprises three stages: a layer of fired-on silver with a high frit content to secure adhesion, a second layer of fired-on silver with a low frit content to give maximum conductivity and an electroplated layer of copper to prevent solder dissolving the silver. The leads on the larger units are thick copper foils fixed to the metallization with solder. The soldering is necessarily a prolonged process to minimize the risk of failure by thermal shock. The edges of the electrodes and the unelectroded portions of the ceramic surface are glazed, and the glaze and the metallization act as an encapsulating layer. Anything thicker would diminish heat dissipation and limit the power which the unit could pass.

Only Class I dielectrics are used in high-frequency (50 kHz–> 100 MHz) power capacitors. Typical ranges of parameter values are as follows: capacitance, 2 pF to 12 nF; peak voltage, 3–25 kV; rated power, 0.25–500 kVA; tan δ, 0.0005–0.001.

PART II PRINCIPAL CERAMIC TYPES AND APPLICATIONS

5.5 Low-Permittivity Ceramic Dielectrics and Insulators

Low-permittivity ($\varepsilon_r < 15$) dielectrics are widely used for straightforward insulation. In this case their mechanical properties may be more important than their dielectric properties, and, when required in tonnage quantities, they must also be based on a low-cost material. When used as substrates for components their dielectric properties become of greater importance. They find some applications as capacitor dielectrics where very small capacitances are required for use at higher frequencies or, as can be the case with capacitors passing high current, where larger size is an advantage because it leads to a higher rate of heat dissipation.

The siliceous ceramics are produced from naturally occurring minerals that have been purified to only a limited extent, for instance by washing out soluble impurities or removing iron-containing contaminants magnetically. The purer oxides used in capacitors require more elaborate processing. The discussion starts with insulating components based on natural minerals.

5.5.1 Electrical porcelains

Clay-based

Ceramic insulators used for the distribution of electrical power at high voltage will be taken as an example of an insulating siliceous clay-based ceramic.

Electrical power is transmitted at high voltage because, for a given power carried on a line, when the voltage is doubled the current is halved. As a result the quantity of conductor material can be reduced to a quarter for the same power loss (I^2R) in the line (see Eq. (2.27)). Therefore, since the cost of conductor materials is high compared with other costs, very significant economies result from an increase in voltage. Materials of high tensile strength such as steel-cored aluminium can be used to replace copper, allowing relatively longer spans and fewer towers. This cost-saving more than offsets the cost of the extra conductor volume required to compensate for the higher resistive materials. Many factors influence the choice of operating voltage, but up to 1000 kV is used.

Transmission by d.c. is sometimes used, since with a.c. systems peak voltages and currents have to be accommodated while only r.m.s. (see Eq. (2.94)) values represent useful power transmission. However, the cost of rectification and conversion back to a.c. of the power at high voltage results in d.c. transmission only being economic for very long (>500 km) lines.

As an example of relative costs, in a conventional 275 kV a.c. line, the towers represent 20–40% of the cost, the conductors 15–35%, while insulators represent only approximately 5–10%.

The electrical porcelain used for such insulators is chosen because of its excellent durability in outdoor conditions, its low cost, its adequate mechanical strength (insulators rated to safely withstand tensile loading up to 90 tonnes are produced) and the ability to produce large monolithic units (the largest produced to date is 12 m long and nearly 2 m in diameter). The dielectric strength (see Section 5.2) is less critical, since the insulators are normally used in air, which limits the electrical stress which can be applied.

The raw materials used in the manufacture of electrical porcelains are clays, fluxes and fillers. Clays are aluminosilicates in the form of small (about 1 μm) platey particles. Kaolinite, with the composition $Al_2Si_2O_5(OH)_4$[†], is the usual clay mineral. Fluxes are commonly feldspars, a group of aluminosilicate minerals containing potassium, sodium and calcium. The potassium form, orthoclase ($KAlSi_3O_8$) is often preferred because of the better electrical and mechanical properties it confers on the porcelain. Fluxes melt at relatively low temperatures to form a glass, which bonds the structure of the fired material, and on cooling crystallizes to give a web of fibrous mullite crystals. Common fillers are finely milled silica (SiO_2) as quartz or cristobalite, or alumina (either as calcined bauxite or refined alumina). Where higher mechanical strength is required silica is replaced by alumina thus limiting the strength-reducing internal micro-cracking resulting from the silica crystal inversions which occur on cooling (Fig. 5.20(c)). The penalties attached to 'aluminous porcelains' are increased cost and relatively poor manufacturing characteristics. Typical compositions lie in the ranges: clays 40–60 wt%: fluxes 15–25 wt% and fillers 25–40 wt%. Control of the fineness of the fluxes and fillers is critical.

The raw materials are blended with water to form a slip from which most of the water is removed by filter-pressing. The 'body' is then homogenized and de-aired in a vacuum extrusion 'pug-mill'. The extruded blanks may be further

† It is common practice to write the chemical formula of kaolinite as $Al_2O_3\,2SiO_2\,2H_2O$ and important to understand that this is a convention (applied to many minerals of interest to the ceramist) and does *not* mean that kaolinite is a mixture of alumina, silca and water!

formed by pressing or by 'jolleying', a process similar to shaping on a potter's wheel (see Section 3.6.3). More usually the blanks are dried to a rigid, but still wet state (~16 wt% moisture) or to a nearly dry (~8 wt% moisture) state, and then machined to shape by turning, milling, etc. Some manufacturers spray-dry the slip and isostatically press blanks from the powder prior to final machining.

After the forming stage the insulator is carefully dried and then coated with a glaze consisting of a water slurry of milled quartz, feldspar and calcium carbonate, plus zircon as an opacifier and, if required, an oxide-based pigment (commonly Mn, Ni, Cr, etc.) to give the desired colour. The glaze formulation is chosen to give a lower thermal expansion than the body to ensure that the surface is in compression after firing. At this stage a grit of similar composition to the body but of lower thermal expansion may be applied to areas where metal parts are to be cemented. This is to effect good adhesion of the cement in the final assembly stage. Following this the insulator is fired to a temperature between 1180 and 1350 °C on a schedule which typically extends from 1 day to 2 weeks, depending on the size of ware and type of body. Firing in oxygen-depleted atmospheres (reduction firing) is occasionally used to achieve vitrification at lower temperatures, but this necessitates raw materials having a low iron content.

In general terms the microstructure (Fig. 5.20(c)) consists of the 'filler' particles embedded in a glass and crystalline matrix (mullite: $Al_6Si_2O_{13}$) derived from the clay and flux. There must be no open porosity and total porosity must be kept to a minimum, typically 4 vol.%, to avoid adverse effects due to ionization of the gas in the pores (see Section 5.2.2).

Most complete insulators comprise metal components cemented to the porcelain insulator to join it to the line structure, or to build up a string of insulators. Various cements are used depending on application, although conventional Portland cement is most common. Galvanized malleable cast iron is preferred for the metal since its thermal expansion is closer to that of porcelain than most metals and it is comparatively inexpensive. None the less, the cost of the metal components when used is often 40–80% of the cost of the insulator. The form of the widely used rod-insulator is shown in Fig. 5.17.

Typical physical properties of the two main types of high voltage electrical porcelain are listed in Table 5.2.

Electrical discharges may occur due to the very high electric fields which exist in the vicinity of a high voltage insulator. For example corona discharge may occur around the insulator–conductor interface, especially in the case of insulators which carry the conductor directly, without intervening metal fittings. This creates interference (RI) to radio and television reception, which is unacceptable in populous areas. Another form of discharge may occur when the insulator surface is wetted under conditions causing the wet layer to become electrically conductive such as when contaminated by sea-spray or polluted atmospheres. The wet layer dries unevenly, forming dry bands of relatively high resistance. Under some conditions the voltage appearing across a band can

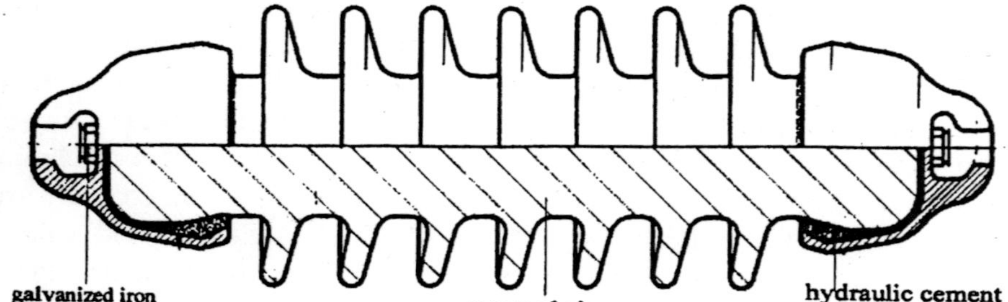

Fig. 5.17 A 36 kV porcelain rod insulator of the variety used to position the overhead power lines for railways. Overall length 500 mm. (Courtesy of New Zealand Insulators, Temuka, NZ.)

Table 5.2 Physical properties of electrical porcelains

Property	*Siliceous*	*Aluminous*
Density/Mg m^{-3}	2.4	2.6–2.8
Cross-breaking strength/MPa	100	150
Coefficient of linear expansion/MK^{-1}	6	5–6
Dielectric Strength/MV m^{-1}	25	25
Volume resistivity/Ω m at 20 °C	$\sim 10^{10}$	$\sim 10^{10}$
Tan $\delta/10^{-4}$ at 20 °C	~ 150	~ 150

exceed the breakdown strength of air ($\sim 3\,\mathrm{MV\,m^{-1}}$) causing an arc to form across the dry band which may grow until 'flash over' occurs from line to earth. This may cause an interruption of power supply to the consumer, or reduced voltage, and can also result in serious damage to the conductors, towers and insulators. Both RI and 'flash-over' can be alleviated by using glazes having surface resistivities in the range 1–100 MΩ/□ (see Section 4.2).

Radio interference is reduced by coating the insulator surface in the line-insulator contact area with a semiconducting glaze to smooth the voltage gradient. Glazes for this purpose usually have a zinc ferrite (see Section 4.4.1) conducting phase, although antimony-doped tin oxide (see Section 4.1.4) is sometimes used. Close control of surface resistivity is not necessary.

The effective use of semiconducting glazes to avoid pollution-induced flash-over is less simple. The glaze needs to have a closely controlled resistivity, designed to pass a leakage current of approximately 1 mA under the service voltage. This ensures that the voltage drop across the dry band does not exceed the air breakdown voltage. The temperature coefficient of resistance must be small (semiconducting glazes usually have a large negative temperature coefficient of resisitivity) and the glaze must be very durable. Where insulators are, broadly speaking, disc-shaped, the current density will vary with the radius leading to a non-linear voltage gradient, and grading of the glaze thickness is needed to compensate for this. Glazes with an antimony-doped tin oxide semiconducting phase are almost invariably used. Although they impart to the insulators better pollution performance than any other type, they are expensive and produced by only a few manufacturers.

Recently insulators comprising a core of glass fibre-reinforced polymer and a sheath of a weather-resistant elastomer ('composite insulators') have become available and now have a promising performance history. In particular, some silicone elastomers have strongly hydrophobic surface characteristics which are maintained for long periods (>20 years) and these give good performance in polluted conditions because the surface never gets wet. Such insulators are replacing porcelain for very high voltage transmission, but at lower voltages (<66 kV) are uneconomic unless pollution conditions are severe. Insulators of toughened glass are also popular, and where a standard design in very large numbers is demanded have a significant price advantage over porcelain. Nevertheless, in excess of 500 000 tonnes of porcelain insulators are manufactured world wide each year.

Varistors play an important role in protecting high voltage distribution equipment (see Section 4.3) against voltage surges caused by lightning strikes, switching operations and similar hazards. Superconducting fault current-limiters (see Section 4.7.4) are also used to protect equipment from current surges.

Low-voltage applications of electrical porcelain exist where an insulating material unaffected by high temperatures or ultraviolet radiation is required. Holders for discharge lamps such as those used for street lighting, and housings for fuses are examples. The porcelain types are typically of similar composition to those discussed above, but the shapes are commonly formed by pressing damp (15 wt.% moisture) body granules in a steel die.

Talc-based

Talc-based ceramics are important electrical porcelains that have major crystalline components of the fired ceramic lying in the ternary phase diagram shown in Fig. 5.18. The principal raw material used for these ceramics is talc ($Mg_3Si_4O_{11}\cdot H_2O$), which is the softest of minerals (no. 1 on the Mohs scale). Talc is also called steatite, and ceramics using it as a raw material are termed 'steatite porcelains' although the talc contained in them has been changed in crystal structure during sintering. Blocks of the mineral steatite, which is also known as soapstone, can readily be machined to shape and, on firing, undergo a change in crystal structure that results in a very small overall expansion accompanied by a large increase in hardness and strength.

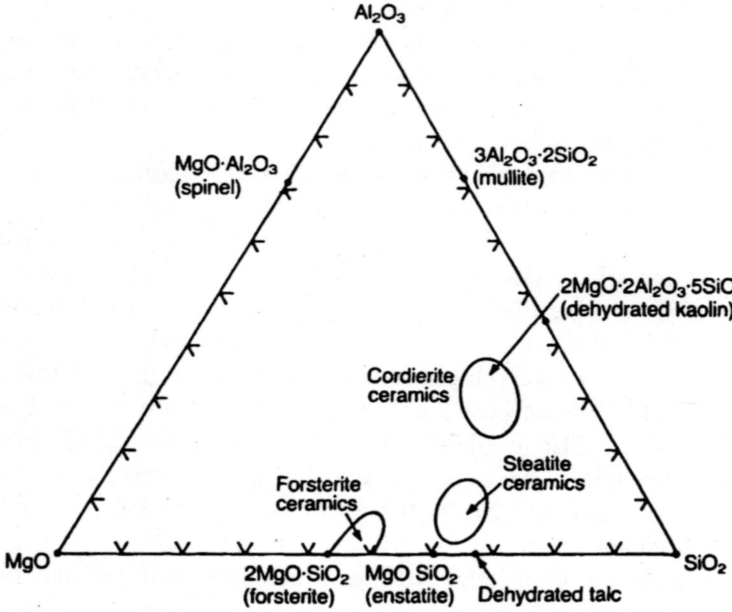

Fig. 5.18 Phases present in the Al_2O_3–MgO–SiO_2 system (in wt%).

Table 5.3 Typical properties of common electrical ceramics

Material	ε_r	$\tan\delta/10^{-4}$ *at* 1 MHz	α_L/MK^{-1} *at* 20–1000 °C	λ/Wm^{-1}K^{-1} *at* 25 °C
'Low-loss' steatite	6.1	7	8.9	3
Cordierite	5.7	80	2.9	2
Forsterite	6.4	2	10.7	3
96Al_2O_3	9.7	3	8.2	35
99.5BeO	6.8	2	8.8	250
AlN	8.8	5–10	4.5	100
Glass	4–15	2–22	0.8–9	0.5–2.0

The various porcelain types fall into three regions of the phase diagram and will be discussed in turn; typical properties are summarized in Table 5.3 and examples of components are illustrated in Fig. 5.19.

Cordierite ceramics are best known for their low expansion coefficient which imparts excellent thermal shock resistance. The major phase developed during firing is cordierite $Mg_2Al_4Si_5O_{18}$ ($2MgO{\cdot}2Al_2O_3{\cdot}5SiO_2$). Clay and talc are the principal ingredients; the talc content is in the range 20–40 wt% but there are many modifications to adjust firing temperature and range. Components are shaped mainly by extrusion and dry pressing, and typical firing temperatures lie in the range 1150–1400 °C. Uses for the ceramic are many and varied, but it is

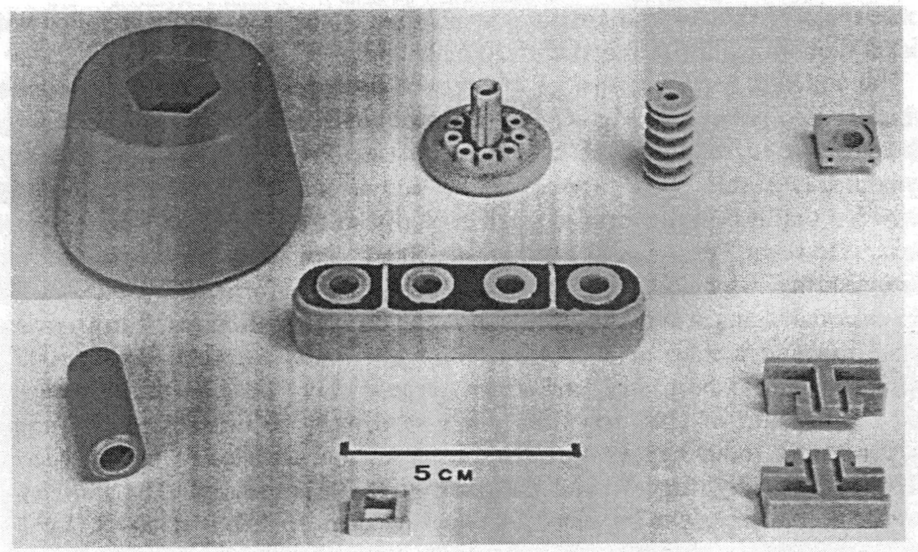

Fig. 5.19 Selection of magnesium silicate parts (parts supplied by Morgan Advanced Ceramics).

particularly suited to applications where there is a need for good thermal shock resistance combined with high electrical resistivity, for example for high-power electrical fuse holders and supports for high-power wire-wound resistors and fan-heater elements.

Steatite ceramics were introduced into electronic components during the 1920s to meet the demands of the rapidly developing radio industry. Their principal attribute is low dielectric losses which are necessary for higher frequencies. As for the cordierite ceramics, the major constituents are clay and talc but with the composition adjusted so that enstatite, a modification of $MgSiO_3$, crystallizes. A typical composition is about 85 wt% talc, 15 wt% clay and 2 wt% calcium carbonate (chalk). The calcium carbonate acts as a flux and is used instead of feldspar to avoid the introduction of alkali metals which would increase dielectric losses. The shrinkage on sintering can be adjusted by calcining up to three-quarters of the talc and varying the proportion of calcined material to compensate for changes in the raw materials. The soft raw talc has a lubricating action that greatly reduces tool wear during fabrication compared with other ceramic powders.

Most steatite ceramics are either dry mixed and pressed, or wet-mixed, spray-dried and pressed, but they can also be wet-mixed and extruded. Typical firing temperatures lie close to 1300 °C. Many small parts are made for the electronics components industry where low dielectric losses are required, for example for tie-bars and other parts for ganged capacitors, small trimmer capacitors, high-power capacitors (see Section 5.6.3), coil formers, lead-throughs and substrates for some types of resistor and circuit. The ceramic is also commonly supplied with nickel metallizing for terminal connector blocks.

The sintered bodies consist of crystals of protoenstatite (or mesoenstatite), which is a polymorph of $MgSiO_3$, a small amount of cordierite and a continuous glassy phase surrounding the crystalline phases. Protoenstatite is the thermodynamically stable phase above 985 °C and is stable at room temperature provided that it is in the form of small crystals ($< 10\ \mu m$) covered with a layer of glass. However, prolonged heating at 500 °C converts protoenstatite into clinoenstatite, which is the thermodynamically stable form below 985 °C. The dimensional changes in the crystals due to the transformation result in an overall expansion and a marked decrease in strength. Badly formulated material, or material that has been overfired so that excessive crystal growth has occurred and the amount of glass has been diminished, may be unstable under humid conditions at room temperature. In bad cases the surfaces of bodies become white and blotchy, they expand and their strength is reduced. The stability of steatite components can be tested by exposing them to steam at 100 °C for 24 h and then checking that their cross-breaking strength is unaltered.

Despite this potential instability, steatite ceramics have been widely used and adequate precautions in manufacture have prevented all but a very few cases of deterioration during component lifetimes of many years in a wide variety of environments.

In forsterite ceramics the mineral forsterite (Mg_2SiO_4) crystallizes. They have excellent low-dielectric-loss characteristics but a high thermal expansion coefficient which imparts poor thermal shock resistance. During the 1960s they were manufactured for parts of rather specialized high-power devices constructed from titanium and forsterite and for which the operating temperature precluded the use of a glass–metal construction. The close match between the thermal expansion coefficients of titanium and forsterite made this possible. Today alumina–metal constructions have completely replaced those based on titanium–forsterite and the ceramic is now manufactured only to meet the occasional special request.

5.5.2 Alumina

Structure, fabrication and properties

Alumina is a widespread component of siliceous minerals. It occurs as single crystals in the form of sapphire, and with chromium 'impurity' as ruby, and in large deposits as the hydrated oxide bauxite ($Al_2O_3{\cdot}2H_2O$). The dehydration of this and other hydrated oxides at temperatures below 1000 °C leads to the formation of γ-Al_2O_3 which is converted to α-Al_2O_3 above 1000 °C. The transformation is irreversible and the α-polymorph is stable from absolute zero to its melting point at 2050 °C.

α-alumina, in powder form, is a byproduct of aluminium production. Bauxite, which is naturally contaminated with other oxides, principally SiO_2 and Fe_2O_3, is purified by the Bayer process. This involves the dissolution of a low-silica-content ore in caustic soda (NaOH) solution under pressure, filtering off the insoluble hydroxides (mostly of iron), and then precipitating $Al(OH)_3$ by diluting the solution at atmospheric pressure and adding a small amount of $Al(OH)_3$ as a nucleating agent. Most of the silica remains in the solution. The hydroxide is washed and then calcined at temperatures in excess of 1000 °C to produce α-Al_2O_3. Because of the involvement of sodium in the extraction process the calcined alumina contains 0.1–0.2% Na_2O. This impurity has important consequences as far as the manufacture of electrical ceramics is concerned and care is exercised to keep it to a minimum. Practically all the powder for the production of alumina ceramics is produced in this way, but a purer product, for instance for growing artificial sapphire crystals, can be made by preparing ammonium alum ($NH_4Al(SO_4)_2 \cdot 12H_2O$) by dissolving 99.999% aluminium metal in sulphuric acid and neutralizing the excess acid with ammonia. The alum is allowed to crystallize, which helps to purify it, and is then calcined below 1000 °C to yield γ-Al_2O_3, which is preferred for some methods of sapphire preparation, or at higher temperatures to give the α form.

The crystallites in alumina ceramics are mainly α-Al_2O_3, mineralogically known as corundum (a synonym for sapphire). Corundum is 9 on the Mohs scale of hardness, i.e. it is next hardest to diamond, at 10. Examples of the microstructures of a high-purity alumina and a debased alumina are shown in Figs 5.20(a) and 5.20(b) respectively. The latter consists of α-Al_2O_3 crystallites embedded in a glass–crystalline matrix usually composed of calcium and magnesium silicates.

The less pure aluminas are blended with silicates so that they can be sintered at 1350 °C or less. The highest-purity materials require a temperature of 1750 °C at atmospheric pressure or hot-pressing. As cost increases with sintering temperature, the grade used in practice is usually the least pure that has adequate properties.

Table 5.4 gives the properties of a range of aluminas and shows the improvement in properties with purity. The most striking change is in thermal

Table 5.4 Typical values of the electrothermal properties of various grades of alumina ceramic

Alumina content/%	85	90	96	99.5	99.9
ε_r at 1 MHz	8.2	8.8	9.0–9.3	9.7	9.8–10.1
$\tan \delta/10^{-4}$ at 1 MHz	9	4	1–3	3	0.4–2
Resistivity/Ω m at 300 °C	4.6×10^8	1.4×10^9	3.1×10^9	2.0×10^9	1.0×10^{13}
Thermal conductivity/ $W m^{-1} K^{-1}$ at 20 °C	14	16	24–35	35	40

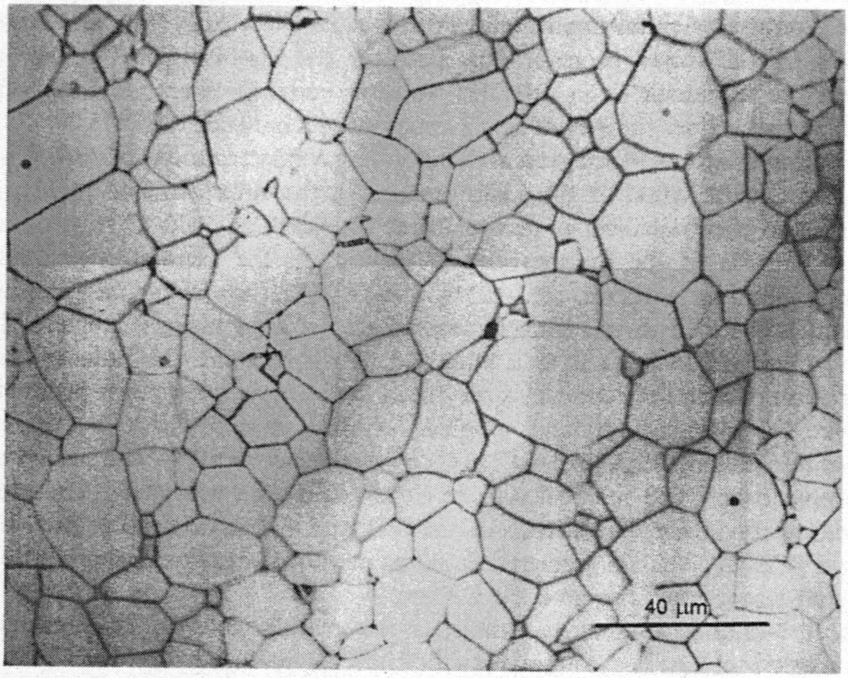

(a)

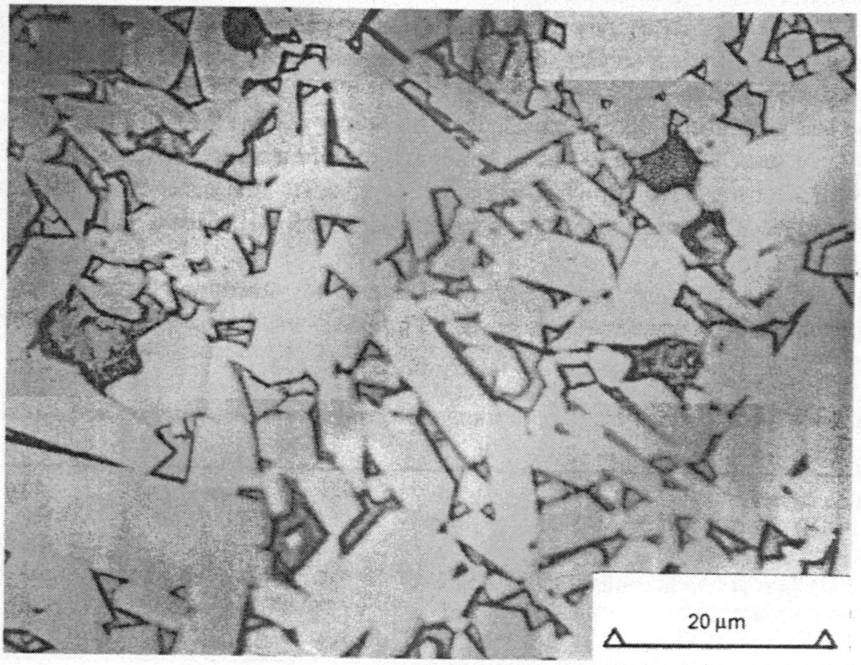

(b)

conductivity which indicates the superiority of the purer grades in applications where the transfer of heat is of importance, e.g. in substrates, discussed below.

The wide range of thermal conductivities reported for 96% Al_2O_3 requires some explanation. Thermal conductivity depends on the transfer of lattice vibrational energy in the form of phonons. Defects in the lattice inhibit this process, and it is found that the incorporation of impurities much in excess of 0.1 mol.% in the lattice appreciably reduces the thermal conductivity. However, if impurities are present as separate phases and have not entered the lattice of the main phase their effect will be in accordance with one of the mixture rules (cf. Section 2.7.4). Which rule will depend on the distribution of the minor phases, for example whether they form continuous layers round each grain of the main phase or are present in separate discrete regions. High conductivity only occurs in 96% alumina when the raw materials and processing conditions are such that the additives to lower the sintering temperature do not enter the Al_2O_3 lattice in more than minor quantities.

There is considerable uncertainty about the room temperature values of the electrical resistivity of good insulators; the best estimates are probably derived by

(c)

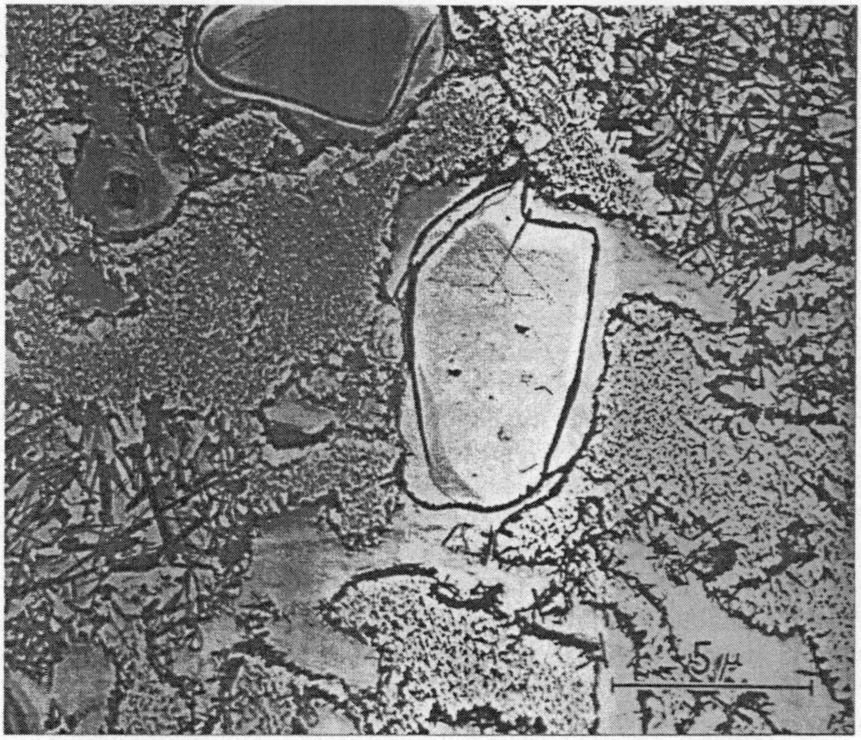

Fig. 5.20 Microstructures of (a) 99.9% Al_2O_3 (Courtesy of E.W. Roberts), (b) 95% Al_2O_3 (Courtesy of R. Morrell), (c) chemically etched electrical porcelain: note the partially dissolved quartz grains and mullite precipitates (Courtesy of S.T. Lundin).

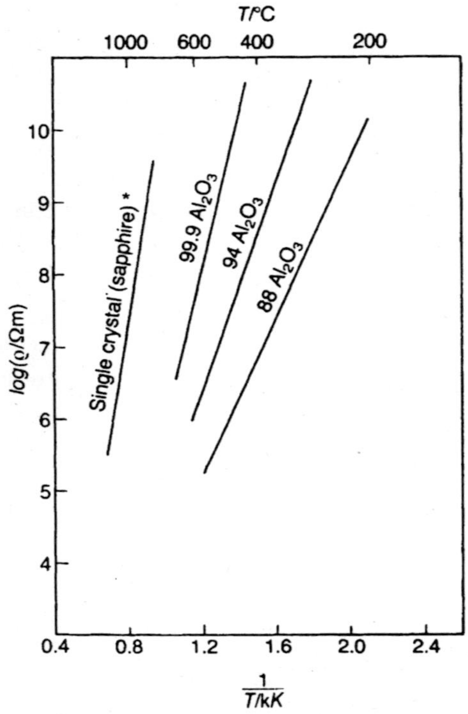

Fig. 5.21 Dependence of resistivity on temperature for various grades of Al_2O_3. *[4]

extrapolation of the linear log ρ–$1/T$ plots of the type shown in Fig. 5.21. Also, the reliable measurement of the resistance of very high resistance specimens is made difficult because of the relatively low resistance of the ceramic surface due to adsorbed impurities, and even that of the surrounding air or other gas. This can be circumvented by using a properly guarded measurement method.

The principle of a guarded measurement is illustrated in Fig. 5.22 which shows a guard ring around one of the measuring electrodes. When the effects of gas conduction have to be guarded against the ring must be extended into a cylinder. The guard ring and one of the centre electrodes are connected to the voltage supply but only the current through the central electrode is measured. Satisfactory measurements can be made in this way at 200 °C and above, but with high resistivities and lower temperatures the polarization and reordering of defects and impurities result in an initially high current that takes many hours to fall to a steady state.

Very careful measurements have been made by F.G. Will *et al.* [4] on high purity (<35 p.p.m. impurities) sapphire over the temperature range 400–1300 °C and at an oxygen partial pressure of 1 μPa ($\sim 10^{-11}$ atm). They find the conduction mechanism to be electronic with negligible ionic contribution. The

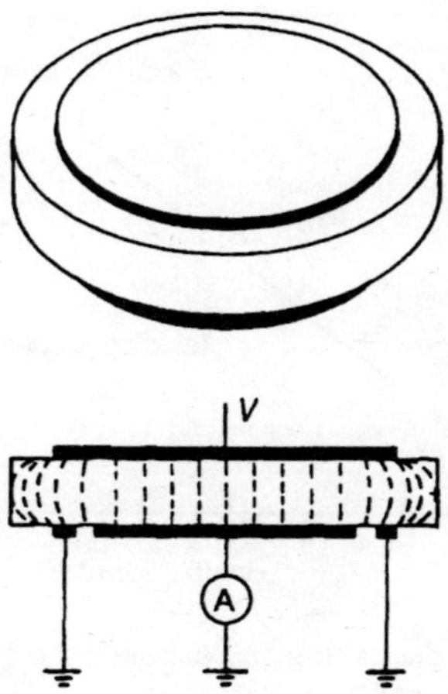

Fig. 5.22 Schematic diagram of a guard ring and circuit for measurements of high resistivity.

measured activation energy of 4.8 eV is consistent with the determined optical band gap of ~9.5 eV.

In the case of polycrystalline alumina the matter is complicated by charge transport along grain boundaries. It is generally found that σ increases with both large and small p_{O_2} values, with a minimum at about 10 mPa (10^{-4} atm) at about 1600 °C. This behaviour seems to be independent of the type of dopant (acceptor or donor) or whether the material is single crystal or polycrystalline. Figure 5.23 illustrates the expected general behavioural pattern, but the positions of the boundaries defining the 'fields' for the various mechanisms is largely speculative and will, of course, depend strongly on the natures and amounts of impurities present.

The effects of deliberately added donors, such as titanium, and acceptors, such as iron and magnesium, on electrical conductivity have been studied. Doping with aliovalent ions affects the concentration of intrinsic defects and, in consequence, the diffusivity of Al and O. In the case of variable-valency dopants, changes in p_{O_2} change the fraction of dopants in the aliovalent state and the nature and concentration of the defects. For example, the dopant Ti substitutes for Al and, in the fully oxidized state, produces the defect $Ti_{Al}^{\cdot}$, compensated by V_{Al}''', so that

$$[Ti_{Al}^{\cdot}] = [3V_{Al}'''] \tag{5.16}$$

and ionic conductivity by transport of V_{Al}''' is encouraged. However, at low p_{O_2} Ti is in the isovalent state, Ti^{3+}, and electron conduction occurs according to

$$Ti_{Al} \rightleftharpoons Ti_{Al}^{\cdot} + e' \tag{5.17}$$

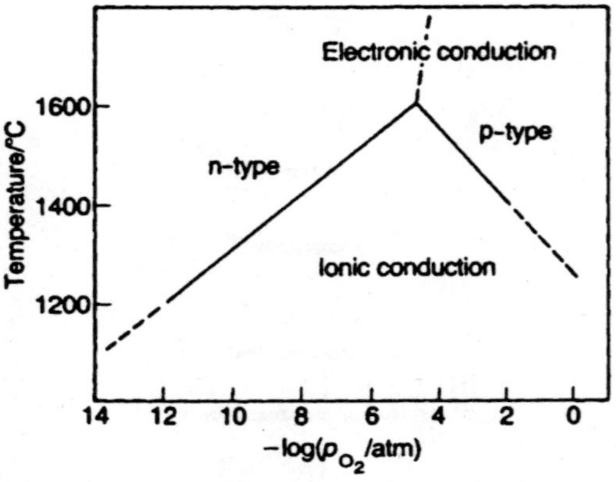

Fig. 5.23 Speculative diagram describing the dependence of charge transport mechanisms in alumina on temperature and oxygen pressure.

The defect concentrations and their dependence on p_{O_2} and temperature are derived from the law of mass action by procedures essentially the same as those outlined in Section 2.6.2. In the case of polycrystalline high-purity alumina, the electronic conductivity increases with decreasing grain size and is attributed to hole transport along grain boundaries.

The conductivity of debased aluminas is dominated by that of the usually continuous silicate phase, and the ideas concerning charge transport through glasses (Section 2.6.3) are therefore appropriate. Debased aluminas are typically more conductive than the high-purity varieties, as illustrated in Fig. 5.21.

The effects of temperature and frequency on the permittivity and dissipation factor of a high-purity alumina ceramic are shown in Fig. 5.24. The discrepancies between the permittivity levels in Fig. 5.24 and values given elsewhere are probably due to differences in microstructure and measurement technique. Reliable room temperature values for ε_r for single-crystal sapphire at 3.4 GHz are 9.39 perpendicular to the c axis and 11.584 parallel to it, which are close to the values measured optically. The average ε_r to be expected for a fully dense ceramic form is therefore 10.12, and values close to this have been determined. Nothwithstanding the uncertainties there is no doubt that the general behavioural pattern indicated by Fig. 5.24 is correct and typical of ceramic dielectrics.

As the temperature increases defects and charge carriers become more responsive to applied fields, causing both permittivity and loss to increase. However their inertia prevents them from responding effectively at high frequencies so that the loss falls, particularly at higher temperatures. An intergranular phase will clearly have a considerable effect on both the resistivity and the dissipation factor.

It is an advantage that alumina can be fired in hydrogen without degradation of the electrical properties since this allows the use of molybdenum heating elements in the high-temperature furnaces required for sintering the high-purity

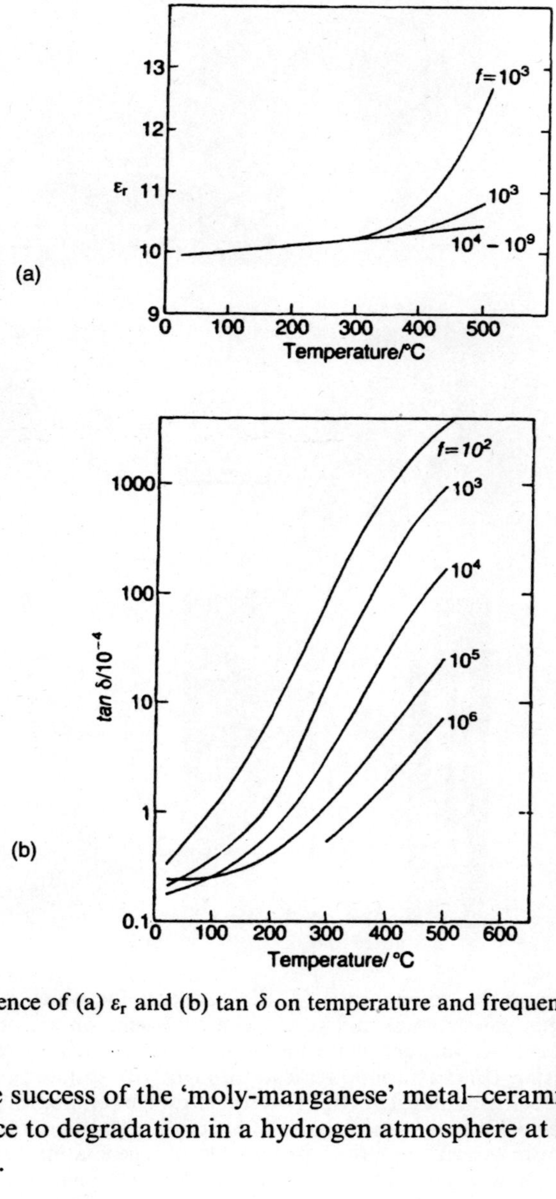

Fig. 5.24 Dependence of (a) ε_r and (b) tan δ on temperature and frequency for a high-purity Al_2O_3 ceramic.

grades. Also, the success of the 'moly-manganese' metal–ceramic joining process rests on resistance to degradation in a hydrogen atmosphere at high temperature (see Section 3.8).

Applications

Alumina ceramics are used wherever exceptionally good dielectric properties, high mechanical strength and high thermal conductivity, coupled with a reliable ceramic–metal joining technology such as the moly-manganese process, are demanded. Examples of components are illustrated in Fig. 5.25.

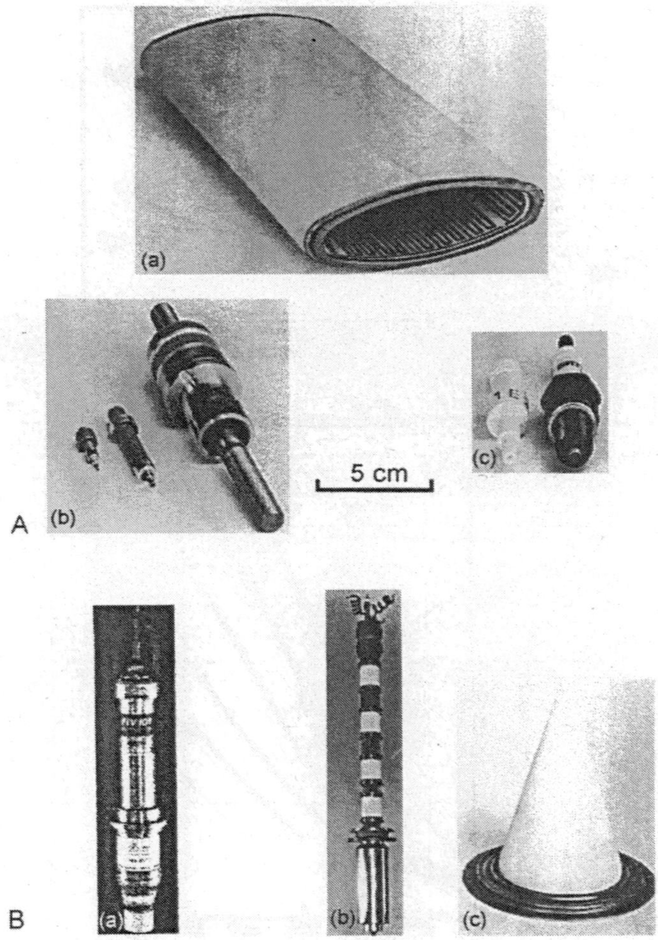

Fig. 5.25 Alumina-metal components: (A) Components for which a combination of high mechanical strength, gas-tightness and good electrical insulation are prime requirements: (a) section of 100 m diameter vacuum-tight ring for an electron synchrotron with internal printed heater for out-gassing; (b) electrical power lead-throughs; (d) isostatically pressed alumina part of spark plug together with plug. (B) High power microwave components: (a) Inductive Output Tube suited to digital TV service and capable of handling 130 kW peak output power, equating to 30 kW mean power (82 cm long); (b) a 4-cavity klystron generating 20 kW continuous power at 470–860 MHz for analogue TV transmission (~125 cm long); (c) a conical microwave 'window' for effecting a vacuum-tight seal between, for example, a microwave generator and antenna (approx. 20 cm high). ((B) Courtesy of 'E2V Technologies' UK.)

The best known use for a 95% alumina ceramic is in spark plugs, mass-produced components which illustrate well the need for a combination of good thermomechanical and electrical properties. They must be strong to withstand not only rough treatment meted out by garage mechanics but also the thermomechanical shocks experienced each time the engine fires, with temperatures and pressures as high as 900 °C and 10 MPa (100 atm) being generated. The alumina must also provide adequate insulation against electrical stresses of the order of 1 MV m^{-1}, and the glazed surface can be kept clean so as to reduce the risk of tracking.

Another important application is in klystrons and magnetrons – devices for the generation of electromagnetic energy, sometimes at very high power levels. For example, the impressive operating characteristics of a large klystron for a 1.3 GHz radar transmitter are as follows: an electron gun voltage of 278 kV and a beam current of 324 A providing a pulse peak power of 30 MW, a pulse length of 35 μs and an average power level of 150 kW. At the other end of the power spectrum are generators for domestic microwave ovens, operating typically at power levels of 1.2 kW at 2.45 GHz. The high gun voltages used in large klystrons necessitate very good insulation and, just as for the electron guns in electron microscopes, it is provided by alumina. A microwave generator, however, makes a further unique demand. Microwave power is passed from the vacuum interior of the device to the outside world through a 'window' as transparent as possible to the radiation. Because of the finite power factor of the window material heat is developed which must be removed to a suitable sink. Therefore requirements for a good microwave window material are a low power factor at the operating frequency, high thermal conductivity, vacuum tightness, high strength and the capability of being joined effectively to a metal. Alumina ceramics meet most of the requirements, but beryllia may be used for the highest power levels because of its unrivalled thermal conductivity. This is so in the case of a particular commercial 'gyrotron' in which a beryllia output window of diameter 2 in passes 200 kW continuously at 60 GHz.

Where the ceramics form part of a precision engineering structure, as in the above examples, it is necessary to bring them to the correct dimensions by diamond machining. They are usually joined to the metal parts by variants of the moly-manganese process. It is worth noting that the manufacture of metal–ceramic components has been so successfully developed that it can replace the metal–glass technology where production costs rather than the technical merits of the alumina ceramic are the deciding factor. The metal–ceramic technology dispenses with the highly skilled and expensive glass blower necessary for the manufacture of devices based on glass envelopes and is suited to mass production.

Alumina ceramics are widely used for thick-film circuit substrates and for integrated circuit packaging. This aspect is discussed in Section 5.5.5.

5.5.3 Beryllia

Beryllia has broadly similar properties to alumina (Table 5.3) but its thermal conductivity is 5–10 times greater. It is therefore used when thermal dissipation combined with electrical isolation is of major importance, e.g. in high-power klystrons and power diodes. Because beryllium is one of the rare elements the oxide is inevitably more expensive than alumina, which limits its use.

The technology developed to produce dense BeO ceramics is similar to that employed for Al_2O_3. Sintering must be carried out at somewhat higher temperatures since the melting point of BeO is higher at 2570 °C. The atmosphere during sintering or other high-temperature (>1000 °C) treatment of BeO must be moisture free since otherwise a volatile hydroxide is formed which is toxic if inhaled as fumes. Sintering and metallization can be carried out in dry hydrogen. Sintered pieces only give rise to a health hazard when they are abraded or heated in moist atmospheres without suitable precautions.

5.5.4 Aluminium nitride

AlN (Table 5.3) has recently been introduced as a substrate because it combines a high thermal conductivity with a thermal expansion coefficient close to that of silicon. Its resistivity is somewhat lower than that of BeO and Al_2O_3 but is sufficient for most substrate applications.

AlN dissociates into its elements at 2500 °C. It is formed by the reaction of aluminium powder with nitrogen and can be sintered in a nitrogen atmosphere.

AlN replaces BeO in many applications since it is competitive in price and does not present the problems of toxicity in manufacture. It may replace Al_2O_3 as a substrate for silicon chips because of its better heat conductivity and its closer match in thermal expansion.

5.5.5 Ceramic 'packaging' technology

Electronic circuits are 'packaged' using substrates of one form or another ranging from the plastic-based laminate through to the sophisticated multilayer ceramic (MLC) substrate. Laminates are dominant in relatively low technologies such as radio, TV and other consumer products, and thick film circuits carried on ceramic (Al_2O_3, BeO, AlN) substrates are extensively used for a wide range of more demanding applications including communications, automotive control and military and space technologies.

The substrates carrying the circuits shown in Fig. 4.5 are a 95–96% alumina. This ceramic has been adopted for its combination of physical and chemical characteristics and, importantly, low cost. It offers a combination of mechanical, thermal and electrical properties which meet the in-service requirements, and compositional and microstructural characteristics suited to thick film printing (see Section 4.2.2). Alumina substrates are manufactured on a very large scale making the unit costs a small fraction of the total circuit cost.

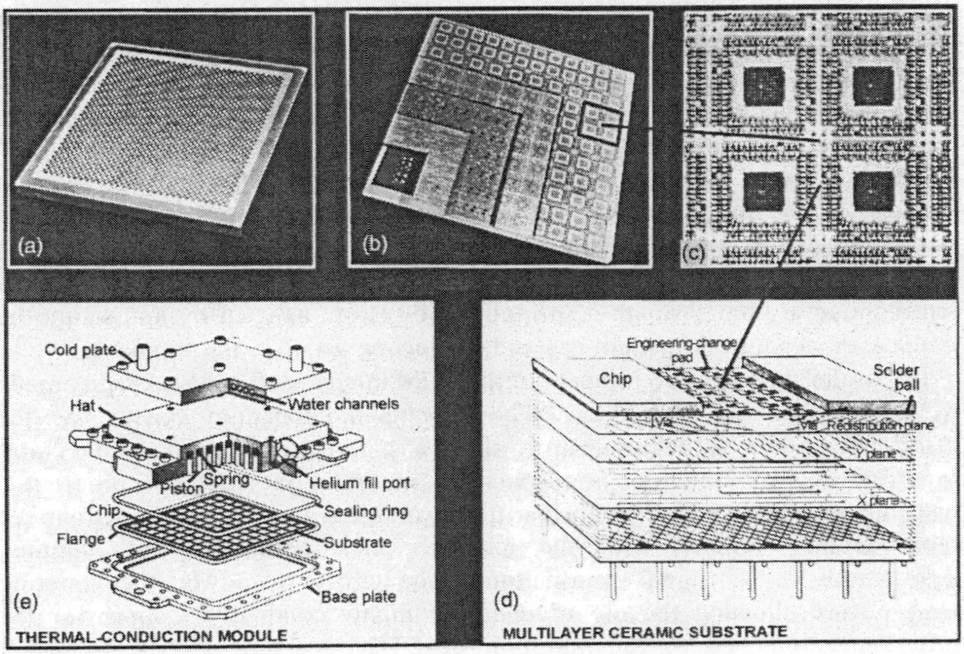

Fig. 5.26 Multilayer ceramic thermal conduction module: (a) underside showing connector pads – substrate size 90 × 90 × 6 mm; (b) multilayer construction; (c) silicon chip mounting positions; (d) schematic diagram illustrating buried interconnections; (e) thermal conduction module. (Courtesy of IBM Corp., East Fishkill.)

Over the past two decades MLC technology has been progressively developed for advanced packaging, especially by main frame computer manufacturers, and it is increasingly exploited for microwave communications. Especially significant is the emergence of low temperature co-fired ceramic (LTCC) technology.

The 'alumina–molybdenum' IBM thermal conduction module (TCM)

In the 1980s computer circuit designers turned to MLC technology in an effort to increase component packing density so reducing interconnect lengths. A major mile-stone was the development of the IBM thermal conduction module (TCM) used in 308X type computers and illustrated in Fig. 5.26. The following discussion outlines the general development trends and reasons for them; the technical details are, presumably, continually changing.

In the manufacture of the multilayer structure for the TCM, square sheets are stamped from a roll of 'green' tape loaded with a debased alumina powder. Many small (about 150 μm diameter) holes ('vias') are punched through the tape at predetermined positions. Electrical 'wiring' patterns with line widths and spacing of about 100 μm are then screen-printed onto each sheet, and the vias filled with a molybdenum-based metallizing ink. After the printing stage the sheets are laminated by warm-pressing and then sintered in a reducing atmosphere. The final structure is an alumina (~92% Al_2O_3) tile of dimensions 90 × 90 × 5 mm made of up to approximately 33 layers of ceramic with buried three-dimensional circuitry, as illustrated in Fig. 5.26. The top and bottom surfaces of the tile are nickel and gold plated with patterns onto which the semiconductor integrated circuits are subsequently bonded.

In the TCM removal of heat is effected directly from the back of each semiconductor chip through a spring-loaded piston, assisted by immersing the multilayer structure in helium gas and by flowing water in the top cover.

In parallel with the development of the TCM intensive efforts were being made to improve on the technology. The alumina–molybdenum system has the disadvantage that the requirement to sinter at high temperature (~1400 °C) and in a low oxygen potential atmosphere to maintain the molybdenum in the metallic state is expensive. A further disadvantage is that the high resistivity of molybdenum, coupled with the relatively high permittivity of alumina, significantly slows signal transit time. The efforts to lower the sintering temperature allowing the use of relatively highly conductive copper for the interconnections led to the adoption of LTCC technology. The following discussion should be read in conjunction with Section 3.7.4.

Low temperature co-fired ceramic technology (LTCC)

Essentially LTCC technology is based on glass-ceramics specially formulated to meet the operational demands. Substrates must be strong enough to withstand mechanical stressing during processing and in service. Ideally thermal conductivity would be high so that heat produced in the components is quickly dissipated without attendant large temperature gradients in the assembly. This is significant in cases where the component density is high or where power components (rather than 'signal') are involved. As the technology has developed thermal conductivity has become far less important, especially in the case of the sophisticated IBM multilayer substrates discussed above in which thermal management does not rely significantly on the ceramic.

As far as electrical properties are concerned most attention is focused on permittivity and dissipation factor. A high permittivity carries two penalties. One is dielectric coupling between closely spaced 'wires' leading to 'cross-talk', and the other signal transit time, the velocity of an electromagnetic wave being inversely proportional to the square root of the relative permittivity (see Eq. (2.120)). Signal transit time is important in so far as it is a factor determining the speed of a computer.

LTCC technology starts with a glass powder carried in an organic- (or water)-based vehicle and cast into a tape. The circuit is printed onto the tape using highly conducting metals such as silver (m.p. ~961 °C), gold (m.p. ~1064 °C) or copper (m.p. ~1084 °C) with line widths and separations of ~75 μm. Whilst in the case of silver and gold sintering is carried out in air, in the case of the copper circuitry, to avoid oxidation of the metal, the co-firing must be in a protective atmosphere, normally wet nitrogen/hydrogen, adding to processing costs. The layers are stacked and laminated and fired in the temperature range 850–950 °C when the glass devitrifies to a glass-ceramic. 'Vias' punched through the tape prior to laminating, and filled with conductor paste effect the electrical connections between layers. The final part is a dense, fully integrated substrate having optimum mechanical and electrical characteristics onto which other components, 'passive' and 'active', can be mounted and interconnected.

For this application the glass-ceramic should completely devitrify during processing to ensure dimensional and shape stability, have a low relative permittivity (<5), and a thermal expansivity close to that of silicon (~3 MK^{-1}) so as to avoid the development of temperature-induced stresses in the semiconductor devices bonded to the finished substrate. By careful choice of composition, thermal expansivity and electrical properties can be tailored, as discussed in Section 3.7.4. Cordierite (see Fig. 5.18) and β-spodumene ($Li_2OAl_2O_34ASi_2O_2$) have suitable properties. The thermal expansivity of cordierite can be tailored by small modifications to composition, for example by adjusting the alumina content and making minor additions of boric oxide and phosphorus pentoxide. To achieve a fully dense final product the glass powder must completely sinter prior to devitrification and so the presence of 'mineralizers' (crystal nucleating agents) are avoided.

Multilayer, LTCC technology is now exploited in high speed digital circuitry for computer main frame equipment, the general form of the modules being still essentially as illustrated in Fig. 5.26. R.R. Tummala [5] and S.H. Knickerbocker *et al.* [6] review the glass-ceramic aspects of IBM TCM technology.

The explosive growth in wireless technologies has been accompanied by a demand for specially tailored 'microwave dielectrics', and very recently by the exploitation of multilayer LTCC-based circuitry (see Section 5.6.5).

5.6 Medium-permittivity Ceramics

Medium-permittivity ceramics are widely used as Class I dielectrics, and in order to be in this category they need to have low dissipation factors. This precludes the use of most ferroelectric compounds in their composition since ferroelectrics have high losses ($\tan \delta > 0.003$), particularly when subjected to high a.c. fields.

Low-loss materials can be obtained with relative permittivities exceeding 500 but accompanied by high negative temperature coefficients, generally exceeding

$-1000\,MK^{-1}$. For most purposes medium-permittivity ceramics have ε_r in the range 15–100.

There are three principal areas in which these dielectrics are applied.

1. High-power transmitter capacitors for the frequency range 0.5–50 MHz for which the main requirement is low loss: a negative temperature coefficient of permittivity is tolerable since it limits the power through the unit when its temperature increases.
2. Stable capacitors for general electronic use: a stability better than $\pm 1\%$ is needed over the operational temperature and voltage ranges, and the frequency lies mainly in the 1 kHz to 100 MHz range.
3. Microwave resonant devices: these operate between 0.5 and 50 GHz and require stabilities of better than $\pm 0.05\%$ over the operational temperature range with dissipation factors better than 2×10^{-4}.

The applications have been listed in order of increasingly stringent specifications so that a material satisfactory for 3 would also satisfy 1 and 2, although it might be unnecessarily expensive.

Medium-permittivity dielectrics are based on interlinked MO_6 groups, where M is either a quadrivalent ion such as Ti, Zr or Sn or a mixture of divalent, trivalent and pentavalent ions with an average charge of 4+. The oxygen octahedra share corners, edges or faces (see Fig. 2.2) in such a way that the O^{2-} ions form a close-packed structure. Sites in the O^{2-} lattice may be occupied by divalent cations that lie in interstices between the MO_6 octahedra, as in perovskite-type materials. TiO_6 octahedra are the most commonly found groups in medium- and high-permittivity dielectrics. The behaviour of rutile (TiO_2) ceramic is typical of these classes of dielectric in several respects.

5.6.1 Rutile ceramic

Titania occurs in three crystalline modifications: anatase, brookite and rutile. Because above approximately 800 °C both anatase and brookite have transformed to rutile it is the only form of significance in the ceramics context and attention here is limited to it.

The rutile structure (Fig. 5.27) is based on nearly close-packed oxygen ions with Ti^{4+} ions occupying half the octahedral sites. The tetragonal unit cell contains two formula units, and the Ti^{4+} is at the centre of a distorted oxygen octahedron.

Rutile is anisotropic, with the values of ε_r at room temperature being approximately 170 and 90 in the *c* and *a* directions respectively. In the polycrystalline ceramic form ε_r averages to intermediate values with a

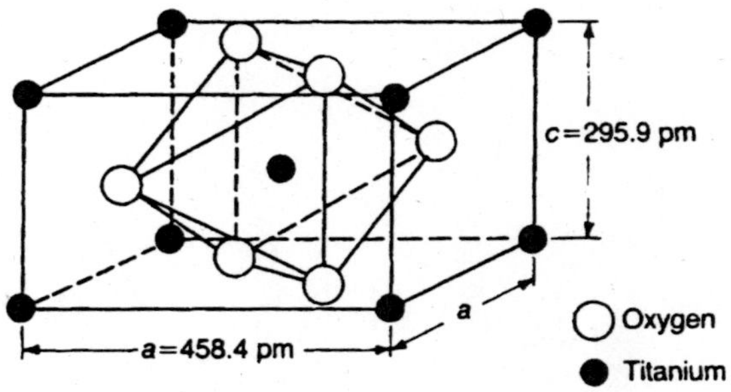

Fig. 5.27 The crystal structure of rutile.

temperature coefficient of approximately $-750\,\mathrm{MK}^{-1}$. The $\varepsilon_r(\omega,T)$ and $\tan\delta(\omega,T)$ characteristics of a titania-based ceramic are shown in Fig. 5.28.

Pure rutile is an excellent insulator at room temperature with an optical band gap between the filled O 2p valence band and the empty Ti 3d conduction band probably in the range 3.5–4.0 eV. A thermal energy of approximately 1.7–2.0 eV can transfer electrons from the valence to the conduction band leading to semiconductivity. Figure 5.29 shows typical conductivity data for a high-purity titania ceramic (>99.95 wt% TiO_2) measured in oxygen at 1 atm.

An important feature of TiO_2 is the extent to which it can be chemically reduced above approximately 900 °C with accompanying significant changes in electrical conductivity, as shown in Fig. 5.30. The fall in resistivity is accompanied by a loss of oxygen and the movement of Ti ions onto interstitial sites, probably the empty octahedral sites in the rutile structure:

$$2\mathrm{O_O} + \mathrm{Ti_{Ti}} \rightarrow \mathrm{O_2(g)} + \mathrm{Ti_I^{\cdot\cdot\cdot\cdot}} + 4\mathrm{e}' \tag{5.18}$$

The law of mass action leads to

$$[\mathrm{Ti_I^{\cdot\cdot\cdot\cdot}}]n^4 = K_n p_{\mathrm{O_2}}{}^{-1} \tag{5.19}$$

and, since $n \approx 4[\mathrm{Ti_I^{\cdot\cdot\cdot\cdot}}]$,

$$n = (4K_n)^{1/5} p_{\mathrm{O_2}}{}^{-1/5} \tag{5.20}$$

Measurements of conductivity at 1000 °C and oxygen pressures below 101 mPa (10^{-10} atm) have confirmed Eq. (5.20) (see Section 2.6.2).

The presence of interstitial Ti ions is confirmed by other research and is consistent with the structural change that occurs with larger oxygen deficiencies (e.g. in $TiO_{1.95}$). This involves a process of crystallographic shear that accommodates the oxygen loss and results in the Ti ions on one side of the shear plane being in interstitial positions relative to those on the other side. Therefore, in reality, a grossly non-stoichiometric phase is made up of stoichiometric regions separated by a series of regularly spaced crystal shear

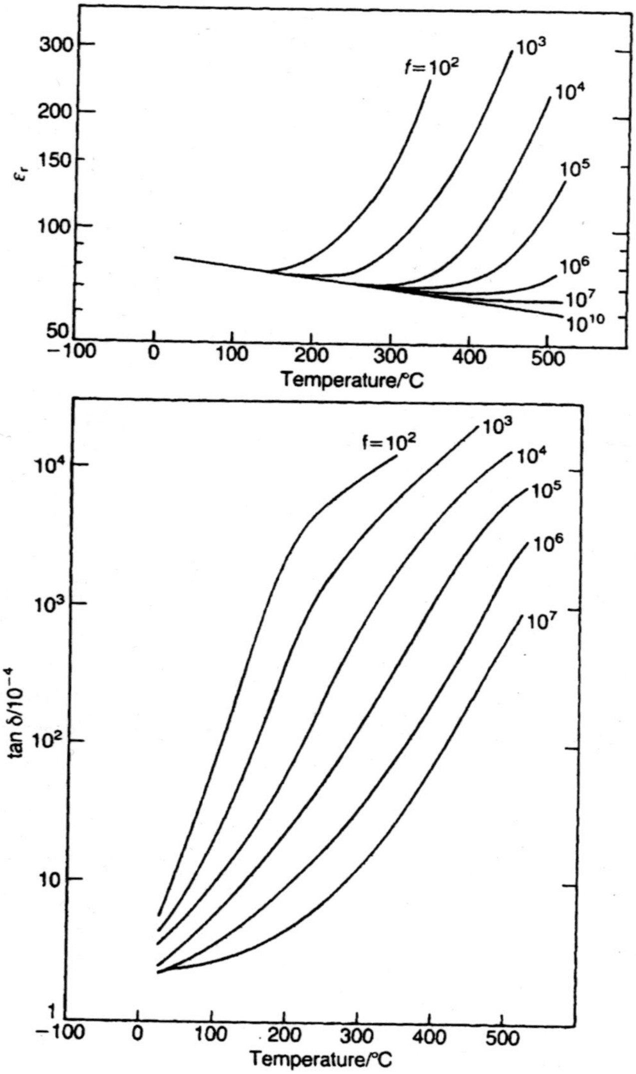

Fig. 5.28 Dielectric properties of titania ceramic as a function of frequency and temperature.

planes, with the whole comprising a so-called Magnéli phase (named after A. Magnéli). In the case of TiO_2, non-stoichiometric compositions can be shown to belong to the stoichiometric series Ti_nO_{2n-1} in which $n = 15, 16, 18, 19, 20, 22, 29, 31$ etc. For example, $TiO_{1.95}$ is actually $Ti_{20}O_{39}$.

The change in conductivity with oxygen pressure can be exploited for gas sensing as discussed in Section 4.6.2. Most compositions containing TiO_2 show similar behaviour when fired in reducing atmospheres, so that sintering in air or oxygen is essential if they are to be used as low-loss dielectrics.

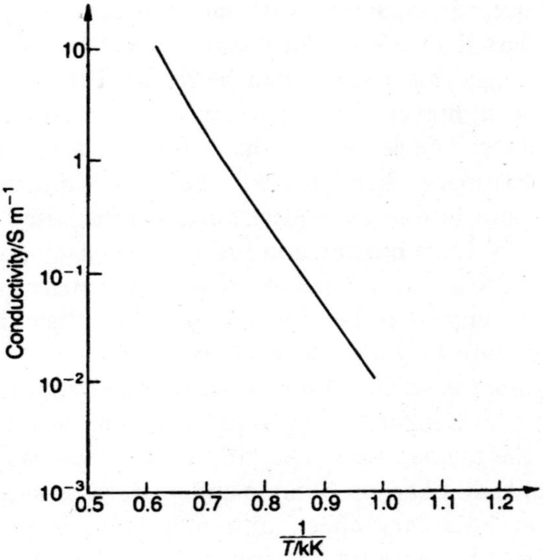

Fig. 5.29 Conductivity of titania ceramic in oxygen (101 kPa; 1 atm) as a function of temperature.

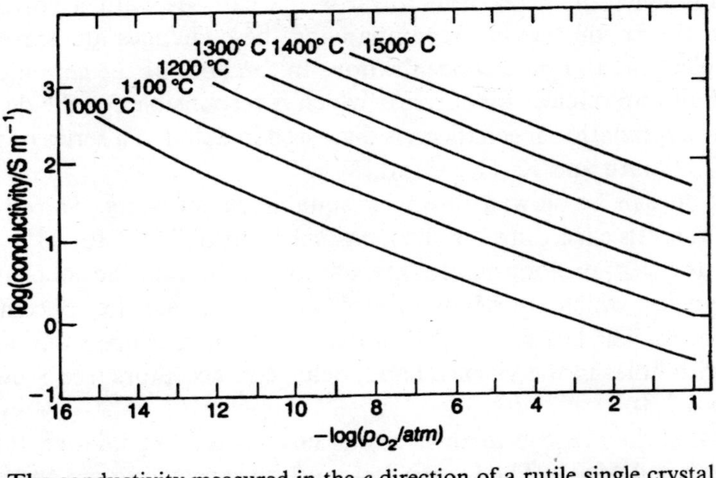

Fig. 5.30 The conductivity measured in the *c* direction of a rutile single crystal as a function of oxygen pressure and temperature (after R.N. Blumenthal *et al.*: see [7] p. 145).

5.6.2 Degradation in titanium-containing oxides

The degradation of capacitor dielectrics has been discussed in general terms in Section 5.2.2 and 5.4.1. Here the topic is amplified with regard to titanium-containing oxides. Deterioration can occur under two different sets of conditions and most probably with differing mechanisms.

In one circumstance, for capacitors with thin dielectrics (approximately 25 μm thick) subjected to less than 5 V d.c. at room temperature, the resistance drops rapidly during prolonged life tests. It can be restored to its initial level by the brief application of a higher (10 ×) voltage, or sometimes just by minor mechanical disturbance. The reason for this behaviour is not known for certain but it seems likely that silver has migrated in the form of a filament which results in a low-resistance path bridging the electrodes. An increased current is able to destroy the filament by Joule heating and fusion. The effect of mechanical shock is less easy to explain. Multilayer capacitors containing structural defects such as cracks or laminations appear to be more likely to be subject to this behaviour. The presence of moisture may also be a factor.

Another circumstance is when deterioration becomes apparent under fields in excess of 0.5 MV m^{-1} at temperatures above 85 °C, and occurs more rapidly the higher the field or the temperature. The fall in resistance has been observed in single crystals of rutile and barium titanate and so must be assumed to be a bulk rather than a grain boundary effect, although there is evidence that grain boundaries play a part in degradation processes in ceramics.

The mechanisms responsible for degradation are complex with $V_O^{\cdot\cdot}$ playing an important role. Because they act as an effective positive charge they migrate under the influence of a d.c. field towards the cathode with a corresponding depletion in the anode region. Accompanying these changes are corresponding changes in electron and hole concentrations in the bulk of the ceramic and it is the steady build-up of electron currents which is responsible for the degradation process. The degradation mechanism is discussed in detail in a series of papers by R. Waser, T. Baiatu and K.-H. Härdtl [8].

Degradation can be slowed down by suitable substituents. The presence of donor ions at levels exceeding 2 mol.%, e.g. substituting Nb^{5+} for Ti^{4+}, La^{3+} for Ba^{2+} or F^- for O^{2-}, prolongs useful life. Donor ions reduce the concentration of oxygen vacancies, which are relatively mobile, and increase the concentration of cation vacancies. The latter have low mobilities at room temperature and, when combined with holes in the valence band, behave as acceptors (see Section 2.6.2, and Section 5.7.3).

Manganese at the 1% level in air-fired dielectrics acts as a palliative. It is present as both Mn^{4+} and Mn^{3+}. The former ion must be expected to act as an effective electron trap since it is readily converted into Mn^{3+}. However, the presence of Mn^{3+} will result in a corresponding concentration of oxygen vacancies.

Humidity is known to accelerate degradation and it is argued that the mechanism involves protons incorporated into the structure, as discussed in the context of hydrogen sensors (see Section 4.6.1). Because an hydroxyl group can occupy a vacant oxygen ion site without distorting the lattice, 'water' may be incorporated into the structure according to the reaction:

$$H_2O + V_O + O^{2-} \rightarrow 2OH^- \qquad (5.21)$$

Proton conduction can then occur via a 'hopping' mechanism between hydroxyl ions.

It is also established that 'water' is able to enter a vitreous phase and, acting as a glass 'network modifier', enhance ionic mobilities. There is no doubt that the nature of the grain boundary phase, coupled with the presence of water vapour, influences the diffusion of silver from electrodes and terminations in multilayer capacitors leading to eventual inter-electrode shorting and component failure [9].

5.6.3 High-power capacitors

The capacitors in the output stage of high-frequency generators provide an application for Class I dielectrics. These units isolate the d.c. voltage component from the external load and the whole output from the generator passes through them. Their values range up to 5000 pF and their frequency of operation ranges up to 50 MHz. They may have to withstand up to 25 kV and pass 150 A. Their shape and fabrication have been outlined in Section 5.4.3. The properties of the principal dielectrics used are given in Table 5.5. The low loss tangents are achieved by selecting raw materials which are low in impurities, especially transition elements, and ensuring that none are introduced during processing.

It is instructive to consider important features of the design of high-power capacitors in some detail, particularly with regard to power dissipation. The average rate $\bar{P}$ at which heat is developed in a dielectric due to the dissipation of electrical energy is

$$\bar{P} = IU \tan\delta \tag{5.22}$$

in which I and U are the root mean square (r.m.s.) current and voltage. Therefore a 500 kW unit with $\tan\delta = 2 \times 10^{-4}$ has to dissipate 100 W as heat. Clearly a particular application requires a particular capacitance value C and, since $I = 2\pi f U C$ and $C = \varepsilon A/h$, Eq. (5.22) becomes

$$\bar{P} = 2\pi f U^2 \varepsilon \frac{A}{h} \tan\delta \tag{5.23}$$

Table 5.5 Properties of dielectrics for 'transmitter' capacitors

Principal constituent	*Relative permittivity*	*tan* $\delta/10^{-4}$	*TCC*/MK^{-1}
Steatite	6	3–5	+100 to +150
$MgTiO_3$	12–15	1–3	+60 to +100
$BaTi_4O_9$	36–40	1–3	−30 to +36
TiO_2	80–90	2–4	−800 to −700

If it is assumed that the rate at which heat can be removed from the unit is proportional to its exposed surface area and to its temperature excess ΔT over the surroundings, we obtain

$$\Delta T \propto fU^2 \frac{\varepsilon}{h} \tan \delta \tag{5.24}$$

Therefore, for a given application, the temperature rise increases with permittivity and frequency, and decreases with the thickness of the dielectric. An allowed ΔT defines the maximum power rating $\bar{P}_{\mathrm{m}}$ for a given capacitor and this, in turn, defines upper and lower bounds on the frequency range over which the unit can be operated at maximum power level.

At low frequencies, say below f_1, the power-handling capability is limited by the high reactance $1/\omega C$ and therefore by the voltage rating U_{m} of the capacitor. Above a frequency f_2 the reactance is low, so that the power capability is limited by the current rating I_{m}. The frequencies are given by

$$f_1 = \frac{\bar{P}_{\mathrm{m}}}{2\pi U_{\mathrm{m}}^2 C} \qquad \text{and} \qquad f_2 = \frac{I_{\mathrm{m}}^2}{2\pi P_{\mathrm{m}} C} \tag{5.25}$$

It is also necessary to consider the power $\bar{P}_{\mathrm{e}}$ dissipated by the electrodes, which is given by

$$\bar{P}_{\mathrm{e}} = R_{\mathrm{e}} I^2 \tag{5.26}$$

where R_{e} is the electrode resistance. Since $\bar{P}_{\mathrm{m}} = IU = I^2/2\pi f C$, Eq. (5.26) can be written

$$\bar{P}_{\mathrm{e}} = R_{\mathrm{e}} 2\pi f C \bar{P}_{\mathrm{m}} \tag{5.27}$$

R_{e} is complicated by the 'skin effect', by which high-frequency currents are concentrated near the surface of a conductor. The effect arises for the following reason. When a direct current flows down a wire the current density is distributed evenly over the cross-section of the wire. The wire can be considered as being made up of elementary current-carrying filaments, each of which has its associated magnetic induction. A little consideration (or consultation of a textbook on electricity) will show that a central filament is linked with more flux than a filament running along the outer surface of the wire. Therefore, when the current alternates, the back e.m.f. is greater along the centre of the cross-section than at the surface. Consequently the current density along the length increases radially from the centre outwards towards the surface. In fact the current density falls off exponentially with depth below the surface, and at a depth δ_{s} its value is $1/e$ of that at the surface. Clearly the effect becomes more pronounced with increasing frequency.

The skin depth δ_s is inversely proportional to the square root of the frequency, and at 1 MHz is 0.064 mm for silver, 0.066 mm for copper and 0.19 mm for a typical solder. A surface resistivity ρ_s can be defined by

$$\rho_s = \rho/\delta_s \tag{5.28}$$

where ρ is the volume resistivity. Alternatively the variation in δ_s with frequency can be introduced explicitly by defining a surface resistivity ρ'_s, which is a material property independent of frequency, such that

$$\rho_s = \rho'_s f^{1/2} \tag{5.29}$$

For silver, copper and solder ρ'_s is $2.5\times10^{-7}\,\Omega\,\mathrm{Hz}^{-1/2}$, $2.6\times10^{-7}\,\Omega\,\mathrm{Hz}^{-1/2}$ and $7.7\times10^{-7}\,\Omega\,\mathrm{Hz}^{-1/2}$ respectively.

In order to evaluate the order of magnitude of the power $\bar{P}_e$ dissipated in the electrode, a value is required for R_e in Eq. (5.27). Because $R_e = \rho_s l/w$, where l/w is the length-to-width ratio, and $l \approx w$, then $R_e \approx \rho_s$. Therefore it follows that

$$\bar{P}_e \approx 2\pi f^{3/2} \rho'_s C \bar{P}_m \tag{5.30}$$

For example, for a 500 kW 500 pF unit with $\rho'_s = 4\times10^{-7}\,\Omega\,\mathrm{Hz}^{-1/2}$, $\bar{P}_e$ is 0.02 W at 0.1 MHz and 7 kW at 500 MHz. Therefore it is evident that below 1 MHz the major contribution to heat generation is dielectric loss, whilst at higher frequencies a significant proportion is due to electrode resistance, and that, because of the skin effect, this resistance cannot be reduced by making the electrodes or leads thicker than a small fraction of a millimetre. However, the thicker the electrodes and leads are the better is the heat transfer from the capacitor.

The TCC is not of the greatest importance for these units since the capacitance values themselves may vary some 10% from nominal in most cases.

A higher-permittivity dielectric based on $CaTiO_3$ is sometimes used. It has a relative permittivity of about 140 and a dissipation factor of 2×10^{-4}. The TCC is about double that of rutile-based dielectrics.

5.6.4 Low-TCC low-loss capacitors

A common function of circuits is the provision of an accurate resonance state. For instance, for a resonance frequency to stay within a tolerance of 0.1% over a temperature range of 100 K a temperature coefficient of less than $10\,\mathrm{MK}^{-1}$ would be required. It might be achieved in the 10–100 kHz range by using a manganese zinc ferrite pot-core inductor (see Section 9.5.1) with a small positive temperature coefficient of inductance combined with a ceramic capacitor having an equal, but negative, temperature coefficient. This is clear from the resonance condition

$$\omega_0 = (LC)^{-1/2}$$

which when differentiated with respect to temperature yields

$$\frac{1}{\omega_0}\frac{\delta\omega_0}{\partial T} = -\frac{1}{2}\left(\frac{1}{L}\frac{\partial L}{\partial T} + \frac{1}{C}\frac{\partial C}{\partial T}\right) \tag{5.31}$$

In most applications a resonance tolerance of 0.1% would only be useful if the resonance were correspondingly sharp, e.g. with a Q in the neighbourhood of 1000 ($\tan\delta = 10^{-3}$). Thus low-TCC capacitors must also be low loss if they are to be of practical value in such applications.

The parameters that contribute to the TCC can be identified by first considering a rectangular parallel-plate capacitor with sides of length x and y and thickness z. Then, since the capacitance is given by

$$C = \frac{\varepsilon x y}{z}$$

differentiation with respect to temperature leads to

$$\begin{aligned}\frac{1}{C}\frac{\partial C}{\partial T} &= \frac{1}{\varepsilon}\frac{\partial \varepsilon}{\partial T} + \frac{1}{x}\frac{\partial x}{\partial T} + \frac{1}{y}\frac{\partial y}{\partial T} - \frac{1}{z}\frac{\partial z}{\partial T}\\ &= \frac{1}{\varepsilon}\frac{\partial \varepsilon}{\partial T} + \alpha_L\end{aligned}$$

or

$$\mathrm{TCC} = \mathrm{TC}_\varepsilon + \alpha_L \tag{5.32}$$

in which TC_ε is the temperature coefficient of permittivity and α_L is the linear expansion coefficient. Eq. (5.32) is derived under the assumption that the expansion coefficients in the x, y and z directions are identical, i.e. the dielectric has isotropic linear expansion characteristics.

The capacitance may change with temperature not only because the dimensions of the capacitor change but also because the permittivity of the dielectric changes. To gain some insight into the sources of the variation in permittivity with temperature, the Clausius–Mosotti equation (Eq. (2.88)) can be differentiated with respect to temperature to give

$$\begin{aligned}\mathrm{TC}_\varepsilon &= \frac{1}{\varepsilon}\frac{\partial \varepsilon}{\partial T}\\ &= \frac{(\varepsilon_r - 1)(\varepsilon_r + 2)}{3\varepsilon_r}\left(\frac{1}{\alpha}\frac{\partial \alpha}{\partial T} + \frac{1}{N}\frac{\partial N}{\partial T}\right)\end{aligned} \tag{5.33}$$

When $\varepsilon_r \geqslant 2$,

$$\frac{(\varepsilon_r - 1)(\varepsilon_r + 2)}{3\varepsilon_r} \approx \frac{\varepsilon_r}{3}$$

and

$$\frac{1}{N}\frac{\partial N}{\partial T} = -\frac{1}{V}\frac{\partial V}{\partial T} = -3\alpha_L$$

where V is the volume containing N polarizable units. Eq. (5.33) therefore reduces to

$$TC_\varepsilon \sim \frac{\varepsilon_r}{3}\left(\frac{1}{\alpha}\frac{\partial \alpha}{\partial T} - 3\alpha_L\right) \tag{5.34}$$

For a number of dielectrics with $\varepsilon_r > 30$, TC_ε is negative and within 15% of $-\alpha_L\varepsilon_r$ as illustrated by the examples given in Table 5.6. Eq. (5.34) suggests that the temperature variation of polarizability is small compared with the volume expansion coefficient in these cases. Lower-permittivity oxides have positive TC_εs and in their case the temperature coefficient of polarizability can be assumed to exceed the volume expansion coefficient. However, the extent to which the Clausius–Mosotti equation can be applied to ionic solids is open to debate.

In order to improve volumetric efficiency, dielectrics combining a small TC_ε with a high permittivity have been sought. Since, in many cases, it is not zero TC_ε but a controlled value that is required, combinations of two components with different TC_εs in a series of ratios has provided ranges of useful dielectrics. A typical combination has end-members with compositions corresponding to $BaTi_3O_7$ ($\varepsilon_r = 35$, $TC_\varepsilon = +35\,MK^{-1}$) and TiO_2 ($\varepsilon_r = 100$, $TC_\varepsilon = -750\,MK^{-1}$) which covers the range of TC_ε and permittivity lying between these two extremes as illustrated in Fig. 5.31.

Combinations of high positive TC_ε with high permittivity and low loss are rare. The antiferroelectric compound $PbZrO_3$ has $\varepsilon_r = 110$ and $TC_\varepsilon = 1400$ but $\tan\delta = 28 \times 10^{-4}$. Sphene ($CaSiTiO_5$), which is also known as titanite, has $\varepsilon_r = 45$, $TC_\varepsilon = 1200$ and $\tan\delta = 5 \times 10^{-4}$. A combination of sphene and rutile gives a dielectric with zero TC_ε, $\varepsilon_r = 60$–70 and low loss. The crystal structure of sphene consists of chains of corner-sharing TiO_6 octahedra interlinked by SiO_4 tetrahedra by corner sharing. The Ti^{4+} ions are displaced from the centres of the octahedra by about 10 pm but in opposite directions in alternate groups;

Table 5.6 Temperature coefficient of permittivity of Class I dielectrics

			TC_ε/MK^{-1}	
Composition	ε_r	α_L/MK^{-1}	*Reported*	$-\varepsilon_r\alpha_L$
TiO_2	100	7.5	−750	−750
$SrTiO_3$	230	10.3	−2400	−2370
$CaTiO_3$	170	10.8	−1600	−1836
$MgTiO_3$	17	7	+100	−119
Al_2O_3	10	8.5	+120	−85
MgO	10	13.5	+190	−135

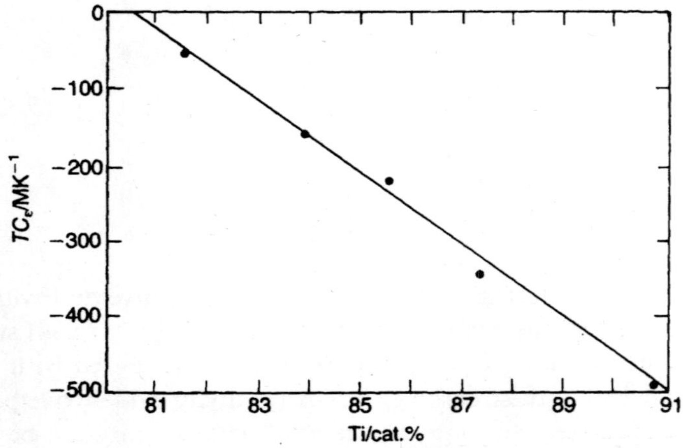

Fig. 5.31 Temperature coefficient versus TiO_2 content for $BaTi_3O_7$–TiO_2 mixtures.

therefore it is an antipolar structure. It is not antiferroelectric since there is no transition to a paraelectric state in which the Ti^{4+} ions have zero displacements.

As pointed out earlier (Chapter 2, Section 2.7.4), Lichtenecker's rule for mixtures leads to the prediction that the TC_ε of a mixture will be equal to the volume average of the TC_εs of its constituents. This presupposes that there is no reaction leading to the formation of new compounds. The permittivity of the mixture can also be predicted approximately by Lichtenecker's rule.

5.6.5 Microwave ceramics

'Wireless technologies' have already penetrated many aspects of modern life and the trend is set to continue. This has in turn been a strong stimulant to the development of specially tailored microwave dielectrics.

Figure 5.32 gives an indication of specific technologies and of the corresponding frequency bands.

Dielectric resonators

The rapid growth of satellite and mobile radio communications has led to a requirement for narrow band, frequency-stable filters and oscillators. Selectivity and stability are necessary to ensure that signals are confined to closely defined allotted frequency bands and to prevent the intrusion of unwanted signals which would interfere with the satisfactory performance of the system. Antennae are also critical components of any wireless communications system. The need for compactness in satellite and hand-held mobile systems is self-evident.

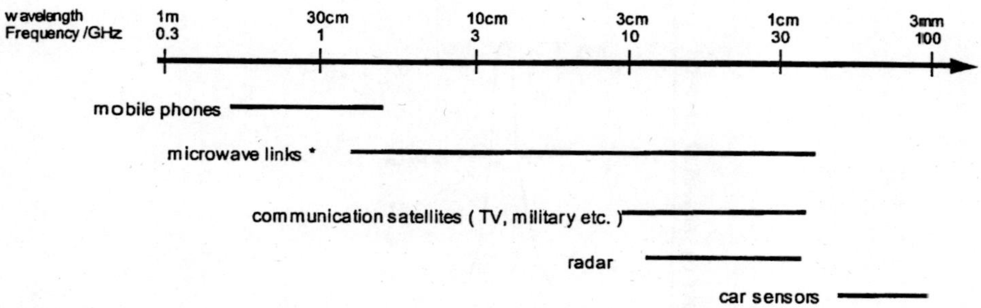

Fig. 5.32 Frequency bands covered by the various wireless technologies (*includes 'Bluetooth'; Harald Blatand (Bluetooth) was a famous King of Denmark c.960 who is known for encouraging communication between people).

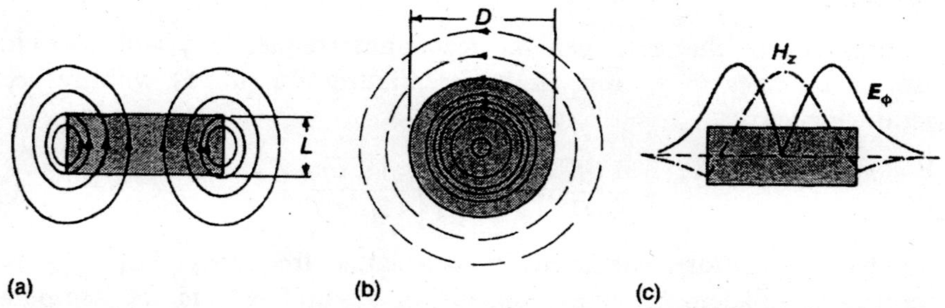

Fig. 5.33 Fields in a microwave resonance dielectric in the simplest standing wave mode: (a) magnetic field; (b) electric field; (c) variation in E_φ and E_z with r at $z = 0$, with reference to cylindrical coordinates (the z axis is perpendicular to the plane of the disc and the origin is at the disc centre).

The solution to providing stable filters and oscillators in the past lay in bulky coaxial and cavity resonators fabricated from the temperature-stable metal alloy Invar. The dielectric resonator (DR) offers a means of miniaturizing the device.

In its simplest form a DR is a cylinder of ceramic of relative permittivity ε_r sufficiently high for a standing electromagnetic wave to be sustained within its volume because of reflection at the dielectric–air interface. The electric and magnetic field components of the fundamental mode of a standing electromagnetic field are illustrated in Fig. 5.33.

The wavelength λ_d of the standing wave in the dielectric approximates to the diameter, D, of the cylinder, i.e. $\lambda_d \sim D$. If the resonance frequency is f_r then in free space $f_r = c/\lambda_0$ where c and λ_0 are respectively the free space velocity and wavelength. In a non-magnetic dielectric medium the velocity $v_d = c/\varepsilon_r^{1/2}$ (see Eq. (2.120)) so that

$$f_r = c/(\lambda_d \varepsilon_r^{1/2}) \sim c/(D\varepsilon_r^{1/2}) \tag{5.35}$$

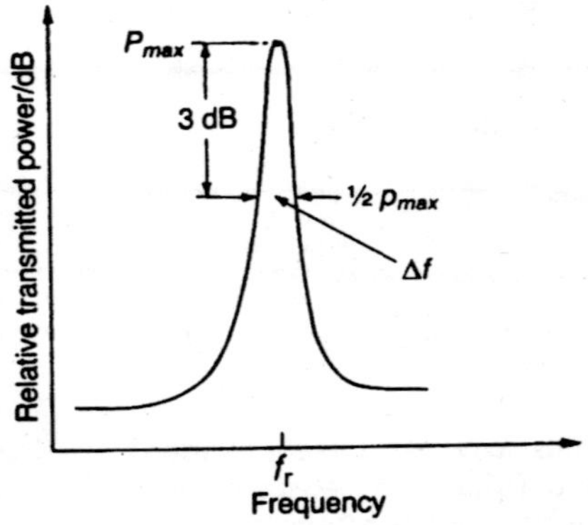

Fig. 5.34 Frequency response of a microwave resonator.

If the temperature changes, then the resonance frequency f_r will also change because of changes in ε_r and D. Differentiating Eq. (5.35) with respect to temperature gives

$$\frac{1}{f_r}\frac{\partial f_r}{\partial T} = -\frac{1}{D}\frac{\partial D}{\partial T} - \frac{1}{2}\frac{1}{\varepsilon_r}\frac{\partial \varepsilon_r}{\partial T} \tag{5.36}$$

$\frac{1}{f_r}\frac{\partial f_r}{\partial T}$ is the temperature coefficient of resonance frequency TC_f, $\frac{1}{D}\frac{\partial D}{\partial T}$ is the temperature coefficient of linear expansion α_L and $\frac{1}{\varepsilon_r}\frac{\partial \varepsilon_r}{\partial T}$ is the temperature coefficient of permittivity TC_ε. Substitution into Eq. (5.36) gives

$$TC_f = -(\tfrac{1}{2}TC_\varepsilon + \alpha_L) \tag{5.37}$$

In the microwave dielectrics context TC_f is usually written τ_f. It follows from Eq. (5.37) that achievement of a temperature-independent resonance frequency, i.e. $\tau_f = 0$, requires balanced control over TC_ε and α_L.

The frequency response of a DR coupled to a microwave circuit is shown in Fig. 5.34. The selectivity Q of the resonator is given by $f_r/\Delta f$ and, under conditions where the energy losses are confined to the dielectric and not to effects such as radiation loss or surface conduction $Q \sim (\tan\delta)^{-1}$ where $\tan\delta$ is the loss factor for the dielectric.

The requirements for DR ceramics are now clear.

1. In order to miniaturize the resonator and to ensure that the electromagnetic energy is adequately confined to the resonator, ε_r must be large and is usually in the range $30 < \varepsilon_r < 100$.
2. To ensure stability against frequency drift with temperature, the temperature coefficient τ_f must be close to zero, which implies control over TC_ε and α_L.
3. To optimize frequency selectivity, $Q = (\tan\delta)^{-1}$ must be maximized and is usually greater than 1000.

Microwave engineers find it helpful to characterize a microwave dielectric by the product Qf (f measured in GHz) which, as a rough 'rule of thumb', is assumed to be constant. The theoretical justification for this is as follows.

Figure 2.38 illustrates that in the case of an ionic solid the optical mode of the lattice vibration resonates at an angular frequency, ω_o, in the region of 10^{13} Hz. In the frequency range from approximately 10^9–10^{11} Hz dielectric dispersion theory shows the contribution to permittivity from the ionic displacement to be nearly constant and the losses to rise with frequency according to

$$\tan\delta = (\gamma\omega_o)^2 f \quad \text{or} \quad Qf = (\gamma\omega_o)^{-2} \sim \text{constant} \tag{5.38}$$

in which γ is a damping factor (see Eq. (2.105)).

It has to be stressed that the relationship is at best a rough guide. There will almost certainly be contributions to dielectric loss occurring at frequencies between the microwave region and the ionic lattice resonance frequency that complicate the $Q(f)$ relationship. For example, such contributions may arise through polarization processes associated with dislocations and microstructural features. There is no alternative to making the measurement if reliable loss data at a particular frequency are required.

Ceramic compositions

The following discussion outlines the historical path taken in the development of microwave ceramics combining high permittivity, high Q and zero τ_f with long term stability and at an economic cost. Key references are given in the overview by R. Freer [10].

Titania (TiO_2) first attracted attention because of its high relative permittivity ($\varepsilon_r \sim 100$) and low loss ($\tan\delta \sim 3 \times 10^{-4}$). However, a very high $\tau_f \sim 400\,MK^{-1}$ renders it totally unsuited.

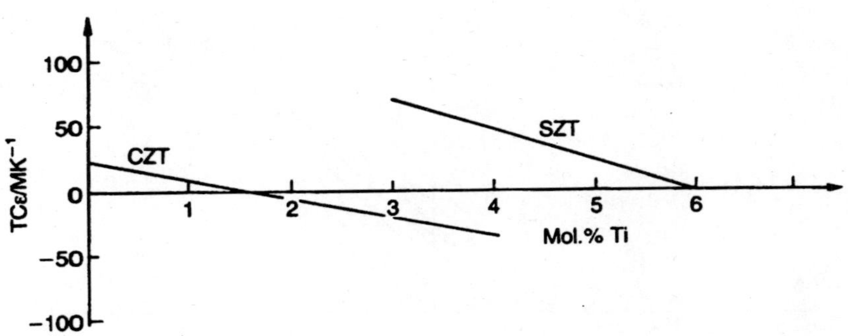

Fig. 5.35 TC_ε of zirconate compositions with varying titanium contents (CZT; $CaZrO_3$–$CaTiO_3$:SZT; $SrZrO_3$–$SrTiO_3$ (measured at 4 GHz).

Table 5.7 Dielectric properties measured at 7 GHz (after [10])

	ε_r	Q	τ_f/MK^{-1}
$MgTiO_3$	17	22 000	−45
$CaTiO_3$	170	1800	800

Interest turned to compositions of high permittivity titanates and zirconates in which one member exhibited a negative and the other a positive TC_ε, the objective being to achieve $\tau_f = 0$. The effect of composition on TC_ε for two such systems is shown in Fig. 5.35. Another system which attracted interest is $MgTiO_3$–$CaTiO_3$ in which the end members have the dielectric properties shown in Table 5.7. The composition $Mg_{0.95}Ca_{0.05}TiO_3$ yields $\tau_f \sim 0$, $\varepsilon_r = 21$ and $Q = 8000$ at 7 GHz.

As early as 1952 the ZrO_2–TiO_2–SnO_2 (ZTS) system was being investigated for possible exploitation in capacitors with low TCCs and in the 1970s [11] it attracted interest for microwave applications. Figure 5.36 shows the region established as single phase ZTS having the orthorhombic $ZrTiO_4$ structure after firing and annealing mixtures of ZrO_2, TiO_2 and SnO_2 powders at 1400 °C. The locus of compositions for which $\tau_f = 0$ is shown.

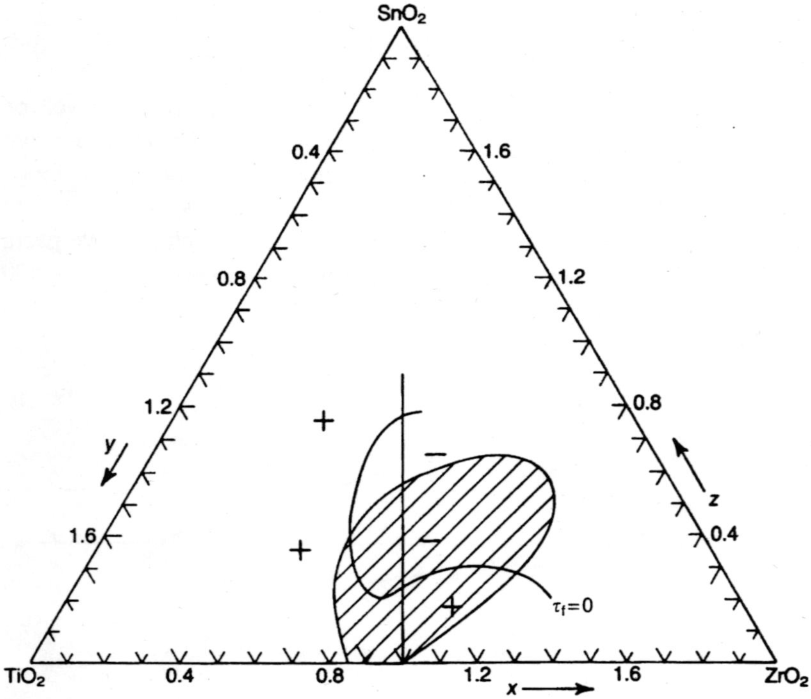

Fig. 5.36 The $Zr_xTi_ySn_zO_4$ (ZTS) system: the shaded area indicates the single-phase microwave ceramic field (after [11]).

In the 1970s barium nonatitanate ($Ba_2Ti_9O_{20}$ or 'B_2T_9') was identified as a candidate. Although not easy to process to a reproducible product because of the existence of many BaO-TiO_2 phases (see Fig. 5.41) it is manufactured on a commercial basis.

The modified perovskite, barium zinc tantalate ($Ba(Zn_{1/3}Ta_{2/3})O_3$) ('BZT'), was reported in the 1980s to offer a combination of high permittivity coupled with exceptionally high Q. It was established that the high Q resulted from long anneals at high temperatures (>1400 °C, 100 h) and that this was accompanied by increased ordering of ions on the B-sites. Unfortunately zinc evaporation and grain-growth also occurred, well illustrating the difficulties encountered in attempting to unambiguously establish correlations between losses and compositional, structural and microstructural changes in electroceramics. Subsequent research established a firm correlation between Q-value and B-site order although the underlying reasons for the correlation remained unclear. A significant advance made in the mid-1980s, especially from the commercial standpoint, was the discovery that small (<5 mol.%) additions of $BaZrO_3$ led to very high Q-values being achieved with relatively short annealing times (~1500 °C, 4 h). There is experimental evidence that the structure consists of B-site ordered domains at the nanometre size scale [12] similar to that occurring in the case of the relaxor ferroelectrics (see Section 5.7.2). The role of the Zr ion in influencing the ordering kinetics is not understood but it is suggested that the increase in Q is, at least in part, a consequence of its segregation to the domain boundaries where it has a stabilizing effect; 'mobile' domain boundaries would be expected to contribute to losses.

The τ_f value for BZT can be tailored to be near zero by the substitution of Ni for some of the Zn so that there are commercial compositions based on $(1-x)Ba[(Zn,Ni)_{1/3}Ta_{2/3}]O_3$–$xBaZrO_3$ system with the Ni:Zn atom ratio approximately 1:7 and $x = \sim 0.03$.

Other microwave dielectrics have been identified and developed into commerical products and those currently exploited are listed in Table 5.8. The modified neodymium titanates are particularly important because of their high permittivity values and the reduction in resonator size this offers.

Broadly speaking microwave dielectrics are all processed in conventional ways, that is by mixing starting materials, calcining, comminution, pressing and firing. Somtimes hot-pressing is used. DRs have to be made to close dimensional tolerances and this requires diamond machining as a final step.

There are many factors which contribute to dielectric loss and in the case of the complex ceramic compounds discussed above, to achieve a satisfying understanding of the relative magnitudes of the various loss mechanisms is challenging. There will be contributions to loss intrinsic to the *idealized* structural chemistry of the material and it is now clear that this is complicated by a domain structure. There will also be contributions of an extrinsic nature, particularly those associated with departures from the ideal structure, point defects and

Table 5.8 Commercial microwave ceramics

Ceramic	ε_r	*Q(2 GHz)*	*Q(f/GHz)*
$Mg_{0.95}Ca_{0.05}TiO_3$	20	–	8000 (7)
$Ba[Sn_x(Mg_{1/3}Ta_{2/3})_{1-x}O_3$	25	–	43 000 (10)
$Ba(Zn_{1/3}Ta_{2/3})O_3$ 'BZT'	30	45 000	14 000 (12)
Zr-modified 'BZT'	30	–	10 000 (10)
$(Zr_xSn_{1-x})TiO_4$ 'ZTS'	35–38	18 500	10 000 (5)
$Ba_2Ti_9O_{20}$ 'B_2T_9'	39	20 000	10 500 (5)
Ca titanate–Nd aluminate 'CTNA'	45	21 000	11 000 (5)
Zr-titanate–Zn-niobate 'ZTZN'	45	21 000	11 000 (5)
$BaNd_2Ti_4O_{12}$	77	3800	4000 (1.8)
(Ba,Pb) $Nd_2Ti_4O_{12}$	90	2000	2000 (1.8)

Ceramic	ε_r	*Q(f/GHz)*	τ_f/MK^{-1}
TiO_2	100	17 000 (3)	400
Al_2O_3	10	33 000 (10)	50

In all cases, through minor compositional changes, the τ_f can be adjusted to $\pm$ a few MK^{-1}. Data for titania and alumina are included for comparison.

'dislocations' and of course, various microstructural features, grain size, grain boundaries and second phases. Even though second phases may in toto be present in extremely small amounts, the manner in which they are distributed can have an effect on properties disproportionate to the amounts (see Chapter 3, Question 19). A clear and unambiguous understanding of dielectric loss mechanisms will only result from research in which impurities are controlled to the p.p.m. level, or lower. All the above variables, particularly the introduction of impurities, depend strongly on all aspects of 'processing', especially those involving high temperatures.

In contrast to dielectric losses permittivity is not, in general, sensitive to small amounts of impurities and for homogeneous dielectrics values can be calculated as described in Section 2.7.1, and the various 'mixture' rules allow good estimates to be made for multiphase dielectrics. For Ba- and Sr-based dielectrics having the perovskite structure the variation of permittivity with temperature, which determines τ_f (see Eq. (5.37)), can be correlated with the 'tolerance factor' t (see Section 2.7.3) [13] providing guidance for tailoring ceramics to have $\tau_f = 0$.

Measurement of dielectric properties

The development of improved microwave ceramics, and especially research into loss mechanisms, necessitates reliable methods for characterizing properties, particularly Q. A popular method is that described by Hakki and Coleman [14]

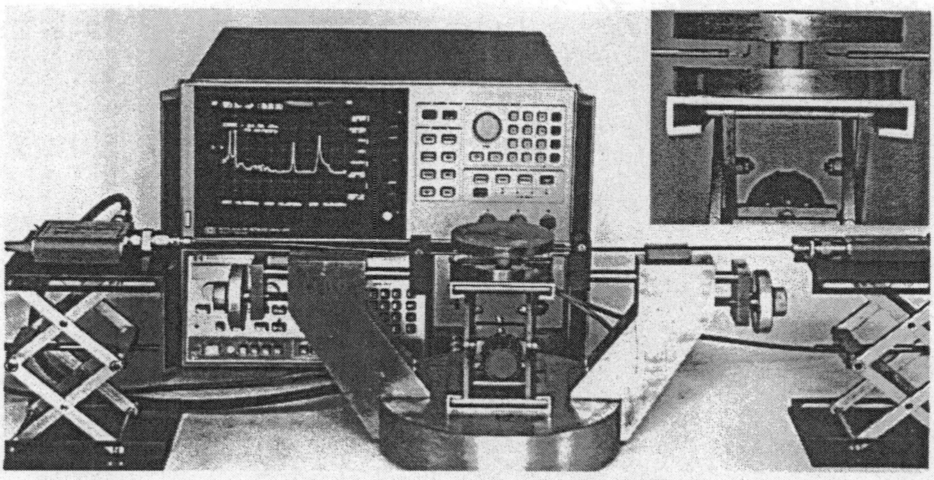

Fig. 5.37 Apparatus for measuring the microwave characteristics of dielectrics. Inset, detail showing specimen between conducting planes, and antennas.

and Courtney [15]. A dielectric cylinder, typically 15 mm in diameter and 5–10 mm long, is placed between two parallel conducting planes forming, in effect, a shorted waveguide (Fig. 5.37). Two antennas, which are loosely electromagnetically coupled to the system, radiate power into and extract power from the cavity. Essentially, ε_r is determined from the resonance frequency f_r, and Q from $f_r/\Delta f$, where Δf is the 3 dB bandwidth. Such measurements are by no means straightforward to make, but are capable of yielding ε_r and Q values with an accuracy to about 1% and about 10%, respectively.

Sometimes the DR carries a fired-on silver coating. This is so with coaxial DRs when the outer and inner surfaces of a hollow cylinder are silvered. The Q of the DR can be written

$$1/Q = 1/Q_d + 1/Q_s + 1/Q_{rad} \tag{5.39}$$

where $1/Q_d$, $1/Q_s$ and $1/Q_{rad}$ are respectively the contributions to total loss from the dielectric, from the conducting surface and from radiated power. Usually $1/Q_{rad}$ can be neglected, and it can be shown that even for a coating of pure silver Q_s is not greater than approximately 1000 at 900 MHz. For such an application there is little point in striving to obtain Q_d values greater than approximately 5000 since they would have a relatively insignificant effect on Q. What may be beneficial is giving the dielectric surface a high polish before being silver coated.

Dielectric resonators are extensively used in mobile communications technology. As far as selectivity and temperature drift are concerned the requirements for hand-held units are far less demanding than for those used in the base stations, and for the more general filter applications. For example, for

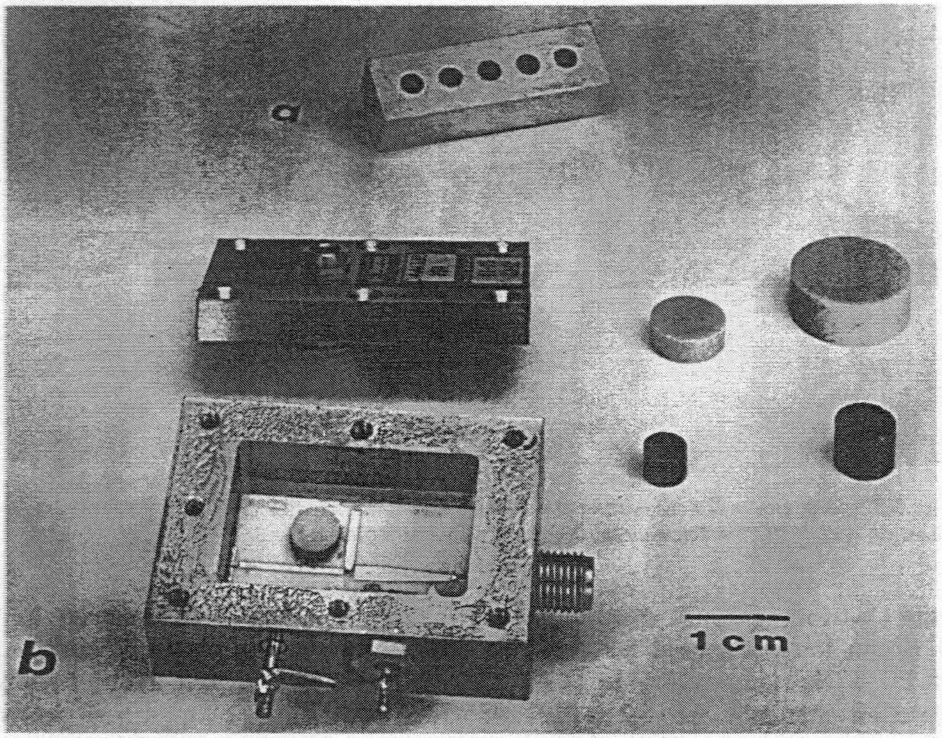

Fig. 5.38 Microwave ceramic components: (a) metallized ceramic 'engine block' for 40 MHz pass band filter at 1.4 GHz; (b) 11.75 GHz oscillator incorporating ceramic dielectric resonator together with various resonator pucks.

hand-held units the resonators are silver plated whereas for base station filters much higher Q-values (e.g. $>20\,000$ at 2 GHz) are essential.

Figure 5.38 shows a microwave oscillator built on to a ceramic substrate and incorporating a dielectric resonator.

LTCC and microwave technology

LTCC technology is extensively exploited in microwave technology and the following discussion should be read in conjunction with Section 5.5.5.

Microwave LTCC materials are essentially established microwave ceramic compositions fluxed to allow processing to a glass-ceramic at temperatures lower than $\sim 900\,°C$. For example, the microwave dielectric $BaNd_2Ti_4O_{12}$ (see Table 5.8) mixed with a glass (B_2O_3 27 mol.%, Bi_2O_3 35 mol.%, SiO_2 6 mol.% ZnO 32%), both in finely powdered form, compacted and processed at 900 °C for 3 h produced a glass-ceramic with values for ε_r, τ_f and Q of approximately 67, $4\,MK^{-1}$ and >1000 (at 6 GHz), respectively [16].

Table 5.9 Dielectric properties of microwave LTCCs [17]

Ceramic	ε_r	*Q(6–7 GHz)*
Commercial $MgTiO_3$–$CaTiO_3$*	21.4	3700
Commercial LTCC	7.4	320
LTCC**	8.5	>1000

*Composition approximately that given in Table 5.8.
**No prior glass melting – see text.

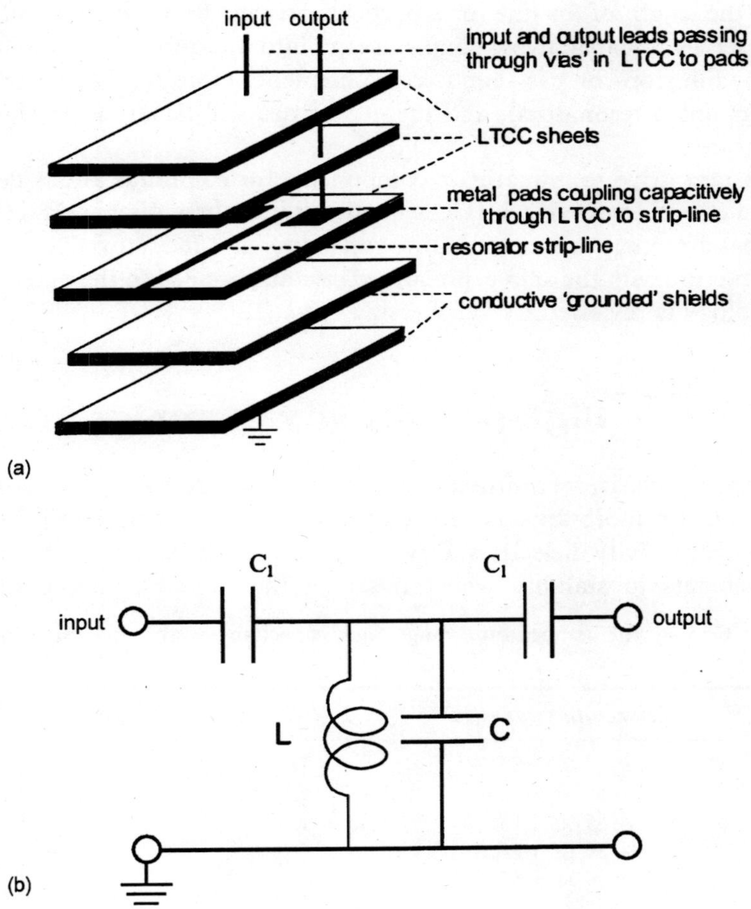

Fig. 5.39 (a) Schematic of a $\lambda/4$ strip line resonator 'pass-band' filter exploiting LTCC technology; the approximate overall dimensions are: $15 \times 10 \times 1$ mm. (b) Equivalent circuit: C_1 is the 'pad-to-strip-line' capacitance and L and C the inductance and capacitance of the strip-line.

In a similar study [17] a commercial microwave dielectric, $MgTiO_3$–$CaTiO_3$, was fluxed with SiO_2, B_2O_2 and ZnO powders avoiding the prior glass-melting step. After firing at 900 °C the final product was a crystalline glass ceramic with no detectable amorphous phase and offering useful dielectric properties (Table 5.9).

Oscillators and filters do not always take the form of the 'puck' as illustrated in Fig. 5.38 but also as 'strip-lines'. Figure 5.39 illustrates a structure for a $\lambda/4$ strip line resonator exploiting LTCC technology.

The principle of the resonator is the same as that of a vibrating string and in the illustrated example one end of the line is maintained at 'ground' potential (the node) and the other is the antinode. The permittivity of the surroundings determine the length of the line for a particular resonance frequency and, in this case, also serves to capacitively couple energy into and out of the resonator.

To tailor the stop- or pass-band to requirements more than one line will be involved (coupled resonators), and the geometries will depart from the simple form illustrated.

The constant drive to miniaturize components for communications devices is stimulating efforts to develop LTCC technology into three dimensional circuitry in which passive components (resistors, capacitors and inductors) are packaged into the structure with the active, integrated circuits, bonded to the outer surface. The technology is reviewed by W. Wersing *et al.* [18].

5.7 High-permittivity Ceramics

Dielectrics with relative permittivities exceeding 1000 are based on ferroelectric materials and are more sensitive to temperature, field strength and frequency than lower-permittivity dielectrics. Development in the past 50 years has resulted in improvements in stability whilst retaining the desirable high-permittivity

Table 5.10 Coding for temperature range and capacitance variation for Class II/III capacitors

EIA Code	*Temperature range/°C*	*EIA Code*	*Capacitance change/%*
X5	−55 to +85	D	±3.3
X7	−55 to +125	E	±4.7
X8	−55 to +150	F	±7.5
Y5	−30 to +85	P	±10
Z5	+10 to +85	R	±15
		S	±22
		T	+22 to −33
		U	+22 to −56
		V	+22 to −82

Example: a capacitor is required for which the capacitance value at 25 °C changes by no more than ±7.5% in the temperature range −30 °C to +85 °C; the EIA Code will be Y5F. EIA codes D–R; Class II: Codes S–V; Class III.

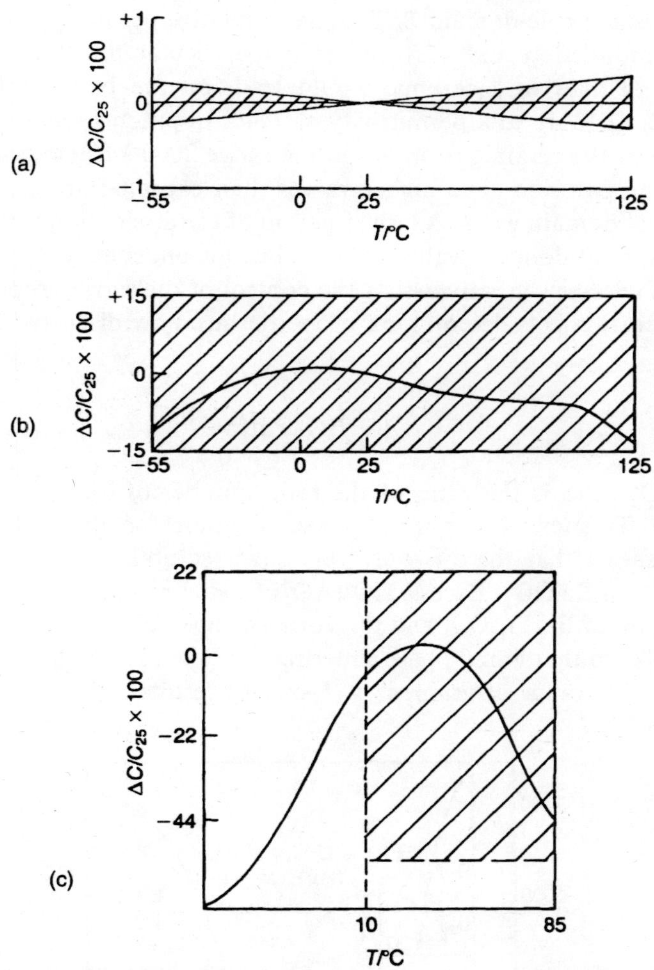

Fig. 5.40 Variation of capacitance with temperature for (a) Class I C0G, (b) Class II X7R and (c) Class III Z5U dielectrics.

feature. The Electronics Industries Alliance (EIA) of the United States has devised a scheme for specifying the variability of capacitance with temperature in the ranges of practical interest. The coding is defined in Table 5.10, and Fig. 5.40 shows typical characteristics compared with those of Class I controlled temperature coefficient dielectrics.

5.7.1 Modified barium titanate dielectrics

The dielectric characteristics of barium titanate ceramics with respect to temperature, electric field strength, frequency and time (ageing) are very dependent on the substitution of minor amounts of other ions for Ba or Ti.

Single-crystal single-domain $BaTiO_3$ has a relative permittivity at 20 °C of 230 in the polar direction and 4770 in the perpendicular directions. The random orientation of axes in a ceramic would lead, on the basis of Lichtenecker's relation (Eq. (2.131)), to a permittivity of 1740. In practice the low-field relative permittivity of the ceramic form lies in the range 2000–4500 and varies with the method of preparation. The higher values than expected are ascribed to small oscillations of domain walls. A large fraction of the dissipation factor can also be accounted for by domain wall motion. Thus an understanding of the domain structure of ceramics greatly assists the control of dielectric properties.

The principal effects determining properties are now discussed in turn.

AO/BO_2 ratio

The AO/BO_2 ratio is the ratio of the total number of ions on Ba sites to the number on Ti sites. The partial phase diagram for the $BaO–TiO_2$ system (Fig. 5.41) shows that there is only very slight solubility for excesses of either BaO or TiO_2 in $BaTiO_3$. Excess $TiO_2(AO/BO_2 < 1)$ results in the formation of a separate phase of $Ba_6Ti_{17}O_{40}$, and this forms a eutectic with $BaTiO_3$ that melts at about 1320 °C so that liquid phase sintering can take place at temperatures below 1350 °C. A wide range of grain sizes (5–50 μm) results.

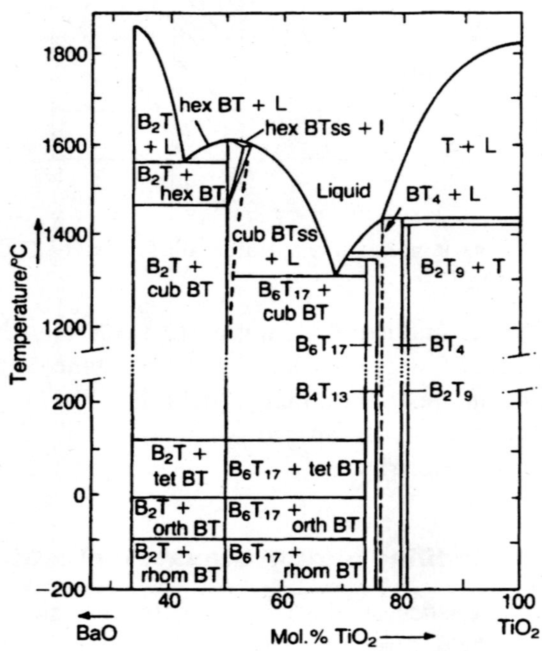

Fig. 5.41 Phase diagram of the $BaO–TiO_2$ system (>34 mol.% TiO_2).

An excess of BaO results in the formation of Ba_2TiO_4 which forms a eutectic with $BaTiO_3$ that melts at about 1563 °C. As is often the case with solid insoluble phases, Ba_2TiO_4 inhibits the grain growth of $BaTiO_3$ sintered at temperatures up to 1450 °C, giving rise to grain sizes in the 1–5 μm range. Excess BaO also lowers the cubic–hexagonal transition from 1570 °C to about 1470 °C in pure $BaTiO_3$. Hexagonal material seldom occurs in sintered ceramics of technical purity because many common substituents, such as strontium for barium, stabilize the cubic form.

The effect of the AO/BO_2 ratio varies with different substituents and additives as discussed in the next section.

Substituents

Uniformly distributed isovalent substituents do not greatly affect the shape of the ε_r–T curve and other characteristics. Their main effect is to alter the Curie point and the lower transitions of $BaTiO_3$.

Barium can be replaced by isovalent ions with r_{12} radii between 130 and 160 pm. As can be seen in Fig. 5.42 the effect on T_c varies considerably among lead, strontium and calcium which enables the transitions to be shifted to suit

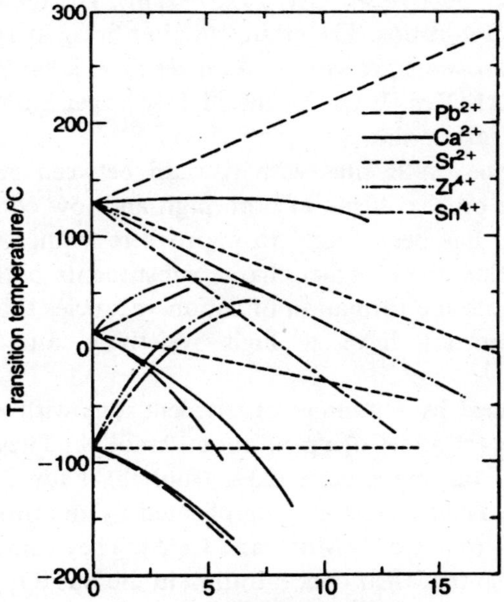

Fig. 5.42 Transition temperature versus concentration of isovalent substituents in $BaTiO_3$: ---, Pb^{2+}; ——, Ca^{2+}; ----, Sr^{2+}; —··—, Zr^{4+}; -·-·-·-, Sn^{4+}. (After B. Jaffe *et al.* (1971) *Piezoelectric Ceramics*, Academic Press, London.)

particular requirements. Ba, Pb and Sr ions can be mixed in any proportions to produce a single-phase perovskite, while the solubility of $CaTiO_3$ is limited to about 20 mol.%. Ti^{4+} can be replaced by isovalent ions with r_6 radii between 60 and 75 pm. Zirconium, hafnium and tin have similar effects on the three transitions, although the solubility of tin may be limited to about 10 mol.%. They reduce T_c but raise the temperature of the other two transitions to such an extent that in the range 10–16 mol.% they almost coincide in the neighbourhood of 50 °C. Particularly high values of permittivity are found for such compositions. The more highly charged B-site ions diffuse less rapidly than the A-site ions so that the effects of inhomogeneities are more often seen for B-site substituents.

Aliovalent ions (see Section 2.6.2) are usually limited in their solubility which may depend on the AO/BO_2 ratio. K^+ can replace Ba^{2+} to which it is very similar in radius. The charge balance can be restored by the simultaneous replacement of O^{2-} by F^-. The effect on dielectric properties is minimal.

A number of trivalent ions with r_{12} radii between 110 and 133 pm, e.g. Bi and La, can substitute on the A site. La^{3+} confers a low resistivity at low concentrations (<0.5 mo.%), and the electrical behaviour is discussed in Section 2.6.2, and Section 4.4.2. It has been studied more widely as a substituent for lead in $PbTiO_3$-based compositions.

Dysprosium with $r_6 \approx 90$ pm and $r_{12} \approx 115$ pm is a rather large ion to be found on a B site and rather small for an A site. At the 0.8 mol.% level it gives optimum properties (ε_r at 25 °C, 3000; resistivity, $7 \times 10^{10}\,\Omega$ m). The resistivity falls to a low value at higher AO/BO_2 ratios. The grain size after firing at 1450 °C in oxygen is 1 μm and few domain walls are visible. As a result ε_r is not greatly affected by applied fields up to 2 MV m^{-1}. Unlike the slightly larger Sm^{3+} ion, it has only a small effect on the Curie point.

Higher-valency ions on B sites with r_6 radii between 58 and 70 pm have similar effects to La on the A site at both high and low concentrations. Nb^{5+} at the 5 mol.% level has been found to improve resistance to degradation. In sufficient concentration these higher-charge substituents both suppress oxygen vacancies and promote the formation of cation vacancies that act as acceptors. The resulting dielectrics have a high resistivity and are resistant to degradation.

Ti^{4+} can be replaced by a number of trivalent ions with r_6 in the range 60–70 pm (Cr, Ga, Mn, Fe and Co) up to about 2 mol.%. They lower T_c and the second transition to varying extents (23 K $(\text{mol.\%})^{-1}$ for T_c for Fe and 10 K $(\text{mol.\%})^{-1}$ for Ga). Their effects are complicated by the presence of oxidation states other than 3+ (e.g. Co^{2+}, Mn^{4+} and Cr^{4+}). They can also dissolve in the intergranular phase so that their concentration in the $BaTiO_3$ phase is difficult to determine. Their main effect is to form acceptors and so to compensate the lowering of resistivity by donors. However, the charge balance may be maintained by the formation of oxygen vacancies which lead to higher ageing rates and to degradation under d.c. fields at lower temperatures.

$BaTiO_3$ sintered with about 3 mol.% Fe_2O_3 at 1300 °C gives a remarkably flat ε_r/T relation with an average ε_r of about 2500. This may be because most of the Fe is present in an intergranular phase that keeps the grain size down to about 1 μm and there is a non-uniform distribution of the Fe within the grains, thus giving regions of differing Curie point that combine to give the flat ε_r–T characteristic. The same composition sintered at 1360 °C gives a normal ε_r–T peak at 60 °C, indicating that the Fe has diffused to a uniform concentration within the grains following further grain growth.

About 0.5 mol.% MnO_2 is frequently added to all classes of dielectric and results in a reduction in the dissipation factor. This may be due to its presence as Mn^{4+} in the sintered bodies with the possibility of trapping carriers by the reactions

$$Mn^{4+} + e' \rightarrow Mn^{3+} \quad \text{and} \quad Mn^{3+} + e' \rightarrow Mn^{2+} \tag{5.40}$$

Mn_{Ti}^{3+} and Mn_{Ti}^{2+} also act as acceptors.

Ti^{4+} can also be replaced by about 2 mol.% of divalent ions with r_6 in the range 60–70 pm such as Ni^{2+} and Zn^{2+}, with similar results to substitution by trivalent ions. Larger divalent ions such as Mn^{2+} ($r_6 = 82$ pm) may be soluble to a lesser extent. Mg^{2+} ($r_6 = 72$ pm) is only soluble in $BaTiO_3$ when AO/BO_2 is greater than unity; otherwise it forms a separate phase of $MgTiO_3$. There is evidence that Ca^{2+} ($r_6 = 100$ pm) may occupy B sites to a limited extent when an excess of Ba is present. All these ions can fulfil an acceptor function and, to varying extents, can prevent $BaTiO_3$-based compositions from becoming conductive when fired in atmospheres low in oxygen (see Section 5.7.3).

Effect of crystal size

The grain size of a ferroelectric ceramic has a marked effect on the permittivity for the size range 1–50 μm (see Fig. 2.48). Below about 1 μm the permittivity falls with decreasing grain size. An important factor leading to this behaviour is the variation in the stress to which a grain is subjected as it cools through the Curie point.

As a single-crystal grain cools through the Curie point it attempts to expand in the *c* direction and contract in the *a* directions, as can be seen in Fig. 2.40(b). It will be constrained from doing so by the surrounding isotropic ceramic. The resulting stresses within the grain can be reduced by formation of an appropriate arrangement of 90° domains and, in large grains, most of the stresses can be relieved by this mechanism. As the grain size decreases the domains become smaller, with the domain width being roughly proportional to the square root of the grain size. The number of domains per grain therefore decreases as the square root of the grain size, and so the smaller the grain the larger is the unrelieved stress. It can be shown using Devonshire's phenomenological theory (cf. Section 2.7.1) that an increase in stress is accompanied by an increase in permittivity, irrespective of any possible contribution from the domain walls *per se*. As the grain size approaches 0.5–0.1 μm, the unrelieved stresses reach values at which they suppress the tetragonality and the permittivity falls to approximately 1000.

In addition to the direct effect of stress described above, a reduction in 90° domain width can enhance permittivity because the domain wall area per unit volume of ceramic increases. The argument outlined below follows that developed by Arlt *et al.* [19].

The decrease in mechanical strain energy resulting from the formation of domains is counterbalanced by the increase in wall energy as the domains develop. At equilibrium the total energy will be a minimum. The strain energy $\mathcal{E}_s$ per unit of volume occupied by 90° domains has been calculated to be

$$\mathcal{E}_s = \frac{dYx^2}{128\pi g} \tag{5.41}$$

where g is the average grain size, d is the domain width, Y is the Young modulus and x $(= c/a - 1)$ is the tetragonality. The domain wall energy $\mathcal{E}_d$ per unit volume is given by

$$\mathcal{E}_d = \frac{\gamma}{d} \tag{5.42}$$

where γ is the surface energy associated with a 90° domain wall. Therefore the total energy $\mathcal{E}_t$ is

$$\mathcal{E}_t = \mathcal{E}_s + \mathcal{E}_d \tag{5.43}$$

Substituting from Eqs (5.41) and (5.42) into Eq. (5.43), differentiating Eq. (5.43) with respect to d and setting $\partial\mathcal{E}_t/\partial d = 0$ yields the condition for minimum total energy:

$$d = \left(\frac{128\pi\gamma g}{Yx^2}\right)^{1/2} \tag{5.44}$$

Substituting the values for barium titanate ($\gamma \approx 3$ mJ m^{-2}, $Y \approx 1.7 \times 10^{11}$ Pa and $x \approx 10^{-2}$) gives

$$d \approx 2 \times 10^{-4} g^{1/2} \tag{5.45}$$

Eq. (5.45) agrees quite well with experimental data for grain sizes between 1 and 10 μm. Above 10 μm more complex domain walls form and the domain width is limited to about 0.8 μm. Below 1 μm the stresses are large enough to reduce the tetragonality and this simple model is no longer valid.

It is easily shown from Eq. (5.45) that, while the domain wall area per grain is proportional to $g^{5/2}$, the domain wall area per unit volume of ceramic is approximately $5000g^{-1/2}$. The part of the permittivity due to domain wall motion will be proportional to the domain wall area and so increases as the grain size diminishes from 10 μm to 1 μm.

Both the direct effect of stress and the changes in the concentration of domain walls appear to contribute to the observed changes due to grain size. The behaviour of domains is markedly affected by dopants and, in consequence, the grain size at which walls become scarce has been reported as varying over a range of values. The importance of grain size is in no doubt, but its optimum value for each composition has to be determined empirically.

The effect on permittivity of applied electric field

The magnitude of the applied electric field has a very marked effect on dielectric properties as can be seen in Fig. 5.43. The effects can be rationalized in very

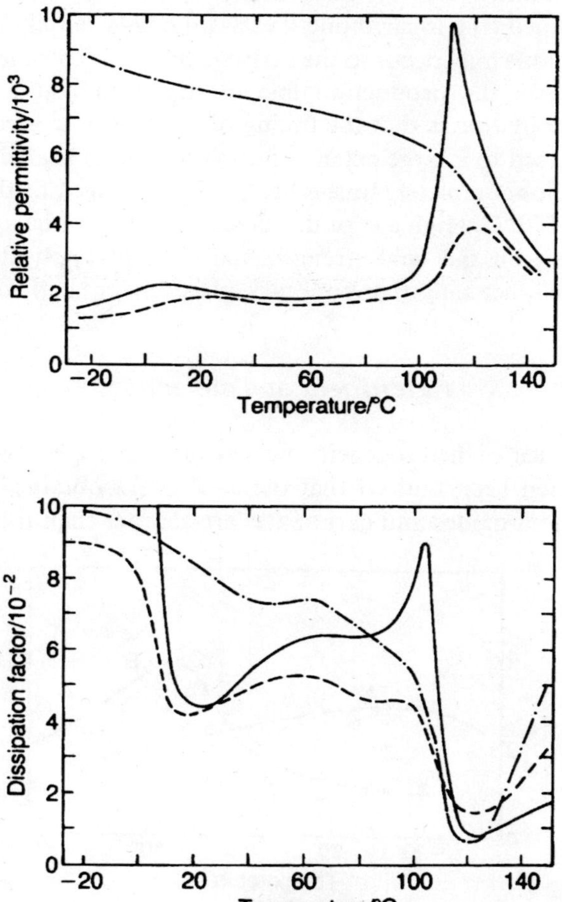

Fig. 5.43 Dielectric properties of technical grade $BaTiO_3$ ceramic under various field conditions: ——, 15 kV m^{-1} peak and 1 kHz; – – –, 15 kV m^{-1} peak and 1 kHz + 1.43 MV m^{-1} d.c.; —·—, 1.6 MV m^{-1} peak and 50 Hz.

general terms by considering the contributions to permittivity and energy dissipation made by domain movement. Reference should also be made to hysteresis effects discussed in Section 2.7.3.

Ageing of capacitors

The mechanism of ageing is discussed in Section 6.3.1, under piezoelectric ceramics. In capacitors it is a minor nuisance and is usually in the 2–5% decade range. However, on automated production lines a capacitance test may take place not many minutes after silvering and soldering, and so may indicate a considerably higher value than that measured a month or so later when the unit is inserted in a circuit. However, once the ageing characteristic of a composition has been determined (the logarithmic law, Eq. (2.124), is only approximate and significant deviations may occur in the early stages), an allowance can be made to relate the value on the production line to that found at a later date. One advantage of automation is that the timing of operations can be very consistent.

Ageing is reversed to a large extent whenever a unit is heated above the Curie point, although not completely unless heating is prolonged or the temperature is increased above 500 °C. High a.c. or d.c. fields also reverse the ageing to a limited extent, depending on the field strength and time of application. Unexpected changes in capacitance may therefore take place under such conditions.

Heterogeneous dielectrics

A possible instance of heterogeneity occurs in iron-doped $BaTiO_3$ mentioned above. It has often been noticed that the ε_r–T peaks obtained with mixed and calcined constituent oxides and carbonates are sharper than those obtained with

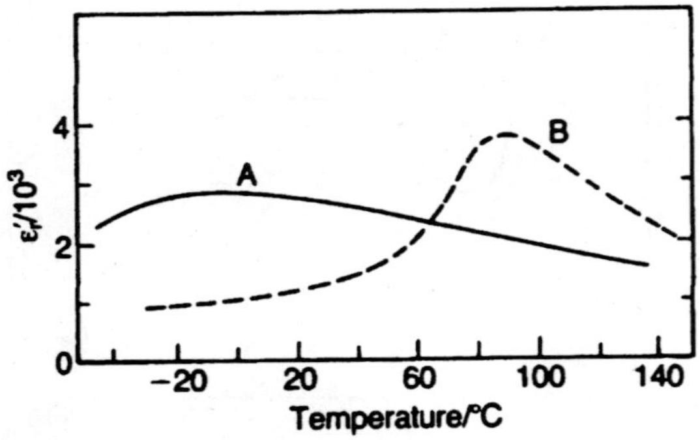

Fig. 5.44 The effect of compositional heterogeneity on dielectric properties: curve A, 43cat.% Ba–7cat.%Bi–49cat.%Ti($BaTiO_3$ and $Bi_4Ti_3O_{12}$ calcined separately); curve B, 46cat.%Ba–5.5cat.%Bi–48cat.%Ti (starting materials calcined together). Both are sintered at 1300 °C for 1 h.

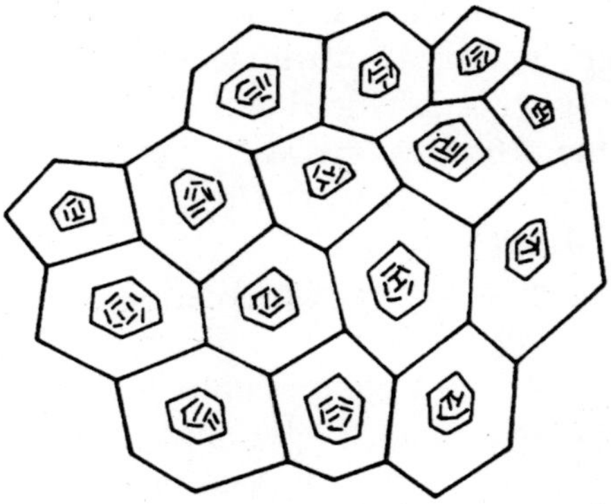

Fig. 5.45 Duplex microstructure of an X7R dielectric ceramic.

mixed and sintered preformed compound oxides. Figure 5.44 shows the effects of adding bismuth as $Bi_4Ti_3O_{12}$ to $BaTiO_3$ rather than as Bi_2O_3 to a mixture of $BaCO_3$ with TiO_2. The flatter characteristics obtained with precalcined compounds are most easily explained as due to the failure of the ions to interdiffuse fully during the subsequent sintering to form a homogeneous composition. The resulting inhomogeneity comprises regions with different Curie points, and the net effect is the flattened characteristic.

In practice the use of this method of controlling ε_r–T relationships depends on the interdiffusion of constituent ions being slow enough to be controlled by practicable furnace schedules. For instance, $(Ba_{1-x}Sr_x)TiO_3$ compositions with different x values, calcined separately and milled to an average particle size of 10 μm, yield a single peak corresponding to their average composition when sintered to full density in air because of the relatively rapid interdiffusion of Ba^{2+} and Sr^{2+} ions. More highly charged ions such as Zr^{4+} diffuse more slowly and can give rise to heterogeneous compositions under sintering conditions that yield satisfactory densities.

The effect of heterogeneity has been observed directly in the case of submicron $BaTiO_3$ powder sintered with a 0.03 molar fraction of $CdBi_2Nb_2O_9$ in air at 1130 °C for 4 h. A schematic diagram of the ceramic (Fig. 5.45) shows grains with a duplex structure. The centre region of each grain exhibits a ferroelectric domain structure which analysis shows to be low in substituent ions. The outer region is high in substituents and appears to have a T_c of −80 °C. The ceramic as a whole has the relatively flat X7R characteristic. Many combinations of additives and $BaTiO_3$ show similar effects, even when the constituents are mixed as simple oxides and carbonates. This type of heterogeneity needs to be distinguished from intergranular phases with widely differing compositions and crystal structures from the main phase. In the present case there is a gradation from a heavily doped to a lightly doped composition with a corresponding gradual change in lattice parameters and dielectric behaviour. It is possible that the additives are initially uniformly distributed in the $BaTiO_3$ lattice but diffuse out as a more perfect crystal is developed during grain growth. Much remains to be understood about the processes involved.

5.7.2 Relaxor ferroelectrics

The dielectric properties of most compositions based on $BaTiO_3$ do not vary greatly with frequency above about 500 Hz until the GHz range is reached.

In contrast, in the case of the relaxor class of dielectric the real part of the permittivity shows a broad peak as a function of temperature with a strong frequency dispersion (see Fig. 5.46). The class comprises a number of perovskites of the type $Pb(B'\,B'')O_3$ for example $Pb(Mg_{1/3}Nb_{2/3})O_3$ (PMN), $Pb(Zn_{1/3}Nb_{2/3})O_3$ (PZN) and $Pb(Sc_{1/2}Ta_{1/2})O_3$ (PST), some solid solutions, for example $Pb_{1-3x/2}La_x(Zr_{1-y}Ti_y)O_3$ (PLZT) and some tungsten bronze structures (see Section 6.4.4), for example $Sr_{5-x}Ba_xNb_{10}O_{30}$ (SBN). The tpyical relaxor characteristics shown for PMN in Fig. 5.46 are evidence of a gradual transition from a macroscopic paraelectric to a ferroelectric phase below the peak permittivity Curie temperature.

G.A. Smolenskii [20] was a pioneering researcher and at the time relaxors were being intensively studied in Russia. Since then they have attracted very considerable interest because of the following attributes. Firstly the low sintering temperatures and high peak permittivity values which are attractive for the

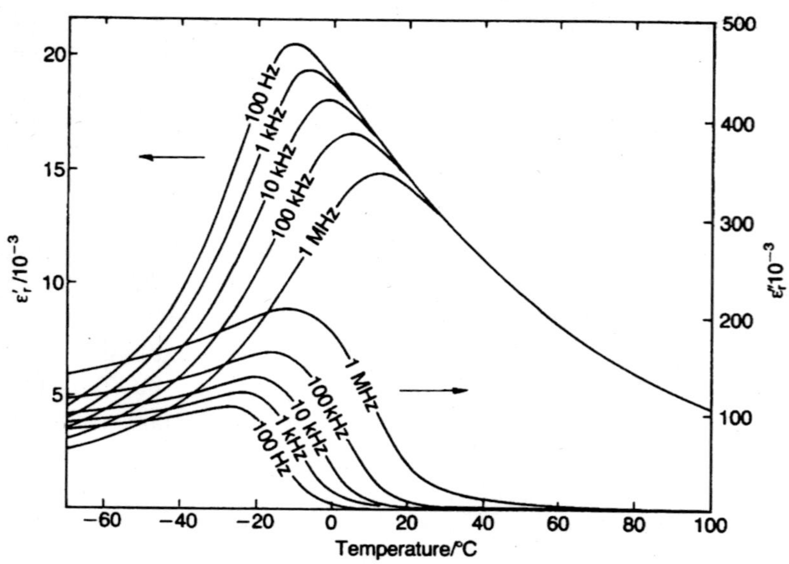

Fig. 5.46 Dielectric properties of $Pb(Mg_{1/2}Nb_{1/2})O_3$ (PMN).

production of MLCCs of low cost and high volumetric efficiencies, and secondly because of their high electrostriction coefficients which make them attractive for actuators (see Section 6.4.3).

It is agreed that B-site crystalline disorder is resonsible for relaxor behaviour although there are uncertainties regarding its precise nature, and it may be that details differ from relaxor to relaxor. In the case of PMN one model [21] proposes the existence of domains of nanometre order size within which the composition is $\{Pb(Mg_{1/2}Nb_{1/2})O_3\}^{-0.5e}$, that is the Mg:Nb ratio is unity. The domains are in a matrix of compositionally disordered material such that the overall composition is stoichiometric, namely Mg/Nb = 0.5. In this model the negative charge the domains carry increases as the domain size increases; it is the associated electrostatic energy which accounts for the observed stability of the domain structure with high temperature annealing.

An alternative model [22] favours a uniform B-site disorder extending throughout the volume and consisting of local clusters of ferroelectric and antiferroelectric ordering with the highly polarizable Pb ion almost certainly playing an important role.

Close parallels can be drawn between the dielectric response of a relaxor and that of a random assembly of electric dipoles – a 'dipolar glass'. The matter is clearly one of very considerable complexity depending as it must upon the sizes, charges and polarizabilities of the ions, and thermal history.

Closely related structural changes occur during the annealing of the microwave dielectric BZT (see Section 5.6.5).

A number of low sintering compositions have been based on $PbFe_{0.55}W_{0.1}Nb_{0.35}O_3$. A typical ε_r–T characteristic is shown in Fig. 5.47. ε_r is greater than 10^4 between −8 and +45 °C but is greatly reduced when d.c. fields are applied. Since in many applications the d.c. field is less than 0.2 MV m^{-1} its effect is not of great importance.

One of the difficulties with most compositions containing lead and niobium is a tendency to form pyrochlore-type rather than perovskite-type structures which results in lower ε_r values. This is particularly the case when Zn ions are present. Pyrochlore is a mineral with a composition approximating RNb_2O_6, where R is a mixture of divalent ions. The pyrochlore-type phase found in lead magnesium niobate has the composition $Pb_{1.83}Nb_{1.71}Mg_{0.29}O_{6.39}$. It has a room temperature relative permittivity of 130 and is paraelectric. The structure contains corner-sharing MO_6 octahedra but they have several different orientations. The effect on the permittivity depends on the amount and distribution of the low-permittivity phase. Small amounts occurring as discrete particles have rather less effect than a corresponding volume of porosity, but they may considerably reduce the permittivity if present as an intergranular phase interposing low-permittivity regions between crystals of the high-permittivity phase. Larger amounts will also result in a significant change in composition, and therefore in properties, of the perovskite-type phase. It is found that the pyrochlore structure forms

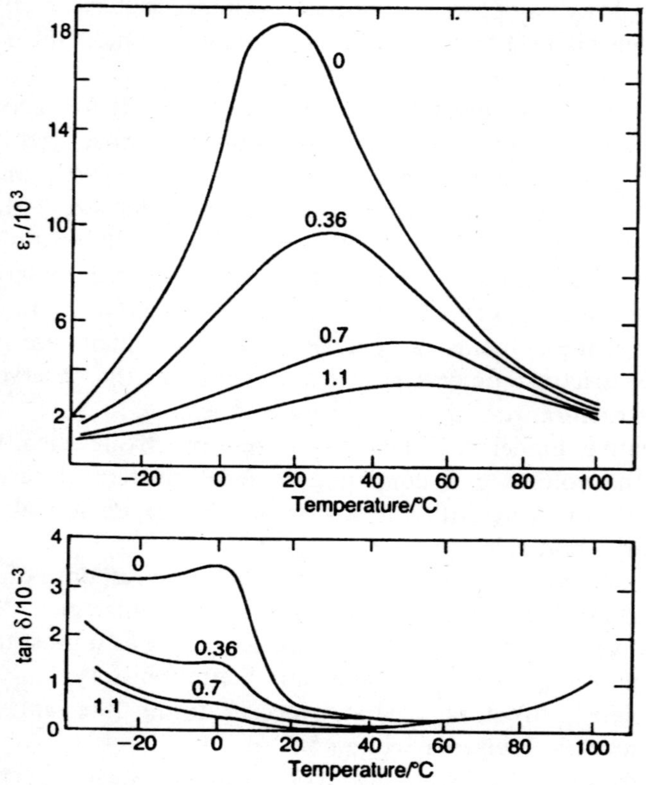

Fig. 5.47 Effect of d.c. bias ($MV\,m^{-1}$) on the dielectric properties of $PbFe_{0.55}W_{0.1}Nb_{0.35}O_3$.

preferentially between 750 and 850 °C whilst perovskite-type structures are formed between 850 and 950 °C. A processing route bypassing the formation of pyrochlore has been developed for PMN [23]; MgO is pre-reacted with the Nb_2O_5 to form $MgNb_2O_6$ which is then reacted with PbO to form the perovskite phase.

A composition approximating to $Pb_{0.85}La_{0.1}Ti_{0.2}Zr_{0.8}O_3$ ($\varepsilon_r \sim 2000$) is only significantly affected by d.c. fields above $2\,MV\,m^{-1}$, as shown in Fig. 5.48. A number of factors contribute to the stability:

1. the grain size (about 1 μm),
2. the relaxor characteristics,
3. a possible antiferroelectric structure between −35 and +65 °C and
4. T_c at 70 °C due to the combined effects of the zirconium and lanthanum contents.

Factor 3 may account for the maintenance of the ε_r level at low temperatures and the small effect of d.c. fields up to 60 °C, and 1 and 2 contribute to a

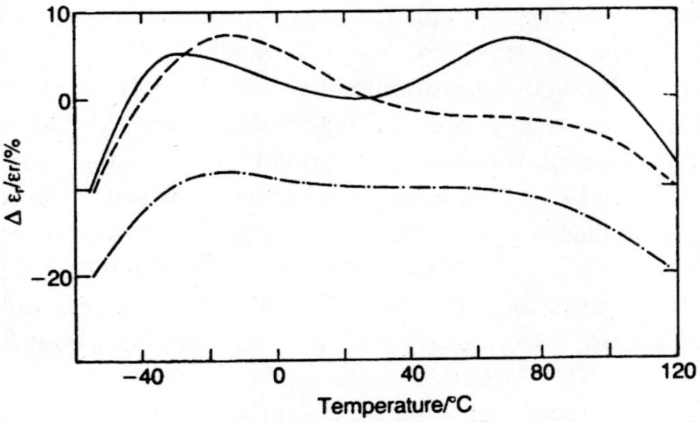

Fig. 5.48 Dependence of ε_r on T and d.c. bias (——, 0; ----, 2 MV m^{-1}; —·—, 4 MV m^{-1}) for $Pb_{0.85}La_{0.1}Ti_{0.2}Zr_{0.8}O_3$ (PLZT). (After G.H. Maher (1983) 33rd Electron. Comp. Conf. (IEEE) 173–7.)

smoothing out of the peaks which might otherwise occur at −40 and +70 °C. Factor 2 has the marginal disadvantage of introducing frequency dependence of ε_r and tan δ below 0 °C. Tan δ measured at 100 kHz rises to above 0.04 at −20 °C.

The composition sinters at 1100 °C which allows some reduction in the palladium content of the Ag–Pd electrodes and reduces the volatility of PbO. The resistivity is very high, leading to a leakage time constant greater than 10^6 s at room temperature and about 10^5 s at 125 °C. There are very few signs of degradation under high d.c. fields up to 150 °C.

The La^{3+} ion has $r_{12} = 132$ pm and $r_6 = 106$ pm and appears to be able to occupy either A or B sites, although a higher lead content may be needed to give appreciable B-site occupancy. The high resistivity and resistance to degradation may be due to the lowering of the concentration of vacant oxygen sites by the replacement of Pb^{2+} by La^{3+}, accompanied by La^{3+} on B sites acting as acceptors for electrons and compensating the La^{3+} ions on A sites.

5.7.3 Multilayer capacitors with base metal electrodes (BME)

The multilayer structure discussed in Section 5.4.3 is applicable to all ceramic dielectrics. It enables the thinnest dielectric plates to be engineered into robust units that are readily attached to the substrates carrying the electronic circuitry by 'surface mount technology' (SMT). The structure also gives minimum electrode inductance permitting use at high frequencies. Very significant progress has been made over the years in reducing the thickness of both dielectric and electrode layers and so increasing volumetric efficiency. An appreciable part of the cost of a multilayer capacitor lies in the palladium-containing electrodes.

With reductions in ceramic sintering temperature, the cost of electrodes has been reduced by the use of Ag–Pd alloys, typically 70 at.% Ag/30 at.% Pd. The reduction in palladium content is limited to about 15 at.% because at lower levels silver migration becomes a problem. The drive to lower cost has led to efforts to avoid it entirely. One approach is based upon the 'injected' or 'fugitive' electrode (see Section 5.4.3 and Fig. 5.12) and the other on the use of the base metals nickel or copper for the electrodes (BME) combined with dielectrics resistant to the reduction firing necessary to retain the electrodes in the metallic state.

Although the fugitive electrode approach is viable it is presently little used, if at all. In contrast BME technology presently accounts for about 50% of MLCC production.

When $BaTiO_3$ is fired under reducing conditions oxygen is lost from the lattice with the formation of doubly ionized oxygen vacancies and electrons in the conduction band (see Section 2.6.2, Eq. (2.48)). On cooling to room temperature a large fraction of the electrons may remain in the conduction band because the ionization energies of V_O and $V_O^{\cdot}$ are low compared to kT at room temperature (see Fig. 2.15). Acceptor ions such as Mn^{3+} on Ti^{4+} sites also give rise to oxygen vacancies but without the liberation of electrons. They have the effect of shifting the minimum in the conductivity–oxygen pressure relation to lower oxygen pressures (Fig. 5.49) so that the concentration of electrons in the conduction band is reduced and positive holes in the valence band become the majority carriers at oxygen pressures above 0.01 Pa (Eq. (2.53)). On cooling to room temperature the positive holes combine with the ionized acceptors since the ionization energies for acceptors lie in the range 1–2 eV. High-resistivity $BaTiO_3$ therefore requires both a sufficient concentration of acceptors and a sufficient pressure of oxygen during firing.

Technical $BaTiO_3$ normally contains sufficient acceptors in the form of Al, Fe etc. to render it insulating after firing at atmospheric pressure, but it requires

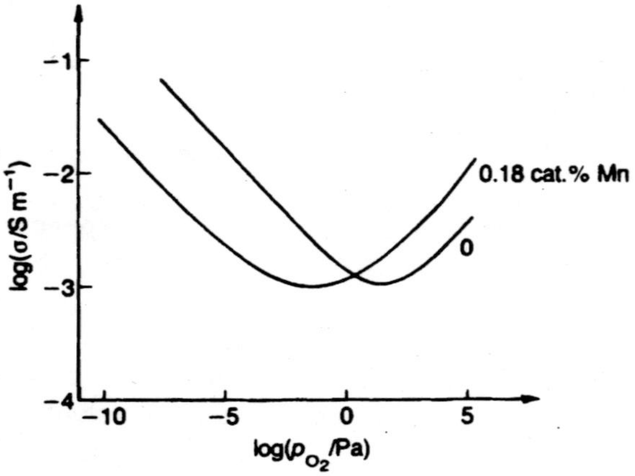

Fig. 5.49 Effect of Mn'_{Ti} on the conductivity–oxygen pressure relation for $BaTiO_3$ at 965 °C.

additional acceptors if it is to be sintered at lower oxygen pressures. The addition of 0.2–0.5 mol.% Mn forms sufficient Mn^{2+} and Mn^{3+} acceptor ions on Ti^{4+} sites on firing at oxygen pressures below 10^{-2} Pa to result in an insulating dielectric at room temperature.

The oxidation reaction for nickel is

$$\mathrm{Ni} + \tfrac{1}{2}\mathrm{O_2(g)} \rightarrow \mathrm{NiO} \tag{5.46}$$

and the standard free-energy change at 1300 °C (ΔG^0 at 1573 K) is approximately -100 kJ. If the activities of Ni and NiO are set equal to unity, the reaction constant K is

$$K = p_{O_2}{}^{-1/2} \tag{5.47}$$

and since $\Delta G^0_{1573} = -RT\ln K = -RT\ln p_{O_2}{}^{-1/2}$,

$$\begin{aligned} p_{O_2} &= \exp\left(\frac{2\Delta G^0_{1573}}{RT}\right) \\ &\approx 2.5 \times 10^{-7}\ \mathrm{bar} \qquad (\approx 10^{-2}\ \mathrm{Pa}) \end{aligned} \tag{5.48}$$

The oxygen potential must be kept below this value at the firing temperature (about 1300 °C) if the nickel electrodes are to remain metallic. This is achieved by firing in nitrogen carrying a controlled H_2O/H_2 mixture. An oxygen meter (see Section 4.6.1) is used to monitor the oxygen potential in the firing zone.

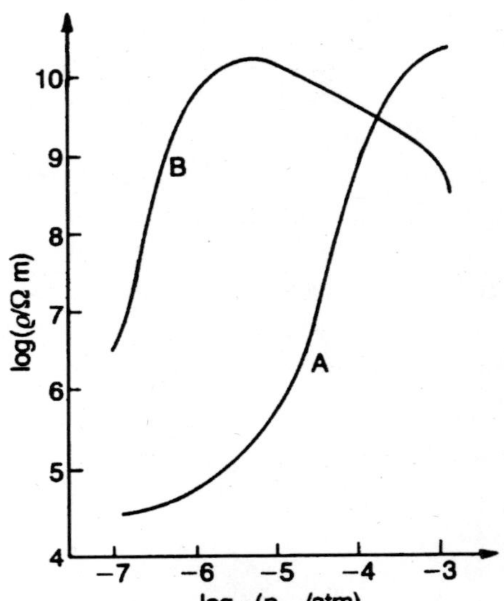

Fig. 5.50 Resistivity at room temperature as a function of oxygen pressure during sintering: curve A, undoped $BaTi_{0.85}Zr_{0.15}O_3$; curve B, $(Ba_{0.85}Ca_{0.15}O)_{1.01}$–$(Ti_{0.9}Zr_{0.1})O_2$. (After [24])

The titanate can also be stabilized against reduction by the substitution of some calcium for barium, and by establishing the ratio $AO/BO_2 > 1$. A suitable composition is $[(Ba_{0.85}Ca_{0.15})O]_{1.01}(Ti_{0.9}Zr_{0.1})O_2$, and its stability against firing under reducing conditions is compared with that for $Ba(Ti_{0.85}Zr_{0.15})O_3$ in Fig. 5.50 [24].

The success of the formulation is probably due to the presence of Ca^{2+} ions on B sites acting as acceptors (Section 5.7.1). Certainly an AO/BO_2 ratio greater than unity has been found to be essential. $CaZrO_3$ is itself very resistant to reducing conditions, and the presence of Ca and Zr ions will confer some measure of increased resistance to reduction on the composition as a whole. The loss of lattice oxygen resulting from the reduction firing, unless compensated for, has the expected adverse effect upon degradation of BME MLCCs and the need to control this has stimulated research into what is undoubtedly a very complex matter.

The BME process cannot be applied to compositions containing lead because of the ease of reduction to metallic lead.

5.7.4 Barrier-layer capacitors (Class IV)

Although barrier layer capacitors are of little or no commercial significance, the principles upon which their manufacture and operation are based, are important. Most materials containing TiO_2, whether as a single phase or in combination with other oxides, become conductive on firing in reducing atmospheres. The ease of reduction is strongly affected by the other ions present: acceptor ions tend to inhibit reduction and donor ions tend to enhance it. In most cases a high resistivity can be restored by annealing in air or oxygen. The mechanism of conduction is discussed in Section 2.6.2.

Barrier-layer capacitors are based on the limited reoxidation of a reduced composition. This results, in the simplest case, in a surface layer of high resistivity and a central portion of conductive material so that the effective dielectric thickness is twice the thickness h_o of a single reoxidized layer and there is an apparent gain in permittivity over that of a fully oxidized unit by a factor of $h/2h_o$, where h is the overall dielectric thickness (Fig. 5.51). Alternatively each

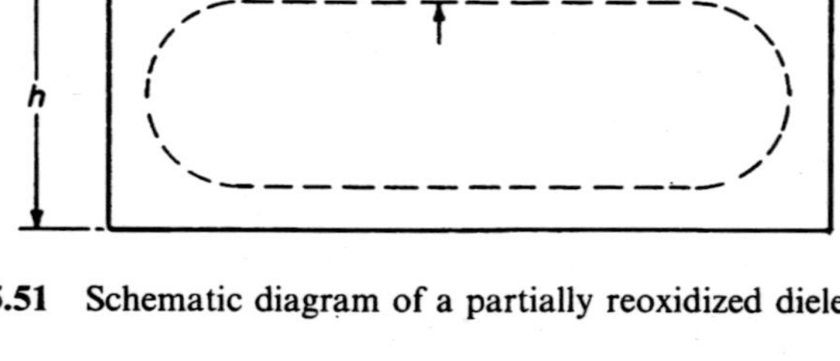

Fig. 5.51 Schematic diagram of a partially reoxidized dielectric.

conductive grain may be surrounded by an insulating barrier layer so that the dielectric property is dispersed throughout the ceramic.

Reoxidized layer units

Units depending on a reoxidized surface layer are generally made by firing $BaTiO_3$ or $SrTiO_3$ discs, approximately 0.5 mm thick, under reducing conditions. A silver electrode paint is applied to the surfaces of the disc and fired on at about 800 °C. The silver paint contains a $PbO–Bi_2O_3–B_2O_3$ glass frit to which is added a small amount (about 1 cat.%) of acceptor ions, e.g. Cu. This leads to the formation of a thin (about 10 μm) insulating layer separating the electrodes from the semiconducting titanate and to an associated very high capacitance. Because the major part of a voltage applied to the capacitor is dropped across the two thin dielectric layers, the working voltage is low, typically about 10 V. Capacitance values per unit electrode area of approximately 10 mF m^{-2} are readily achieved.

Internal barrier layers

The thinnest reoxidized layers, which result in very large effective permittivities, have properties similar in some respects to those of varistors as described in Section 4.3. They contain Schottky barriers in the semiconducting surfaces of the grains which result in properties similar to those of two back-to-back diodes (see Fig. 2.22). Their working voltages are therefore limited to the range within which the current is low. In order to withstand higher voltages it is necessary to have a ceramic structure that comprises a number of such barrier layers in series between the electrodes. There must also be an intergranular component that allows the diffusion of oxygen and dopant ions to the crystallite surfaces during oxidation.

The grain size in these units averages about 25 μm. Crystallites smaller than about 10 μm have a large fraction of their volume taken up by Schottky barriers and associated space charges that increase their resistivity, and if the crystals are larger than 50 μm there is the possibility that only a few crystals will separate the electrodes in some places, resulting in a greater likelihood of breakdown. Additions of small amounts (about 1%) of silica and alumina provide an intergranular layer that allows ionic movement and access to oxygen at high temperatures. Dysprosium or other donor ions are added to assist in the reduction process.

Discs or other shapes are first fired in air to remove organic matter and then sintered in air to obtain the required level of crystal growth. A reducing atmosphere of carbon monoxide or hydrogen is then introduced. This is found to inhibit crystal growth and so cannot usually be combined with the sintering stage. After cooling, a boric oxide frit containing acceptor ions such as Cu, Mn, Bi or Tl is painted on the surfaces of the pieces which are then reheated in air to 1300–1400 °C. The acceptor ions diffuse along the grain boundaries and modify the surface properties of the crystallites in much the same way as the acceptors that protect dielectrics for base-metal-electroded multilayer capacitors. However, their precise behaviour has not been established since it is extremely difficult to determine the structure of the thin intergranular layers and their interfaces with the crystallites.

Model calculations

The structure of internal-barrier-layer units can be represented by the simplified model shown in Fig. 5.52. The overall capacitance can be calculated as follows. The capacitance C_i of an individual element, assuming $t_g \gg t_b$, is given by

$$C_i = \frac{\varepsilon_r \varepsilon_0 t_g^2}{t_b} \tag{5.49}$$

and that of a series connected column by

$$C_i/(\text{no. of elements in column}) = \frac{\varepsilon_r \varepsilon_0 t_g^2 / t_b}{t/t_g} \tag{5.50}$$

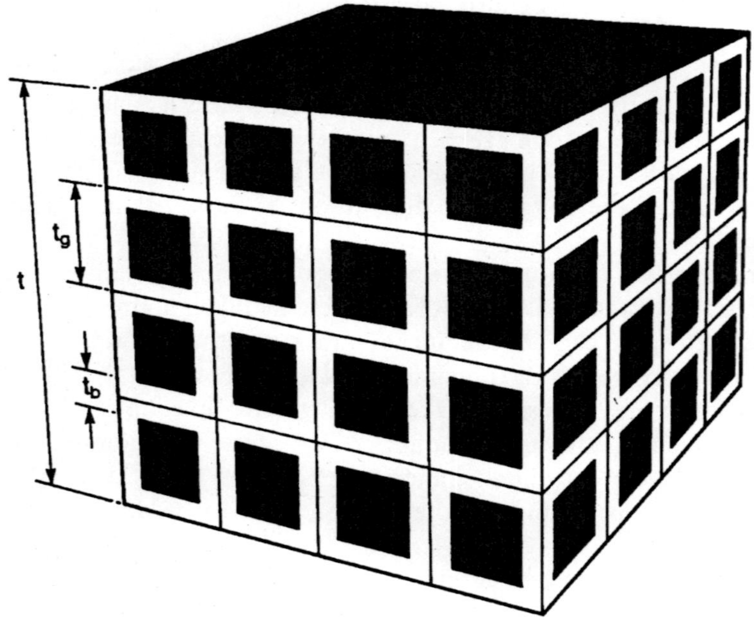

Fig. 5.52 Schematic diagram of a section through an internal-barrier-layer capacitor.

The capacitance C per unit area is

$$C = \frac{\varepsilon_r \varepsilon_0 t_g}{t t_b} \tag{5.51}$$

since there are $1/t_g^2$ columns per unit area. It follows that the effective relative permittivity ε_{re} of the composite dielectric is

$$\varepsilon_{re} = \frac{\varepsilon_r t_g}{t_b} \tag{5.52}$$

Assuming $t_g = 50\,\mu m$, $t_b = 0.2\,\mu m$ and $\varepsilon_r = 200$, we obtain $\varepsilon_{re} = 50\,000$. Values as high as this can be achieved in practice with tan δ values of typically 0.03.

Units based on $SrTiO_3$ are more stable with respect to field and temperature than those based on $BaTiO_3$. Their capacitance is only reduced by 5% under maximum d.c. field and their variation with temperature can be kept within $\pm 20\%$ over a -20 to $+85\,°C$ range; their effective permittivity is 10 000–20 000. $BaTiO_3$ units have effective permittivities of up to 50 000.

5.7.5 Ferroelectric memories

No text covering electroceramics can omit some discussion of 'ferroelectric memories' even though the engineering of the ceramics into the memory elements of a computer is so highly specialized that it is arguably more properly dealt with in the context of semiconductor technologies. Certainly the topic demonstrates very clearly the interdisciplinary nature of electroceramics. The 'materials' aspects – especially the production of very well defined precursors for fabricating thin ferroelectric films, characterizing them and improving understanding of their basic physics are pursuits well within the ceramist's field of interest. There are evident common interests between those working to maximize the volumetric efficiency of a multilayer capacitor by reducing dielectric layer thickness and those engaged on fabricating and optimizing the properties of thin ferroelectric films for memories.

The topics very briefly touched upon here are dealt with in greater depth in the monograph by J.F. Scott [25] which is recommended to those wishing or needing to pursue the subject; an apt starting point is to quote from the introduction to the monograph.

> Since ferroelectricity was discovered in 1921 it has been obvious to many scientists and engineers that the two stable polarization states +P and −P could be used to encode the 1 and 0 of the Boolean algebra that forms the basis of memory and logic circuitry in all modern computers. Yet until very recently this has been unsuccessful. In fact, although ferroelectric materials are used in a wide variety of commercial devices, it has until now always been the case that some other property of the material – especially pyroelectricity or piezoelectricity – is the characteristic actually employed. Ironically, no devices using ferroelectrics have actually required ferroelectricity to work.

For non-volatile random access memories (NV-RAMs; in which the stored information is retained even if power to the chip is interrupted), ferroelectrics serve not just as capacitors (as in the case of the DRAM, described below), but as the memory element itself. Their principal advantages in this application are low-voltage (1.0 V) operation, small size (about 20% of a conventional EEPROM* cell – and cost is proportional to size once high-yield production is achieved), radiation hardness (not just for military applications but also for satellite communications systems) and very high speed (60 ns access time in commercial devices, sub-nanosecond in laboratory tests on single cells).

Additional advantages offered by a NV-RAM over a conventional EEPROM are fast write-access (<30 ns) and high number of write cycles ($>10^{13}$).

The principle on which a computer memory is based is illustrated in Fig. 5.53. The binary-coded information is stored in an array of cells addressed by voltage pulses applied to the 'word' and 'bit' lines. The response of a particular cell, measured by the associated electronic circuitry, determines its state.

Figure 5.53(a) indicates the situation for a 'passive' FeRAM (ferroelectric random access memory). If the capacitor is in the +P state (say a '1') and the application of two V/2 voltages is sufficient to switch the polarization to the −P state, there is a resulting 'large' current pulse. However were the cell in the −P state (a '0') then the current pulse would be very much smaller. If the state of the cell is such that the polarization is switched then a follow-on pulse must be applied to return it to its original state.

In practice there are complications because the ferroelectric elements do not have a sufficiently well defined switching voltage. This can lead to unintentional switching of elements other than the one addressed and, consequently, to an unacceptable error rate. The problem can be overcome by coupling each memory element with a transistor which provides the sharply defined switching voltage. In its simplest form each node of such an 'active' matrix would then take the form shown in Fig. 5.53(b). In fact, for a variety of reasons associated with the changes which occur in the physical properties of ferroelectrics with time, the transistor-memory cell circuitry at each node is more complex than outlined above, but the general principles are the same.

In the case of dynamic random access memories (DRAMs) the cell is a thin film capacitor. The state of the cell ('0' or '1') is read by the current pulse following an addressing voltage pulse. Because the charge stored in the capacitor leaks away in time ($\tau = RC$) then, for the information to be retained it must be periodically refreshed, the rate of leakage determining the necessary refresh interval.

In the conventional DRAM the capacitor dielectric is SiO_2 ($\varepsilon_r \sim 5$) or a combination of SiO_2 and Si_3N_4 ('ONO' film) and is formed by controlled oxidation or nitridation of the silicon crystal. It is necessary for the memory capacitor to have a capacitance value of approximately 30 fF (i.e. 30×10^{-15} F)

* EEPROM, electrically erasable programmable read-only memory – a 'FLASH' memory.

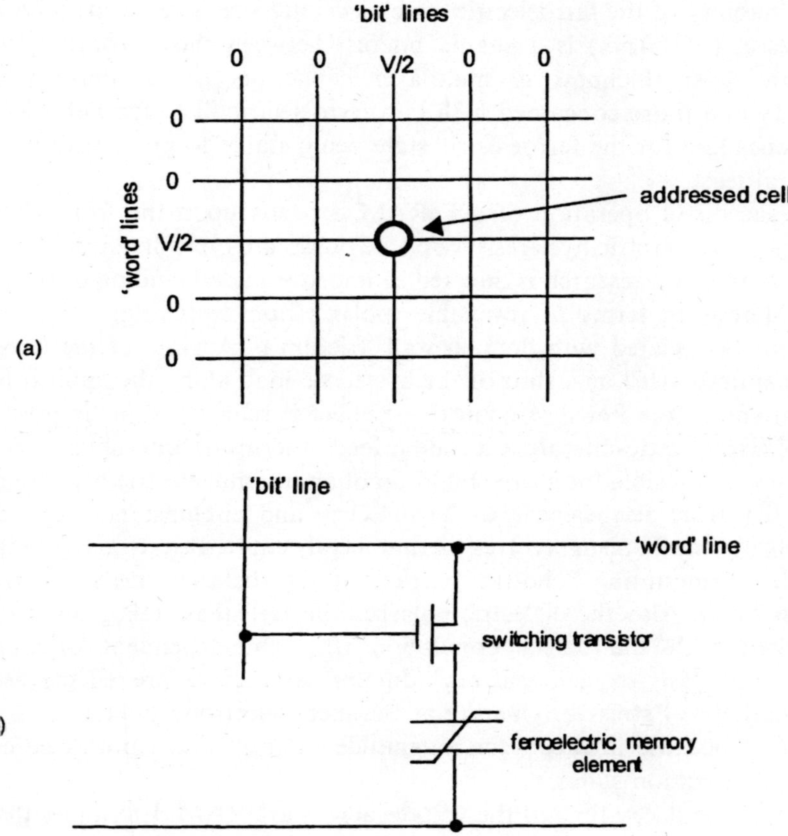

Fig. 5.53 Schematic of FeRAM matrix: (a) a matrix of memory cells (b) each memory cell coupled with a switching transistor to more closely define the threshold switching voltage.

and this places a lower limit on its size. A lower limit on the thickness of the dielectric is necessary to retain adequate resistance to electrical breakdown arising from 'tunnelling' currents. Capacitor size, and therefore component density, can be reduced by increasing the permittivity of the dielectric. A significant reduction in capacitor size could be achieved by replacing the silica with Ta_2O_5 ($\varepsilon_r \sim 25$) and this step seems likely in the foreseeable future. If a 'ferroelectric' capacitor (ε_r of order 10^2 to 10^3) could be used then a bit density increase of about ten-fold might be achieved. This is the stimulus for considerable research and development effort, especially on $SrTiO_3$ ($\varepsilon_r > 200$) and $Ba_xSr_{(1-x)}$ TiO_3 e.g. $Ba_{0.7}Sr_{0.3}TiO_3$ ($\varepsilon_r > 300$), the composition being chosen so that the material is in the paraelectric phase at room temperature.

For FeRAMs materials attracting interest are $PbZr_{(1-x)}Ti_xO_3$ (with $x \sim 0.6$) and the Aurivillius layer-structured perovskite, $SrBi_2Ta_2O_9$ (SBT).

The stability of the ferroelectric state as crystal size is reduced to typical film thicknesses (<100 nm) is a shared interest between those working to reduce dielectric layer thickness in multilayer capacitors to maximize volumetric efficiency and those concerned with thin ferroelectric films for FeRAMs. There is evidence [26] for the ferroelectric state being stable to grain sizes as small as 40 nm, at least.

The successful operation of a FeRAM depends upon the form of the high frequency ferroelectric hysteresis loop (the pulses are typically in the nanosecond period range) and research is directed to improve understanding of the origin of the hysteresis in terms of reversible (polarization switching) and irreversible processes (associated with domain wall 'friction'). 'Ageing' of the ferroelectric state manifests itself as a shift of the hysteresis loop along the applied field axis (x-axis) and in the FeRAM context the effect is referred to as 'imprint'; in the general ferroelectrics literature it is described in terms of 'internal bias field'. The processes responsible for a slow build-up of an internal electric field are complex and differ from ferroelectric to ferroelectric and circumstance. The evidence suggests one mechanism involves thermionically emitted electrons from the metal cathode, surmounting Schottky barriers at the dielectric–metal interface (see Section 2.6.5). Also the dielectric polarization itself may 'relax' due to changes associated with the defect population (the time-dependent orientation of 'acceptor dopant–oxygen vacancy' dipoles) and there are effects associated with mechanical stress originating at the metal-electrode interfaces. Imprint is important because it affects the magnitude of the voltage pulses necessary to switch polarization states.

Because the decay time of the voltage across a DRAM determines the refresh time it is important to understand the mechanisms responsible for current leakage. These may be related to the voltage distribution across the dielectric and to charge-injection from the electrodes.

In all cases there is the risk of dielectric deterioration and although the operating voltages are low, the electric fields in the dielectric are high. The possible mechanisms leading to failure are outlined in Section 5.6.2 and discussed in some detail in [25].

There is a growing industrial demand for powders specially tailored for DRAM and FeRAM technologies. These are produced by 'chemical routes' (see Section 3.4) with tight control exercised over composition. The general experimental approaches to forming thin films are summarized in Section 3.6.9 and in [25].

Ferroelectric memories appear set to penetrate many market sectors especially 'smart' cards for direct, 'contactless' purchasing, for example at petrol stations, for recording medical histories and for automatic motorway tolling, etc. They will also be widely exploited in many 'consumer' products including digital cameras, 'Walkman', washers/dryers, etc. However, they have to be seen in the context of other competing technologies, for example MRAMS (magnetic random access memories) which are based on thin film magnetic alloy and the dependence of tunnelling current on the direction of magnetization.

The up-to-date (2000) situation with regard to the technology and commercial exploitation of ferroelectric memories is summarized by J.F. Scott [27] and H. Takasu [28].

Problems

1. Describe the general principles that have guided the development of Class I ceramic dielectrics offering a range of TC_ε values.

   A parallel-plate capacitor at 25 °C comprises a slab of dielectric of area 10^{-4} m^2 and thickness 1 mm carrying metal electrodes over the two major surfaces. If the relative permittivity, temperature coefficient of permittivity and linear expansion coefficient of the dielectric are respectively 2000, $-12\,\mathrm{MK}^{-1}$ and $8\,\mathrm{MK}^{-1}$, estimate the change in capacitance which accompanies a temperature change of +5 °C around 25 °C. [Answer: -0.035 pF]

2. Given that a 10 μF multilayer capacitor is made up of 100 active 15 μm thick layers of dielectric of relative permittivity 10^4, make a realistic estimate of the overall size of the capacitor. [Answer: very approximately 5 mm × 5 mm × 2 mm]

3. A 100 F double layer carbon capacitor has a working voltage of 3 V and overall dimensions 30 × 30 × 30 mm. What is the volumetric efficiency of the capacitor?

   It is required to use a pack of such capacitors to operate at 300 V. Estimate the volumetric efficiency of the 300 V pack.

   Compare the energy densities (J/kg) of the single cell and cell pack.
   [Answer: ~3.7 F cm^{-3}; ~370 μF cm^{-3}]

4. A double layer carbon capacitor consists of an anode and cathode of 1 g each of carbon cloth (S.S.A. = 2500 m^2 g^{-1}), the electrolyte being sulphuric acid. With reference to [3] make and justify an estimate of the capacitance of the cell.
   [Answer: ~100 F]

5. An X7R type 0.022 μF capacitor has a dissipation factor (1 kHz, 1 V (r.m.s.)) of 1.5% and a self-resonance at 30 MHz. Given that the e.s.r. at resonance is 0.056 Ω estimate (a) the contribution to the impedance from the resistance of the leads and (b) the inductance of the capacitor. [Answers: (a) 0.052 Ω; (b) 1.28 nH]

   Sketch curves showing the expected form of the following relationships:

   (i) $\Delta C/C$ versus d.c. bias (0–50 V);

   (ii) tan δ versus d.c. bias (0–50 V);

   (iii) $\Delta C/C$ versus r.m.s. voltage (0–10 V);

(iv) $\tan\delta$ versus r.m.s. voltage (0–10 V);

(v) $|Z|$ versus frequency (1–100 MHz).

In each case explain, using no more than 50 words, the reasons for the forms of the curves.

6. Describe the various approaches made to reduce the manufacturing costs of ceramic multilayer capacitors.

7. A 3000 pF power capacitor has maximum power and voltage ratings of 200 kV A and 15 kV respectively. Its dissipation factor can be assumed to be constant at 0.001. Under particular ambient conditions, and operating at 5 kV (r.m.s.) and 100 kHz, its temperature is 10 °C above the 30 °C of the surroundings. Assuming the same ambient conditions, estimate the temperature of the unit when it is operating at (i) 5 kV (r.m.s.) and 300 kHz and (ii) 10 kV (r.m.s.) and 150 kHz. Calculate the reactive current carried by the capacitor, the e.s.r. and the dissipated power under each of the two operating conditions. [Answers: (i) 60 °C; current, 28.3 A; e.s.r., 0.17 Ω; dissipated power, 141 W; (ii) 90 °C; current, 28.3 A; e.s.r. 0.35 Ω; dissipated power, 283 W]

8. Describe the essentials of the design principles of a dielectric resonator (DR) and the advantages offered over cavity resonators.

   A ceramic of relative permittivity 37 is in the form of a cylindrical DR for use at 1 GHz. Estimate the overall dimensions of the DR. The ceramic has a temperature coefficient of linear expansivity of $5\,\text{MK}^{-1}$ and a temperature coefficient of permittivity of $-16\,\text{MK}^{-1}$. Estimate by how much the resonance frequency will change for a 5 °C change in temperature. [Answer: diameter 2.5 cm; 15 kHz]

9. A thin porcelain disc of diameter 500 mm and having a smoothly rounded edge carries a semiconducting glaze. At the centres of top and bottom surfaces there are 20 mm diameter metal terminals making good electrical contact with the glaze surface. With 10 kV applied between the terminals a current of 1 mA is required to flow. If the resistivity of the glaze is 600 Ω m calculate the necessary glaze thickness at the electrode edges and how the thickness must be graded to ensure a constant voltage gradient along the radii.

   Estimate the steady state surface temperature above that of the air ambient assuming a rate of loss of heat of $10\,\text{W}\,\text{m}^{-2}\,\text{K}^{-1}$, and comment on how this might affect performance in a polluted atmosphere. [Answers: ~1.5 mm; ~60 μm; ~2.5 K; reduces risk of condensation on the glaze surface and so risk of 'flash-over']

10. Offer outline explanations of why the permittivity of polycrystalline $BaTiO_3$ increases with decreasing grain size. Show that the 90° domain wall area is proportional to (grain size)$^{5/2}$ and that the domain wall area per unit volume is approximately $5 \times 10^3\, g^{-1/2}$. Numerical data can be found in Section 5.7.1.

11. A disc of reduced semiconducting rutile crystal 2 cm in diameter and 2 mm thick is heated in air for 10 s at 300 °C. After cooling, circular electrodes, 1 cm in diameter, are applied symmetrically to the two major surfaces. The chemical diffusion coefficient $\tilde{D}$ for the oxidation reaction in reduced single-crystal TiO_2 is given by

$$\tilde{D} = 2.61 \times 10^{-7} \exp\left(-\frac{Q}{RT}\right) \mathrm{m^2\,s^{-1}}, \text{ with } Q = 38.5\,\mathrm{kJ\,mol^{-1}}$$

Estimate the capacitance value measured between the electrodes. [Answer: ~1 nF]

12. Estimate the resonance frequency of the strip-line resonator shown in Fig. 5.39 given that the length of the line is 5 mm and the LTCC properties are as given in Table 5.9 [17].

By reference to the equivalent circuit explain why the device is a pass-band filter and suggest a design modification which would more closely define the band, and why. [Answer: ~5 GHz]

13. Estimate the missing 2 GHz Q-values in Table 5.8. Explain the basis for your estimate making clear the circumstances which might invalidate it.

14. Given brief reasoned justifications for your choice of material for the manufacture of (i) a 10 kV insulator for outdoor use, (ii) the insulating parts of a precision adjustable high-Q air capacitor, (iii) a substrate for carrying a microwave circuit and (iv) a high-power fuse holder.

15. Briefly trace how the demands of electronic circuit technology have impacted on the development of ceramic 'packaging' technology.

An alumina substrate of dimensions 1 cm × 1 cm × 0.5 mm carries a device dissipating 20 W. If the substrate is bonded to a metallic heat sink, estimate the steady state difference in temperatures between the surface carrying the device and the heat sink. The thermal conductivity of alumina may be taken to be $35\,\mathrm{W\,m^{-1}\,K^{-1}}$. [Answer: ≈3 °C]

16. Discuss the effect of grain size over the range 50 nm to 50 μm on the real part of the permittivity of bulk polycrystalline $BaTiO_3$. What is the evidence for the ferroelectric state being stable at the smallest grain size and explain why this is relevant to FeRAM technology. Discuss possible reasons why lower permittivity values are found for $BaTiO_3$ in thin film form even when the grain size is the same as that in bulk ceramic form. (See [26]).

Bibliography

1. Buchanan, R.C. (ed.) (1991) *Ceramic Materials for Electronics*, 2nd edn, Marcel Dekker, New York.

2. Herbert, J.M. (1985) *Ceramic Dielectrics and Capacitors*, Gordon and Breach, London.
3. Kötz, T. and Carlen, M. (2000) Principles and applications of electrochemical capacitors. *Electrochim. Acta* **45**, 2483–98.
4. Will, F.G. *et al.* (1992) Conduction mechanism in single crystal alumina. *J. Am. Ceram. Soc.* **75**, 295–304.
5. Tummala, R.R. (1991) Ceramic and glass-ceramic packaging in the 1990s. *J. Am. Ceram. Soc.* **74**, 895–908.
6. Knickerbocker, S.H., Kumar, A.H. and Herron, L.W. (1993) Cordierite glass-ceramics for multilayer ceramic packaging. *Am. Ceram. Soc. Bull.* **72**, 90–5.
7. Kofstad, P. (1972) *Nonstoichiometry, Diffusion and Electrical Conductivity in Binary Metal Oxides*, J. Wiley and Sons, Inc., Chichester.
8. Waser, R., Baiatu, T. and Härdtl, K.-H. (1990) d.c. Electrical degradation of perovskite-type titanates: I, ceramics, II, single crystals and III, A model of the mechanisms, *J. Am. Ceram. Soc.* **73**, 1645–73.
9. Kanai, H. *et al.* (1998) Effects of microstructure on insulation resistance degradation of relaxors, in *Advances in Dielectric Ceramic Materials*, Vol. **88**, The Am. Ceram. Soc., 295–9.
10. Freer, R. (1993) Microwave dielectric ceramics – an overview. *Silicates Industriels* **9–10**, 191–7.
11. Wolfram, G. and Göbel, H.E. (1981) Existence range, structural and dielectric properties of $Zr_x\ Ti_y\ Sn_z\ O_4$ ceramics $(x+y+z)=2$. *Mater. Res. Bull.*, **16**, 1455–63.
12. Davies, P.K., Tong, J. and Negas, T. (1997) Effect of ordering-induced domain boudaries on low-loss Ba $(Zn_{1/3}Ta_{2/3})O_3$-$BaZrO_3$ perovskite microwave dielectrics. *J. Am. Ceram. Soc.*, **80**, 1727–40.
13. Reaney, I.M., Colla, E.L. and Setter, N. (1994) Dielectric and structural characteristics of Ba- and Sr-based complex perovskites as a function of tolerance factor. *Jpn. J. Appl. Phys.* **33**, 3984–90.
14. Hakki, B.W. and Coleman, P.D. (1960) A dielectric resonator method of measuring inductive capacities in the millimeter range, *I.R.E. Trans. Microwave Theory Tech.* **8**, 402–10.
15. Courtney, W.E. (1970) Analysis and evaluation of a method of measuring the complex permittivity and permeability of microwave insulators, *IEEE Trans. Microwave Theory Tech.* **MTT-18**, 476–85.
16. Dernovsek, O. *et al.* (2001) LTCC glass-ceramic composites for microwave application, *J. Eur. Ceram. Soc.* **21**, 1693–7.
17. Jantunen, H. *et al.* (2000) Preparing low-loss low-temperature cofired ceramic material without glass addition, *J. Am. Ceram. Soc.* **83**, 2855–7.
18. Wersing, W. *et al.* (1999) Integrated passive components using low temperature cofired ceramics, *SPIE* **3582**, 193–9.
19. Arlt, G., Hennings, D. and de With, G. (1985) Dielectric properties of fine-grained barium titanate ceramics, *J. Appl. Phys.* **58**, 1619–25.
20. Smolenskii, G.A. (1970) Physical phenomena in ferroelectrics with diffused phase transition, *J. Phys. Soc. Japan* **28**, 26–37.
21. Chen, J., Chan, H.M. and Harmer, M.P. (1989) Ordering structure and dielectric properties of undoped La/Na-doped $Pb(Mg_{1/3}Nb_{2/3})O_3$, *J. Am. Ceram. Soc.* **72**, 593–8.

22. Gosula, V. *et al.* (2000) X-ray scattering study of the transition dynamics in relaxor ferroelectric $Pb(Mg_{1/3}Nb_{2/3})O_3$, *J. Phys. Chem. Solids* **61**, 221–7.
23. Swartz, S.L. and Shrout, T.R. (1982) Fabrication of perovskite lead magnesium niobate, *Mater. Res. Bull.* **17**, 1245–50.
24. Sakabe, Y., Minai, K. and Wakino, K. (1981) High-dielectric constant ceramics for base metal monolithic capacitors, *Jpn. J. Appl. Phys.* **20** *Suppl.*, 147–50.
25. Scott, J.F. (2000) *Ferroelectric Memories*, Springer, Berlin.
26. Frey, M.H. *et al.* (1998) The role of interfaces on an apparent grain size effect on the dielectric properties for ferroelectric barium titanate ceramics, *Ferroelectrics*, **206–207**, 337–53.
27. Scott, J.F. (2000) Ferroelectric memories today, *Ferroelectrics*, **236**, 247–58.
28. Takasu, H. (2000) The ferroelectric memory and its applications, *J. Electroceram.*, **4**, 327–38.

6

PIEZOELECTRIC CERAMICS

The texts by B. Jaffe, W.R. Cook and H. Jaffe [1], J.M. Herbert [2], J.C. Burfoot and G.W. Taylor [3] and L.M. Levinson (ed.) [4] are recommended to supplement the discussions.

6.1 Background Theory

All materials undergo a small change in dimensions when subjected to an electric field. If the resultant strain is proportional to the square of the field it is known as the electrostrictive effect. Some materials show the reverse effect – the development of electric polarization when they are strained through an applied stress. These are said to be piezoelectric (pronounced 'pie-ease-oh'). To a first approximation the polarization is proportional to the stress and the effect is said to be 'direct'. Piezoelectric materials also show a 'converse' effect, i.e. the development of a strain x directly proportional to an applied field.

The phenomenon of electrostriction is expressed by the relationship

$$x = QP^2 \tag{6.1}$$

where Q is the electrostriction coefficient.

For materials with $\varepsilon_r \gg 1$ $P \sim D$ and so Eq. (6.1) may be written

$$x = QD^2 = Q\varepsilon^2 E^2 \tag{6.2}$$

The electrostriction coefficient is a fourth-rank tensor because it relates a strain tensor (second rank) to the various cross-products of the components of $\boldsymbol{E}$ or $\boldsymbol{D}$ in the x, y and z directions.

Among the 32 classes of single-crystal materials, 11 possess a centre of symmetry and are non-polar. For these an applied stress results in symmetrical ionic displacements so that there is no net change in dipole moment. The other 21

Electroceramics: Materials, Properties, Applications. 2nd Edition. Edited by A. J. Moulson and J. M. Herbert.
 ISBN 0 471 49747 9 (hardback) 0 471 49748 7 (paperback)

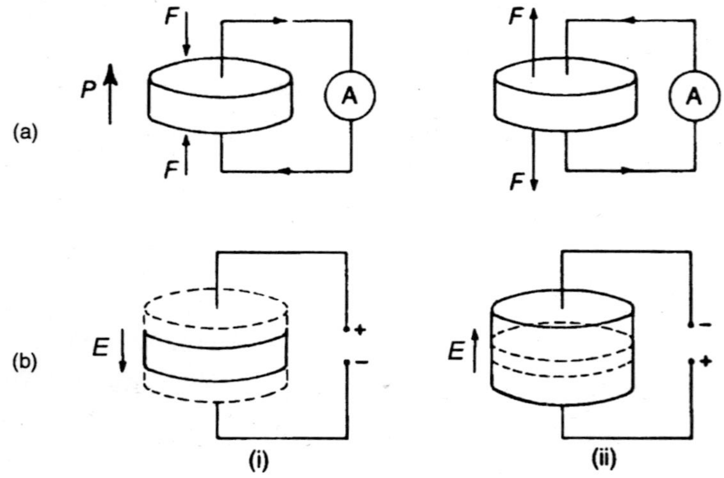

Fig. 6.1 (a) The direct and (b) the indirect piezoelectric effects: (i) contraction; (ii) expansion. The broken lines indicate the original dimensions.

crystal classes are non-centrosymmetric, and 20 of these exhibit the piezoelectric effect. The single exception, in the cubic system, possesses symmetry characteristics which combine to give no piezoelectric effect.

If a piezoelectric plate (Fig. 6.1), polarized in the direction indicated by P, carries electrodes over its two flat faces, then a compressive stress causes a transient current to flow in the external circuit; a tensile stress produces current in the opposite sense (Fig. 6.1(a)). Conversely, the application of an electric field produces strain in the crystal, say a negative strain; reversal of the field causes a positive strain (Fig. 6.1(b)). The changes in polarization which accompany the direct piezoelectric effect manifest themselves in the appearance of charges on the crystal surface (see Eq. (2.71)) and, in the case of a closed circuit, in a current.

Polycrystalline materials in which the crystal axes of the grains are randomly oriented all behave electrostrictively whatever the structural class of the crystallites comprising them. If the crystals belong to a piezoelectric class and their crystal axes can be suitably aligned, then a piezoelectric polycrystalline ceramic becomes possible.

A polar direction can be developed in a ferroelectric ceramic by applying a static field; this process is known as 'poling'. There is, of course, no question of rotating the grains themselves, but the crystal axes can be oriented by reversal, or by changes through other angles that depend on the crystal structure involved, so that the spontaneous polarization has a component in the direction of the poling field. Electrodes have to be applied to the ceramic for the poling process and these also serve for most subsequent piezoelectric applications. The exception is when the deformation is to be a shear. In this case the poling electrodes have to be removed and electrodes applied in planes perpendicular to the poling electrodes.

It should be noted that a poling process is often necessary with single-crystal ferroelectric bodies because they contain a multiplicity of randomly oriented domains. There is therefore a sequence of states of increasing orderliness: polycrystalline ferroelectric ceramics, poled ferroelectric ceramics, single-crystal ferroelectrics and single-domain single crystals.

Piezoelectric properties are described in terms of the parameters D, E, X and x, where X is stress. The electrical response due to the direct effect can be expressed in terms of strain by

$$D = ex \qquad \text{or} \qquad E = hx \tag{6.3}$$

and the converse effect can be expressed by

$$x = g^*D \qquad \text{or} \qquad x = d^*E \tag{6.4}$$

The piezoelectric coefficients e, h, g^* and d^* are tensors and are defined more precisely later. As in the case of the electrostriction coefficients, their values depend on the directions of E, D, x and X so that the relationships between them are sometimes complex. If x and D are assumed to be collinear, differentiation of Eq. (6.2) gives

$$\delta x = 2Q\varepsilon^2 E\,\delta E \tag{6.5}$$

This indicates that, if a biasing field E_b is applied to an electrostrictive material, there will be a direct proportionality between changes in strain and small changes in field ($E_b \gg \delta E$).

From Eq. (6.5)

$$\delta x/\delta E = 2Q\varepsilon^2 E = 2Q\varepsilon D \sim 2Q\varepsilon P \tag{6.6}$$

and from the differential form of Eq. (6.4) it follows that

$$d^* = 2Q\varepsilon P \tag{6.7}$$

Ferroelectric materials above their Curie point behave electrostrictively and comparison of the electrostriction coefficient with $d^*/2\varepsilon P$ shows them to be of similar magnitude. This suggests that the large d-coefficients shown by some ferroelectric materials are due to a combination of large electrostriction coefficients and large spontaneous polarization and permittivity values.

The various direct-effect coefficients are properly defined by the following partial derivatives, where the subscripts indicate the variables held constant (T is the temperature):

$$\begin{aligned} \left(\frac{\partial D}{\partial X}\right)_{E,T} &= d \\ -\left(\frac{\partial E}{\partial X}\right)_{D,T} &= g \\ \left(\frac{\partial D}{\partial x}\right)_{E,T} &= e \\ -\left(\frac{\partial E}{\partial x}\right)_{D,T} &= h \end{aligned} \tag{6.8}$$

The converse-effect coefficients are properly defined similarly:

$$\begin{aligned} \left(\frac{\partial x}{\partial E}\right)_{X,T} &= d^* \\ \left(\frac{\partial x}{\partial D}\right)_{X,T} &= g^* \\ -\left(\frac{\partial X}{\partial E}\right)_{x,T} &= e^* \\ -\left(\frac{\partial X}{\partial D}\right)_{x,T} &= h^* \end{aligned} \tag{6.9}$$

It is instructive to outline the thermodynamic argument proving that $d = d^*, g = g^*, e = e^*$ and $h = h^*$. The First Law states that the addition of an increment of heat $\mathrm{d}q$ to a system produces a change in the internal energy $\mathrm{d}U$ and causes the system to do work $\mathrm{d}W$ on its surroundings. Thus

$$\mathrm{d}q = \mathrm{d}U + \mathrm{d}W \tag{6.10}$$

If the changes are performed reversibly, $\mathrm{d}q = T\mathrm{d}S$, and

$$T\mathrm{d}S = \mathrm{d}U + \mathrm{d}W \tag{6.11}$$

where S is the entropy. In the case of an isolated piezoelectric material the mechanical and electrical work is done *on* the system, and hence

$$T\mathrm{d}S = \mathrm{d}U - \mathrm{d}W = \mathrm{d}U - X\mathrm{d}x - E\mathrm{d}D \tag{6.12}$$

It is possible to define the thermodynamic state of a system in terms of groups of certain independent variables, and T, X, x, D and E are available in the present case. For example, the equilibrium could be expressed in terms of the independent variables (T, x, D) when the appropriate starting point would be the energy function

$$A = U - TS \tag{6.13}$$

Now

$$\mathrm{d}A = \mathrm{d}U - T\mathrm{d}S - S\mathrm{d}T$$

which, on substituting for $\mathrm{d}U - T\mathrm{d}S$ from Eq. (6.12), becomes

$$\mathrm{d}A = X\mathrm{d}x + E\mathrm{d}D - S\mathrm{d}T \tag{6.14}$$

Because $\mathrm{d}A$ is a perfect differential it follows that

$$\left(\frac{\partial X}{\partial D}\right)_{x,T} = \left(\frac{\partial E}{\partial x}\right)_{D,T} \tag{6.15}$$

Similarly, if attention is focused on the independent variables (T, X, E), the appropriate function is

$$G = U - TS - Xx - ED \tag{6.16}$$

which leads to

$$\mathrm{d}G = -S\mathrm{d}T - x\mathrm{d}X - D\mathrm{d}E \tag{6.17}$$

and to the relationship

$$\left(\frac{\partial x}{\partial E}\right)_{X,T} = \left(\frac{\partial D}{\partial X}\right)_{E,T} \tag{6.18}$$

Similarly, the function

$$G_1 = U - TS - ED \tag{6.19}$$

leads to

$$\left(\frac{\partial X}{\partial E}\right)_{x,T} = -\left(\frac{\partial D}{\partial x}\right)_{E,T} \tag{6.20}$$

and

$$G_2 = U - TS - Xx \tag{6.21}$$

leads to

$$\left(\frac{\partial x}{\partial D}\right)_{x,T} = -\left(\frac{\partial E}{\partial X}\right)_{D,T} \tag{6.22}$$

Thus the equalities $d = d^*$ etc. are proved, and the relationships can be summarized as follows:

$$\left(\frac{\partial D}{\partial X}\right)_{E,T} = \left(\frac{\partial x}{\partial E}\right)_{X,T} = d \tag{6.23}$$

$$-\left(\frac{\partial E}{\partial X}\right)_{D,T} = \left(\frac{\partial x}{\partial D}\right)_{X,T} = g \tag{6.24}$$

$$\left(\frac{\partial D}{\partial x}\right)_{E,T} = -\left(\frac{\partial X}{\partial E}\right)_{x,T} = e \tag{6.25}$$

$$\left(\frac{\partial E}{\partial x}\right)_{D,T} = \left(\frac{\partial X}{\partial D}\right)_{x,T} = -h \tag{6.26}$$

6.2 Parameters for Piezoelectric Ceramics and their Measurement

The discussion is now restricted to matters relevant to the technology of piezoceramics. If it is assumed that the quantities d, g, ε, and s are constants and, furthermore, the tensor nature of the various parameters (see below) is not taken into account, then equations of state can be written as follows.

$$D = \mathrm{d}X + \varepsilon^X E \tag{6.27a}$$

$$x = s^E X + \mathrm{d}E \tag{6.27b}$$

$$E = -gX + D/\varepsilon^X \tag{6.28a}$$

$$x = s^D X + gD \tag{6.28b}$$

Where s^E, s^D are the elastic compliances and the superscripts denote the parameters held constant; similarly ε^X denotes the permittivity measured at constant stress (the usual condition); to measure the permittivity at constant strain ε^x is more difficult to accomplish.

Because ferroelectric ceramics have non-linear characteristics the effects are more correctly described by the equations,

$$\delta D = d\delta X + \varepsilon^X \delta E \tag{6.29a}$$

$$\delta x = s^E \delta X + d\delta E \tag{6.29b}$$

$$\delta E = -g\delta X + \delta D/\varepsilon^X \tag{6.30a}$$

$$\delta x = s^D \delta X + g\delta D \tag{6.30b}$$

$$\delta E = -h\delta x + \delta D/\varepsilon^x \tag{6.31a}$$

$$\delta X = c^D \delta x - h\delta D \tag{6.31b}$$

$$\delta D = e\delta x + \varepsilon^x \delta E \tag{6.32a}$$

$$\delta X = c^E \delta x - e\delta E \tag{6.32b}$$

in which the coefficients are recognized as being functions of the independent variables, that is $d(X,E)$, $g(X,D)$, etc.

The various coefficients which appear in electromechanical equations of state such as (6.29a) to (6.32b) are not all independent. For example, from Eqs (6.23) and (6.24)

$$d/g = (\partial x/\partial E)_X/(\partial x/\partial D)_X = (\partial D/\partial E)_X = \varepsilon^X \tag{6.33}$$

It is understood that in all cases the temperature is maintained constant.

In a similar way, from Eqs (6.24), (6.25) and (6.26), it follows that

$$e/h = \varepsilon^x \quad \text{and} \quad g/h = s^D \tag{6.34}$$

The electromechanical coupling coefficient (k) is a measure of the ability of a piezoelectric material to transform mechanical energy into electrical energy, and vice versa. It is defined [5] by

$$k^2 = W_{12}^2/W_1 W_2 \tag{6.35}$$

in which W_{12}, W_1 and W_2 are respectively the piezoelectric, mechanical and electrical energy densities. In the piezoceramics' context an effective coupling coefficient k_{eff}, defined below, is in common usage.

There are important relationships which follow from the above equations of state. When a piezoelectric material is stressed, electrically or mechanically, the developed energy densities are, from Eqs (6.29a) and (6.29b).

$$\tfrac{1}{2}\delta D\,\delta E = \tfrac{1}{2}d\,\delta X\delta E + \tfrac{1}{2}\varepsilon^X(\delta E)^2 \tag{6.36}$$

$$\tfrac{1}{2}\delta x\,\delta X = \tfrac{1}{2}s^E(\delta X)^2 + \tfrac{1}{2}d\,\delta E\delta X \tag{6.37}$$

The terms involving $(\delta E)^2$ and $(\delta X)^2$ are the electrical and mechanical energy densities, and those involving $\delta E \delta X$ the piezoelectric energy densities. It follows from Eq. (6.35) that

$$k^2 = \{d\,\delta X\,\delta E\,d\,\delta E\,\delta X/s^E(\delta X)^2\ \varepsilon^X(\delta E)^2\}^{1/2} \tag{6.38}$$

or

$$k^2 = d/(s^E\varepsilon^X)^{1/2} \tag{6.39}$$

The elastic compliance s, (and stiffness c) has very different values depending upon whether the electric field E within the material is maintained at zero (the component is 'short-circuited'), or whether the electric displacement D remains constant (the component is 'open-circuited'). The 'short circuit' value is specified by the superscript E and the 'open circuit' condition by the superscript D. Similarly the permittivity can be measured with the specimen free to deform, that is at constant stress ε^X, or in the clamped state, ε^x.

Important relationships between s^E and s^D (and between c^E and c^D) can be derived as follows.

Equating Eqs (6.29b) and (6.30b) leads to

$$s^E\delta X + d\,\delta E = s^D\delta X + g\,\delta D$$

After substituting for δD and δE from Eqs (6.29a) and (6.30a) and noting from Eq. (6.33) that $g = d/\varepsilon^X$ it follows, after a little manipulation that

$$s^D = s^E(1 - d^2/s^E\varepsilon^X)$$

or, from Eq. (6.39)

$$s^D = s^E(1 - k^2) \tag{6.40}$$

Also, because $c = 1/s$

$$c^E = c^D(1 - k^2) \tag{6.41}$$

In a similar way, but now equating Eqs (6.30a) and (6.31a) and substituting for δx from Eq. (6.30b), it follows that

$$\varepsilon^x \sim \varepsilon^X(1 - k^2) \tag{6.42}$$

[the approximation rests on $k^4 \ll 1$]

Rearranging Eq. (6.42) leads to $k^2 = (\varepsilon^X - \varepsilon^x)/\varepsilon^X$ or to

$$k^2 = (\tfrac{1}{2}E^2\varepsilon^X - \tfrac{1}{2}E^2\varepsilon^x)/\tfrac{1}{2}E^2\varepsilon^X \tag{6.43}$$

When an electric field is applied to an unconstrained piezoelectric body then both electrical and mechanical energies are stored. If the body is mechanically clamped then only electrical energy is stored. Eq. (6.43) is the basis of the definition of the 'effective coupling coefficient' k_{eff}, namely

$$k_{\text{eff}}^2 = \frac{\text{input electrical energy converted into mechanical energy}}{\text{input electrical energy}} \tag{6.44}$$

Similarly from Eq. (6.40)

$$k_{\text{eff}}^2 = \frac{\text{input mechanical energy converted into electrical energy}}{\text{input mechanical energy}} \tag{6.45}$$

In practice the energy transfer electrical-to-mechanical (or vice versa) occurs in a complex 3-dimensional way. The strains caused by applied electrical or mechanical stresses have components in three orthogonal directions necessitating the description of the piezoelectric effect in terms of tensors, as outlined below.

The state of strain in a body is fully described by a second-rank tensor, a 'strain tensor', and the state of stress by a stress tensor, again of second rank. Therefore the relationships between the stress and strain tensors, i.e. the Young modulus or the compliance, are fourth-rank tensors. The relationship between the electric field and electric displacement, i.e. the permittivity, is a second-rank tensor. In general, a vector (formally regarded as a first-rank tensor) has three components, a second-rank tensor has nine components, a third-rank tensor has 27 components and a fourth-rank tensor has 81 components.

Not all the tensor components are independent. Between Eqs (6.29a) and (6.29b) there are 45 independent tensor components, 21 for the elastic compliance s^E, six for the permittivity ε^X and 18 for the piezoelectric coefficient d. Fortunately crystal symmetry and the choice of reference axes reduces the number even further. Here the discussion is restricted to poled polycrystalline ceramics, which have ∞-fold symmetry in a plane normal to the poling direction. The symmetry of a poled ceramic is therefore described as ∞mm, which is equivalent to 6mm in the hexagonal symmetry system.

The convention is to define the poling direction as the 3-axis, as illustrated in Fig. 6.2. The shear planes are indicated by the subscripts 4, 5 and 6 and are perpendicular to directions 1, 2 and 3 respectively. For example, d_{31} is the coefficient relating the field along the polar axis to the strain perpendicular to it, whilst d_{33} is the corresponding coefficient for both strain and field along the polar axis. Shear can only occur when a field is applied at right angles to the polar axis so that there is only one coefficient, d_{15}. There are also piezoelectric coefficients corresponding to hydrostatic stress, e.g. $d_{\text{h}} = d_{33} + 2d_{31}$ (see Section 6.4.6).

Since the elastic compliance (s) and stiffness (c) are tensor properties, in general $c_{jk} \neq 1/s_{jk}$; the exact relations are given in Eq. (6.54) and in the Appendix. Since $s_{jk} = s_{kj}$ and $c_{jk} = c_{kj}$, only six terms (s_{11}, s_{12}, s_{13}, s_{33}, s_{44}, s_{66} or c_{11}, c_{12}, c_{13}, c_{33}, c_{44}, c_{66}) are needed for poled ceramics. Of these, s_{66} and c_{66} are irrelevant because shear in the plane perpendicular to the polar axis produces no

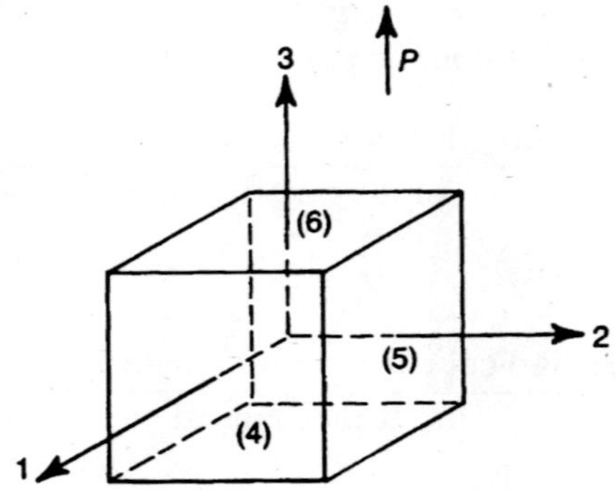

Fig. 6.2 Labelling of reference axes and planes for piezoceramics.

piezoelectric response. Two sets of elastic constants are needed, one for short-circuit and the other for open-circuit conditions, as indicated by superscripts E or D respectively.

The Poisson ratios ν are given

$$\nu_{12} = -s_{12}^E/s_{11}^E \qquad \nu_{13} = -s_{13}^E/s_{33}^E \tag{6.46}$$

Following the above conventions, Eqs (6.33) and (6.34) become

$$\frac{d_{ij}}{g_{ij}} = \left(\frac{\partial x_j}{\partial E_i}\right)_{X,T} \bigg/ \left(\frac{\partial x_j}{\partial D_i}\right)_{X,T} = \left(\frac{\partial D_i}{\partial E_i}\right)_{X,T} = \varepsilon_{\mathrm{ii}}^X \tag{6.47}$$

For example,

$$\frac{d_{31}}{g_{31}} = \varepsilon_{33}^X \tag{6.48}$$

Similarly

$$\frac{e_{ij}}{h_{ij}} = \varepsilon_{ii}^x \tag{6.49}$$

Also, Eq. (6.29a) is expanded to

$$\delta D_1 = d_{15}\delta X_5 + \varepsilon_1^X \delta E_1 \tag{6.50a}$$

$$\delta D_2 = d_{15}\delta X_4 + \varepsilon_1^X \delta E_2 \tag{6.50b}$$

$$\delta D_3 = d_{31}(\delta X_1 + \delta X_2) + d_{33}\delta X_3 + \varepsilon_3^X \delta E_3 \tag{6.50c}$$

The coefficients e_{33} and e_{31} can be expressed in terms of d_{33} and d_{31} from Eq. (6.50c):

$$e_{33} = \left(\frac{\partial D_3}{\partial x_3}\right)_E = 2d_{31}c_{13}^E + d_{33}c_{33}^E \tag{6.51}$$

$$e_{31} = \left(\frac{\partial D_3}{\partial x_1}\right)_E = d_{31}(c_{11}^E + c_{12}^E) + d_{33}c_{13}^E \tag{6.52}$$

A combination of Eqs (6.48), (6.49), (6.51) and (6.52) enables the coefficients e and h to be calculated from the more usually quoted coefficients d and g.

The indirect effect given in Eq. (6.29b) expands to

$$\delta x_1 = s_{11}^E \delta X_1 + s_{12}^E \delta X_2 + s_{13}^E \delta X_3 + d_{31}\delta E_3 \tag{6.53a}$$

$$\delta x_2 = s_{11}^E \delta X_2 + s_{12}^E \delta X_1 + s_{13}^E \delta X_3 + d_{31}\delta E_3 \tag{6.53b}$$

$$\delta x_3 = s_{13}^E(\delta X_1 + \delta X_2) + s_{33}^E \delta X_3 + d_{33}\delta E_3 \tag{6.53c}$$

$$\delta x_4 = s_{44}^E \delta X_4 + d_{15}\delta E_2 \tag{6.53d}$$

$$\delta x_6 = 2(s_{11}^E - s_{12}^E)\delta X_6 \tag{6.53e}$$

The stiffness coefficients can be derived from the compliances using the following relations which hold for either open- or closed-circuit conditions and also with the symbols c and s interchanged:

$$\begin{aligned}
c_{11} &= \frac{s_{11}s_{33} - s_{13}^2}{f(s)} \\
c_{12} &= -\frac{s_{12}s_{33} - s_{13}^2}{f(s)} \\
c_{13} &= -\frac{s_{13}(s_{11} - s_{12})}{f(s)} \\
c_{33} &= \frac{s_{11}^2 - s_{12}^2}{f(s)} \\
c_{44} &= \frac{1}{s_{44}}
\end{aligned} \tag{6.54}$$

where $f(s) = (s_{11} - s_{12})\{s_{33}(s_{11} + s_{12}) - 2s_{13}^2\}$.

For convenience the various property relationships are summarized in an appendix to this chapter.

The values of the piezoelectric properties of a material can be derived from the resonance behaviour of suitably shaped specimens subjected to a sinusoidally varying electric field. To a good approximation the behaviour of the piezoelectric

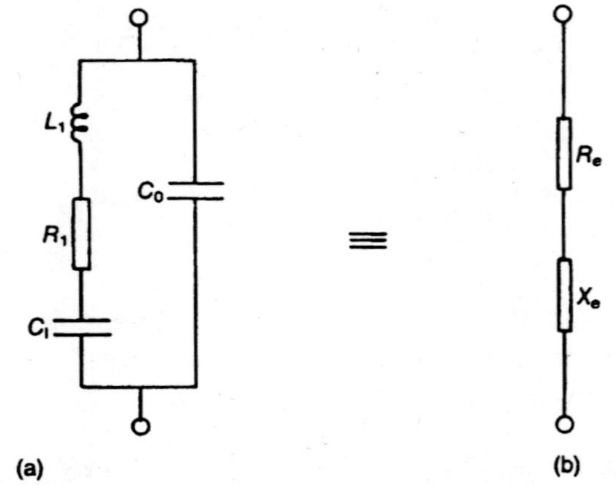

Fig. 6.3 (a) Equivalent circuit for a piezoelectric specimen vibrating close to resonance; (b) the equivalent series components of the impedance of (a).

specimen close to its fundamental resonance can be represented by the equivalent circuit shown in Fig. 6.3(a) [6].

There is an equivalence between the differential equations describing a mechanical system which oscillates with damped simple harmonic motion and driven by a sinusoidal force, and the series L_1, C_1, R_1 arm of the circuit driven by a sinusoidal e.m.f. The inductance L_1 is equivalent to the mass (inertia) of the mechanical system, the capacitance C_1 to the mechanical stiffness and the resistance R_1 accounts for the energy losses; C_o is the electrical capacitance of the specimen. Fig. 6.3(b) is the equivalent series circuit representing the impedance of the parallel circuit.

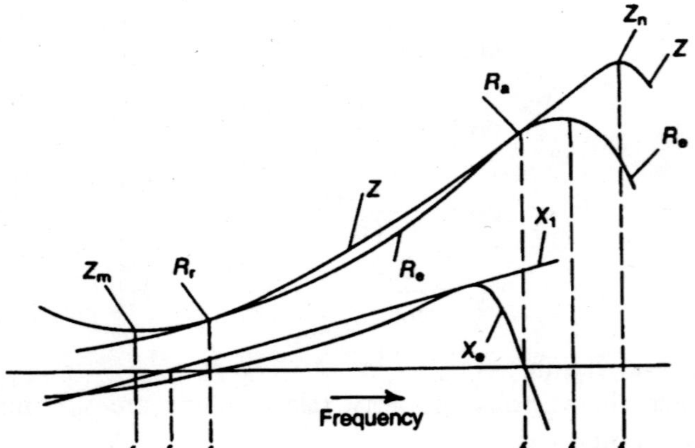

Fig. 6.4 The characteristic frequencies of the equivalent circuit exaggerating the differences between f_m, f_s and f_r and between f_a, f_p and f_n.

The form of the frequency response of the electrical circuit, shown in Fig. 6.4, has the following characteristic frequencies:

1. f_m and f_n when the impedance is, respectively, a minimum and a maximum, and
2. the resonance frequency f_r and antiresonance frequency f_a when the reactance of the circuit is zero.

The reactance of the mechanical arm is zero at the 'series' resonance frequency f_s when $j\omega L_1 = 1/j\omega C_1$; the parallel resonance f_p occurs when the currents flowing in the two arms are in antiphase, that is when $j(\omega L_1 - 1/\omega C_1)^{-1} = -1/j\omega C_0$, or when $\omega = \{(C_0 + C_1)/L_1 C_0 C_1\}^{1/2}$.

It follows from Eq. (6.44) and the equivalent circuit that

$$k_{\text{eff}}^2 = C_1/(C_1 + C_0) \tag{6.55}$$

and it can be shown that

$$k_{\text{eff}}^2 = C_1/(C_1 + C_0) = (f_p^2 - f_s^2)/f_p^2 \sim (f_a^2 - f_r^2)/f_a^2 \sim (f_n^2 - f_m^2)/f_n^2 \tag{6.56}$$

Provided that the mechanical quality factor (see Fig. 5.34 for analogy) for the resonator is sufficiently high ($Q_m \geqslant \sim 100$) the approximations apply.

Values for f_n and f_m are readily measured using an impedance analyser.

Specific ('33') coupling coefficient, compliance and permittivity values can be measured on appropriately dimensioned specimens from which the corresponding d and g coefficients can be calculated; the following examples illustrate how this is accomplished.

In the case of a piezoelectric ceramic rod (~ 5 mm diam. and 15 mm long) electroded on its end faces and poled along its length, for the fundamental resonance condition when there is a standing elastic longitudinal wave along the length with antinodes at the ends and the centre a node, it can be shown that

$$k_{33} = \frac{\pi}{2}\frac{f_s}{f_p}\tan\left(\frac{\pi}{2}\frac{f_p - f_s}{f_p}\right) \tag{6.57}$$

The appropriate permittivity ε_{33}^X is easily determined from the capacitance C of the specimen at a frequency well below resonance:

$$\varepsilon_{33}^X = \frac{Cl}{A} \tag{6.58}$$

where l and A are the length and cross-sectional area of the rod respectively. The elastic compliance s_{33}^D is related to the parallel resonance frequency by

$$\frac{1}{s_{33}^D} = 4\rho f_p^2 l^2 \tag{6.59}$$

where ρ is the density of the material.

The other '33' parameters can be calculated from the measured values of k_{33}, s_{33}^D and ε_{33}^X and using Eqs (6.33), (6.39), (6.40) and (6.42).

A more common geometry is a thin disc of diameter d electroded over both faces and poled in a direction perpendicular to the faces. The resonance on which attention is focused is that of a radial mode, excited through the piezoelectric effect across the thickness of the disc. In this case the route from the resonant frequencies to the coefficients d and g is the same as in the case of the rod, although the expressions are more complex.

Figure 6.5 shows the typical frequency response for such a disc. The lower frequency responses are radial resonances and the higher frequency response the thickness mode resonance.

The planar coupling coefficient k_p is related to the parallel and series resonant frequencies by

$$\frac{k_p^2}{1-k_p^2} = f\left(J_0, J_1, \nu \frac{f_p - f_s}{f_s}\right) \tag{6.60}$$

where J_0 and J_1 are Bessel functions and ν is Poisson's ratio. The relationship between k_p and $(f_p - f_s)/f_s$ is shown in Fig. 6.6 for the case in which $\nu = 0.3$. Fortunately, the curve is very insensitive to ν, and in the range $0.27 < \nu < 0.35$ the

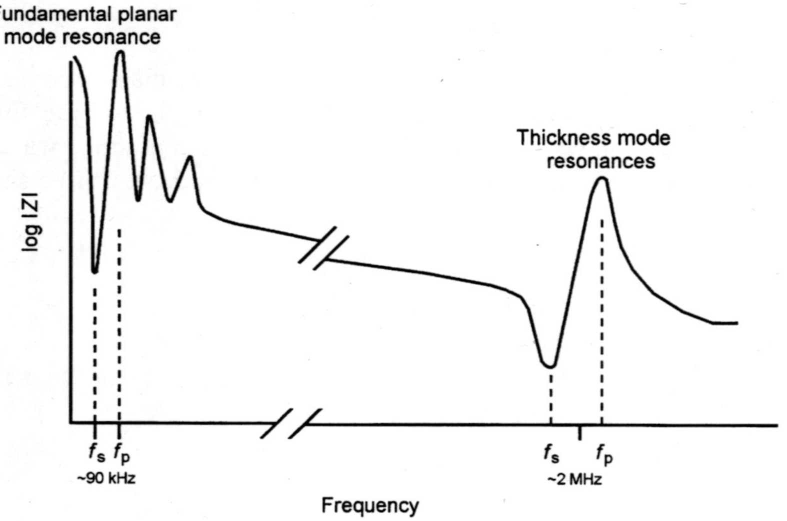

Fig. 6.5 A typical 'impedance–frequency' characteristic for a piezoelectric disc specimen having diameter ~ 25 mm and thickness ~ 1 mm and poled through the thickness.

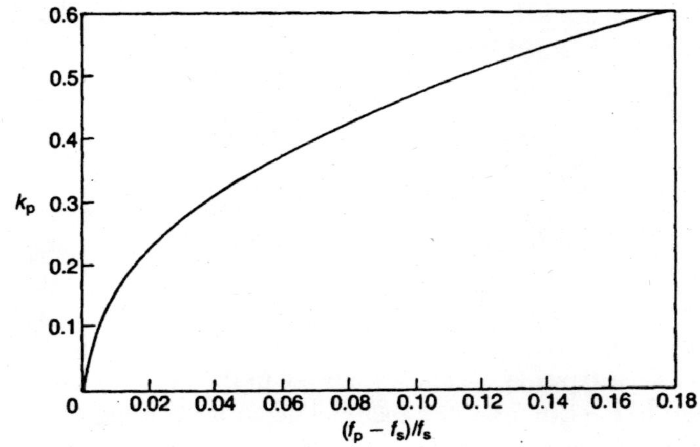

Fig. 6.6 The planar coupling coefficient k_p as a function of $(f_p - f_s)/f_s$.

differences cannot easily be discerned; for the common piezoelectric ceramics $0.28 < \nu < 0.32$.

With an assumed value for ν k_{31} can be calculated from

$$k_{31}^2 = \frac{1-\nu}{2} k_p^2 \tag{6.61}$$

Alternatively, using Eq. (6.62), k_{31} may be determined directly from the resonance and antiresonance frequencies of a bar poled across its thickness and using Eq. (6.62)

$$k_{31}^2/(1 - k_{31}^2) = (\pi/2)(f_p/f_s) \tan \{(\pi/2)[(f_p - f_s)/f_s]\} \tag{6.62}$$

The measured k_{31} and k_p values, together with Eq. (6.61), then permit the *calculation* of ν.

The compliance, s_{11}^E can be calculated using

$$\frac{1}{s_{11}^E} = \frac{\pi^2 d^2 f_s^2 (1 - \nu^2)\rho}{\eta_1^2} \tag{6.63}$$

where η_1 is a root of an equation involving the Bessel functions and the Poisson ratio; it has a value of approximately 2. Then (cf. Eq. (6.39))

$$d_{31} = k_{31}(\varepsilon_{33}^X s_{11}^E)^{1/2} \tag{6.64}$$

and (cf. Eq. (6.48))

$$g_{31} = \frac{d_{31}}{\varepsilon_{33}^X} \tag{6.65}$$

If the minimum impedance $|Z_m|$ at resonance is known, the mechanical Q factor Q_m can be obtained from the following equation:

$$\frac{1}{Q_m} = 2\pi f_s |Z_m| (C_0 + C) \frac{f_p^2 - f_s^2}{f_p^2}$$
$$\approx 4\pi \Delta f \, |Z_m| \, (C_0 + C) \qquad (6.66)$$

The dielectric Q factor is the reciprocal of the dielectric dissipation factor $\tan\delta$ (see Section 2.7.2).

Figure 6.5 shows the thickness resonances from which the coupling coefficient (k_t) can be calculated. This usually has a value in the range 0.45 to 0.5, that is much smaller than typical k_{33} values which lie in the range 0.65–0.75. The reason for this is that in the case of a disc the thickness ('longitudinal') vibrations are in effect partially clamped since they must be accompanied by planar distortions. The longitudinal coupling coefficient k_{33} is measured on more appropriately dimensioned rods (see Eq. (6.57)) to minimize this clamping.

6.3 General Characteristics and Fabrication of 'PZT'

6.3.1 Effects of domains

The first piezoceramic to be developed commercially was $BaTiO_3$, the model ferroelectric discussed earlier (see Section 2.7.3). By the 1950s the solid solution system $Pb(Ti,Zr)O_3$ (PZT), which also has the perovskite structure, was found to be ferroelectric and PZT compositions are now the most widely exploited of all piezoelectric ceramics. The following outline description of their properties and fabrication introduces important ideas for the following discussion of the tailoring of piezoceramics, including PZT, for specific applications. It is assumed that the reader has studied Sections 2.3 and 2.7.3.

The $Pb(Zr_{1-x}Ti_x)O_3$ phase diagram [1] is shown in Fig. 6.7. The morphotropic phase boundary (MPB) defines the composition at which there is an abrupt structural change, the composition being almost independent of temperature. That is the phase boundary between the high temperature rhombohedral and tetragonal forms is, practically speaking, a vertical line. As shown in Fig. 6.8, the piezoelectric activity peaks in the region of the MPB composition and considerable effort has been directed to elucidating the reason(s) for this technically very important phenomenon.

The current understanding is that the MPB is not a sharp boundary but rather a temperature-dependent compositional range over which there is a mixture of tetragonal and monoclinic phases. At room temperature (300 K) the two phases coexist over the range $0.455 \leqslant x \leqslant 0.48$ [7]. The enhanced piezoelectric activity of

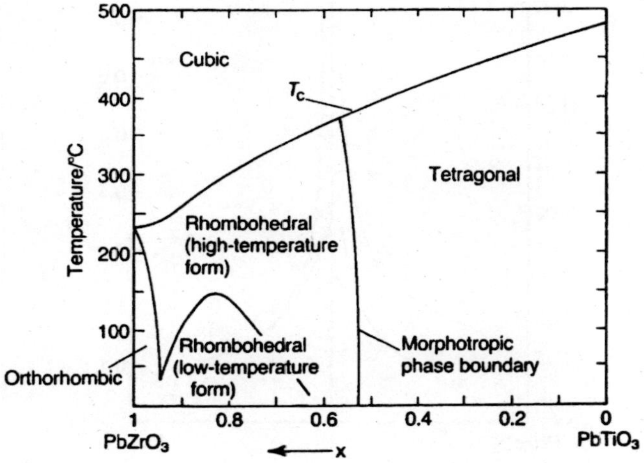

Fig. 6.7 Phase stabilities in the system $Pb(Ti_{1-x}Zr_x)O_3$. (After Jaffe *et al.* [1].)

the commercial compositions ($x \sim 0.48$) can be rationalised in terms of the relatively large ionic displacements associated with stress (electrical or mechanical) – induced rotation of the monoclinic polar axis [8].

As a ferroelectric perovskite in ceramic form cools through its Curie point it contracts isotropically since the orientations of its component crystals are random. However, the individual crystals will have a tendency to assume the anisotropic shapes required by the orientation of their crystal axes. This tendency will be counteracted by the isotropic contraction of the cavities they occupy. As a consequence a complex system of differently oriented domains that minimizes the elastic strain energy within the crystals will become established.

The application of a sufficiently strong field will orient the 180° domains in the field direction, as nearly as the orientation of the crystal axes allows (see Section 2.7.3). The field will also have an orienting effect on 90° domains in tetragonal material and on 71° and 109° domains in the rhombohedral form, but the response will be limited by the strain situation within and between the crystals. There will be an overall change in the shape of the ceramic body with an expansion in the field direction and a contraction at right angles to it. When the field is removed the strain in some regions will cause the polar orientation to revert to its previous direction, but a substantial part of the reorientation will be permanent.

An externally applied stress will affect the internal strain and the domain structures will respond; this process is termed the ferroelastic effect. Compression will favour polar orientations perpendicular to the stress while tension will favour a parallel orientation. Thus the polarity conferred by a field through 90° domain changes can be reversed by a compressive stress in the field direction. Stress will not affect 180° domains except in so far as their behaviour may be coupled with other domain changes.

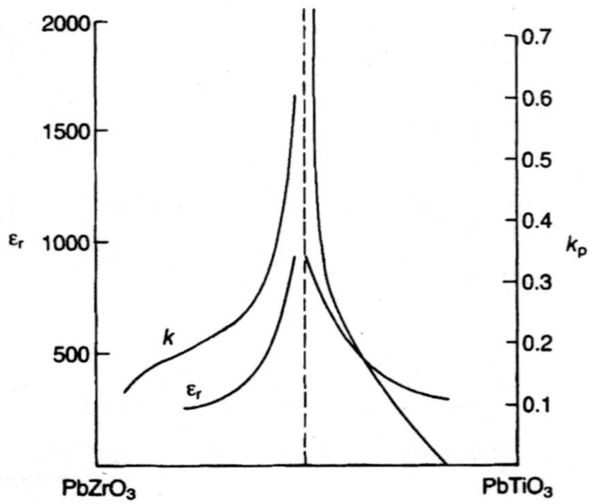

Fig. 6.8 Coupling coefficient k_p and permittivity ε_r values across the PZT compositional range. (After Jaffe *et al.* [1].)

A polar axis can be conferred on the isotropic ceramic by applying a static field of 1–4 MV m^{-1} for periods of several minutes at temperatures usually somewhat above 100 °C when domain alignment occurs. More rapid domain movement takes place at higher temperatures but the maximum alignment of 90° domains takes some time to be established. The temperature is limited by the leakage current which can lead to an increase in the internal temperature and to thermal breakdown (see Section 5.2.2); the field is limited by the breakdown strength. If the applied voltage exceeds about 1 kV it is necessary to ensure very clean surfaces between the electrodes and to immerse the pieces to be poled in an insulating oil. Surface breakdown takes place very readily between electrodes on the surface of a high-permittivity material when it is exposed to air.

Depoling can be achieved by applying a field in the opposite direction to that used for poling or, in some cases, by applying a high a.c. field and gradually reducing it to zero, but there is a danger of overheating because of the high dielectric loss at high fields. Some compositions can be depoled by applying a compressive stress (10–100 MPa). Complete depoling is achieved by raising the temperature to well above the Curie point and cooling without a field.

Alternating fields cause domain walls to oscillate. At low fields the excursions of 90°, 71° or 109° walls result in stress–strain cycles that lead to the conversion of some electrical energy into heat and therefore contribute to the dielectric loss. When peak fields are sufficient to reverse the spontaneous polarization the loss becomes very high, as shown by a marked expansion of the hysteresis loop (Fig. 6.9).

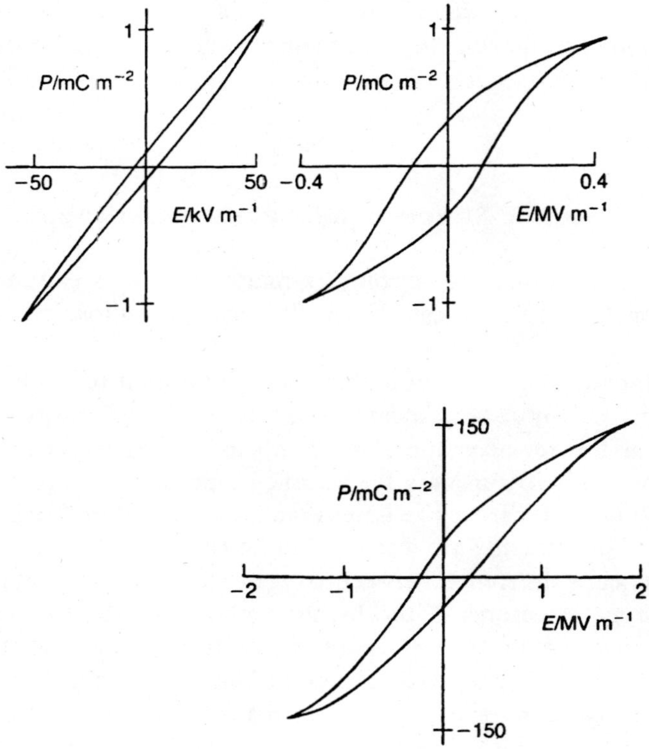

Fig. 6.9 *P–E* loops with increasing peak field strengths at 50 Hz for a ceramic ferroelectric.

'Ageing' (see Section 2.7.3) affects many of the properties of piezoelectric ceramics. Most of the piezoelectric coefficients fall by a few per cent per decade although the frequency constant increases. Ageing can be accelerated and properties stabilized by heating to temperatures of about 80 °C in the case of $BaTiO_3$ and rather higher for PZT.

Ageing effects are known to be significantly changed when the concentration of vacant oxygen sites is increased either by doping or by heating in mildly reducing atmospheres. It has been suggested that dipoles are formed between negatively charged defects (e.g. Co^{3+} on Ti^{4+} sites) and the positively charged $V_O^{\cdot\cdot}$, and that these are aligned in the polar direction by movement of the $V_O^{\cdot\cdot}$ as a result of the combined action of the local field associated with P_s and thermal energy. The dipoles then provide an internal field stabilizing the domain configuration thereby reducing ageing rate.

An alternative mechanism is based on the local stresses that are caused by the development of axial anisotropy within the grains as they cool below T_C. Lower-energy domain configurations may be generated by phonon action at room temperature and might be assisted by a high concentration of V_O. Another suggestion has been that defects such as V_O may diffuse to the domain walls and stabilize them. The first of these mechanisms, dipole formation and rotation, is assumed in the next section.

6.3.2 Effects of aliovalent substituents

The effect of substituents is a complex matter but, with caution, a number of important generalizations can be made regarding aliovalent substituents in perovskites.

Donor dopants, i.e. those of higher charge than that of the ions they replace, are compensated by cation vacancies; acceptors, i.e. dopants of lower charge than that of the replaced ions, are compensated by oxygen vacancies. Each dopant type tends to suppress the vacancy type that the other promotes. The common dopants in perovskite-type ceramics are listed in Table 6.1. The effects of aliovalent substituents are discussed in Section 2.6.2.

The significant difference between oxygen vacancies and cation vacancies in perovskite-type structures is the higher mobility of the former. Cations and cation vacancies tend to be separated by oxygen ions so that there is a considerable energy barrier to be overcome before the ion and its vacancy can be interchanged. Oxygen ions, however, form a continuous lattice structure so that oxygen vacancies have oxygen ion neighbours with which they can easily exchange.

Typical concentrations of dopants (0.05–5 at.%) must result in the formation of dipolar pairs between an appreciable fraction of the dopant ions and the vacancies, e.g. $2La^{\cdot}_A$–V''_A or $2Fe^{3+\prime}_B$–$V^{\cdot\cdot}_O$. Donor–cation vacancy combinations can be assumed to have a stable orientation so that their initially random state is unaffected by spontaneous polarization or applied fields. Acceptor–oxygen vacancy combinations are likely to be less stable and thermally activated reorientation may take place in the presence of local or applied fields. The dipoles, once oriented in a common direction, will provide a field stabilizing the domain structure. A reduction in permittivity, dielectric and mechanical loss and an increase in the coercive field will result from the inhibition of wall movement. Since the compliance is affected by the elastic movement of 90° walls under stress, it will also be reduced by domain stabilization.

Table 6.1 Common aliovalent substituents

A-site donors	La^{3+}, Bi^{3+}, Nd^{3+}
B-site donors	Nb^{5+}, Ta^{5+}, Sb^{5+}
A-site acceptors	K^+, Rb^+
B-site acceptors	Co^{3+}, Fe^{3+}, Sc^{3+}, Ga^{3+}, Cr^{3+}, Mn^{3+}, Mn^{2+}, Mg^{2+}, Cu^{2+}

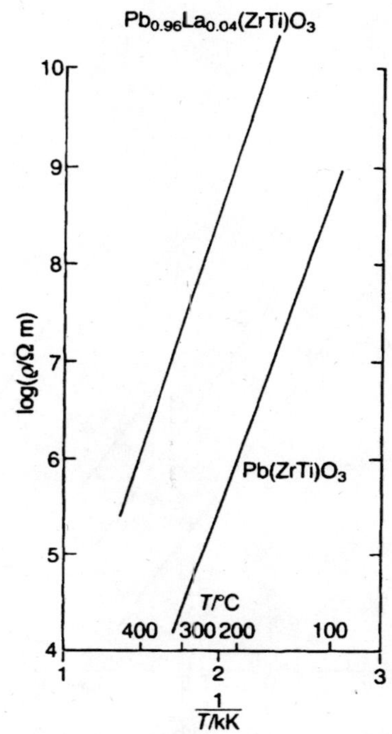

Fig. 6.10 The effect of lanthanum addition on the resistivity of PZT. (After J.C. Jaffe *et al.* [1].)

Donor doping in PZT would be expected to reduce the concentration of oxygen vacancies, leading to a reduction in the concentration of domain-stabilizing defect pairs and so to lower ageing rates. The resulting increase in wall mobility causes the observed increases in permittivity, dielectric losses, elastic compliance and coupling coefficients, and reductions in mechanical Q and coercivity.

The introduction of oxygen vacancies through acceptor doping also leads to a slight reduction in unit-cell size, which tends to reinforce the effects referred to above.

Some indication of the real complexities is afforded by the different responses of $BaTiO_3$ and PZT to the addition of donors such as La^{3+} (see Section 2.6.2). In PZT PbO lost by volatilization during sintering can be replaced in the crystal by La_2O_3. For example, if the excess positive charge of the La^{3+} is balanced by lead site vacancies,

$$0.01La_2O_3 + Pb(Ti,Zr)O_3 \rightarrow (Pb_{0.97}La_{0.02}V_{0.01})(Ti,Zr)O_3 + 0.03PbO \uparrow$$

Thus the lanthanum dopant is vacancy compensated and no electronic charge carriers are generated. As shown in Fig. 6.10, La substitution results in a marked increase in the resistivity of PZT.

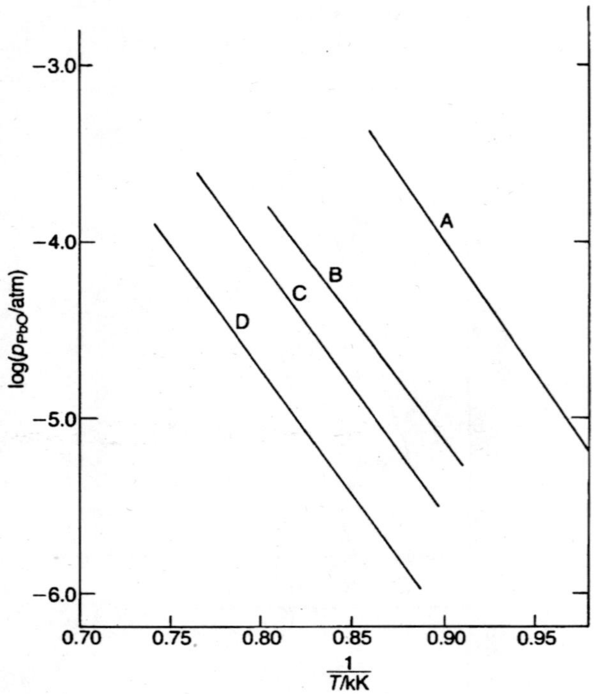

Fig. 6.11 Vapour pressure of PbO over lead-containing ceramics: curve A, solid PbO; curve B, $PbZrO_3$; curve C, $Pb(Ti_{0.45}Zr_{0.55})O_3$; curve D, $PbTiO_3$.

Lead lanthanum zirconate titanates (PLZT) containing 3–12 mol.% La and 5–30 mol.% Ti form a class of ceramics with important dielectric, piezoelectric and electro-optic properties. They may contain vacancies on B as well as A sites and have a remarkable facility for changing their polar states under the influence of applied fields.

In contrast, in $BaTiO_3$, in which the volatility of the constituents during sintering is low, small concentrations (< 1 mol.%) of donors are compensated by electrons in the conduction band with an accompanying change of colour from pale yellow to grey or black. However, higher donor concentrations lead to two effects: a decrease in crystal size in ceramics sintered to full density, and an increase in resistivity above that for undoped $BaTiO_3$ together with a reversion to the pale colour. The enhancement of permittivity and piezoelectric coupling coefficient that occurs with PZT does not occur with donor-doped $BaTiO_3$. It appears that at higher donor concentrations the formation of cation vacancies becomes energetically favoured. At present there is no widely accepted explanation of this very pronounced difference in behaviour between PZT and $BaTiO_3$.

6.3.3 Fabrication of PZT

Normal powder technology is used in the fabrication of piezoelectric ceramics. The highest values of the coefficients are obtained when the composition is near stoichiometric, the content of fluxing reagents and impurities is minimal, and the density is as high as possible. Contamination during milling is kept low by using zirconia-based milling media.

Most of the compositions used at present contain PbO as a major constituent. Despite its volatility above 800 °C (Fig. 6.11), the PbO must be retained during sintering at temperatures up to 1300 °C. A.I. Kingon and B. Clark [9] address in some detail the matter of 'atmosphere control' with regard to firing PZT, and the effect of PbO content on densification kinetics. Calcination is usually carried out in lidded alumina pots. For the final sintering the 'work' is surrounded by a lead-rich powder, such as $PbZrO_3$, and again placed in closed saggars. Because of the limited access to the atmosphere that results and the ease with which PbO is reduced to metallic lead, all organic constituents must be removed before sintering by a preliminary firing at about 600 °C in air. The final firing is usually carried out in batch-type electric kilns well filled with 'work'. Despite precautions, there is normally a loss of 2–3% of the initial PbO content which is compensated by an addition to the starting materials.

An alternative to firing in closed vessels is very rapid firing in which the material, carried on a moving belt, is exposed to a high peak temperature for a brief period in a special kiln. This procedure is applicable to thin sheets of material.

A further difficulty with compositions containing substantial amounts of ZrO_2 results from the low reactivity of some grades of the material. The TiO_2 powders, developed as pigments, react rapidly with PbO, and the resulting titanates only take up Zr^{4+} ions slowly from unreacted ZrO_2. Grades of ZrO_2 are available which have not been calcined at high temperatures during their preparation and which therefore react more rapidly with PbO so that a mixed $Pb(Zr,Ti)O_3$ is formed at an early stage. The problem can also be alleviated by prolonged milling during the mixing stage and by a second calcination after milling the product of an initial firing.

All the constituents of a PZT composition can be precipitated from nitrate solutions to yield highly homogeneous reactive powders. Such powders can also be prepared by calcining citrates that contain the A- and B-site ions in a 1:1 ratio. It is possible to sinter at lower temperatures using these materials and, apart from the economy in energy, this has the advantage of lessening the loss of PbO through volatilization. However, sintered bodies adequate for the majority of piezoelectric applications are obtained by the lower-cost route starting from a mixture of oxides or carbonates.

Simple shapes are formed by die-pressing, long bodies of uniform section are formed by extrusion, thin plates are formed by band-casting or calendering, and large rings and more intricate shapes are formed by slip-casting.

The sintered product usually has a density higher than 95% of theoretical and a crystallite size in the 5–30 μm range. Both an excess and a deficiency of PbO result in inferior piezoelectric properties so that careful control of all aspects of the manufacturing process is essential.

Electrodes are applied after any necessary machining to shape or finishing. For the majority of applications a suitably formulated silver-bearing paint is fired on at 600–800 °C. In the case of thin pieces, Ni–Cr or gold electrodes can be applied by sputtering or evaporation.

Poling takes place with the specimens immersed in transformer oil at 100–150 °C and with an applied field of 1–4 MV m^{-1}. In the case of $BaTiO_3$ the field must be maintained whilst cooling to some 50 °C below the Curie point. For PZT the temperature and voltage are optimized to give the maximum piezoelectric coefficients without allowing the leakage current to reach levels that could result in thermal runaway and electrical breakdown. Higher fields can be used if they are applied as a succession of short pulses.

Corona poling can also be used. In this case voltages of order 10^4 V are applied to either a single needle or an array of needles with their points a few millimetres from the ceramic surface; the opposite side of the ceramic is earthed. In this way a high field is established in the ceramic. The advantage that corona poling offers is a diminished risk of electrical breakdown. This is because the poling charge cannot be quickly channelled to a 'weak spot', as it could be if it were to be carried on a metallic electrode.

6.4 Important Commercial Piezoceramics

Compositions, composites and devices have been developed for the four main uses of piezoceramics:

1. the generation of voltages;
2. electromechanical actuation;
3. frequency control;
4. the generation and detection of acoustic and ultrasonic energy.

The specific requirements to optimize performance for the various applications are discussed in the following sections.

6.4.1 Barium titanate

$BaTiO_3$ was the first material to be developed as a piezoceramic and was used on a large scale for application 4 above. For most commercial purposes it has been superseded by PZT, although its 'lead-free' composition is attractive from the 'health and safety' perspective.

The structural transitions which occur in $BaTiO_3$ (see Section 2.7.3) are accompanied by changes in almost all electrical and mechanical properties. The transition temperatures can be altered by substitutions on the A and B sites, and for many applications it is necessary to move them away from the working temperature ranges so that the associated large temperature coefficients are avoided. The cubic–tetragonal and orthorhomic–rhombohedral transitions occur well away from normal working temperatures, but the tetragonal–orthorhombic transition occurs close to them. The substitution of Pb and Ca for Ba lowers this transition temperature and has been used to control piezoelectric properties around 0 °C which is important for underwater detection and echo sounding.

The substitution of Zr or Sn for Ti raises both the tetragonal–orthorhomic and orthorhomic–rhombohedral transitions. Raising the former to above the working-temperature range yields enhanced piezoelectric properties provided that the poling is effected below this temperature range and that it is not subsequently exceeded. In the past such compositions have found widespread use as bimorphs for record-player pick-up cartridges.

Technically pure $BaTiO_3$ (Fig. 6.12), or $BaTiO_3$ doped with isovalent substituents, generally has too high a loss at the high field strengths (0.2–0.4 $MV\,m^{-1}$) required to generate useful ultrasonic powers. The dielectric loss arises largely from the movement of domain walls. The control of the high-field loss is therefore a matter of controlling domain wall movement.

As discussed above, the introduction of acceptor dopants leads to domain wall clamping. Manganese reduces the low-field loss (see Section 5.7.1) but has little effect at high fields. Co^{3+} substituted for Ti at the 1–2at.% level is particularly effective, as shown in Fig. 6.12. Compositions containing cobalt must be fired in a fully oxidizing atmosphere since Co^{3+} is easily reduced to Co^{2+}. This change in oxidation state is accompanied by a change in colour from blue-black to green and ultimately to yellow, and an almost complete loss of piezoelectric properties.

Since the polar axes in barium titanate and PZT (see Fig. 2.40(b) and Fig. 2.44) are longer than the perpendicular axes, ceramics expand in the polar direction during poling. The application of a high compressive stress in the polar direction to a poled ceramic causes depoling since the 90° domains switch direction as a result of the ferroelastic effect and the polar directions of the crystallites become randomized.

In $BaTiO_3$ less than 10% of 90° domains are permanently altered in their polar direction by poling, whereas some 40%–50% of 90°, 71° and 109° domains are affected in PZT compositions. It is therefore understandable that $BaTiO_3$ shows a greater resistance than PZT to depoling by compressive stresses, and this resistance is particularly strong in cobalt-doped material, especially after a period of ageing. This further illustrates the clamping effect of Co^{3+} ions on domain walls (see Section 6.3.2). Although iron doping has a similar effect in

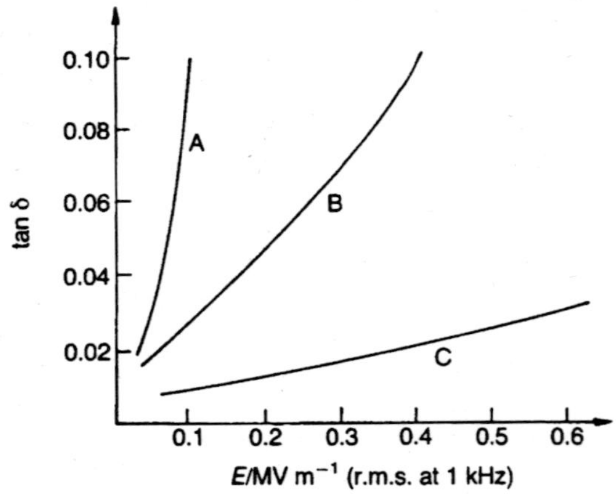

Fig. 6.12 Dependence of tan δ on the applied field strength for barium titanates: curve A, technically pure barium titanate; curve B, barium titanate containing lead and 1–2 at.% Ti^4 replaced by Co^{3+}, unpoled; curve C as B but poled.

PZT it has not been found possible to obtain a PZT ceramic as resistant to high pressures as $BaTiO_3$. For this reason the use of cobalt-doped $BaTiO_3$ for producing high acoustic powers has continued in some applications, despite its inferior piezoelectric activity.

6.4.2 Lead zirconate–lead titanate ('PZT')

$PbTiO_3$ (Curie point, 495 °C) has a similar tetragonal structure to $BaTiO_3$ but with a c axis approximately 6% longer than its a axis at room temperature. The high internal stresses developed on cooling a sintered ceramic through Tc cause disintegration. $PbZrO_3$ (Curie point 234 °C) is orthorhombic with a structure similar to that of orthorhombic $BaTiO_3$ but is antiferroelectric, i.e. the dipoles due to a displacement of the Zr^{4+} ions from the geometric centre of the surrounding six O^{2-} ions are alternatively directed in opposite senses so that the spontaneous polarization is zero.

The replacement of zirconium by isovalent hafnium has little effect apart from shifting the morphotropic boundary (see Fig. 6.7) to 52mol.% $PbTiO_3$. Replacement by tin causes a slight loss in piezoelectric activity, shifts the morphotropic boundary to 42 mol.% $PbTiO_3$ and lowers the Curie point from 370 to 250 °C.

The isovalent A-site substituents barium, strontium and calcium lower the Curie point and have a small influence on the morphotropic composition. At the 5–10 mol.% level they enhance the permittivity and piezoelectric properties. $Pb_{0.94}Sr_{0.06}Ti_{0.47}Zr_{0.53}$ for instance has a relative permittivity of 1300, a k_p of 0.58

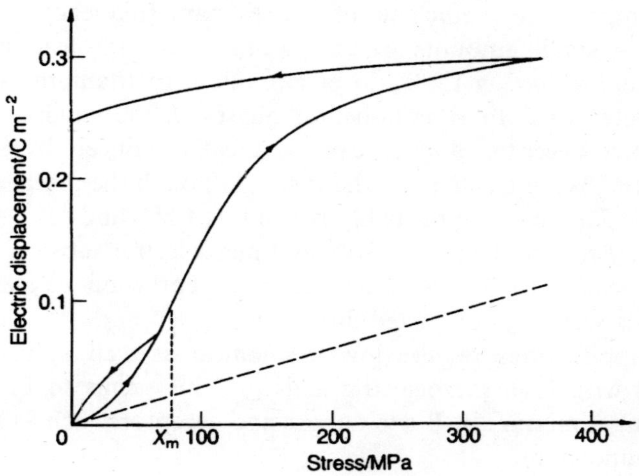

Fig. 6.13 Electric displacement as a function of compressive stress for donor-doped PZT: - - -, output expected from a low-stress piezoelectric constant.

and a Curie point of 328 °C compared with values of 730, 0.53 and 386 °C for the unsubstituted composition.

The mechanism whereby aliovalent substituents may affect properties has already been discussed. Doping PZT with iron restricts domain movement in much the same way as cobalt in $BaTiO_3$ does and results in the low loss at high fields necessary in the generation of high-energy vibrations. Uniaxial compressive stresses below the level that cause depoling raise the permittivity and dissipation factor in iron-doped PZT, presumably as a result of the response of the additional 90° walls generated by the applied stress.

Figure 6.13 shows a plot of charge against compressive stress for donor-doped PZT. The output is considerably in excess of what would be expected from the piezoelectric coefficient determined at low stress. The excess charge is to be attributed to the switching of 90° domains. Up to a limiting stress X_m, the charge is reabsorbed on releasing the stress, indicating that the 90° domain wall movement is reversible, but at higher stresses only a part is reabsorbed and the low-stress piezoelectric coefficient is correspondingly progressively diminished. The longer the time a stress is applied, the greater is the charge released and the smaller is the fraction of it that is reabsorbed. The extra output due to the reversible movement of 90° walls is made use of in high-voltage generators used for the ignition of gas by sparks (see Section 6.5.1).

Donor-doped PZTs have higher permittivities and *d* coefficients than acceptor-doped materials and are therefore more suitable for converting mechanical into electrical vibrations. They have higher dissipation factors than acceptor-doped materials and are therefore not as suitable for wave filters. If this were not the case, their low ageing coefficients would be an advantage.

Low temperature coefficients of the resonant frequency can be obtained by substituting small amounts of an alkaline earth for lead in acceptor-doped material and adjusting the ratio of zirconium to titanium, which governs the ratio of tetragonal to rhombohedral phases in the ceramic. The ageing that accompanies acceptor doping can be greatly reduced by substituting small amounts (<1%) of Cr or U on the B site, although the mechanism has not been elucidated. The ions are probably present as Cr^{4+} and U^{4+} or U^{6+} and, since they slightly reduce the permittivity and piezoelectric activity, it can be assumed that they accelerate the orientation of the acceptor ion–oxygen vacancy dipoles so that it is virtually completed during poling.

Some applications require low mechanical as well as low dielectric losses combined with high piezoelectric activity. This seems to be best achieved by substituting a mixture of B-site donors and acceptors such as Nb and Mg in 2:1 atomic proportions.

6.4.3 Lead-based relaxor piezoelectric and electrostrictive ceramics

There are many factors which determine the suitability of a particular piezoceramic for a given transducer application. As a general rule the higher the Curie point (T_c) the lower are the room temperature values of permittivity and charge (or strain) coefficient. However, the low transition temperatures, which favour high piezoelectric coefficients, carry with them the risk of depoling during the processing steps, for example machining, electroding and soldering, usually necessary to engineer the material into a device.

The lead-based perovskite relaxor ferroelectrics, $Pb(B'B'')O_3$, (see Section 5.7.2) have exceptionally high permittivity values and therefore, from Eq. (6.7) are expected to be strongly electromechanically active and especially suited to transducer applications.

The search for piezoceramics having optimum electromechanical properties has been extended to cover MPB compositions of $PbTiO_3$ with various relaxors. The search is assisted by using a relationship established between the Curie point of the MPB composition and the tolerance factor (see Section 2.7.3) for the relaxor component [10].

The article by S.-E. Park and T.R. Shrout [11] provides a valuable supplement to the discussion.

Polycrystalline ceramics

The ternary diagram (Fig. 6.14) is essentially the PZT binary diagram extended to take account of additions of a relaxor of the type $Pb(B'\ B'')O_3$.

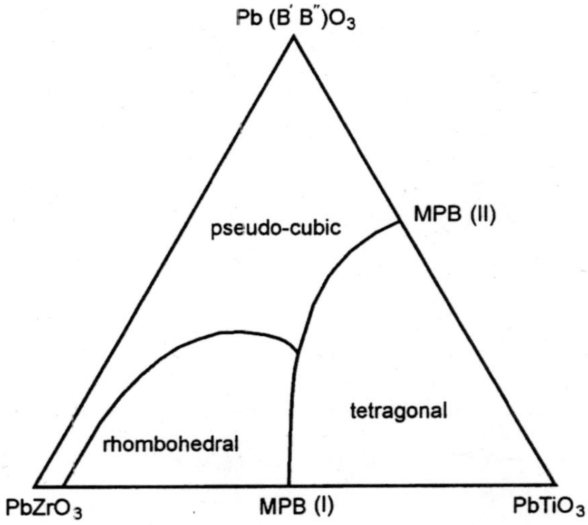

Fig. 6.14 Room temperature ternary showing structural phase-fields in the PZ–PT–relaxor system.

The system now shows two morphotropic phase boundaries, the one already discussed on the PZ–PT join, and the other on the PT–$Pb(B'B'')O_3$ join. Compositions close to that of the MPB (II) have attracted considerable interest because of their potential for electromechanical applications. The properties of some selected systems are given in Table 6.2 together with those for the 'PZT' system for comparison.

As discussed in Section 5.7.2, in the case of niobium-containing perovskites it is necessary to avoid the formation of pyrochlore-type phases if reproducible and optimum dielectric and piezoelectric properties are to be achieved. G. Roberts *et al.* [13] developed a modification of the B-site precursor ('columbite' process) route to produce high quality PNN–PZT ceramics.

Table 6.2 Piezoelectric properties of some 'relaxor based' compositions close to the MPB(II) (data taken from [11])

Composition	x *at MPB(II)*	k_{33}	$d_{33}/pC\ N^{-1}$	$\varepsilon_r^X/10^3$	$T_c/°C$
$(1-x)$ PMN–xPT	0.33	0.73	690	5	160
$(1-x)$ PST–xPT	0.45	0.73	655	4	205
$(1-x)$ PSN–xPT	0.43	0.76	504	2.5	248
0.5 PNN–0.345 PT–0.155PZ[a]	0.40	0.80	900	7	(~90[b])
$(1-x)$ PZ–xPT[c]	0.48 MPB (I)	0.64–0.75	220–600	1–3.4	195–365

PMN = $Pb(Mg_{1/3}Nb_{2/3})O_3$; PST = $Pb(Sc_{1/2}Ta_{1/2})O_3$; PSN = $Pb(Sc_{1/2}Nb_{1/2})O_3$;
PNN = $Pb(Ni_{1/3}Nb_{2/3})O_3$; PZ = $PbZrO_3$; PT = $PbTiO_3$.
(a) Taken from [12].
(b) Estimated.
(c) The quoted property value ranges are based on data taken from a leading manufacturer's brochure.

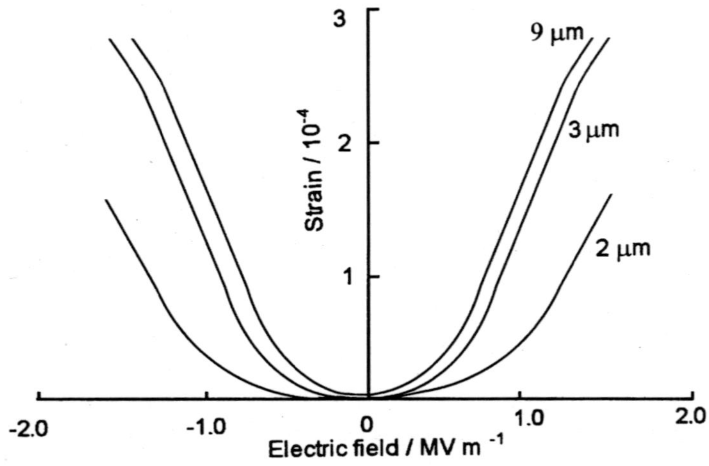

Fig. 6.15 Electrostriction in 0.93 PMN 0.07 PT showing the dependence on grain size (after T.R. Shrout *et al.* [14]).

Although the polycrystalline relaxor-based compositions have useful piezoelectric characteristics, taking their properties overall into account they do not offer significant advantages over the well established PZT system. Their high electrostriction coefficients make them attractive for certain actuator applications (see Section 6.5.2), but the potentially important advance has been the production of single crystals.

Through additions of $PbTiO_3$ to the prototype relaxor $Pb(Mg_{1/3}Nb_{2/3})O_3$ the temperature at which the permittivity peaks can be adjusted and, close to the composition 0.9 PMN–0.1 PT, it occurs at around room temperature. This is a composition suited to actuator applications. The dependence of electrostriction on grain size (Fig. 6.15) [14] is believed to be linked to the dependence of the size of the nanosized ordered regions on grain size (T. Mishima *et al.* [15]).

Single crystals

Efforts to grow usefully sized PZT single crystals with the prospect of their offering significantly enhanced electromechanical properties compared to those of the polycrystalline counterparts, have not been successful. However, in the case of the 'relaxor-PT' systems large single crystals can be flux-grown (see Section 3.11) and these do offer exceptionally good electromechanical properties (Table 6.3) which are, of course, dependent upon the 'cut' of the piece with respect to the crystallographic axes of the single crystal.

Table 6.3 Electromechanical properties of some relaxor-based single crystals (data taken from [11])

Crystal	*Cut*	ε_r	k_{33}	$d_{33}/pC\ N^{-1}$†
$Pb(Zn_{1/3}Nb_{2/3})O_3$	111	900	0.38	83
	001	3600	0.85	1100
PZN–8 mol.% PT*	111	2150	0.40	82
	001	4200	0.94	2070
PMN–35 mol.% PT	001	3100	0.92	1240

*The MPB lies at ~10 mol.% PT. †Effective d-coefficient.

6.4.4 Lead niobate

The ferroelectric polymorph of lead niobate ($PbNb_2O_6$) is metastable at room temperature. Above 1200 °C the structure is tetragonal tungsten bronze. On cooling slowly below 1200 °C it transforms to an orange-brown rhombohedral form that is paraelectric at room temperature. Rapid cooling from 1200 to 700 °C, combined with additions such as 2 wt% $ZrTiO_4$, allows the tetragonal form to persist at lower temperatures. It is bluish-green in colour with a Curie point at 560 °C below which it undergoes a small orthorhombic distortion in the *c* plane and becomes ferroelectric. The tungsten bronze structure is typified by certain alkali metal compounds of tungstic oxide, e.g. $K_{5.7}W_{10}O_{30}$.

The spaces between the MO_6 (M≡W, Ta, Nb, Ti etc.) form tunnels. In the section shown in Fig. 6.16 the largest tunnels, A2, have five O^{2-} ions round their periphery. The somewhat smaller tunnels, A1, have four peripheral O^{2-} ions, while the smallest, A3, have three. Lower-valency cations are located in sites in these tunnels above and below the plane shown. The MO_6 groups provide two distinguishable B sites shown as B1 and B2. The general formula for a tungsten bronze is

$$[(A1)_2(A2)_4A3][(B1)_2(B2)_8]O_{30}$$

The A3 sites are vacant unless small (r_6=50–70 pm) ions such as Li^+ or Mg^{2+} are present. A large number of compounds are known in which one in six of the A1 and A2 sites is also vacant, giving rise to the simpler formula AB_2O_6 of which $PbNb_2O_6$ is an example.

In the majority of ferroelectric tungsten bronzes the polar axes are parallel to the A-site tunnels. $PbNb_2O_6$ is an exception; below the Curie point it undergoes an orthorhombic distortion in the plane perpendicular to the A-site tunnels as indicated in Fig. 6.16. It possesses four possible polar directions and can be poled successfully in ceramic form.

Although it is difficult to obtain a ceramic with less than 7% porosity and the piezoelectric coefficients are smaller than those of $BaTiO_3$, $PbNb_2O_6$ has advantages in certain applications. Its Q_m value is approximately 11, which is

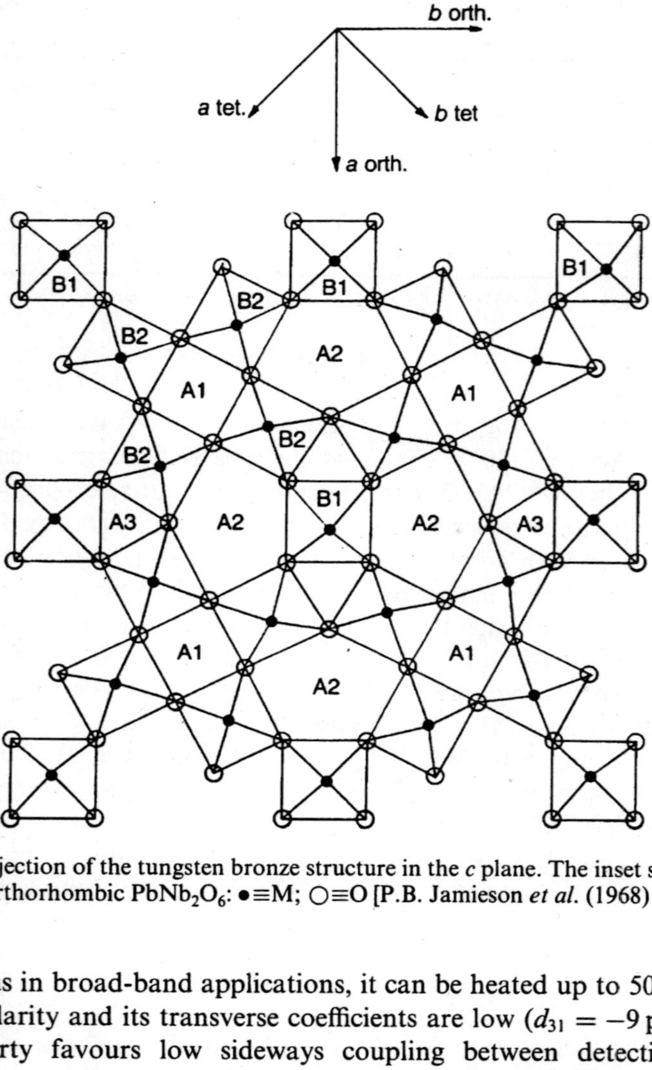

Fig. 6.16 Projection of the tungsten bronze structure in the *c* plane. The inset shows the polar directions in orthorhombic $PbNb_2O_6$: ●≡M; ○≡O [P.B. Jamieson *et al.* (1968) *J. Chem. Phys.* **48**, 5048–57].

advantageous in broad-band applications, it can be heated up to 500 °C without losing its polarity and its transverse coefficients are low ($d_{31} = -9\,\mathrm{pC\,N^{-1}}$). The latter property favours low sideways coupling between detecting elements formed side by side on the same piece of ceramic and results in a hydrostatic coefficient d_h of useful magnitude ($d_h = d_{33} + 2d_{31}$). It also has the advantage of a lower PbO vapour pressure during sintering so that guarding against lead loss is less of a problem than with PZT.

The substitution of lead by barium enhances the piezoelectric properties which peak when Pb/Ba ≈ 1. At this and higher barium contents the structure changes to tetragonal with the polar axis parallel to the A-site tunnels. There is a morphotropic boundary, similar to that found in PZT compositions, and peak values of piezoelectric properties are found near the $Pb_{1/2}Ba_{1/2}Nb_2O_6$ composition. d_{33} rises to $220\,\mathrm{pC\,N^{-1}}$ and d_{31} rises to $-90\,\mathrm{pC\,N^{-1}}$, while Q_m increases to 300 and the Curie point falls to 250 °C; thus most of the features peculiar to

$PbNb_2O_6$ are lost. The properties are intermediate between those of $BaTiO_3$ and PZT but there has been little attempt to exploit the material commercially.

6.4.5 Lithium niobate and lithium tantalate

Lithium niobate and lithium tantalate have similar properties and their crystal structure is similar to that of ilmenite. It consists of corner-sharing MO_6 groups (M≡Nb or Ta) that share one face with an LiO_6 group and an opposite face with an empty O_6 octahedron. Figure 6.17 shows diagrammatically the positions of Li^+ and M^{5+} ions between layers of hexagonally close-packed O^{2-} ions. The M^{5+} and Li^+ ions are displaced from the centres of their octahedra in opposite directions in the ferroelectric phases. When the polarization is reversed both cations move in the same direction, the M^{5+} ion to the other side of the midplane between the O^{2-} ions and the Li^+ ion through to the other side of the adjacent O^{2-} plane. There are only two possible polar directions and they are at 180° to one another. As a consequence only limited piezoelectric activity can be induced in ceramic preparations by poling.

Large crystals of both $LiNbO_3$ and $LiTaO_3$ can be grown by the Czochralski method (see Fig. 3.11). Their principal properties are given in Table 6.5 below.

The high melting point of $LiTaO_3$ necessitates the use of an iridium crucible when growing crystals. This in turn makes it necessary to have a nitrogen atmosphere to prevent the formation of IrO_2 which is volatile at 1650 °C. As a result the crystals are dark in colour and conductive as grown and must be embedded in $LiTaO_3$ powder and annealed at 1400 °C in air to render them colourless and of high resistivity. The $LiTaO_3$ powder prevents the volatilization

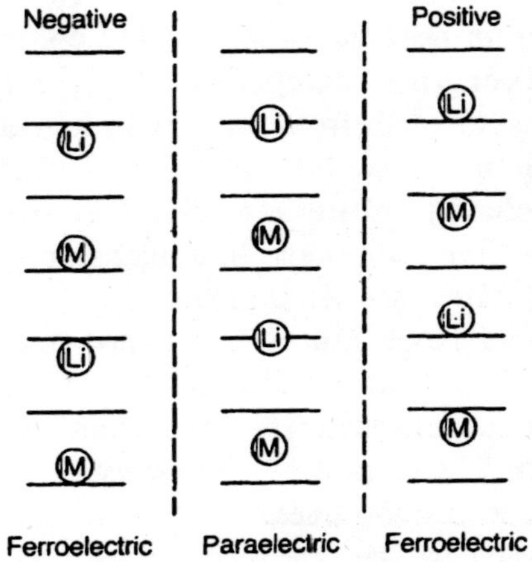

Fig. 6.17 The position of Li^+ and M (Nb^{5+}, Ta^{5+}) ions relative to the O^{2-} planes (horizontal lines) in $LiNbO_3$ and $LiTaO_3$ in the two ferroelectric states and the paraelectric state.

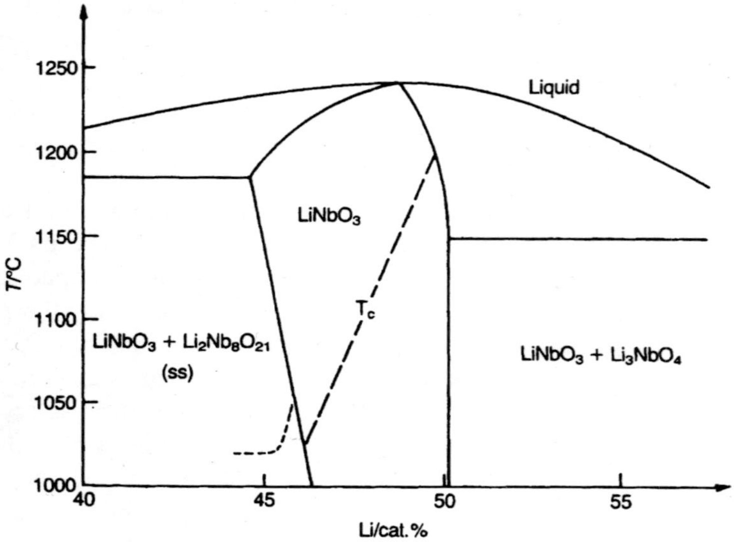

Fig. 6.18 Phase diagram for $LiNbO_3$ as determined for crystal growth.

of Li_2O from the crystals. They are poled by applying pulsed fields of about $1\,kV\,m^{-1}$ whilst cooling through the Curie point.

$LiNbO_3$ is grown in a similar manner but using a pure platinum crucible in air. Rhodium, with which platinum is usually alloyed to improve its high-temperature strength, dissolves to some extent in $LiNbO_3$. Poling can be carried out by passing a current of a few milliamperes through the crystal during growth or by applying a field of about $0.3\,kV\,m^{-1}$ at 1190 °C for 10 min.

The congruent melt composition for both $LiTaO_3$ and $LiNbO_3$ is low in lithium. The phase diagram near the stoichiometric composition for $LiNbO_3$ is shown in Fig. 6.18. The congruent composition $Li_{0.95}TaO_{2.975}$ is satisfactory for piezoelectric applications. Growth from melts with higher lithium contents has the problem that the lithium concentration in the melt increases during growth. However, as the solid solubility of Li_2O in crystals is low, highly homogeneous crystals with Li/Ta and Li/Nb ratios close to unity can be grown from melts with high Li_2O contents. 1% MgO is often added to $LiNbO_3$ melts as it assists the growth of crack-free crystals and improves their non-linear optical behaviour.

Suitably oriented slices cut from $LiTaO_3$ crystals have low-temperature coefficients of shear mode resonant frequency and have been made into high-stability wave filters. The surface of single-crystal $LiNbO_3$, which is readily prepared with a high degree of smoothness, can be used as the substrate for surface acoustic wave devices such as the intermediate frequency stages in television receivers. The high Curie point of $LiNbO_3$ makes it a valuable high-temperature vibration detector, for instance in the heat exchangers of nuclear reactors.

$LiNbO_3$ possesses a useful combination of piezoelectric and electro-optical properties that enable the development of devices using visible or near-infrared radiation for the logical processing of information and the routing of channels in telecommunications (see Section 8.3.5).

6.4.6 Piezoceramic–polymer composites

General background

The following discussion should be supplemented by reference to the article by T.R. Gururaja *et al.* [4], one of the contributing authors being R.E. Newnham, a pioneer in the field of piezoelectric composites.

It is instructive to compare the basic properties of the piezoelectric polymer, polyvinylidene fluoride (PVDF) with those of 'PZT'. The flexibility and low density of the polymer contrasts with the stiffness, brittleness and high density of 'PZT'. On the other hand the piezoelectric '*d*' coefficient for PVDF is relatively small ($\sim -30\,\mathrm{pC\,N^{-1}}$; the mechanisms by which the polarisation in PVDF changes differ from those in PZT accounting for the different algebraic sign). However, because the relative permittivity of PVDF is low (~10) compared to that of 'PZT' the voltage coefficients are similar in value.

Composite technology in general sets out to combine materials in such a way that the properties of the composite are the optimum for a particular application. The property, whether mechanical, thermal, electrical etc., is determined by the choice of component and their relative amounts and, most importantly, the 'connectivity', that is the manner in which the components are interconnected.

Piezoceramic–polymer composites are a relatively recent addition to the range of composite materials and have been developed principally because their properties offer advantages, especially for sonar and medical ultrasonic imaging technologies, over those of the piezoceramics alone. For these applications the transducer is usually interfacing with water or soft tissue, for example body skin. The advantages include relatively good acoustic matching between transducer and medium, improved electromechanical coupling coefficient and higher bandwidth, the last allowing the engineering of well-defined ultrasonic pulses and in consequence good temporal and spatial resolution.

Sonar and imaging transducers transmit into and receive ultrasound energy from the surrounding medium and so are generating and experiencing hydrostatic pressure changes. The corresponding coefficients of significance are d_h and g_h, the former being a measure of the effectiveness of the transducer as a transmitter and the latter as a receiver. Because the transducer is required to be effective in both modes it is customary to introduce the product $d_h g_h$ as a 'figure of merit' useful for comparing materials for this application. There are other critical considerations including the sharpness of the rise-time of the emitted pulse and the extent of the 'ringing' following the pulse. Unacceptably long 'ringing' would interfere with receiving a well-defined echo. The low mechanical Q of a polymer plays an important role in achieving good pulse characteristics.

It follows from a consideration of Fig. 6.2, assuming cylindrical symmetry about the poling direction, that

$$d_h = d_{33} + 2d_{31} \tag{6.67}$$

and, from Eq. (6.33)

$$g_h = d_h/\varepsilon \tag{6.68}$$

It is apparent from Eq. (6.67) and the data in Table 6.5 that because of the strong 'cross-coupling' (this term describes the coupling of the piezoelectric effects in the '3' and '1' (or '2') directions) and consequently the magnitude and sign of the d_{31} coefficients relative to those of the d_{33}, the hydrostatic coefficients for the common 'PZT' family of piezoceramics are small. It is also clear that because of the large values of permittivity the g_h coefficients also tend to be small. However, through appropriate composite design it is possible to very significantly increase the figure of merit, $d_h g_h$.

The other important consideration concerns the transmission of ultrasound (and other forms of energy) from one medium to another and the importance of 'impedance matching'. When wave energy is transferred from one medium to another then a part is transmitted and the rest reflected. The ratio of reflected to transmitted energies depends on the characteristic impedances of the two media and the transmission is total if these are matched. In the case of acoustic waves the specific impedance (Z) of a medium is given by the product of the density ρ and the velocity of sound v*. that is

$$Z = \rho v \tag{6.69}$$

Z is measured in 'rayl' (after Lord Rayleigh) and for water it has the value approximately 1.5 Mrayl ($\rho = 1000\,\mathrm{kg\,m^{-3}}$and $v \sim 1500\,\mathrm{m^2\,s^{-1}}$); in the case of a typical 'PZT', Z is approximately 35 Mrayl and so the two are poorly matched. In contrast, for polymers Z is typically approximately 3.5 Mrayl and so the acoustic match with water is far better. A further disadvantage of PZT is the high Q_m which is unsuited to the detection of sharp pulses because of the associated and troublesome 'ringing' effects.

The concept of 'connectivity' is discussed in Section 2.7.4 and in [4]. There are 10 different ways of connecting the phases in a two-phase composite, each described by two numbers, the first defining (in the present context) how the active ceramic phase is connected and the second how the passive polymer phase

* It is customary for transducer engineers to use c for the velocity of sound; in the present text c is reserved for the velocity of light.

is connected. Here attention is confined to the two most commonly encountered connectivities, 0–3 and 1–3.

A 0–3 composite consists of discrete piezoceramic particles dispersed in a polymer; the isolated particles have zero connectivity and the polymer matrix 3-dimensional connectivity. A 1–3 composite consists of piezoceramic rods extending from electrode to electrode and embedded in a polymer, as shown in Fig. 6.19. The rods have one-dimensional connectivity and the polymer again 3-dimensional connectivity. Some important considerations concerning the two types are summarized below.

0–3 composites

Probably the major positive attributes of 0–3 composites are their ease of manufacture and conformability, and in the latter respect they are similar to PVDF. As the volume fraction of piezoceramic particles, poled in the same direction, is increased so too, of course the d coefficient of the composite increases. However the increase in permittivity accompanying the loading of high permittivity particles in a low permittivity matrix leads to a progressive decrease in the g coefficient. Consequently there is an optimum particle loading as far as maximizing $d_h g_h$ is concerned.

Poling a high permittivity particle embedded in a low permittivity matrix presents a technical problem. When an electric field is applied to such a composite the local field is high in the low permittivity component and low in the high permittivity component (the student should recall that the normal component of the electric displacement vector **D** (ε**E**) does not change in value as a boundary between two different dielectrics is crossed). For a 'PZT' particle in a polymer matrix the electric field experienced by the particle – the poling field – might well be between a factor of a 100 and 1000 (depending upon its shape and orientation with respect to the field) lower than that in its surroundings. Ways have been found to overcome this problem [4].

Although 0–3 composites are relatively little used they can be exploited for low technology applications such as pressure-sensitive pads.

1–3 composites

Composites having 1–3 connectivity are successfully exploited in sonar ('sound navigation and ranging'), medical imaging technologies, NDT, etc. Important matters crucial to the satisfactory functioning of a 1–3 composite are briefly summarized below where, to illustrate the ideas, it is assumed the composite is operating as a sonar device, that is under 'hydrostatic' conditions. The subject matter is discussed in detail by W.A. Smith [16].

1. The composite is one part of a system and has to interface with and properly match the 'drive' and 'sensing' circuitry. This requires the electrical capacitance of the composite component to be tailored.

2. Because the piezoceramic rods are continuous from electrode to electrode (Fig. 6.19(a)) poling presents no problem.

3. Because of the very large differences in mechanical compliance between the two phases, the hydrostatic pressures experienced by the polymer phase transfer forces to the rods magnifying the stress in them in the poled direction. This magnified stress increases the charge induced on to the electrode tending to compensate for the inactive polymer. As a result, in typical practical cases, the piezoelectric charge coefficient (d_{33}) is not very sensitive to volume fraction of 'PZT'.

   Also, because the polymer bears a fraction of the forces due to the lateral pressure, the 'negative' d_{31}-effect is reduced compared to that which would be experienced by the ceramic alone and subjected to the same hydrostatic pressure. To enhance this advantage ways are found to engineer an increase in stiffness of the composite in the '1' and '2' directions.

   A further significant related benefit derives from the fact that the rods are long and thin. As discussed in Section 6.2, a thin rod is not as laterally self-clamped as a cylinder of smaller length-to-diameter ratio (in the extreme, a disc).

4. In contrast to d_{33}, the permittivity of the composite is strongly dependent upon volume fraction of 'PZT' (see Eq. 2.127) and therefore so too is the voltage coefficient (g_{33}).

5. It follows that the 'figure of merit' $d_h g_h$ can be significantly increased in comparison to that for the PZT alone.

6. The size-scale of the structure is critical for two reasons:

   (a) it influences the effectiveness of the stress-transfer referred to in 3 above, and

   (b) it determines the way in which the vibrational energy is distributed within the transducer and how this depends upon frequency. (The reader should have in mind that the attenuation (and hence range) of a sound wave propagating in water depends upon the frequency; also the resolution with regard to detecting a reflecting object depends upon wavelength, and so too on frequency.)

With regard to 6(a), it is self evident that at a large rod-spacing much of the energy received by the composite is wasted deforming the inactive polymer. In the case of 6(b) the matter is more complex. When the composite is active in the 'receive' mode, that is when vibrational energy is incident on the composite, or when it is energized in the 'transmit' mode, ideally the energy is located entirely in the thickness vibration, the composite disc vibrating in a piston-like manner. But as the rods vibrate longitudinally they excite the propagation of unwanted transverse (shear) waves in the polymer and it is necessary to consider how these waves interact with a periodic structure. (The student might recognize that the essential underlying physics is the same as that describing X-ray diffraction, and also the propagation of electron waves in a crystal.) As expected, there are frequencies for which the waves cannot propagate in the plane of the composite because of interference effects. The periodicity of the lattice of rods introduces frequency 'stop-bands', that is frequency ranges over which the shear waves in the polymer cannot be sustained. If the rod spacing is d then the first stop-band occurs at a wave number π/d the second at $2\pi/d$ and so on. 'Wave number' is simply the reciprocal of the wavelength and so the first stop band occurs for a wavelength d/π and the second at half this wavelength, and so on. The dimensions of the transducer, and the periodicity of the array of rods, are chosen so that the shear waves are suppressed.

Preventing the excitation of the transverse waves not only reduces the wastage of acoustic energy but, more importantly, prevents the development of spurious signals.

It follows therefore that the higher the operating frequency of the composite plate, the finer must be the size scale of the rod array.

7. Sonar and imaging technologies are based on sending out ultrasound pulses and receiving the echoes. The resolution capabilities depend upon the emitted pulses being sharp, and the sharper the pulse the more Fourier components it has. Therefore, for a transducer to emit and detect a sharp pulse it must be able to respond to a wide frequency range – that is it must have the 'broad band' characteristic associated with low mechanical Q. The very 'lossy' nature of the polymer phase endows the composite with this ($Q<10$).

Fabrication of piezoceramic–polymer composites

The fabrication of a 0–3 composite is straightforward involving essentially the dispersal of piezoceramic particles in a polymer and forming the mix into a shape by casting, moulding, extrusion, tape-casting, calendering, etc. However, the variables are many and include composition of the ceramic particle, its shape and volume fraction; in the case of non-equiaxed particles then texture (the orientation of the particles with respect to the poling direction) is very significant and so too is the incorporation of porosity to tailor compliance in a controlled way. These matters are discussed in reference [4].

In the case of 1–3 composites the variety of approaches to fabrication depend very much on the imagination and ingenuity of the fabrication technologist. For

(a)

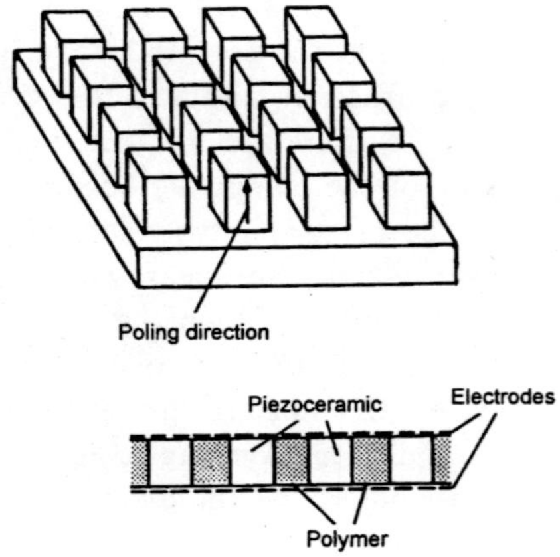

(b)

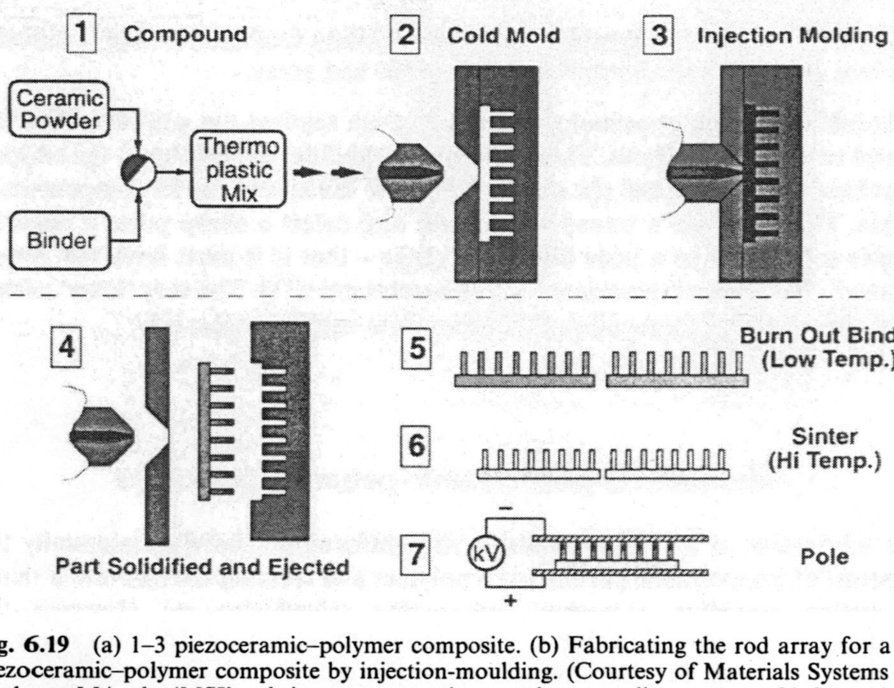

Fig. 6.19 (a) 1–3 piezoceramic–polymer composite. (b) Fabricating the rod array for a 1–3 piezoceramic–polymer composite by injection-moulding. (Courtesy of Materials Systems Inc. Littleton MA; the 'MSI' website: www.matsyinc.com is an excellent source of information.)

Table 6.4 Comparison of the properties of piezoelectric materials

Composite/piezoceramic	*Connectivity*	*Relative permittivity*	$g_h/mV\,m\,N^{-1}$	$d_h/pC\,N^{-1}$	$d_h\,g_h/10^{-15}\,Pa^{-1}$
'PZT' (Navy type VI)	–	3200	1.5	45	68
$PbTiO_3$ ceramic *	–	230	23	47	1080
$PbNb_2O_6$ ceramic *	–	225	33	67	2200
15 vol.%PZT rods–polyurethane	1–3	460	66	268	17 700
$PbTiO_3$–rubber	0–3	45	100	40	4000
PVDF	–	10	150	14	2100

* Modified compositions.

a comprehensive discussion the reader is referred to the paper by V.E. Janas and A. Safari [17].

An obvious fabrication approach is the 'dice-and-fill'. A piezoceramic plate is diced into an array of square-section pillars standing on the undiced base of the plate. The structure is then in-filled with the chosen polymer after which the base is ground away to produce the structure shown in Fig. 6.19(a). Typically rods of thickness down to approximately 100 μm can be diamond-machined in this way.

In another approach extruded and sintered rods in reasonable lengths (~100 mm) are arranged in a jig, in-filled with the polymer the composite then being sliced into individual transducer plates. Again the lower limit on the rod diameter is approximately 100 μm.

Fine-scale arrays (<50 μm) have been produced by first making a metal master by hard X-ray lithography. This constitutes the mould from which a polymer 'negative' array is made. The piezoceramic powder slurry or paste is formed in the 'sacrificial' polymer mould which is subsequently burnt away in a presintering heating stage.

The most successful approach and the one suited to economic mass production is by injection-moulding (see Section 3.6.8), as illustrated in Fig. 6.19(b), and composites having rod diameters down to 100 μm are produced by this route.

The high values of the '$d_h\,g_h$' figure of merit offered by piezoceramic–polymer composites can be appreciated from Table 6.4.

6.4.7 Summary of properties

The useful properties of some of the important single-phase piezoceramics are summarized in Table 6.5. The values for $BaTiO_3$ and the PZT range are taken from manufacturers' data sheets where precise compositions are, of course, not revealed; other values are from the literature.

Table 6.5 Typical values of the properties of some piezoelectric materials

Property	*Unit*	*α-Quartz*[a]	*$BaTiO_3$*	*PZT A*[b]	*PZT B*[b]	*$PbNb_2O_6$*	*$Na_{1/2}K_{1/2}NbO_3$*	*$LiNbO_3$*[a]	*$LiTaO_3$*[a]	*$PbTiO_3$*[c]
Density	$Mg\,m^{-3}$	2.65	5.7	7.9	7.7	5.9	4.5	4.64	7.46	7.12
T_c	°C		130	315	220	560	420[d]	1210	665	494
ε^X_{r33}		4.6	1900	1200	2800	225	400	29	43	203
ε^X_{r11}			1600	1130	–	–	600	85	53	–
$\tan\delta$	10^{-3}		7	3	16	10	10	–	–	22
k_p			0.38	0.56	0.66	0.07	0.45	0.035	0.1	–
k_{31}			0.21	0.33	0.39	0.045	0.27	0.02	0.07	0.052
k_{33}			0.49	0.68	0.72	0.38	0.53	0.17	0.14	0.35
k_{15}			0.44	0.66	0.65	–	–	0.61	–	0.36
k_{jk}		(11)[e] 0.1 (14)[e] 0.05								
d_{31}			-79	−119	−234	−11	−50	−0.85	−3.0	−7.4
d_{33}	$pC\,N^{-1}$		190	268	480	80	160	6	5.7	47
d_{15}			270	335	–	–	–	69	26	–
d_{jk}		(11)[e] 2.3 (14)[e] 0.67								
Q_m		$>10^6$	500	1000	50	11	240	–	–	326
s^E_{11}		12.8	8.6	12.2	14.5	29	9.6	5.8	4.9	11
s^E_{12}	$\mu m^2\,N^{-1}$	−1.8	−2.6	−4.1	−5.0	–	–	−1.2	−0.52	–
s^E_{13}		−1.2	−2.9	−5.8	−6.7	−5 to −8	–	−1.42	−1.28	–
s^E_{33}		9.6	9.1	14.6	17.8	25	10	5.0	4.3	11
s^E_{44}		20.0	23	32	–	–	–	17.1	10.5	–

[a]Single crystals.
[b]PZT A and PZT B are two typical PZT materials illustrating, in particular, the wide range of achievable Q_m values.
[c]+5 mol.% $Bi_{2/3}Zn_{1/3}Nb_{2/3}O_3$.
[d]Depoles above 180°C.
[e]Numbers in parentheses are *jk* values.

Fig. 6.20 (a) Piezoceramic parts for a range of applications including accelerometers, underwater acoustics, pressure and liquid level sensors, medical diagnostics and therapeutics and NDT; the ultrasound focusing bowls are for medical imaging and for producing high-intensity focused ultrasound.

High-quality, pore-free microstructures of PZ – PT piezoceramic, (b) and (c), are essential for reliable, high-performance applications, e.g. composites and arrays where very small elements are cut from larger pieces (e.g. see Fig. 6.36); (d) the layer–structured bismuth titanate ferroelectric ($Bi_4Ti_3O_{12}$); $T_c \sim 650°C$; the crystal structure consists of 'perovskite layers' separated by bismuth oxide layers) is exploited in high-temperature applications, including accelerometers and flow-meters (reproduced with permission of Ferroperm Piezoceramics A/S, Denmark).

6.5 Applications

Devices exploiting piezoceramics have been developed for the generation of high voltages, electromechanical actuators and sensors, frequency-control and the generation and detection of acoustic and ultrasonic energy. In the following sections some typical examples are chosen for detailed descriptions, not only because of their technical importance but also because they provide the opportunity to illustrate important ideas. Reference to the monographs by J.M. Herbert [2] and K. Uchino [18] is recommended to supplement the discussions.

A selection of piezoceramic pieces and components is shown in Fig. 6.20.

6.5.1 Generation of voltages

Gas igniter

The principle of a gas igniter is illustrated in Fig. 6.21(a). It is usual to use two poled cylinders back to back, rather than one, so as to have twice the charge available for the spark. It is necessary to apply the force F quickly otherwise the voltage generated disappears as charge leaks away through the piezoceramic, across its surfaces and via the apparatus.

The following argument is developed for a single element of cross-sectional area A and length L, as shown in Fig. 6.21(b). The material parameters are d_{33}, g_{33}, ε_{33}^{X}, ε_{33}^{x} and k_{33}. The generation of a spark can be seen as consisting of two stages. Firstly the compressive force F on an area A will alter the length L of the cylinder by $-\delta L$. This happens under open-circuit conditions and so

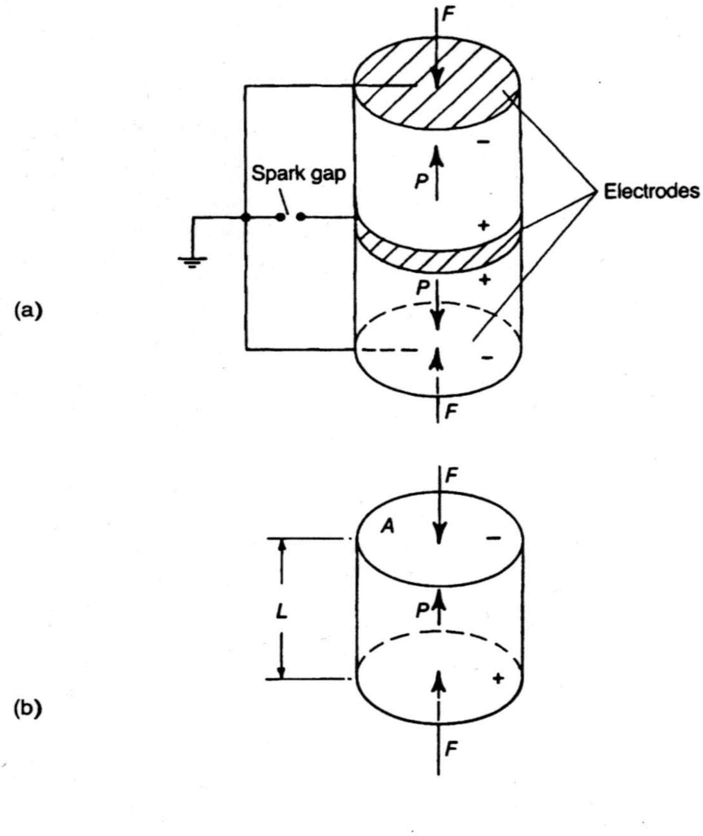

Fig. 6.21 (a) A piezoelectric spark generator. (b) A piezoceramic cylinder under axial compressive force.

$$x = -\frac{\delta L^D}{L} = -s_{33}^D \frac{F}{A} \tag{6.70}$$

The mechanical work done is

$$w_m = \tfrac{1}{2} F \,\delta L^D = \tfrac{1}{2} s_{33}^D \frac{F^2 L}{A} \tag{6.71}$$

The available electrical energy w_{el} is, from the definition of k^2,

$$w_{el} = \tfrac{1}{2} k_{33}^2 s_{33}^D \frac{F^2 L}{A} \tag{6.72}$$

At this stage the cylinder has an electric potential difference U between its ends and is in a clamped state so that the electric potential energy is (C is the capacitance)

$$\begin{aligned} w_{el} &= \tfrac{1}{2} C U^2 \\ &= \tfrac{1}{2} U^2 \varepsilon_{33}^x \frac{A}{L} \end{aligned} \tag{6.73}$$

Since $k_{33}^2 = g_{33}^2 \varepsilon^X / s_{33}^E$, $\varepsilon_{33}^x \approx \varepsilon_{33}^X (1 - k_{33}^2)$ and $s_{33}^D = s_{33}^E\,(1 - k_{33}^2)$, Eqs (6.72) and (6.73) yield the expected result:

$$U = g_{33} \frac{FL}{A} \tag{6.74}$$

For a spark to occur U must exceed the breakdown voltage of the gap. It is assumed that the voltage given by Eq. (6.74) is established when breakdown occurs. At this stage the field in the ceramic is in the same direction as the poling field and assists in maintaining the domain orientations in the poled sense.

When the spark gap breaks down the second stage in energy generation is initiated. The spark discharge results in a change from open- to closed-circuit conditions with the voltage dropping to a lower level. The compliance changes from s_{33}^D to s_{33}^E and the drop in field strength now allows 90°, 71° and 109° domain wall movement. If only the first of these effects is considered, the increase in compliance will allow a further compression of the ceramic and a movement $\delta L^E - \delta L^D$ of the applied force. δL^E is the movement that would have occurred had the force been applied under short-circuit conditions in the first place. Therefore

$$\delta L^E - \delta L^D = (s_{33}^E - s_{33}^D) \frac{FL}{A} = k_{33}^2\, s_{33}^E \frac{FL}{A} \tag{6.75}$$

The additional strain will result in an increase w_{ex} in the electrical energy available:

$$w_{ex} = \tfrac{1}{2} k_{33}^4\, s_{33}^E \frac{F^2 L}{A} \tag{6.76}$$

Therefore the total energy that can be dissipated in the spark is

$$w_{\mathrm{T}} = w_{\mathrm{el}} + w_{\mathrm{ex}} = \tfrac{1}{2}k_{33}^2\,(k_{33}^2\,s_{33}^E + s_{33}^D)\frac{F^2L}{A} \tag{6.77}$$

or

$$w_{\mathrm{T}} = \tfrac{1}{2}k_{33}^2\,s_{33}^E\,\frac{F^2L}{A} = \tfrac{1}{2}d_{33}\,g_{33}\,\frac{F^2L}{A} \tag{6.78}$$

In practice, with most PZT compositions, Eq. (6.78) may underestimate the energy released because domain reorientation associated with ferroelasticity will contribute significantly to the charge developed (cf. Fig. 6.13). However, the magnitude and duration of the applied force must be such that the changes in polarization due to ferroelasticity are reversible if the output of the igniter is not to deteriorate with usage.

Consider the following simple example. A force of 1000 N acts on a cylinder 6 mm in diameter and 15 mm long with $d_{33} = 265\,\mathrm{pC\,N^{-1}}$, $\varepsilon_{33}^X = 1500\varepsilon_0$ and $k_{33} = 0.7$. In this case $U = 10.4\,\mathrm{kV}$, $w_{\mathrm{el}} = 0.72\,\mathrm{mJ}$ and $w_{\mathrm{ex}} = 0.66\,\mathrm{mJ}$. The total energy from a single cylinder is 1.38 mJ so that, if a pair is used, the energy available for ignition should be ample since, under average conditions, about 1 mJ is sufficient. The stress applied is 35 MPa, which will result in a significant enhancement in output from reversible domain reorientation.

Transformer

The transfer of energy from one set of electrodes to another on a piezoceramic body can be used for voltage transformation. In Fig. 6.22(a) a flat plate carries electrodes on half its major surfaces and on an edge. The regions between the larger electrodes, and between them and the edge electrode, are poled separately. A low a.c. voltage is applied to the larger-area electrodes at a frequency which excites a length mode resonance. A high-voltage output can then be taken from the small electrode and the common of the larger ones. The transformation can, very approximately, be in the ratio of the input to output capacitances but depends strongly on the output load, as does the efficiency. The ceramic needs to combine high k_{31} and k_{33} coupling coefficients with a high Q_{m}, preferably exceeding 1500. Very significant gains in voltage and amplification can be made by adopting a multilayered structure for the primary, as illustrated in Fig. 6.22(b), and for both primary and secondary in (c) when the efficiency depends upon k_{33}^2.

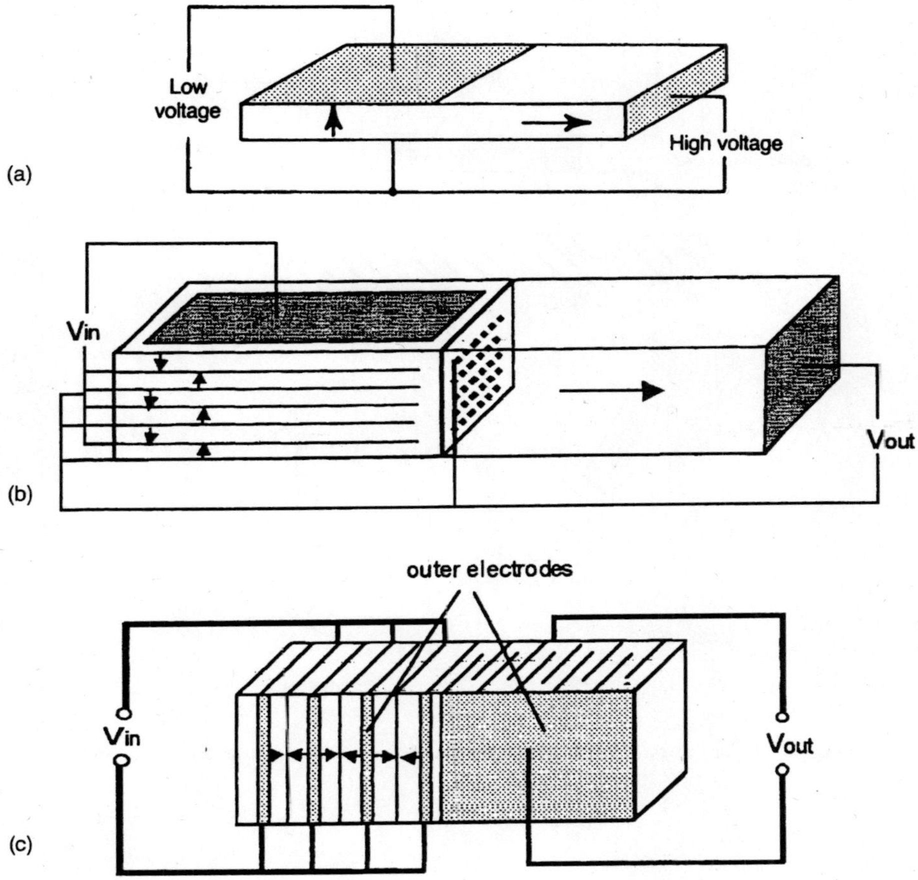

Fig. 6.22 (a) Simple form of piezoelectric transformer; (b) transformer incorporating multilayered primary; (c) transformer with multilayered primary and secondary.

In addition to suitability to cheap mass production, piezoelectric transformers offer the advantage of a low profile when mounted on printed circuit boards. Over recent years they have found a commercially significant application for powering fluorescent lamps for back-lighting the screen in lap-top computers and electronic notebooks with liquid crystal displays. For these applications the output voltage and power are typically approximately 1 kV (rms) and 5 W, respectively.

The literature contains [19] impressive performance data for multilayer transformers, made more impressive by the low sintering temperature (960 °C) of the 'PZT' fluxed with small amounts (3 wt.%) of mixes of B_2O_3, Bi_2O_3 and CdO. Open circuit voltage step-up ratios as high as 9000 are claimed.

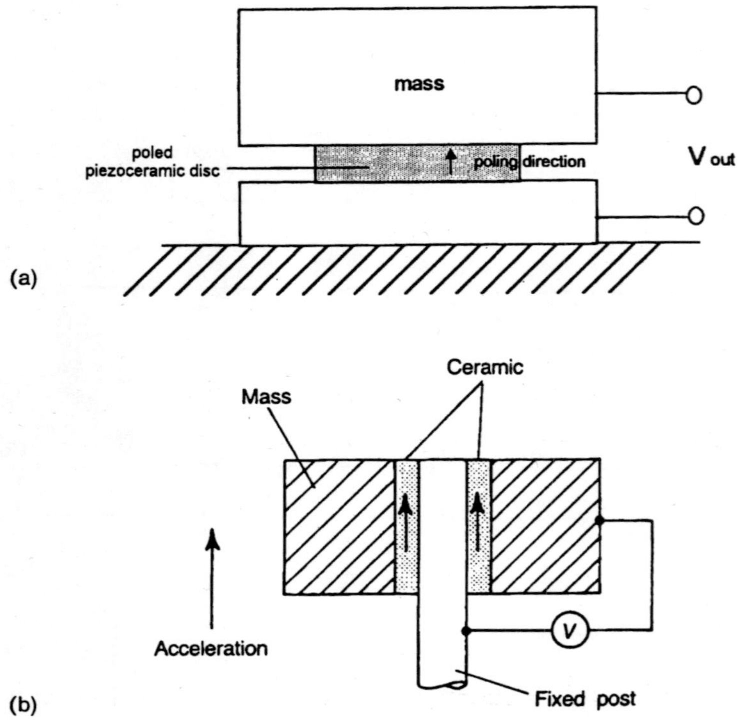

Fig. 6.23 (a) Simple form of accelerometer; (b) shear-mode accelerometer.

A simple form of accelerometer is shown in Fig. 6.23(a). During acceleration in the direction of poling the mass exerts a force, F ($= M \times \text{acc}^{\text{n}}$) on the piezoceramic disc and therefore a stress of F/A where A is the area of the disc. The measured voltage output is proportional to the acceleration.

Accelerometers have been designed which use ceramics in the shear mode. The transducer is a cylinder poled along its axis but with the poling electrodes removed and sensing electrodes applied to the inner and outer major surfaces. The cylinder is cemented to a central post and has a cylindrical mass cemented to its outer surface (Fig. 6.23(b)). The cylinder is subjected to a shearing action between the mass and the supporting post when accelerated axially. Motion in perpendicular directions has very little effect, since first the average stress in the ceramic will be small, and second the d_{11} coefficient is very close to zero. The device is therefore very highly directional.

Cantilever bimorphs (see Section 6.5.2 below) with small masses attached to their free end can be used as accelerometers provided that the frequency of the vibrations to be detected is well below the resonance frequency of the transducer.

There are many applications for accelerometers, an important one being to control the stability of a car during emergency braking and steering. Another is to sense the onset of a car crash when appropriate measures need to be activated, for example the release of air-bags.

6.5.2 Generation of displacement – 'actuators'

Overview

In advanced precision engineering there is a need for a variety of types of actuator. For example, in the fabrication of semiconductor chips, circuit components have to be precisely positioned for the various processing steps. In optical equipment lenses and mirrors require micropositioning, and even the shapes of mirrors are adjusted to correct image distortions arising from, for example, atmospheric effects. In autofocusing cameras there are actuators capable of producing precise rotational displacements. Actuators are also required for ink-jet printers, for positioning videotape-recording heads and for micromachining metals. The certainty that the range of applications and demand for actuators will grow has stimulated intensive research into piezoelectric and electrostrictive varieties.

When compared with piezoelectric/electrostrictive transducers, the electromagnetically driven variety suffer shortcomings in 'backlash', in the relatively low forces that can be generated and in their relatively low response speeds. Piezoelectric/electrostrictive actuators are capable of producing displacements of $10 \pm 0.01\,\mu m$ in a time as short as $10\,\mu s$, even when the transducer is subjected to high (~ 100 kgf) opposing forces.

There is an increasing interest in materials showing a strong electrostrictive effect for actuators. In the case of electrostriction the sign of the strain is independent of the sense of the electric field, and the effect is exhibited by all materials. In contrast, with the piezoelectric effect the strain is proportional to the applied field and so changes sign when the field is reversed. Generally speaking, the piezoelectric effect is significantly greater than the electrostrictive effect. However, in the case of high-permittivity materials, and especially ferroelectrics just above their Curie points, the electrostrictive effect is large enough to be exploited.

Electrostrictive materials offer important advantages over piezoelectric ceramics in actuator applications. They do not contain domains (of the usual ferroelectric type), and so return to their original dimensions immediately a field is reduced to zero, and they do not age. Figure 6.24(a) shows the strain–electric field characteristic for a PLZT (7/62/38) piezoelectric and Fig. 6.24(b) the absence of significant hysteresis in a PMN ($0.9\,Pb(Mg_{1/3}Nb_{2/3}O_3$–$0.1\,PbTiO_3)$) electrostrictive ceramic.

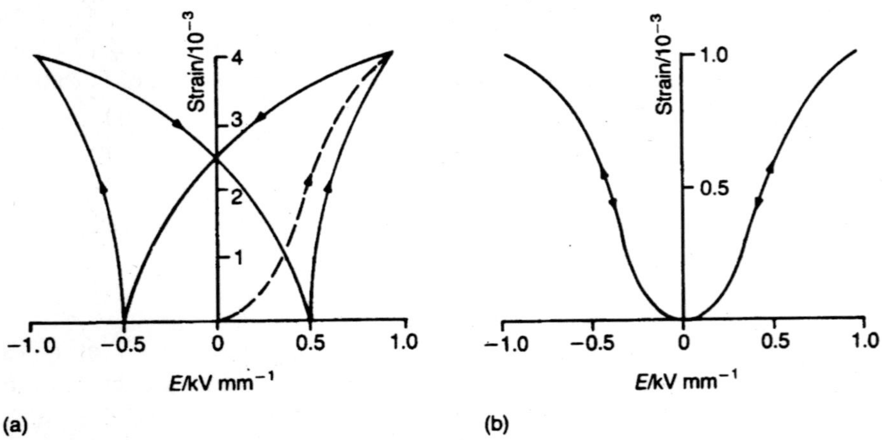

Fig. 6.24 Dependence of strain on electric field for (a) a readily poled and depoled piezoelectric PLZT and (b) electrostrictive PMN.

Recently single crystal relaxors have become available offering very high electrostrictive strains (see Table 6.3). There are many forms of actuator exploiting the different types of piezoelectric/electrostrictive response and the following choice is made to illustrate principles.

Stack actuator and 'Moonie'

Multilayer stack actuators can be made in much the same way as described for a multilayer capacitor (see Section 5.4.3). However, because of the design of a MLCC each layer is clamped at its edges resulting in the development of electric field-induced mechanical stresses which can be a cause of eventual failure. In the case of the actuator the objective is to achieve maximum strain and if the same construction were applied as in the case of the MLCC dangerously large stresses would be developed which not only would detract from the desired movement but would constitute a serious failure risk. The problem is avoided by extending the electrodes to the edge of each layer and devising a means of connecting up alternate layers without the risk of shorting to adjacent layers. This can be achieved in a variety of ways and one strategy is illustrated in Fig. 6.25.

The attractive features of a piezoelectric/electrostrictive actuator are mentioned above, but a penalty is the small displacements generated. For a d_{33} value of $500\,\mathrm{pC\,N^{-1}}$ (or, more appropriately in the present context, $\mathrm{pm\,V^{-1}}$) a voltage of 1000 V applied across a 1 mm thick poled disc would produce a displacement of only 0.5 μm. A multilayer stack of the same piezoceramic comprising 10 layers each 100 μm thick would produce a displacement of 5 μm for the same applied voltage. Alternatively, the multilayer stack allows lower operating voltages for a given displacement.

Various designs have been developed to magnify the small displacements, at the cost of reduced blocking force. A well known design, the 'Moonie', is illustrated in Fig. 6.26. The brass end-caps not only follow the d_{33}-controlled displacement but also redirect the d_{31} displacement into the 3-direction. This magnified displacement can be as much as 20 μm on a 3 mm thick actuator for an applied voltage of as little as 50 V.

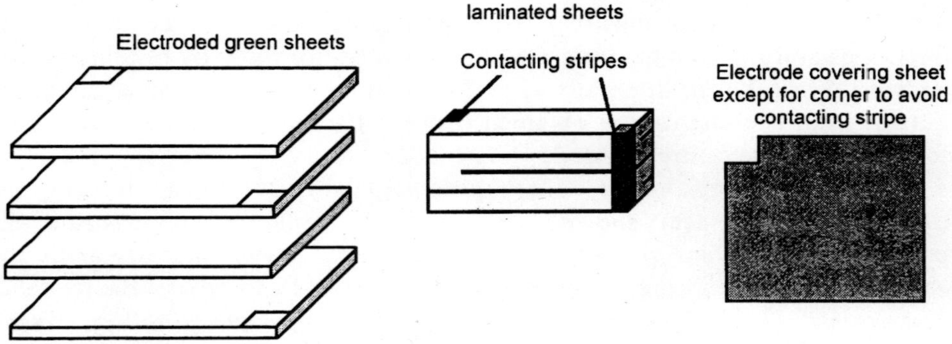

Fig. 6.25 The multilayer stack actuator showing one strategy for electroding to avoid clamping stresses.

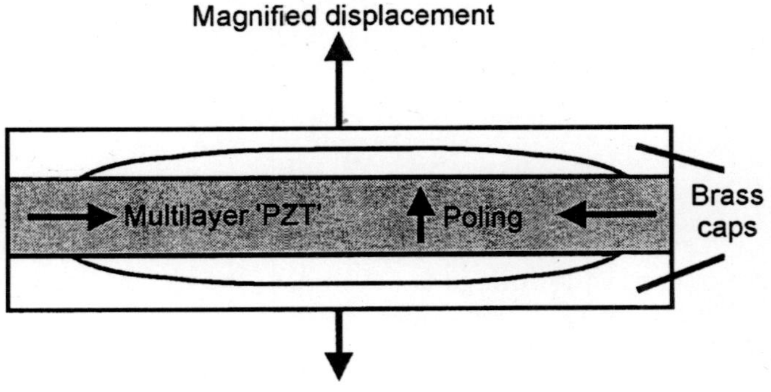

Fig. 6.26 The 'Moonie' actuator.

Bi-morph

Piezoceramics have high Young moduli so that large forces are required to generate strains that produce easily measured electrical responses from solid blocks of material. Compliance can be greatly increased by making long thin strips or plates of material and mounting them as cantilevers or diaphragms, the latter the basis of the widely exploited 'buzzer'.

Bending a plate causes one half to stretch and the other to compress so that, to a first approximation, there can be no electrical output from a homogeneous body by bending. This is overcome in the bimorph by making the two halves of separate beams with an intervening electrode, as well as electrodes on the outer surfaces, as shown in Fig. 6.27. If the beams are poled from the centre outwards (or from the surfaces inwards) the resulting polarities will be in opposite directions and the voltages generated on the outer electrodes by bending will be additive (Fig. 6.27(a)). Alternatively, if the beams are poled in the same direction, output from bending can be obtained between the outer electrodes, connected together, and the centre electrode (Fig. 6.27(b)).

In order to calculate the voltage generated when the device is bent, the cantilever arrangement shown in Fig. 6.28(a) is chosen to illustrate the principles. The force F produces a bending moment varying linearly along the length of the beam, being a maximum at the fixed end and zero at the free end. The central plane OO′ is the neutral plane and, as is evident from Fig. 6.28(b), the strain in a filament a distance y from it is y/R, and is tensile in the upper half and compressive in the lower half.

The radius of curvature R at a distance l from the fixed end is related to the end deflection δz by

$$R = \frac{L^3}{3(L-l)\delta z} \tag{6.79}$$

Therefore the average strain x_l at l in the upper half of the beam is

$$x_l \frac{H}{4R} = \frac{H}{4}\frac{3(L-l)}{L^3}\delta z \tag{6.80}$$

Since the open-circuit voltage is developed, the electromechanical effects take place at constant D. Therefore, from

$$\left(\frac{\partial E_3}{\partial x_1}\right)_D = -h_{31} \tag{6.81}$$

the field $E_3(l)$ developed at l is given by

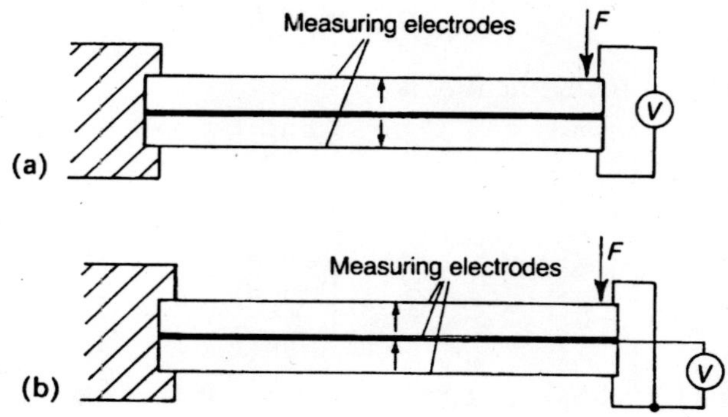

Fig. 6.27 Cantilever bimorphs showing (a) series connection and (b) parallel connection of beams.

$$E_3(l) = h_{31} x_l$$

or

$$E_3(l) = h_{31} \frac{3}{4} \frac{H}{L^3} (L - l) \delta z \tag{6.82}$$

The total charge density $\sigma(l)$ appearing on each surface of the bimorph half is

(a)

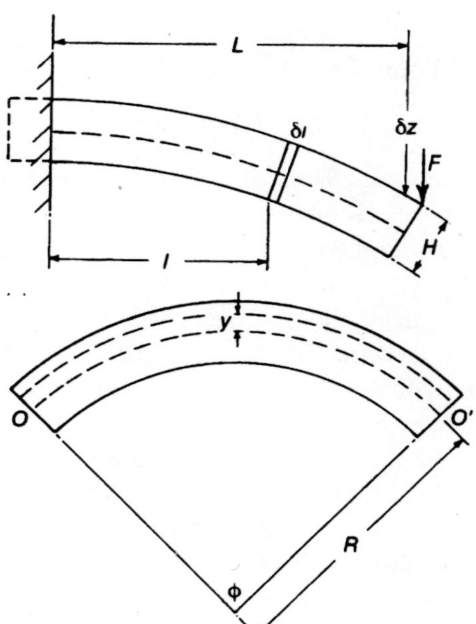

(b)

Fig. 6.28 Strains in a bent rectangular section cantilever of width w and depth H.

$$\sigma(l) = \varepsilon_{33}^x E_3(l) \tag{6.83}$$

The charge $Q_{\delta l}$ on the element δl is

$$Q_{\delta l} = \varepsilon_{33}^x E_3(l) w \delta l \tag{6.84}$$

and the total charge is

$$Q_T = \int_0^L Q_{\delta l} \mathrm{d}l \tag{6.85}$$

From Eqs (6.83), (6.85) and (6.86)

$$Q_T = \frac{3}{4}\frac{Hw}{L^3} \delta z \varepsilon_3^x h_{31} \int_0^L (L-l)\mathrm{d}l \tag{6.86}$$

so that

$$Q_T = \frac{3}{8}\frac{Hw}{L} \varepsilon_3^x h_{31} \delta z \tag{6.87}$$

If C^x is the clamped capacitance of one of the bimorph plates, then the voltage across it is $U = Q_T/C^x$ and that across two oppositely poled plates connected in series is $U_T = 2U$. Therefore

$$U_T = \frac{3}{8}\left(\frac{H}{L}\right)^2 h_{31} \delta z \tag{6.88}$$

If $L = 10\,\mathrm{mm}$, $H = 1\,\mathrm{mm}$, $\delta z = 1\,\mu\mathrm{m}$ and $h_{31} = -6.2 \times 10^8\,\mathrm{V\,m^{-1}}$,

$$U_T = 2.3\ V$$

Although, by the converse effect, a bimorph will bend when a voltage is applied to it, this will not be according to the inverse of Eq. (6.88) because the deformation due to an applied field is governed by $d = (\partial x/\partial E)_X$ which is not the inverse of $h = (\partial E/\partial x)_D$. Furthermore, the applied field will produce uniform strains along the length of the bimorph so that it will be bent in the form of a circular arc, in contrast with the more complex shape given by Eq. (6.79).

Under an applied voltage the stress-free filaments in a bent bimorph will be in two planes, one above and one below the central plane and at equal distances a from it. The centre plane is strain free because the stresses on opposite sides of the joining layer are equal and opposed to one another. The steps in the argument are illustrated in Fig. 6.29 and the stress system is similar to that set up in a heated bimetallic strip, as analysed by S. Timoshenko [20].

If y in Fig. 6.28 is replaced by a, the strain at the neutral planes is a/R which, from Fig. 6.21, is Ed_{31}, i.e.

$$a/R = Ed_{31} \tag{6.89}$$

For a circular arc the depression of the free end of a bimorph relative to the clamped end is

$$\delta z = \frac{L^2}{2R} = \frac{L^2 E}{2a} d_{31} \tag{6.90}$$

a can be determined by taking moments in a cross-section of the bimorph and, provided that the thickness of the material joining the two plates is less than a fifth of the beam thickness, $a \approx H/3$, so that

$$\delta = \frac{3}{2}\left(\frac{L^2}{H}\right) E d_{31} \tag{6.91}$$

In terms of the applied voltage,

$$\delta z = \frac{3}{2}\left(\frac{L}{H}\right)^2 d_{31} U \tag{6.92}$$

With the same dimensions as before, and with $d_{31} = -80\,\text{pC}\,\text{N}^{-1}$, a voltage of 84 V must be applied to produce a deflection of 1 μm.

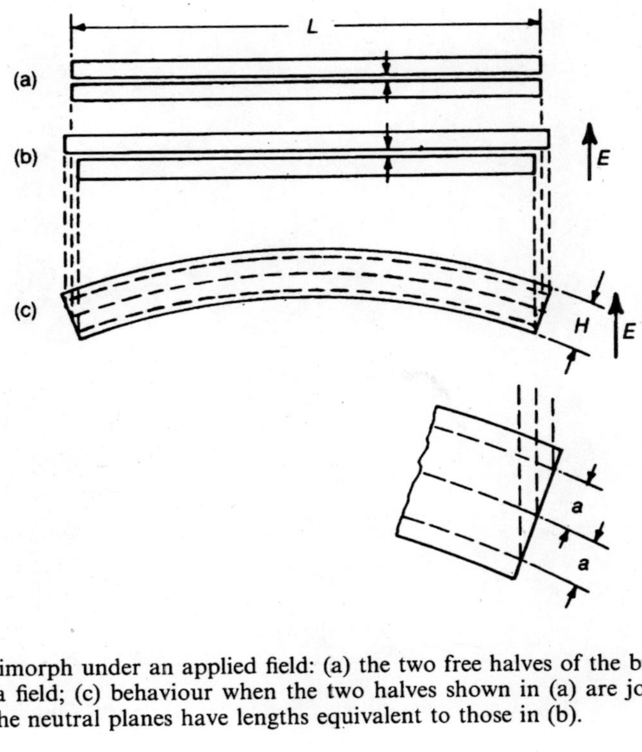

Fig. 6.29 A bimorph under an applied field: (a) the two free halves of the bimorph; (b) the free halves in a field; (c) behaviour when the two halves shown in (a) are joined to form a bimorph and the neutral planes have lengths equivalent to those in (b).

It is clear that useful movements for most purposes can only be obtained for moderate voltages if L/H is made large, but this leads to a fragile component. One solution is to replace one of the ceramic plates by a metal strip. This will at least halve the deflection per volt, but it can also be arranged that the bending always compresses the ceramic and the sense of the applied field will then coincide with the poling direction so that depoling effects are avoided. Higher maximum voltages can be used since the ceramic is stronger in compression than in tension. Since the fields must be high to obtain large movements it is advantageous to use electrostrictive materials, and this also eliminates the hysteretic behaviour to which piezoelectric materials are prone.

Very recently the viscous polymer processing route (see Section 3.6.7) has been exploited by D.H. Pearce and co-workers [21] to fabricate a helical form of the bimorph, the 'Helimorph', illustrated in Fig. 6.30. Both axial and rotational actuations are developed which have potential for a variety of applications in the mechanical, optical and acoustical fields.

Fig. 6.30 'Helimorph'© actuators. (Courtesy of '1Limited' Cambridge, UK.)

Rotary motion

The direct generation of rotary motion can be achieved by the use of surface wave transducers based on the generation of acoustic waves confined to a thin layer near to the surface of a solid. The basic theory of surface acoustic waves (SAWs) was established by Lord Rayleigh in 1885. The waves are due to a combination of longitudinal and shear motions governed by the stress-free boundary conditions at the surface. The particles of the surface layer of the solid describe elliptical motions as indicated in Fig. 6.31(a), with the amplitude decreasing with distance from the surface until, at a depth of approximately 1 μm, it is negligible. The velocity of SAWs is approximately 300 m s^{-1}, a little lower than the velocity of bulk waves, and so they require times of the order of microseconds to travel 1 cm.

Because of the horizontal component of the motions of the surface particles the 'slider' in contact with the surface experiences a force tending to move it in a direction opposite to that of the travelling wave, as indicated in Fig. 6.31(a). Rotary motion is achieved by propagating a SAW round an elastic ring driving a contacting ring slider. Correctly phased vibrations are propagated into the elastic ring from an appropriately poled and energized piezoelectric ring (Fig. 6.31(b)).

The SAW is generated by superimposing two standing waves of equal amplitude but differing in phase by $\pi/2$ with respect to both time and space. The time phase difference results from one wave being generated by the voltage $U_0\sin(\omega t)$ and the other by $U_0\cos(\omega t)$; the spatial phase difference results from the $3\lambda/4$ and $\lambda/4$ gaps between the two poled segments. The two standing waves can be written

$$y_1 = y_0 \cos(n\theta)\cos(\omega t) \tag{6.93}$$

and

$$y_2 = y_0 \cos\left(n\theta - \frac{\pi}{2}\right)\cos\left(\omega - \frac{\pi}{2}\right)$$

in which y_1 and y_2 represent the vertical displacements of the ring surface, θ is the angular displacement round the ring and n is the number of wavelengths accommodated round the ring. The resultant wave is given by

$$\begin{aligned} y = y_1 + y_2 &= y_0\{\cos(n\theta)\cos(\omega t) + \sin(n\theta)\sin(\omega t)\} \\ &= y_0 \cos(\omega t - n\theta) \end{aligned} \tag{6.94}$$

which represents a surface wave travelling with velocity ω/n.

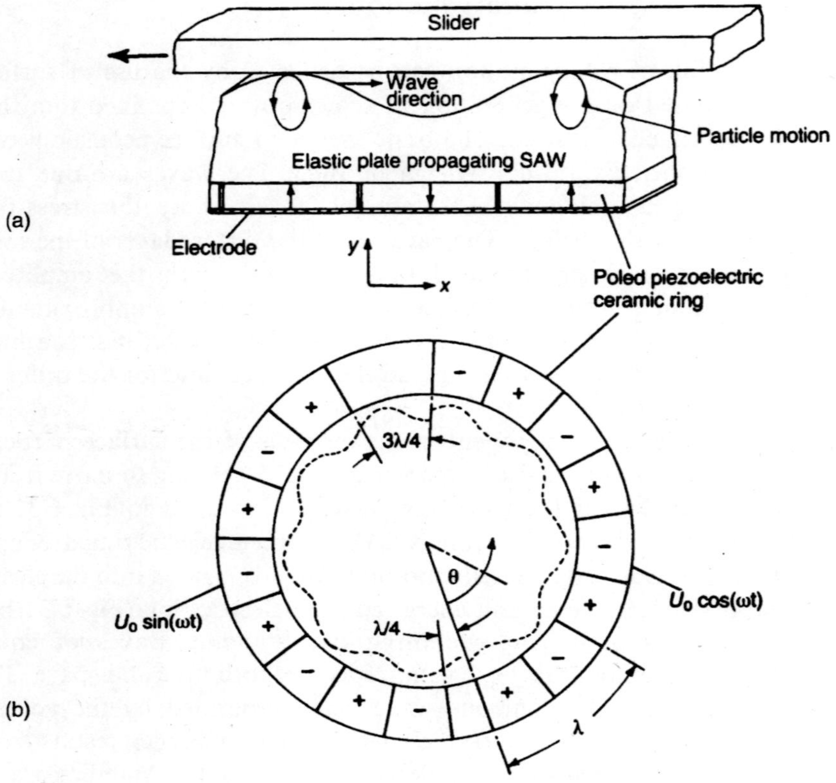

Fig. 6.31 Principle of the rotary actuator: (a) side view; (b) plan view showing poled segments and how temporal and spatial phase differences are established.

Miscellaneous

There are a variety of actuators which are the products of the ingenuity of the engineer, such as the 'Inchworm', which develops precise linear motion and the 'walking' variety. These and others are described by K. Uchino [18].

6.5.3 High frequency applications

These applications can be divided into those which demand high power (e.g. ultrasonic cleaner) intermediate power (e.g. 'Tweeter') and signal power (e.g. Delay line). Some examples are briefly described below.

Ultrasonic cleaner

For high-power ultrasonic applications, such as ultrasonic cleaning, the transducers are operated at frequencies typically in the range 20–50 kHz. A PZT disc having its fundamental thickness mode frequency in this range would be large; for example, the thickness for 20 kHz would be about 12 cm! The production of such large pieces would present difficulties, and so too would coping with the energy dissipation. The problems are overcome by adopting a composite structure comprising the piezoceramic together with carefully chosen metal end-pieces.

The structure of a typical composite power transducer for ultrasonic cleaning is illustrated in Fig. 6.32. The 'PZT' toroids are kept in compression (stress ~25 MPa) by the bolt to reduce the risk of their fracturing under the high drive fields. The 'PZT' will be of the 'hard' variety ('acceptor'-doped; see Section 6.3.2) so as to minimize the risk of depolarization under the high mechanical stresses experienced. The associated high Q_m value and low electrical tan δ ensure that the losses are kept within acceptable limits. The structure has the added advantage that any heat developed in the ceramic can be dissipated through the massive metal end pieces.

The device must be dimensioned so that at resonance a half-wavelength is accommodated in the distance L. The amplitude of motion at the outer metal surface will be related to that of the ceramic through the ratio q of their acoustic impedances:

$$q = \frac{\rho_c v_c A_c}{\rho_m v_m A_m} \tag{6.95}$$

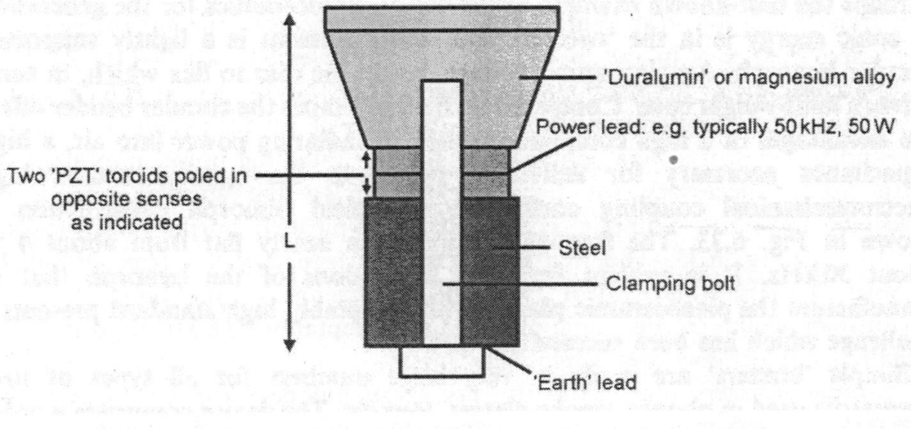

Fig. 6.32 The essentials of a power transducer for ultrasonic cleaning (the component is cylindrically symmetrical about the vertical axis).

where ρ_c, ρ_m, v_c, v_m, A_c and A_m are the density, sound velocity and sectional areas of the metal and the ceramic respectively. By making one end-piece of magnesium, for which $\rho_m v_m = 8.4 \times 10^6\,\mathrm{kg\,m^{-2}\,s^{-1}}$ (8.4 Mrayl) and the other of steel with $\rho_m v_m = 41$ Mrayl, and suitably adjusting the sectional areas of the components, about 15 times the amplitude of motion can be obtained at the magnesium surface than at the steel surface, so that the major part of the acoustic energy is emitted in only one direction.

A final benefit of this type of structure is that the acoustic behaviour is primarily governed by the properties and dimensions of the metal parts and only to a minor extent by those of the ceramic, so that small variations in the ceramic properties become unimportant.

For efficient transfer of power from the generator to the medium, usually water, the two must be acoustically matched. The discontinuity can be smoothed by fixing a $\lambda/4$ thick layer of material having an acoustic impedance intermediate between that of the radiating surface material and water, and polymers having impedances of about 3.5 Mrayl are readily available. The velocity of sound in them is approximately $2500\,\mathrm{m\,s^{-1}}$ so that the thickness required at 50 kHz is about 12 mm. In practice the transducer is often bonded to an ultrasonic cleaning tank and then the tank and water become a complicating part of the transducer.

'Tweeter'

Perhaps the best-known example of the use of piezoceramics for the generation of sonic energy is in the 'tweeter'. The active element is a lightly supported circular bimorph. An alternating voltage causes the disc to flex which, in turn, drives a light-weight cone. Compared with other shapes the circular bender offers the advantages of a high compliance suited to radiating power into air, a high capacitance necessary for delivering power to the speaker and a high electromechanical coupling coefficient. A typical bimorph construction is shown in Fig. 6.33. The frequency response is nearly flat from about 4 to about 30 kHz. It is evident from the dimensions of the bimorph that to manufacture the piezoceramic plates to an acceptably high standard presents a challenge which has been successfully met.

Simple 'buzzers' are made in very large numbers for all types of tone generators used in phones, smoke alarms, toys etc. The device comprises a poled PZT disc cemented to a thin metal disc which vibrates at the frequency of the applied voltage.

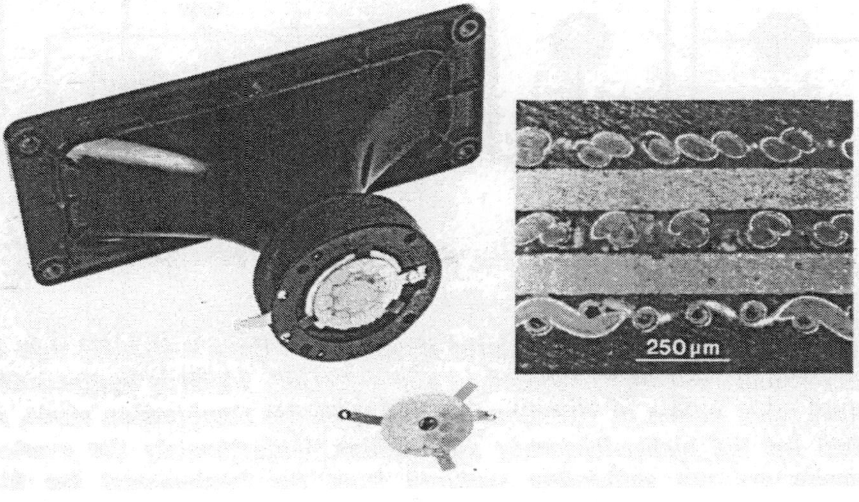

Fig. 6.33 Hi-fi 'tweeter' showing ceramic bimorph plate element. Inset, cross-section of element showing piezoceramic plates sandwiched between metallized glass fibre mesh.

Wave filters and delay lines

Piezoelectric crystals, notably quartz, are used to control or limit the operating frequency of electrical circuits. A well-known example is their use in 'quartz clocks'. The fact that a dielectric body vibrating at a resonant frequency can absorb considerably more energy than at other frequencies provides the basis for piezoelectric wave filters. The equivalent circuit for a piezoelectric body vibrating at frequencies close to a natural frequency is given in Fig. 6.3. At resonance the impedance due to L_1 and C_1 falls to zero and, provided that R_1 is small, the overall impedance is small.

A filter is required to pass a certain selected frequency band, or to stop a given band. The passband for a piezoelectric device is proportional to k^2, where k is the appropriate coupling coefficient. The very low k value of about 0.1 for quartz only allows it to pass frequency bands of approximately 1% of the resonant frequency. However, the PZT ceramics, with k values of typically about 0.5, can readily pass bands up to approximately 10% of the resonant frequency. Quartz has a very high Q_m (about 10^6) which results in a sharp cut-off to the passband. This, coupled with its very narrow passband, is the reason why the frequency of quartz oscillators is very well defined. In contrast PZT ceramics have Q_m values in the range 10^2–10^3 and so are unsuited to applications demanding tightly specified frequency characteristics.

The simplest of resonators is a thin (about 400 μm) disc electroded on its plane faces and vibrating radially. For a frequency of 450 kHz the diameter needs to be

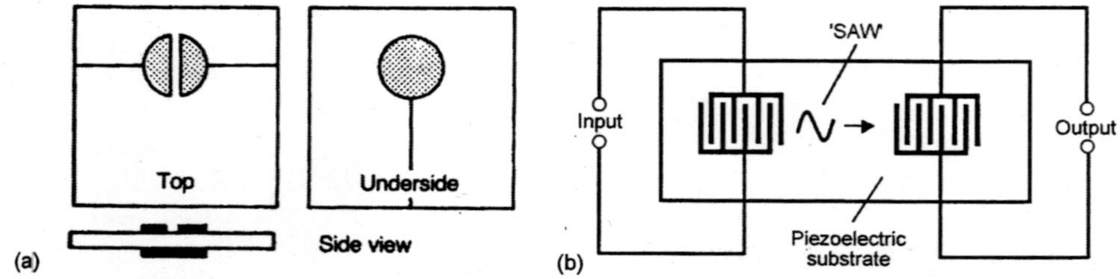

Fig. 6.34 (a) Trapped-energy filter. (b) Surface acoustic wave (SAW) filter.

about 5.6 mm. However, if the required resonant frequency is 10 MHz then the diameter would need to be reduced to about 250 μm, which is impracticable. Therefore other modes of vibration, e.g. the thickness compression mode, are exploited for the higher-frequency applications. Unfortunately the overtone frequencies are not sufficiently removed from the fundamental for filter applications, but the problem is solved by exploiting the 'trapped-energy' principle to suppress the overtones.

The plate is partly covered with electrodes of a specific thickness. Because of the extra inertia of the electrodes the fundamental frequency of the thickness modes beneath the electrodes is less than that of the unelectroded thickness. Therefore the longer wave characteristic of the electroded region cannot propagate in the unelectroded region; it is said to be trapped. In contrast, the higher-frequency overtones can propagate away into the unelectroded region where their energy is dissipated. If the top electrode is split, coupling between the two parts will only be efficient at resonance. A trapped-energy filter of this type is illustrated in Fig. 6.34(a).

Filters based on this principle made from PZT ceramic have been widely used in the intermediate frequency stages of FM radio receivers. More stable units suitable for telecommunication filters are made from single-crystal $LiTaO_3$ and quartz.

Filters and delay lines of the form shown in Fig. 6.34(b) are made which exploit the surface acoustic wave (see Section 6.5.2 above). The SAW is propagated in the direction normal to the overlap of the interdigitated electrodes, the wavelength launched being related to the electrode spacing and width. For a combined space and half-width of 15 μm ($\lambda_{surface} = 30\,\mu$m) the structure will propagate a centre frequency of 100 MHz. That is it operates as a filter.

The transit time of the SAW between the launch and receiving electrodes can be tailored to introduce a specific time delay into a microwave circuit.

A delay line can also be formed from a slice of a special glass designed so that the velocity of sound is as nearly as possible independent of temperature. (The term 'isopaustic' has been coined to describe such a material.) PZT ceramic transducers are soldered on two 45° metallized edges of the slice. The input transducer converts the electrical signal to a transverse ('shear') acoustic

Fig. 6.35 A signal delay device.

wave which travels through the slice and is reflected at the edges as shown in Fig. 6.35. At the output transducer the signal is reconverted into an electrical signal delayed by the length of time taken to travel around the slice. Unwanted reflections inside the slice are suppressed by the damping spots which are screen-printed epoxy resin loaded with tungsten powder. By grinding the slice to precise dimensions the delay time can be controlled with an accuracy to better than 1 in 10^4.

Such delay lines are used in colour television sets to introduce a delay of approximately 64 μs, the time taken for the electron beam to travel once across the screen. The delayed signal can be used to average out variations in the colour signal that occur during transmission and so improve picture quality. Similar delay lines are used in videotape recorders.

The intravascular catheter

The intravascular catheter, used on a routine basis to image arterial tissue, is an essential medical tool enabling the surgeon to identify disease type and to take appropriate steps to combat it, and the demand is for very large numbers (currently approximately 175 000 p.a.).

In a typical design the transducer head, illustrated in Fig. 6.36(b), comprises an array of 64 'PZT' bars (approximate dimensions $500 \times 30 \times 70\,\mu$m) arranged around the catheter tip.

To form the array a 70 μm thick poled sheet (5×0.6 mm) of PZT is bonded between two metallized polymer sheets. The PZT is sliced into the 64 elements by diamond-cutting, the cut just penetrating the bottom polymer layer so allowing the whole to be wrapped around the catheter tip. The outer polymer film, about 20 μm thick, serves as a $\lambda/4$ acoustic matching layers and also carries the thin copper tracks to electrically address each element.

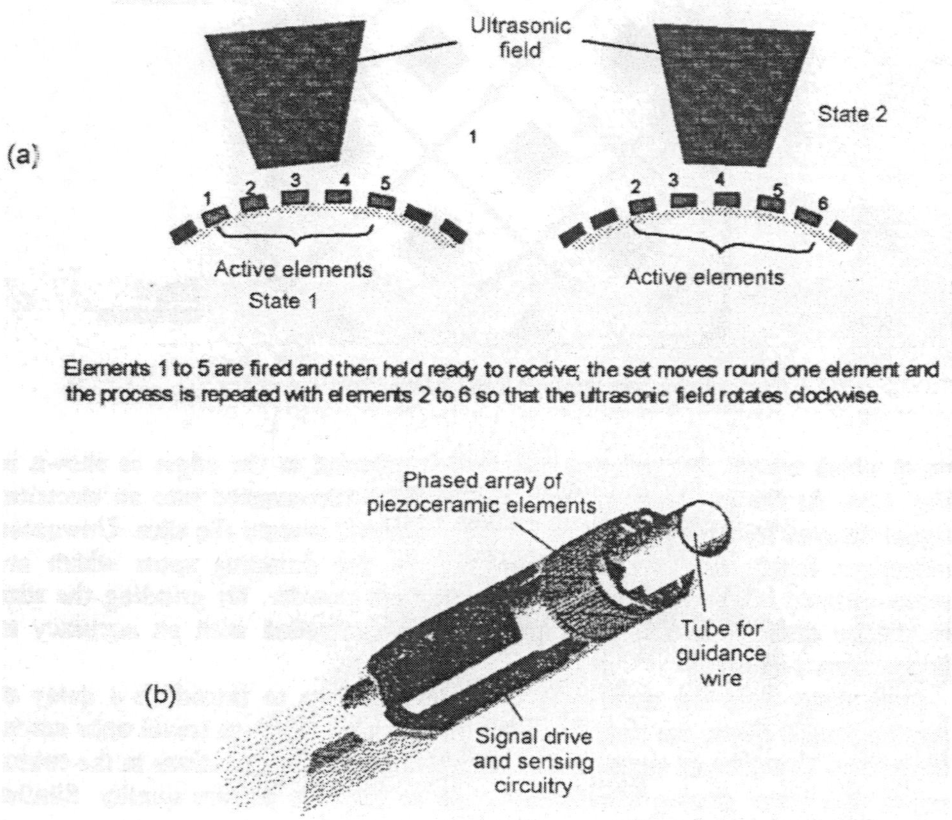

Fig. 6.36 The intravascular catheter: (a) the 'phased array' principle; (b) a schematic of the catheter head (the guide-wire tube is 1.47 mm o.d.). The catheter as a whole is passed through the artery along a pre-inserted guide-wire ((b) reproduced with permission of Jomed Inc.).

Many development problems have been encountered and overcome, including the risk of de-poling during processing (cutting, bonding, welding, sterilization) and of fractures due to stresses imposed on the elements during bending the array to conform to the catheter surface. Because both types of fault can degrade the image, the achievement of optimum mechanical strength is of the utmost importance demanding a very high quality microstructure free from any stress-raising defects. Typically the grain sizes of the various candidate piezoceramics lie in the range 1 to $\sim 5\,\mu$m. E.L. Nix *et al.* [22] have carried out a detailed analysis of the microstructures and mechanical and piezoelectric properties of a range of candidate piezoceramics, including a single crystal material.

6.5.4 Piezoceramic–polymer composites

Because of the special relationship between the structure of piezocomposites, especially the 1–3 variety, and the major application – for underwater technologies – the topic is dealt with in Section 6.4.6.

6.5.5 Summary

The discussion has highlighted a selection of applications chosen to illustrate the two major functions of piezoceramics namely to generate voltages or displacements. The choice, necessarily limited, also spans the displacement and power levels commonly encountered.

Other applications include the following, listed according to the principal functional modes.

- *Small voltage generators (sensors)*: car crash sensors to activate air-bags, noise cancellation, active suspension for cars, anti-'knock sensor' for i.c. engines, liquid level sensing, etc.
- *Displacement actuators*: deformable mirrors, computer-controlled knitting machine actuators, fuel-injection, ink-jet printer, cooling fans, electrical relays, etc.
- *High frequency generators at both 'power' and 'signal' levels*: ultrasonic welding of polymers, lithotripters, humidifiers, 'buzzers' ('sounders') of all types, non-destructive testing devices, depth gauges for boats, under water and various medical imaging technologies, etc.

The fascinating and instructive article, 'Smart Electroceramics' by R.E. Newnham and G.R. Ruschau [23] is strongly recommended.

Appendix: Piezoelectric Relations for Ceramics poled in the 3 Direction

Strains as functions of stresses in terms of compliances:

$$x_1 = s_{11}X_1 + s_{12}X_2 + s_{13}X_3 \tag{6A.1}$$

$$x_2 = s_{12}X_1 + s_{11}X_2 + s_{13}X_3 \tag{6A.2}$$

$$x_3 = s_{13}X_1 + s_{13}X_2 + s_{33}X_3 \tag{6A.3}$$

$$x_4 = s_{44}X_4 \tag{6A.4}$$

$$x_5 = s_{44}X_5 \tag{6A.5}$$

$$x_6 = 2(s_{11} - s_{12})X_6 \tag{6A.6}$$

Stresses as functions of strains in terms of stiffness coefficients:

$$X_1 = c_{11}x_1 + c_{12}x_2 + c_{13}x_3 \tag{6A.7}$$

$$X_2 = c_{12}x_1 + c_{11}x_2 + c_{13}x_3 \tag{6A.8}$$

$$X_3 = c_{13}x_1 + c_{13}x_3 + c_{33}x_3 \tag{6A.9}$$

$$X_4 = c_{44}x_4 \tag{6A.10}$$

$$X_5 = c_{44}x_5 \tag{6A.11}$$

$$X_6 = 2(c_{11} - c_{12})x_6 \tag{6A.12}$$

Relations between the c and s coefficients:

$$c_{11} = \frac{s_{11}s_{33} - s_{13}^2}{f(s)} \tag{6A.13}$$

$$c_{12} = -\frac{s_{12}s_{33} - s_{13}^2}{f(s)} \tag{A.14}$$

$$c_{13} = -\frac{s_{13}(s_{11} - s_{12})}{f(s)} \tag{6A.15}$$

$$c_{33} = \frac{s_{11}^2 - s_{12}^2}{f(s)} \tag{6A.16}$$

$$c_{44} = \frac{1}{s_{44}} \tag{6A.17}$$

$$f(s) = (s_{11} - s_{12})\{s_{33}(s_{11} + s_{12}) - 2s_{13}^2\} \tag{6A.18}$$

The Poisson ratios:

$$\nu_{12} = -s_{12}^E/s_{11}^E \tag{6A.19}$$

$$\nu_{13} = -s_{13}^E/s_{33}^E \tag{6A.20}$$

Relationships between clamped and free permittivities:

$$\varepsilon_1^x = (1 - k_{15}^2)\varepsilon_1^X \tag{6A.21}$$

$$\varepsilon_3^x = (1 - k_p^2)(1 - k_t^2)\varepsilon_3^X \approx (1 - k_{33}^2)\varepsilon_3^X \tag{6A.22}$$

In Eqs (6A.23)–(6A.34) ε represents the permittivity ε^X at constant stress and s represents the compliance s^E under short-circuit conditions.

The equations of state for the direct effect:

$$D_1 = \varepsilon_1 E_1 + d_{15} X_5 \tag{6A.23}$$

$$D_2 = \varepsilon_1 E_2 + d_{15} X_4 \tag{6A.24}$$

$$D_3 = \varepsilon_3 E_3 + d_{31}(X_1 + X_2) + d_{33} X_3 \tag{6A.25}$$

The equations of state for the converse effect:

$$x_1 = s_{11} X_1 + s_{12} X_2 + s_{13} X_3 + d_{31} E_3 \tag{6A.26}$$

$$x_2 = s_{11} X_2 + s_{12} X_1 + s_{13} X_3 + d_{31} E_3 \tag{6A.27}$$

$$x_3 = s_{13}(X_1 + X_2) + s_{33} X_3 + d_{33} E_3 \tag{6A.28}$$

$$x_4 = s_{44} X_4 + d_{15} E_2 \tag{6A.29}$$

$$x_6 = 2(s_{11} - s_{12}) X_6 \tag{6A.30}$$

The coupling coefficients are related to the d coefficients by

$$k_{33} = \frac{d_{33}}{(s_{33}\varepsilon_3)^{1/2}} \tag{6A.31}$$

$$k_{31} = -\frac{d_{31}}{(s_{11}\varepsilon_3)^{1/2}} \tag{6A.32}$$

$$k_p = \frac{-d_{31}}{\{(s_{11} + s_{12})\varepsilon_3/2\}^{1/2}} = k_{31}\left(\frac{2}{1 - \nu_{12}}\right)^{1/2} \tag{6A.33}$$

$$k_{15} = \frac{d_{15}}{(s_{44}\varepsilon_1)^{1/2}} \tag{6A.34}$$

$$k_t = \frac{e_{33}}{(\varepsilon_3^x c_{33}^D)^{1/2}} \tag{6A.35}$$

The open- and short-circuit compliances are related by

$$s_{11}^D = s_{11}^E(1 - k_{31}^2) \tag{6A.36}$$

$$s_{12}^D = s_{12}^E - k_{31}^2 s_{11}^E \tag{6A.37}$$

$$s_{33}^D = s_{33}^E(1 - k_{33}^2) \tag{6A.38}$$

$$s_{44}^D = s_{44}^E(1 - k_{15}^2) \tag{6A.39}$$

The piezoelectric coefficients are related by

$$g_{31} = d_{31}/\varepsilon_3^X \tag{6A.40}$$

$$g_{33} = d_{33}/\varepsilon_3^X \tag{6A.41}$$

$$g_{15} = d_{15}/\varepsilon_1^X \tag{6A.42}$$

$$e_{31} = \varepsilon_3^x h_{31} = d_{31}(c_{11}^E + c_{12}^E) + d_{33}c_{13}^E \tag{6A.43}$$

$$e_{33} = \varepsilon_3^x h_{33} = 2d_{31}c_{13}^E + d_{33}c_{33}^E \tag{6A.44}$$

$$e_{15} = \varepsilon_1^x h_{15} = d_{15}c_{44}^E \tag{6A.45}$$

$$h_{31} = g_{31}(c_{11}^D + c_{12}^D) + g_{33}c_{13}^D \tag{6A.46}$$

$$h_{33} = 2g_{31}c_{13}^D + g_{33}c_{33}^D \tag{6A.47}$$

$$h_{15} = g_{15}c_{44}^D \tag{6A.48}$$

$$d_{31} = e_{31}(s_{11}^E + s_{12}^E) + e_{33}s_{13}^E \tag{6A.49}$$

$$d_{33} = 2e_{31}s_{13}^E + e_{33}s_{33}^E \tag{6A.50}$$

$$d_{15} = e_{15}s_{44}^E \tag{6A.51}$$

Problems

1. Quote from p. 341

> This suggests that the large *d*-coefficients shown by some ferroelectric materials are due to a combination of large electrostriction coefficients and large spontaneous polarization and permittivity values.

Explain the reasonableness (or otherwise) of the statement in terms of the elementary physics of the processes involved.

2. (a) Show that the two ways of expressing the units of the piezoelectric d-coefficient are dimensionally equivalent.

(b) From the electrostriction data given in Fig. 6.15 estimate an effective maximum d_{33} coefficient. [Answer: $200\,\mathrm{pC\,N^{-1}}$]

(c) Suggest an explanation for the dependence of the electrostriction coefficient on grain size shown in Fig. 6.15. [Answer: $\sim 200\,\mathrm{pm\,V^{-1}}$]

3. A rectangular plate of piezoceramic of dimensions $10\,\mathrm{mm} \times 3\,\mathrm{mm} \times 1\,\mathrm{mm}$ is electroded over the $10\,\mathrm{mm} \times 3\,\mathrm{mm}$ faces. Calculate the new dimensions when a potential difference of 100 V is applied between the electrodes and the field and poling directions coincide ($d_{33} = 525\,\mathrm{pC\,N^{-1}}$; $d_{31} = -220\,\mathrm{pC\,N^{-1}}$). [Answer: $9.99978\,\mathrm{mm} \times 2.999934\,\mathrm{mm} \times 1.0000525\,\mathrm{mm}$]

4. A piezoceramic disc, thickness t and diameter D, is electroded over its faces and poled. A constant voltage is applied in the sense that t increases in an unrestrained way. From a consideration of the energies involves in the definition of k_{33} derive the relationship:

$$k_{33} = d_{33}/(s_{33}^{E}\varepsilon_{33}^{X})^{1/2}$$

in which k_{33} is the electromechanical coupling coefficient, s_{33}^{E} the elastic compliance measured at constant applied electric field and ε_{33}^{X} the permittivity at constant mechanical stress.

For a p.d. of 1.5 kV applied between the electrodes calculate the extension under no restraining force, and estimate the force necessary to block the extension, the 'blocking force' ($d_{33} = 500\,\mathrm{pC\,N^{-1}}$, $t = 1\,\mathrm{mm}$, diam.$= 10\,\mathrm{mm}$, $s_{33}^{E} = 20 \times 10^{-12}\,\mathrm{m^2\,N^{-1}}$). [Answer $\sim 300\,\mathrm{kgf}$]

5. A uniform tensile stress of 50 MPa acts along the axis of a piezoceramic cylindrical rod of length 12 mm and diameter 6 mm. Calculate the potential difference developed between the ends of the rod. Given that the coupling coefficient is 0.5, calculate the total stored energy in the rod ($\varepsilon_{33}^{X} = 700\varepsilon_0$; $d_{33} = 350\,\mathrm{pC\,N^{-1}}$). [Answer: 33.9 kV; 25 mJ]

6. A cylindrical rod of ceramic $BaTiO_3$ is poled along an axis perpendicular to its electroded end-faces. The cylinder stands on one flat end and carries a load of 10 kg uniformly distributed over the upper face. Calculate the open-circuit voltage difference developed between the end-faces, the mechanical strain and the length change.

In a separate experiment, a potential difference is applied between the ends of the unloaded cylinder. Calculate the voltage necessary to obtain the same strain as developed previously. Explain why the developed and applied voltages differ.

Cylinder dimensions: height, 10 mm; diameter, 10 mm. Electromechanical properties of ceramic: $s_{33}^{D} = 6.8 \times 10^{-12}\,\mathrm{m^2\,N^{-1}}$; $h_{33} = 1.5\,\mathrm{GN\,C^{-1}}$; $d_{33} = 190\,\mathrm{pC\,N^{-1}}$. [Answer: 130 V, 8.6×10^{-6} and $0.086\,\mu\mathrm{m}$; 453 V]

7. Using the data given in Problem 6 analyse the energetics, assuming a coupling coefficient of 0.5 and $\varepsilon_r \approx 1600$.

8. A piezoceramic bimorph is loaded in a symmetrical 'three-point' arrangement, in which the loads are applied via small steel rods of circular section parallel to each other and perpendicular to the length of the bimorph. Neglecting the effects of the metal shim, calculate the voltage developed when the centre is displaced $2\,\mu$m from the plane containing the outer loading lines (distance between outer loading points, 20 mm; bimorph width, 3 mm; bimorph depth, 1 mm; $g_{31} = -4.7 \times 10^{-3}\,\mathrm{V\,m\,N^{-1}}$; $g_{33} = 11.4 \times 10^{-3}\,\mathrm{V\,m\,N^{-1}}$; $c_{11}^D = 168 \times 10^9$ Pa; $c_{12}^D = 78.2 \times 10^9$ Pa; $c_{13}^D = 71.0 \times 10^9$ Pa). [Answer: ~ 2.6 V].

9. The following electroded pieces fabricated from the same piezoceramic are available: (i) a disc – diameter 21 mm, thickness 0.8 mm, fully poled through its thickness, (ii) a square section rod 20×3×3 mm fully poled along its length, and (iii) a rod of the same dimensions as in (ii) but fully poled perpendicular to its length.

   An impedance analyser is also available.

   Outline a strategy for determining ε_3^X, ε_3^x, s_{33}^D, s_{33}^E, k_{33}, k_3, d_{33}, g_{33}, k_p, and calculate the various parameters given the following.

   Mass of disc is 2.15 g and the planar resonance and antiresonance frequencies are 102 kHz and 120 kHz.

   The longitudinal resonance (the lengthwise poled rod supported at its centre (node)) and antiresonance frequencies are 71.8 kHz and 88.1 kHz.

   The resonance and antiresonance frequencies for the bar poled across its thickness are 72.6 kHz and 76.9 kHz.

   The capacitance of the disc measured at 1 kHz is 6.1 nF.

   From the data make an estimate of the Poisson ratio and comment on its realism.

   [Answers: $\varepsilon^x = 1592\varepsilon_0\,\mathrm{F\,m^{-1}}$; $s_{33}^D = 10.4 \times 10^{-12}\,\mathrm{m^2\,N^{-1}}$; $k_{33} = 0.62$; $k_{31} = 0.37$; $s_{33}^E = 16.9 \times 10^{-12}\,\mathrm{m^2\,N^{-1}}$; $d_{33} = 303\,\mathrm{pC\,N^{-1}}$; $g_{33} = 21.5 \times 10^{-3}\,\mathrm{V\,m\,N^{-1}}$; $k_p =$ ~ 0.60]

10. Discuss the design of the bimorph for a hi-fi 'tweeter' and describe a fabrication route.

11. Regarding the components of a 0–3 composite as characterised by their resistivity values rather than permittivities, suggest strategies for poling the composite.

12. An experimental 1–3 piezoceramic/polymer composite disc transducer operating at the fundamental thickness mode resonance frequency is intended to detect objects in water having dimensions of approximately 10 mm. It is firmly fixed to a backing.

    The piezoelectric ceramic rods have a density of $7500\,\mathrm{kg\,m^{-3}}$, the polymer $1200\,\mathrm{kg\,m^{-3}}$ and the volume fraction of the ceramic is 0.25. The elastic compliance of the ceramic (s_{33}^D) is $10 \times 10^{-12}\,\mathrm{m^2\,N^{-1}}$ and that of the polymer $115 \times 10^{-12}\,\mathrm{m^2\,N^{-1}}$.

    Sketch out the form the composite should take defining the disc thickness and the dimensions and spacing of the rods, briefly justifying steps in the argument.

Given that the appropriate relative permittivities of the ceramic and the polymer are respectively 1500 and 3.5, and that the d_{33} coefficient for the ceramic is $375\,pC\,N^{-1}$, calculate a value for the hydrophone figure of merit and show that the units are $m^2\,N^{-1}$. For this estimate it may be assumed that the composite has been structurally modified so that the d_{31} contribution is negligible. Comment on the realism or otherwise of the calculated value.

Calculate the capacitance per unit area of the composite, and its acoustic impedance.

Describe a production route suited to manufacturing the composite in large numbers.

[Answer: $\sim 40\,000$ (units of $10^{-15}\,m^2\,N^{-1}$); $\sim 1\,\mu F\,m^{-2}$; ~ 9 Mrayl]

Bibliography

1. Jaffe, B., Cook, W.R. and Jaffe, H. (1971) *Piezoelectric Ceramics*, Academic Press, London.
2. Herbert, J.M. (1982) *Ferroelectric Transducers and Sensors*, Gordon and Breach, London.
3. Burfoot, J.C. and Taylor, G.W. (1979) *Polar Dielectrics and their Applications*, Macmillan, London.
4. Levinson, L.M. (ed.) (1988) *Electronic Ceramics*. Marcel Dekker, New York.
5. IRE Standards on Piezoelectric Crystals: determination of the elastic, piezoelectric and dielectric constants – the Electromechanical Coupling Factor, 1958, Proceedings IRE April 1958, 764–78.
6. IRE Standards on Piezoelectric Crystals: the piezoelectric vibrator: definitions and methods of measurement, 1957, Proceedings IRE, March 1957, 353–8.
7. Noheda, B. *et al.* (2000) Stability of the monoclinic phase in the ferroelectric perovskite Pb $Zr_{1-x}Ti_xO_3$, *Phys. Rev. B*, **63**, 14103–1 to 9.
8. Guo, R. *et al.* (2000) Origin of the high piezoelectric response in Pb $Zr_{1-x}Ti_xO_3$, *Phys. Rev. Lett.*, **84**, 5423–6.
9. Kingon, A.I and Clark, B. (1983) Sintering of PZT Ceramics: I, Atmosphere Control and Sintering of PZT Ceramics: II, Effect of PbO content on densification kinetics, *J. Am. Ceram. Soc.*, **66**, 253–60.
10. Eitel, R.E. *et al.* (2001) New high temperature morphotropic phase boundary piezoelectrics based on $Bi(Me)O_3$–$PbTiO_3$ ceramics, *Jpn. J. Appl. Phys.*, **40**, 5999–6002.
11. Park, S.-E. and Shrout, T.R. (1997) Characteristics of relaxor-based piezoelectric single crystals for ultrasonic transducers, *IEEE Trans. Ultrasound, Ferroelectrics and Frequency Control*, **44**, 1140–7.
12. Kondo, M. *et al.* (1997) Piezoelectric properties of $PbNi_{1/3}Nb_{2/3}O_3$–$PbTiO_3$–$PbZrO_3$ ceramics, *Jpn. J. Appl. Phys.*, **36**, 6043–5.
13. Roberts, G. *et al.* (2001) Synthesis of lead nickel niobate–lead zirconate titanate solid solutions by a B-site precursor method, *J. Am. Ceram. Soc.*, **84**, 2869–72.

14. Shrout, T.R. *et al.* (1987) Grain size dependence of dielectric and electrostriction (properties) of $Pb(Mg_{1/3}Nb_{2/3})O_3$ -based ceramics, *Ferroelectrics*, **76**, 479–87.
15. Mishima, T. *et al.* (1997) Lattice image observations of nanoscale ordered regions in $Pb(Mg_{1/3}Nb_{2/3})O_3$, *Jpn. J. Appl. Phys.*, **36**, 6141–4.
16 Smith, W.A. (1989) The role of piezocomposites in ultrasonic transducers, IEEE, Ultrasonics Symposium, Montreal, Canada. **2**, 755–66.
17. Janas, V.E. and Safari, A. (1995) Overview of fine-scale piezoelectric ceramic/polymer composite processing, *J. Am. Ceram. Soc.*, **78**, 2945–55.
18. Uchino, K. (1997) *Piezoelectric Actuators and Ultrasonic Motors*, Kluwer Academic Publishers, London.
19. Longtu, Li. *et al.* (1990) Lead zirconate titanate ceramics and monolithic piezoelectric transformer of low firing temperature, *Ferroelectrics*, **101**, 193–200.
20. Timoshenko, S. (1925) Analysis of bi-metal thermostats, *J. Opt. Soc. Am.* **11**, 233.
21. Pearce, D.H. *et al.* (2002) On piezoelectric Super-Helix actuators, *Sensors and Actuators: A*, **100**, 281–6.
22. Nix, E.L. *et al.* (2001) Piezoelectric transducer arrays for intravascular ultrasound. In *Ferroelectrics UK 2001*, I.M. Reaney and D.C. Sinclair (eds), Maney Publishing, London, pp. 101–12.
23. Newnham, R.E. and Ruschau, G.R. (1991) Smart Electroceramics, *J. Am. Ceram. Soc.*, 463–80.

7

PYROELECTRIC MATERIALS

The discussion draws on the review articles by R.W. Whatmore [1] and by R.W. Whatmore and R. Watton [2]. The now classical text by R.A. Smith, F.E. Jones and R.P. Chasmar [3] is recommended for supplementary reading.

7.1 Background

True pyroelectricity results from the temperature dependence of the spontaneous polarization $\boldsymbol{P}_s$ of polar materials and is therefore shown by ferroelectric materials whether they are single-domain single crystals or poled ceramics. Because a change in polarization in a solid is accompanied by a change in surface charges (see Section 2.7.1) it can be detected by an induced current in an external circuit. If the pyroelectric material is perfectly electrically insulated from its surroundings, the surface charges are eventually neutralized by charge flow occurring because of the intrinsic electrical conductivity of the material. Effective neutralization occurs in a time approximately equal to the electrical time constant $\rho\varepsilon$ of the material.

When an electric field $\boldsymbol{E}$ is applied to a polar material the total electric displacement $\boldsymbol{D}$ is given by

$$\begin{aligned}\boldsymbol{D} &= \varepsilon_0\boldsymbol{E} + \boldsymbol{P}_{\text{total}} \\ &= \varepsilon_0\boldsymbol{E} + (\boldsymbol{P}_s + \boldsymbol{P}_{\text{induced}})\end{aligned} \tag{7.1}$$

which, from Eqs (2.76) and (2.81), is

$$\boldsymbol{D} = \varepsilon\boldsymbol{E} + \boldsymbol{P}_s \tag{7.2}$$

Therefore

$$\frac{\partial \boldsymbol{D}}{\partial T} = \frac{\partial \boldsymbol{P}_s}{\partial T} + \boldsymbol{E}\frac{\partial \varepsilon}{\partial T}$$

Electroceramics: Materials, Properties, Applications. 2nd Edition. Edited by A. J. Moulson and J. M. Herbert.
 ISBN 0 471 49747 9 (hardback) 0 471 49748 7 (paperback)

(assuming constant $\boldsymbol{E}$), and

$$\boldsymbol{p}_{\mathrm{g}} = \boldsymbol{p} + \boldsymbol{E}\frac{\partial \varepsilon}{\partial T} \tag{7.3}$$

where $\boldsymbol{p} = \partial \boldsymbol{P}_{\mathrm{s}}/\partial T$ is the true pyroelectric coefficient and $\boldsymbol{p}_{\mathrm{g}}$ is sometimes referred to as a generalized pyroelectric coefficient. Since a temperature change ΔT produces a change in the polarization vector, the pyroelectric coefficient has three components defined by

$$\Delta P_i = p_i \Delta T \qquad i = 1, 2, 3 \tag{7.4}$$

Therefore the pyroelectric coefficient is a vector but, because in practical applications the electrodes that collect the pyrocharges are positioned normal to the polar axis, the quantities are usually treated as scalars, and this is done in the following discussion.

The contribution $E(\partial \varepsilon/\partial T)$ (Eq. (7.3)) can be made by all dielectrics, whether polar or not, but since the temperature coefficients of permittivity of ferroelectric materials are high, in their case the effect can be comparable in magnitude with the true pyroelectric effect. This is also the case above the Curie point and where, because of the absence of domains, the dielectric losses of ferroelectrics are reduced, which is important in some applications. However, the provision of a very stable biasing field is not always convenient.

Since pyroelectric materials are polar, they are also piezoelectric, and the strain resulting from thermal expansion will result in the development of a surface charge. However, this is a small effect that seldom exceeds 10% of the primary pyroelectric effect.

Because P_{s} falls to zero at the Curie point, ferroelectric materials are likely to exhibit high pyroelectric coefficients just below their transition temperatures. The various ways in which P_{s} falls as the Curie point is approached from below are shown in Fig. 7.1. High pyroelectric coefficients are observed for ferroelectrics that exhibit second-order transitions, such as triglycine sulphate with a transition temperature of 49 °C and a pyroelectric coefficient of at least 280 μC m^{-2} K^{-1} at 20 °C. There are difficulties in exploiting materials which exhibit first-order transitions, first because they exhibit hysteresis – the transition occurs at a higher temperature when the temperature is rising than when it is falling – and second because it would be difficult in most applications to keep the pyroelectrics in a sufficiently constant temperature environment. A number of materials are used at temperatures well below their Curie points where, although the pyroelectric coefficients are smaller, they vary less with the ambient temperature.

For practical purposes, the very small signals generated by pyroelectric elements must be amplified. The most widely used first stage consists of a field effect transistor (FET) which responds to electric potential rather than to charge. In this case, it is advantageous for the material to have a low permittivity to match the low input capacitance of the FET. Therefore the compositions with high

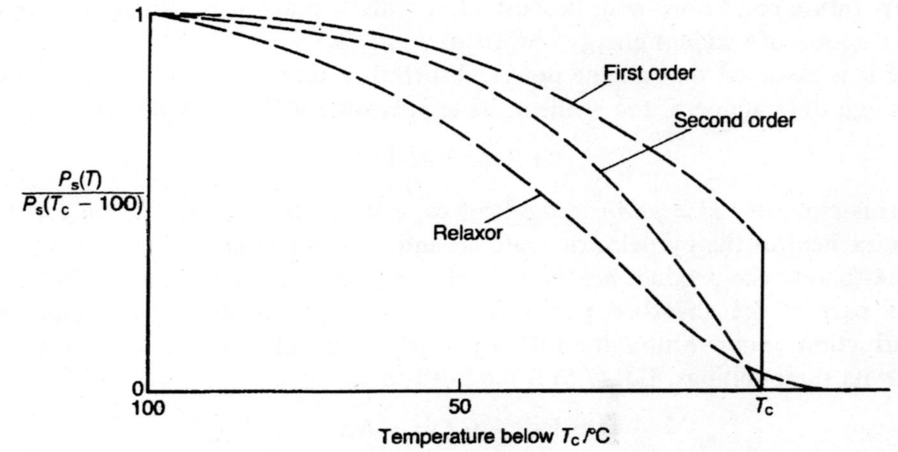

Fig. 7.1 Form of $P_s(T)$ for various classes of ferroelectric.

permittivities which are exploited as dielectrics and piezoelectrics are unsuitable, and special materials have been developed for pyroelectric applications.

7.2 Infrared Detection

Pyroelectric materials are used mainly for the detection of infrared radiation. The elements for the detectors are typically thin slices of material (e.g. $1.0 \times 1.0 \times 0.1$ mm) coated with conductive electrodes, one of which is a good absorber of the radiation.

Figure 7.2 illustrates a detector at a temperature T above its surroundings. If radiation at a power density W_i/A is incident on the face for a time $\mathrm{d}t$, the energy absorbed is $\eta W_i\,\mathrm{d}t$. The emissivity η is for the particular surface, wavelength and

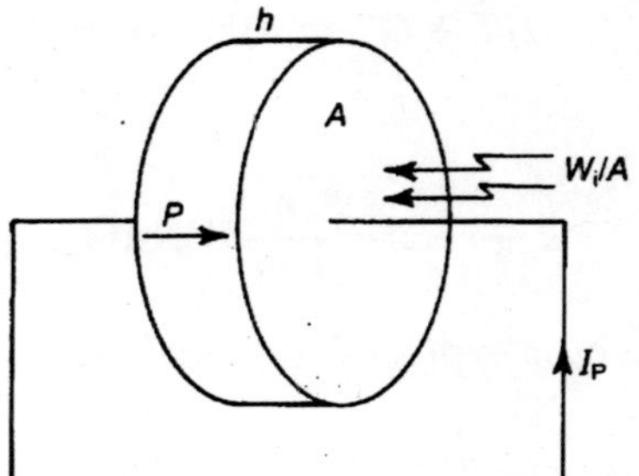

Fig. 7.2 A pyroelectric detector element.

temperature conditions and, because of Kirchhoff's law, it is also a measure of the fraction of incident energy absorbed.

If it is assumed that all the power absorbed in time $\mathrm{d}t$ is rapidly distributed through the volume of the element, its temperature will rise by $\mathrm{d}T$ where

$$\eta W_i \mathrm{d}t = H\,\mathrm{d}T \tag{7.5}$$

In this equation $H = \rho c A h$ is the heat capacity of the element where c is the specific heat of the pyroelectric material and, in this context, ρ is its density. In what follows the product ρc, the volume specific heat, is given the symbol c'.

If part of the absorbed power is lost to the surroundings by reradiation, conduction or convection at a rate G per unit temperature excess of the element over its surroundings, Eq. (7.5) is modified to

$$\eta W_i \mathrm{d}t - GT\,\mathrm{d}t = H\mathrm{d}T$$

or

$$H\dot{T} + GT = \eta W_i \tag{7.6}$$

in which the dot notation signifies a time derivative. If the incident power is shut off at $t = t_0$,

$$H\dot{T} + GT = 0 \tag{7.7}$$

and

$$T = T_0 \exp\left(-\frac{t}{\tau_\mathrm{T}}\right) \tag{7.8}$$

where T_0 is the temperature excess at $t = t_0$ and $\tau_\mathrm{T} = H/G$ is the thermal time constant.

To obtain a continuous response from a pyroelectric material the incident radiation is pulsed, and this situation is analysed by assuming that the energy varies sinusoidally with frequency ω and amplitude W_0. Equation (7.6) then becomes

$$H\dot{T} + GT = \eta W_0 \exp(\mathrm{j}\omega t)$$

or

$$\dot{T} + \frac{1}{\tau_\mathrm{T}} T = \frac{\eta}{G\tau_\mathrm{T}} W_0 \exp(\mathrm{j}\omega t) \tag{7.9}$$

Using the integrating factor $\exp(t/\tau_\mathrm{T})$ and integrating gives

$$T = \left\{G\tau_\mathrm{T}\left(\frac{1}{\tau_\mathrm{T}} + \mathrm{j}\omega\right)\right\}^{-1} \eta W_i \tag{7.10}$$

The pyrocurrent I_p collected from the electrodes is given by

$$I_p = \dot{Q} = \dot{T}\frac{dQ}{dT} \tag{7.11}$$

where Q is the total instantaneous charge. If it is assumed that the element is operating in a polar state

$$dQ = A\,dP_s = ApdT$$

and

$$I_p = pA\dot{T} \tag{7.12}$$

which, after substituting $\dot{T}$ from Eq. (7.10), becomes

$$I_p = j\omega\left\{(G\tau_T)\left(\frac{1}{\tau_T} + j\omega\right)\right\}^{-1} pA\eta W_i \tag{7.13}$$

The 'current responsivity' r_I is defined as the modulus of I_p/W_i, so that

$$r_I = \left|\frac{I_p}{W_i}\right| = \left|j\omega pA\eta\left\{G\tau_T\left(\frac{1}{\tau_T} + j\omega\right)\right\}^{-1}\right|$$

which, after algebraic manipulation, becomes

$$r_I = \frac{pA\eta\omega}{G}(1 + \omega^2\tau_T^2)^{-1/2} \tag{7.14}$$

A common arrangement for a detecting system where the voltage u from the pyroelectric element is fed to the gate of an FET with a high input impedance is shown in Fig. 7.3. The resistor R_G correctly biases the FET, and C_A and R_A are respectively the input capacitance and resistance of the amplifying and associated system. The voltage output is I_p/Y where the admittance Y is given by

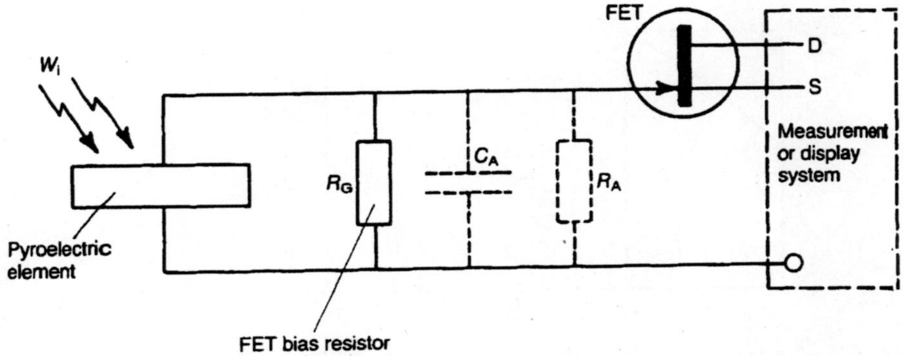

Fig. 7.3 A common pyroelectric detecting system.

$$Y = \frac{1}{R_G} + \frac{1}{R_A} + j\omega(C_E + C_A) \tag{7.15}$$

and C_E is the capacitance of the element. Usually, although not always, $R_A \gg R_G$ and $C_A \ll C_E$, so that

$$Y \approx R_G^{-1} + j\omega C_E$$

and

$$|Y| = R_G^{-1}(1 + \omega^2 \tau_E^2)^{1/2} \tag{7.16}$$

where $\tau_E = R_G C_E$ is the electrical time constant of the circuit. Therefore the voltage responsivity r_V is given by

$$r_V = \left|\frac{u}{W_i}\right| = \left|\frac{I_p}{YW_i}\right| = \frac{r_I}{|Y|} \tag{7.17}$$

or

$$r_V = \frac{R_G p A \omega \eta}{G(1 + \omega^2 \tau_T^2)^{1/2}(1 + \omega^2 \tau_E^2)^{1/2}} \tag{7.18}$$

Eq. (7.18) shows that, for maximum sensitivity at low frequency, G should be minimized by isolating the element to reduce the loss of heat. τ_T can be minimized by reducing the thickness h of the detecting element so as to reduce its thermal capacity. r_V is shown as a function of frequency in Fig. 7.4. There is a maximum at $\omega^2 = 1/\tau_E\tau_T$ with only a small variation with frequency between $1/\tau_T$ and $1/\tau_E$. τ_E and τ_T usually lie in the range 0.1–10 s for high sensitivity. The maximum value of r_V is

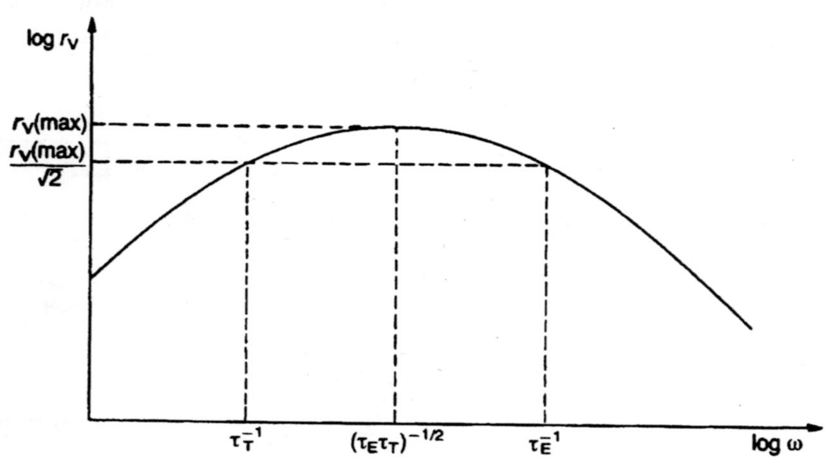

Fig. 7.4 Variation in voltage responsivity r_V with frequency. Note that τ_r^{-1} etc. denote ω values.

$$r_V(\max) = \frac{pA\eta R_G}{G(\tau_E + \tau_T)} \tag{7.19}$$

At high frequencies, when $\omega^2\tau_T^2 \gg 1$, $\omega^2\tau_E^2 \gg 1$ and $C_E \gg C_A$, Eq. (7.18) reduces to

$$r_V = \frac{p\eta}{c'A\varepsilon\omega} \tag{7.20}$$

in which ε is the permittivity of the material.

Eq. (7.20) suggests a figure of merit,

$$F_V = \frac{p}{c'\varepsilon} \tag{7.21}$$

which describes, in terms of material properties only, the effectiveness of a pyroelectric element used under defined conditions. Figures of merit only apply under given defined operational conditions and care has to be exercised in making use of them. As shown later, the appropriate figure of merit when the intrinsic 'noise' arising from dielectric loss is a dominating influence is

$$F_D = \frac{p}{c'\varepsilon^{1/2}\tan^{1/2}\delta} \tag{7.22}$$

7.3 Effects of Circuit Noise

All signal detectors are required to detect the signal against a background of 'noise'. Therefore, the signal-to-noise ratio must be optimized or, put another way, for maximum sensitivity the noise has to be minimized. The sensitivity of any detector is determined by the noise level in the amplified output signal. In the case of a pyroelectric detector and its associated circuitry, the principal sources of noise are Johnson noise, amplifier noise and thermal fluctuations.

The noise level can be expressed in terms of the power incident on the detector necessary to give a signal equivalent to the noise. If the noise voltage is ΔV_N then the 'noise equivalent power' (NEP) is defined by

$$\text{NEP} = \frac{\Delta V_N}{r_V} \tag{7.23}$$

Because it is preferable to have a quantity that increases in value as the performance of the system is improved, the 'detectivity' D is defined as

$$D = \frac{1}{\text{NEP}} \tag{7.24}$$

Johnson noise and thermal fluctuations are briefly discussed below, Johnson noise because it is usually the dominant noise and thermal fluctuations because they set a lower limit on the achievable noise level.

7.3.1 Johnson noise

Johnson noise arises because the random thermal motion of electrons in an isolated resistor produces random fluctuations in voltage between its ends, covering a broad frequency band. It can be shown that

$$\Delta\overline{I_J^2} = 4kTg\Delta f \tag{7.25}$$

where $\Delta\overline{I_J^2}$ is the mean square Johnson current covering a bandwidth Δf, k is Boltzmann's constant, T is the absolute temperature and g is the conductance of the noise generator. In the case of the pyroelectric detector system discussed above,

$$g = \frac{1}{R_G} + \omega C_E \tan\delta \tag{7.26}$$

where $\omega C_E \tan\delta$ is the AC conductance of the pyroelectric element.

The root mean square (r.m.s.) noise voltage ΔV_J for unit bandwidth is $\Delta I_J/|Y|$, where ΔI_J is the r.m.s. noise current for unit bandwidth and Y is the admittance. Therefore

$$\Delta V_J = \frac{(4kTg)^{1/2}}{|Y|} \tag{7.27}$$

At high frequencies, when $\omega C_E \tan\delta \gg 1/R_G$,

$$\Delta V_J \approx \left(\frac{4kT}{\omega C_E}\right)^{1/2} \tan^{1/2}\delta \tag{7.28}$$

Therefore $D = 1/\mathrm{NEP} = r_V/\Delta V_J$ which, on substitution from Eqs (7.20) and (7.28), gives

$$D = \frac{\eta}{(Ah)^{1/2}}(4kT\omega)^{-1/2}\left(\frac{p}{c'}\right)(\varepsilon\tan\delta)^{-1/2} \tag{7.29}$$

Under the conditions defined the material parameter to be maximized is p, with c', ε and $\tan\delta$ minimized as required by F_D (Eq. (7.22)).

7.3.2 Thermal fluctuations

Thermal fluctuations arise even when a body is in thermal equilibrium with its surroundings through radiation exchange only. Calculation of the mean-square value of the power fluctuations can be accomplished by the methods of either classical statistical mechanics or, when the radiation is considered to be quantized into photons, quantum mechanics.

Both approaches give

$$\overline{\Delta W_T^2} = 16k\sigma\eta T^5 A\Delta f \quad (7.30)$$

for the mean square power fluctuations covering a bandwidth Δf for a body of surface area A and emissivity η at equilibrium at temperature T; σ is the Stefan constant. If the body is assumed to be 'black' ($\eta = 1$), for unit area and unit bandwidth

$$\overline{\Delta W_T^2} = 16k\sigma T^5 \quad (7.31)$$

The r.m.s. value at 300 K is $5.5 \times 10^{-9}\,\mathrm{m^{-1}\,Hz^{-1/2}\,W}$, placing an upper limit of $1.8 \times 10^8\,\mathrm{m\,Hz^{1/2}\,W^{-1}}$ on the detectivity D. The highest detectivities achieved in practice are between one and two orders of magnitude below this.

In ferroelectrics the major contributor to tan δ is domain wall movement which diminishes as the amplitude of the applied field diminishes; the value applicable to pyroelectric detectors will be that for very small fields. The permittivity is also very sensitive to bias field strength, as is its temperature coefficient. The properties of some ferroelectrics – the 'relaxors' – are also frequency dependent. It is important, therefore, to ensure that when assessing the suitability of a ferroelectric for a particular application on the basis of measured properties that the measurements have been made using values of the parameters (frequency, field strength etc.) appropriate to the application. This is not always done.

7.4 Materials

The properties and figures of merit of a range of pyroelectric materials (at 20 °C except where indicated) are given in Table 7.1. The table omits a number of secondary characteristics that determine suitability in particular applications. Triglycine sulphate (TGS) has high figures of merit but is a rather fragile water-soluble single-crystal material. It can be modified to withstand temperatures in excess of its Curie point without depoling, but it cannot be heated in a vacuum to the temperatures necessary for outgassing without decomposing. It is difficult to handle and cannot be used in devices where it would be subjected to either a hard vacuum or high humidity. In contrast, polyvinylidene fluoride (PVDF) has poor figures of merit but is readily available in large areas of thin film. It is considerably more stable to heat, vacuum and moisture than TGS. It is mechanically robust and, as indicated by Eq. (7.18), can have its voltage sensitivity enhanced by the use of a large area. It also has low heat conductivity

Table 7.1 Properties of some pyroelectric materials. (Data should be regarded as approximate)

Material and form	$p/\mu C\,m^{-2}\,K^{-1}$	ε_r'	$tan\delta$	$c'/MJ\,m^{-3}\,K^{-1}$	$T_c/°C$	$F_V/m^2\,C^{-1}$	$F_D/TPa^{-1/2}$*
TGS, (single crystal; 35 °C)	280	38	0.01	2.3	49	0.36	66
DTGS (single crystal; 40 °C)	550	43	0.02	2.6	61	0.53	83
$LiTaO_3$ (single crystal)	230	47	0.001	3.2	665	0.17	110
$(SrBa)Nb_2O_6$ (single crystal)	550	400	0.001	2.3	121	0.07	70
modified PZ (polycrystalline**)	400	290	0.003	2.5	230	0.06	58
		300†	0.014†			0.06	26
modified PT (polycrystalline)	350	220	0.01	2.5	250	0.07	32
		220†	0.03†			0.07	18
PVDF (film)	27	12	0.01	2.4	80	0.1	10
P(70VDF/30TrFE)*** (film)	33	7.4	0.017	2.3	121	0.22	13.6
0.25PZ 0.75PT (sol-gel thin film)	220	350†	0.008†	2.7	–	2.5	17
(sputtered)	450	300	0.01	2.7	–	2.6	14
'BST (a) $(0.67Ba\ 0.33Sr)TiO_3$ (ceramic; 0.6 V μm^{-1})	7000	8800	0.004	–	25	–	124
'PST (b) $Pb(Sc_{1/2}Ta_{1/2})O_3$ (sputtered on Si; 10 V μm^{-1}; 50°C)	850	600	0.0025	–	10	–	80
'PST' (c) (ceramic; 4 V μm^{-1}; 40°C)	3500	2000	0.005	2.7	25	0.06	110

*F_D is also expressed in the equivalent units, $10^{-6}\,Pa^{-1/2}$: TGS = triglycine sulphate: DTGS = deuterated TGS: **'PZFNTU' see text: *** co-polymer of vinylidene fluoride and trifluoroethylene.

†Measured at 33 Hz; all other dielectric data measured at 1 kHz.

Properties measured at 25°C, unless indicated in parentheses. Where appropriate bias electric fields in parentheses.

(a) From [2].

(b) After P. P. Donohue, M.A. Todd *et al.* (2001) *Integrated Ferroelectrics*, **41**, 25–34.

(c) After N. M. Shorrocks *et al.* (1990) *Ferroelectrics*, **106**, 387–92.

and low permittivity so that both thermal and electrical coupling between neighbouring elements on the same piece of material are minimized. Its high $\tan\delta$ is a disadvantage.

Lithium tantalate is a single-crystal material that is produced in quantity by the Czochralski method (see Section 3.11) for piezoelectric applications and is therefore readily available. It is stable in a hard vacuum to temperatures that allow outgassing procedures. It is insensitive to humidity. It is widely used where precise measurements are to be made.

Strontium barium niobate is a single-crystal material with the tungsten bronze type of structure which is made by the Czochralski method but has yet to find a major use. It has relaxor characteristics of the type shown in Fig. 7.1 which give it a high pyroelectric coefficient and detectivity, but its high permittivity lowers the figure of merit F_V.

The ceramic based on lead zirconate (PZ) has intermediate figure of merit values. Its final form as a thin plate is obtained by sawing up a hot-pressed block and lapping and polishing to the required dimensions. Similar methods must be applied to single crystals and so the costs of manufacture are not very different.

The adaptability of ceramics is well illustrated by the development of PZ for pyroelectric devices. It was found initially that the addition of about 10 mol% lead iron niobate ($PbFe_{1/2}Nb_{1/2}O_3$) to lead zirconate converted it from an antiferroelectric to a ferroelectric material with a relatively low permittivity and a high pyroelectric coefficient. However, this material was found to undergo a transition between two ferroelectric rhombohedral forms at 30–40 °C which rendered its pyroelectric output unstable. The replacement of about 5% of the B-site atoms by Ti shifted this transition to around 100 °C without significantly increasing the permittivity. For some devices a somewhat lower resistivity is advantageous since it eliminates the need for a high-value, and expensive, external resistor. It was found that replacement of 0.4–0.8% of the B-site atoms by uranium controlled the resistivity in the 10^8–10^{11} Ω m range, as shown in Fig. 7.5; at the same time the permittivity and loss were reduced. The final composition $Pb_{1.02}(Zr_{0.58}Fe_{0.20}Nb_{0.20}Ti_{0.02})_{0.994}U_{0.006}O_3$ (PZFNTU), although complex, can be manufactured reproducibly.

A pyroelectric response is shown by any dielectric to which an electric field is applied provided there is a variation in permittivity with temperature (see Eq. 7.3). The permittivity of ferroelectrics is strongly temperature-dependent close to the Curie point and high pyroelectric figures of merit can be obtained by operating them in the region of T_c and with a biasing electric field. This is because of a combination of high p and low $\tan\delta$. A little below T_c a sufficiently strong biasing field can sweep the domains from the material therefore removing the contribution to losses from domain wall movements. Table 7.1 includes data for 'BST' and 'PST' operated in this mode.

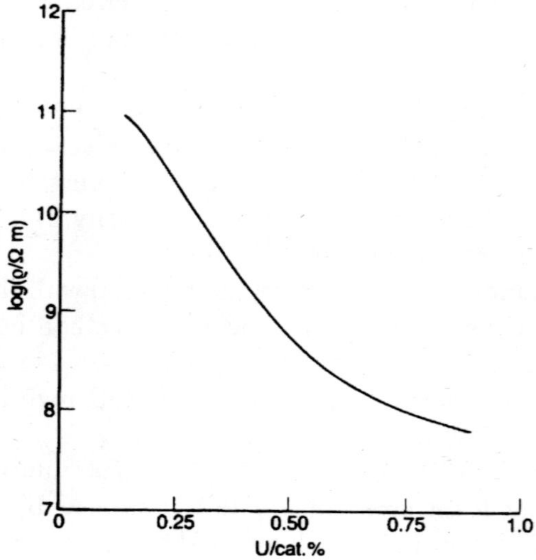

Fig. 7.5 Variation in the resistivity of PZFNTU with the uranium content.

7.5 Measurement of the Pyroelectric Coefficient

The pyroelectric coefficient can be measured in a variety of ways but a direct method is illustrated in Fig. 7.6. From Eq. (7.12)

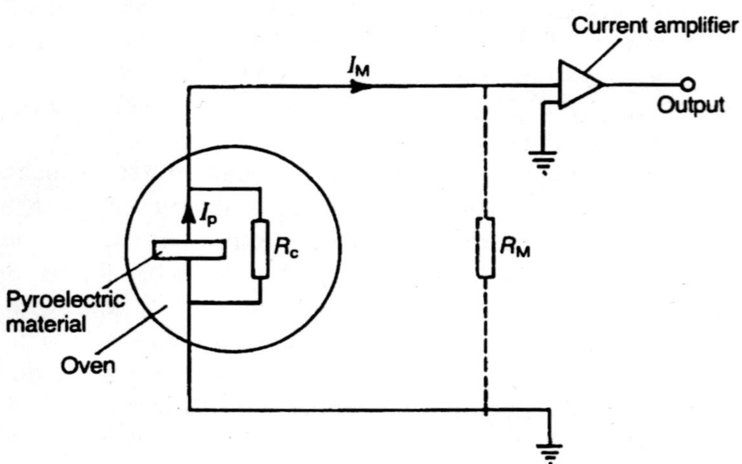

Fig. 7.6 Circuit for the measurement of the pyroelectric coefficient.

$$p = \frac{I_p}{A\dot{T}} \tag{7.32}$$

where

$$I_p = I_M\left(1 + \frac{R_M}{R_c}\right)$$

in which I_M is the measured current and R_M and R_c are the input resistance of the amplifier and the leakage resistance of the crystal respectively. Provided that $R_M \ll R_c$, which is usually the case, then $I_M \approx I_p$ and p can be determined by changing the temperature of the crystal at a known rate.

7.6 Applications

Pyroelectric materials respond to changes in the intensity of incident radiation and not to a temporally uniform intensity. Thus humans or animals moving across the field of view of a detector will produce a response as a result of the movement of their warm bodies which emit infrared radiation ($\lambda \approx 10\,\mu$m). To obtain a response from stationary objects requires the radiation from them to be periodically interrupted. This is usually achieved by a sector disc rotating in front of the detector and acting as a radiation chopper.

All pyroelectric materials are piezoelectric and therefore develop electric charges in response to external stresses that may interfere with the response to radiation. This can largely be compensated for by the provision of a duplicate of the detecting element that is protected from the radiation by reflecting electrodes or masking, but which is equally exposed to air and mounting vibrations. The principle is illustrated in Fig. 7.7. The duplicate is connected in series with the detector and with its polarity opposed so that the piezoelectric outputs cancel. This results in a small reduction in sensitivity (<3 dB) but compensation is an

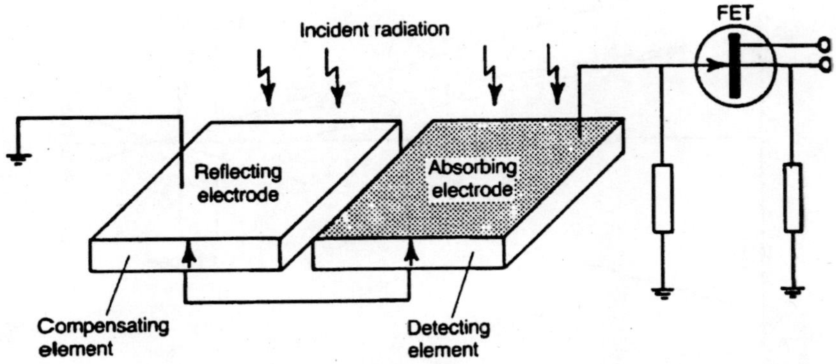

Fig. 7.7 'Dummy' element to compensate for piezoelectrically induced currents.

essential feature in many cases. Using a compensating element of larger area than that of the detecting element, e.g. larger by a factor of 3, does not affect the piezoelectric voltage cancellation but results in a response and detectivity closer to that of the detecting element alone. It may also be necessary to ensure that the detector and compensator are so packaged as to reduce the transmission of vibrations to them.

7.6.1 Radiometry

The radiant power $\mathrm{d}W$ emitted into a solid angle $\mathrm{d}\Omega$ from a small area $\mathrm{d}A$ varies with the angle θ measured from the normal to $\mathrm{d}A$ and is given by

$$\mathrm{d}W = \frac{\sigma\eta T^4}{\pi}\mathrm{d}A\,\mathrm{d}\Omega\cos\theta \tag{7.33}$$

The situation is illustrated in Fig. 7.8 where $\mathrm{d}A$ is a small part of a uniformly radiating surface AB of emissivity η. Provided that θ is small, which is also the condition that r is approximately constant for all radiating elements, Eq. (7.33) becomes

$$\mathrm{d}W \approx \frac{\eta\sigma T^4}{\pi}\mathrm{d}A\mathrm{d}\Omega$$

or

$$\frac{W}{A\mathrm{d}\Omega} \approx \frac{\eta\sigma T^4}{\pi} \tag{7.34}$$

If the radiometer is designed so that the imaged uniformly emitting surface covers the detector surface, then both $A = \int \mathrm{d}A$ and the solid angle $\mathrm{d}\Omega$ subtended

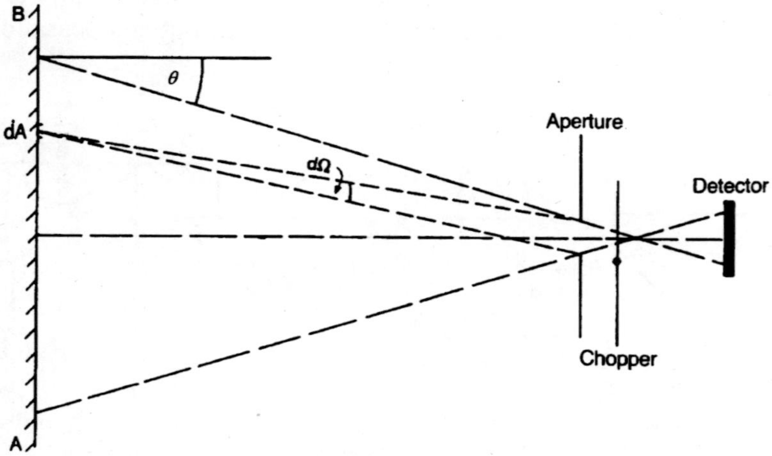

Fig. 7.8 Radiometer optics.

by the aperture at a point on the emitting surface are known. Therefore if W is measured and η is known, T is determined.

The radiation has to be chopped, and so the detector receives pulses of radiation from the source and the chopper alternately. Therefore the temperature is measured relative to the temperature of the chopper surface. In practice, radiometers are calibrated so that the temperature can be read directly from a scale. A particular commercial model which operates over the temperature range 0–600 °C uses a $LiTaO_3$ single crystal as the pyroelectric detector.

7.6.2 Pollutant control

The amount of specific impurities in a gas can be monitored by passing radiation from an infrared source through a tube containing a gas sample and then through a narrow passband optical filter corresponding to a frequency at which the pollutant absorbs radiation to a greater extent than any other constituent of the gas. For example, CO_2 absorbs strongly at about 4.3 μm. The experimental arrangement is illustrated in Fig. 7.9. The response from the infrared detector can be compared with that obtained when using a different filter corresponding to a wavelength to which the gas is transparent. A variety of pollutants can be detected by changing filters.

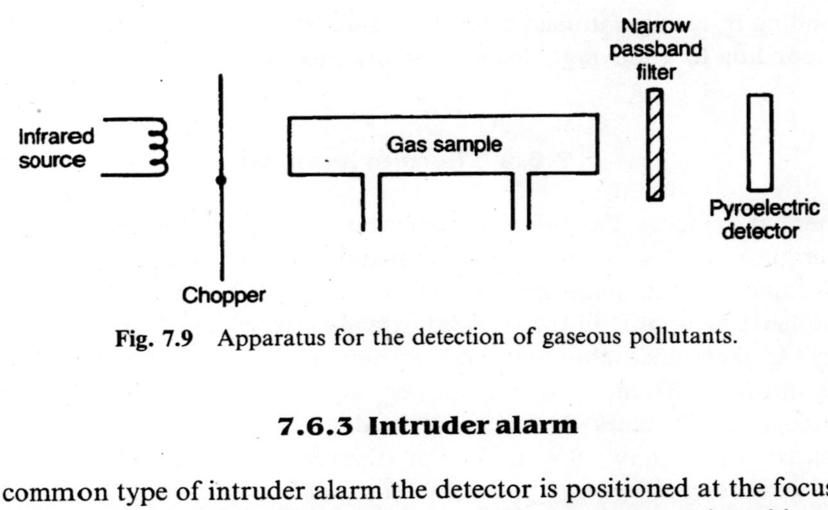

Fig. 7.9 Apparatus for the detection of gaseous pollutants.

7.6.3 Intruder alarm

In a common type of intruder alarm the detector is positioned at the focus of an encircling set of parabolic mirrors as shown in Fig. 7.10. A moving object causes a succession of maxima and minima in the radiation from it reaching the detector.

The wavelength λ_m of radiation emitted with maximum power from a black body at temperature T is given by Wien's displacement law:

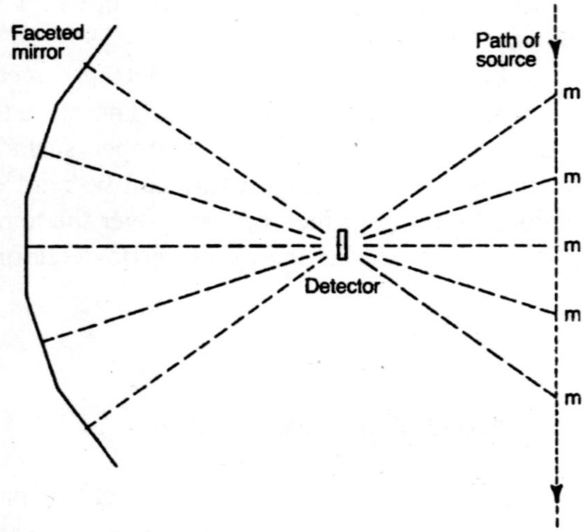

Fig. 7.10 Faceted mirror in an intruder alarm: m indicates the positions of the source giving radiation maxima at the detector.

$$\lambda_m T = 2944\,\mu\text{m K} \tag{7.35}$$

For a black body at 300 K, $\lambda_m = 9.8\,\mu$m. A filter is included which cuts out radiation of wavelength shorter than about 5 μm and so prevents the device from responding to changes in background lighting levels. Such a detector is capable of responding to a moving person up to distances of 100 m.

7.6.4 Thermal imaging

In thermal imaging the infrared radiation emitted by surfaces at different temperatures is focused onto a sensitive plate in an exactly analogous manner to the formation of a photographic image using visible light. There are two atmospheric 'windows' in the infrared extending from 3 to 5 μm and from 8 to 14 μm. As mentioned above, the power radiated from a black body at 300 K peaks around 10 μm, and the power density in the 8–14 μm band is approximately 25 times that radiated in the 3–5 μm band (about 150 W m^{-2} compared with about 6 W m^{-2}). For this reason infrared imaging using pyroelectrics exploits energy in the 8–14 μm band.

The pyroelectric elements used in the devices described so far are commonly square plates with sides about a millimetre long and thicknesses around 30 μm. Because entire scenes are focused onto the plates in thermal imaging, they have to be larger, typically squares of side about 1 cm; the thicknesses are the same as for the simpler devices.

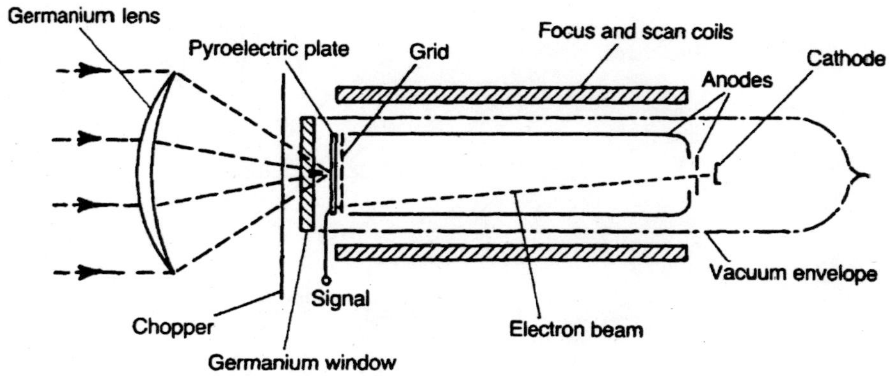

Fig. 7.11 Pyroelectric vidicon.

The effect of a spatially variable radiation flux falling on a pyroelectric plate is to produce a corresponding charge distribution. In imaging devices this charge distribution is detected by a process giving rise to a current which is amplified and electronically processed to produce a television picture corresponding to the infrared radiation from the original scene. A widely used device is the infrared vidicon shown schematically in Fig. 7.11. It is used in many systems, both civil and military. A UK-manufactured camera is used for locating survivors buried in collapsed building rubble.

The radiation from the scene is chopped and focused, using germanium optics, onto an infrared absorbing electrode carried on the front face of the pyroelectric plate. The voltages developed across the plate are typically a few millivolts. The metal grid, positioned close to the detector, collects most of the electrons as it is scanned. Some electrons pass through the grid; the number depends on the local potential at the surface of the plate. The charge deposited on the plate produces a signal at the front electrode which is processed into the television picture.

In the particular mode of operation outlined, when the radiation is interrupted by the chopper, the detector cools, giving negative signals and causing the local surface potential to go negative with respect to the cathode. In order for the detector to be properly addressed by the electron beam it is biased to a potential of approximately +300 mV. This is achieved by the deposition of charge from gas molecules deliberately introduced into the tube and ionized by the electron beam. Circuitry in synchronism with the chopper reverses the sense of the negative output signal current when the chopper is closed and adds this to the positive signal obtained when the chopper is open. This both removes any flicker in the picture due to detector or bias non-uniformity and increases the sensitivity for imaging moving objects.

The charge on the surface of the plate must be removed before the electron-beam scan is repeated. This will occur automatically if the bias is high enough to ensure that sufficient electrons can reach the plate to neutralise the pyroelectric

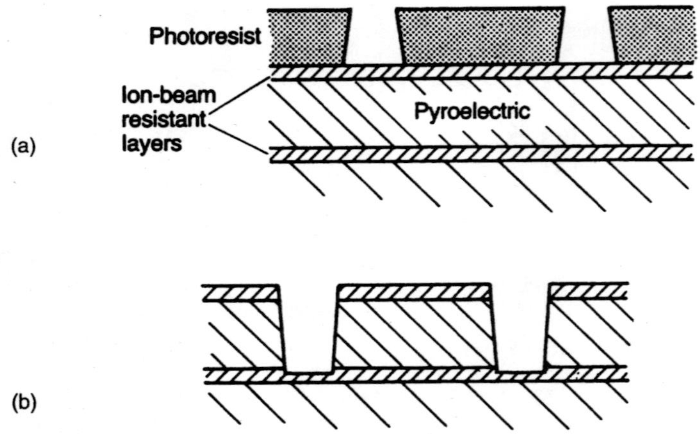

Fig. 7.12 Reticulation to improve thermal isolation: (a) before milling; (b) after milling.

charges. Sufficient ions must be created in the residual gas to provide the required bias.

Lateral flow of heat in the pyroelectric detector plate tends to even out the temperature differences, blurring the charge pattern and hence the final image. This process will depend on the thickness h of the plate and the thermal diffusivity $\Lambda = \lambda/c'$ of the plate material. Vidicon performance improves as both h and Λ are reduced in value, and a suitable figure of merit is

$$F_{\text{vid}} = \frac{p}{c'\varepsilon\Lambda h} = \frac{p}{\lambda\varepsilon h} = \frac{F_{\text{V}}}{\Lambda h} \tag{7.36}$$

According to Eq. (7.36) PVDF is an attractive candidate for vidicon targets, and it has been used, although it does not approach the performance obtained using TGS and other members of that family.

The effect of thermal diffusion can be countered by reticulating the target. Reticulation involves the formation of grooves that cut through the thickness of the target almost completely (or completely) and divide it into small square regions which are nearly (or completely) separate, as illustrated in Fig. 7.12. Suitable patterns for the grooves can be formed by depositing a metal, or some other material resistant to ion-beam etching, on the target surface and then partially removing it in the required pattern by photo-engraving. The target is then exposed to an ion beam that etches away the bare pyroelectric material far more rapidly than the metal or other protective layer. This method has been used

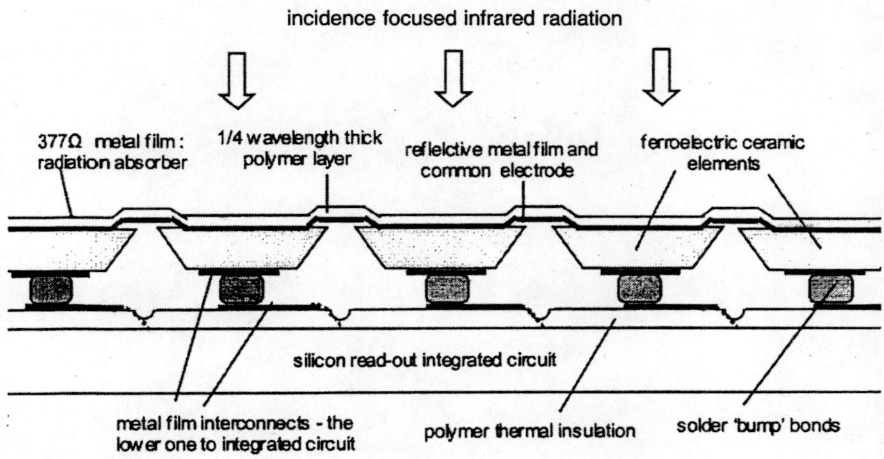

Fig. 7.13 Ferroelectric pyroelectric imaging array structure (after R.W. Whatmore and R. Watton [2]).

to divide TGS crystals some 30 μm thick into islands 20 μm wide separated by 5 μm wide grooves.

As an attractive alternative to the vidicon the array of elements can be connected directly to individual voltage detectors on a silicon integrated circuit (usually referred to as a ROIC: 'read-out integrated circuit') via solder 'bumps'. The resulting signals can then be electronically processed to yield a picture in the standard way. Fig. 7.13 shows a section through part of such an imaging array.

Arrays can be fabricated in the following way. A wafer, $\sim 15\,\mu$m thick by ~ 75 mm diameter is prepared from a hot-pressed ceramic block (e.g. 'PST' or 'BST') by diamond sawing, lapping and polishing. Both faces are coated with metal films by sputtering after which a polymer film is 'spun' onto one face. The coated wafer is then bonded to a glass support for the reticulation process carried out by laser-assisted etching under potassium hydroxide. This gives an element pitch of down to $\sim 40\,\mu$m with groove widths down to $\sim 10\,\mu$m. The structure is then flip-chip bonded to the integrated circuit after which the glass support layer is removed to expose the $\lambda/4$ polymer/metal film radiation-absorbing structure.

To reduce expense, efforts are made to exploit integrated thin film technologies. For example, arrays have been produced via thin film deposition of the pyroelectric onto a sacrificial layer, e.g. a suitable metal or polysilicon, which is then selectively etched away. Thermal isolation of the pyroelectric element is achieved through engineering a gap between it and the ROIC silicon wafer. Vias in the supporting layer permit electrical connections to be made between the detector and the wafer via solder bonds. Imaging arrays have been produced in this way incorporating sputtered PST and sol-gel formed PZT films.

Fig. 7.14 A thermal image produced by an infrared camera exploiting uncooled pyroelectric ceramic detection technology. (Courtesy of QinetiQ Limited.)

The ideal route would be one in which the pyroelectric detector material is laid down in thin film form by a route compatible with the production of the silicon ROIC. There are obvious parallels with the development of FeRAMS (see Section 5.7.5) and the substantial effort now devoted to their development will have a positive impact on the manufacture of pyroelectric arrays. Challenges lie in the requirement to process the deposited films at temperatures not too high for the underlying integrated circuit, and the need to engineer the temperature diffusion characteristics within the element and its surroundings so as to optimise image definition.

The type of infrared imager described is now widely exploited for both civilian and military use and typical image quality currently achieved is illustrated in Fig. 7.14.

Pyroelectric infrared detectors are inferior in detectivity by one or two orders of magnitude compared with photoconductors such as cadmium mercury telluride, as shown in Fig. 7.15. However, such materials require temperatures of 200 K for efficient operation and generally respond to rather narrow bands at the infrared wavelengths. Pyroelectric devices can discriminate temperature differences of 0.1 K but find many useful applications in which the discrimination is limited to about 0.5 K. They have the great practical advantage of operating at normal ambient temperatures.

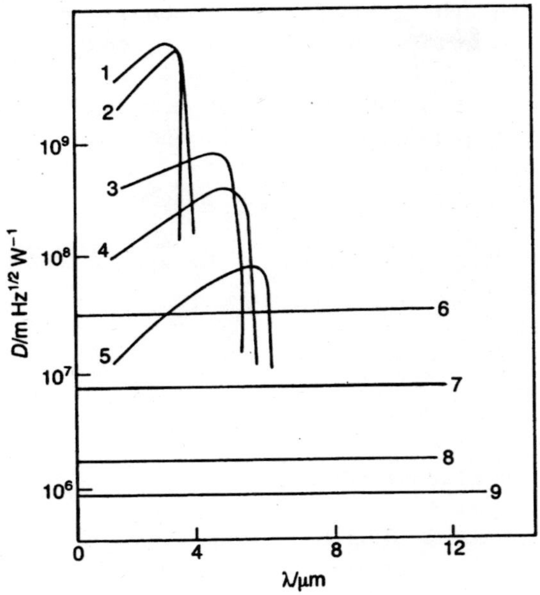

Fig. 7.15 Detectivities D of some photoconductive and pyroelectric detector materials: curve 1, PbS; curve 2, InAs; curve 3, HgCdTe; curve 4, PbSe; curve 5, InSb; curve 6, L-alanine-doped TGS; curve 7, $LiTaO_3$; curve 8, PZ; curve 9, PVDF. (Curves 1–5 measured at 200 K; curves 6–9 measured at 300 K.)

The uncooled thermal imaging technology exploiting ferroelectric ceramics is being challenged by other technologies, significantly by one based upon 'resistance bolometer' materials such as vanadium oxide (VO_x) and amorphous silicon.

Problems

1. Explain why it is that the sun appears to the eye as a uniformly bright disc. Would it appear equally bright were its distance from us doubled?

2. Check by integration that Eq. (7.33) is consistent with the expression $W = \sigma T^4$ for the total power radiated from unit area of a black body.

3. Make an order of magnitude estimate of the time taken for a plate of pyroelectric material 30 μm thick to achieve a uniform temperature if the face is suddenly illuminated with infrared radiation ($\lambda \approx 4.0\,\mathrm{W\,m^{-1}\,K^{-1}}$; $\rho = 7.45\,\mathrm{Mg\,m^{-3}}$; $c = 0.33\,\mathrm{kJ\,kg^{-1}\,K^{-1}}$). [Answer: ≈ 0.1 ms]

4. A TGS plate of uniform thickness (30 μm) receives a uniform radiation flux for 1 ms at normal incidence. Using data given in Table 7.1, calculate the lower radiation intensity

threshold, given that a change in voltage of 1 nV on the pyroelectric element can be detected. [Answer: $\approx 2.8\,\mu W\,m^{-2}$]

5. (a) Describe and justify a suitable route for preparing a ceramic for use as a pyroelectric detector. The description should include co-precipitation, hot pressing, control of porosity and grain size. (b) Describe a route for the construction of a thin film PST-based pyroelectric array for thermal imaging.

6. Explain concisely the underlying physics of the $\lambda/4$ radiation absorbing structure shown in Fig. 7.13.

7. A ceramic pyroelectric material has the formula

$$PbZr_{0.8}Fe_{0.1}Nb_{0.1}Ti_{0.02}U_{0.0021}O_3$$

What do each of these cations contribute?

8. What is the objection to using PZT compositions of the type developed for piezoelectric applications in a pyroelectric infrared detector? What are the properties needed and how have they been achieved?

9. A pyroelectric ceramic has a tan δ of 0.005 arising solely due to dielectric losses and $\varepsilon_r = 250$ at 100 Hz. What is the minimum resistivity that the ceramic can have without incurring a 20% increase in tan δ. What will be the consequent effect on the noise-related figure of merit of a 20% increase? [Answers: $7.2 \times 10^8\,\Omega\,m$; 10% decrease]

Bibliography

1. Whatmore, R.W. (1986) Pyroelectric devices and materials, *Rep. Prog. Phys.*, **49**, 1335–86.
2. Whatmore, R.W. and Watton, R. (2001) Pyroelectric Materials and Devices. In: *Infrared Detectors and Emitters: Materials and Devices*, Kluwer Academic, Chapter 5, pp. 99–148.
3. Smith, R.A., Jones, F.E. and Chasmar, R.P. (1957) *The Detection and Measurement of Infra-red Radiation*, Clarendon Press, Oxford.

8

ELECTRO-OPTIC CERAMICS

The last decade has seen an intensification of research and development effort into 'photonics', a term embracing the many related optical technologies harnessed principally for communications and the processing of information. Specifically, it is concerned with generating, amplifying, detecting, guiding, modulating, or modifying optical radiation and covers a range of interests including lasers, opto-electronics, fibre-optics, non-linear optics, acousto-optics, optical data storage and electro-optics. The focus of attention here is mainly on the electro-optical effects in polar ceramics. However, the last decade has also seen a strong interest grow in the non-linear properties of poled glasses and the potential they have for modifying and controlling optical signals. The discussion includes a brief summary of this development.

The article by G.H. Haertling [1], a pioneer in the field of electro-optical ceramics, and the monographs by J.M. Herbert [2] and F. Aguilló-López *et al.* [3], which cover much of the same ground as the present text but in greater detail, are recommended to supplement the discussion. The classical work by M.E. Lines and A.M. Glass [4] and the monograph by J.C. Burfoot and G.W. Taylor [5] are invaluable. The text by R.M. Wood [6] and the review of 'poled glasses' by W. Margulis *et al.* [7] are especially suited to the 'materials' student.

There are many books covering the background optics and that by A. Yariv and P. Yeh [8] can be recommended.

8.1 Background Optics

To appreciate properly how electro-optic ceramics function, it is first necessary to consider the nature of light and its interaction with dielectrics.

James Clerk Maxwell (1831–1879), against a background of experimental and theoretical work by André Ampère (1775–1836). Karl Gauss (1777–1855) and Michael Faraday (1791–1867), developed the electromagnetic wave theory of

Electroceramics: Materials, Properties, Applications. 2nd Edition. Edited by A. J. Moulson and J. M. Herbert.
 ISBN 0 471 49747 9 (hardback) 0 471 49748 7 (paperback)

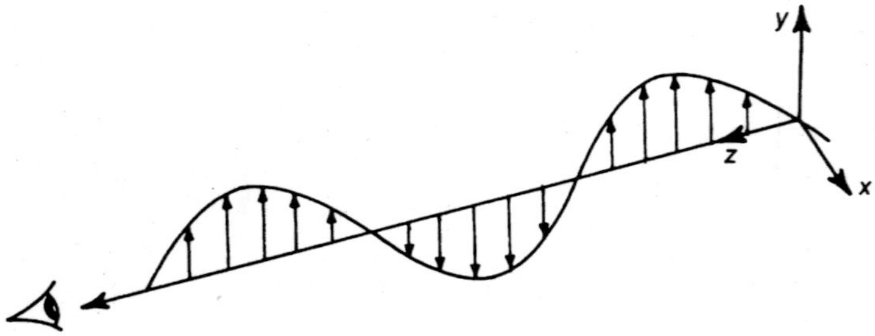

Fig. 8.1 *E* vector of a wave polarized in the *y–z* plane and propagating in the *z* direction towards the observer.

light. Maxwell's equations describe how an electromagnetic wave originates from an accelerating charge and propagates in free space with a speed of $2.998 \times 10^8\,\mathrm{m\,s^{-1}}$. An electromagnetic wave in the visible part of the spectrum may be emitted when an electron changes its position relative to the rest of an atom, involving a change in dipole moment. Light can also be emitted from a single charge moving at high speed under the influence of a magnetic field: because the charge follows a curved trajectory it is accelerating and, as a consequence, radiating. The radiation, termed 'synchrotron radiation', is emitted naturally from regions of the Universe, e.g. the Crab Nebula. An electromagnetic wave in free space comprises an electric field $\boldsymbol{E}$ and a magnetic induction field $\boldsymbol{B}$ which vibrate in mutually perpendicular directions in a plane normal to the wave propagation direction. The $\boldsymbol{E}$ vector of a single sinusoidal plane-polarized wave propagating in the z direction is shown in Fig. 8.1. The wave is polarized in the y–z plane, and the convention is to describe the plane to which $\boldsymbol{E}$ is confined as the plane of polarization.

Radiation from a single atom, say in an electric light bulb filament, only persists in phase and polarization for a time of the order of 10^{-8} s, so that light from such sources contains a random mixture of polarizations and phases as well as a wide range of wavelengths. Such radiation is said to be unpolarized, incoherent and white, but by various means (particularly the use of laser sources) light can be obtained with a specified polarization, a coherence persisting for times in the neighbourhood of 10^{-5} s and a linewidth of less than 10 Hz at a frequency of 10^{14} Hz. For many purposes the polarization is the simplest feature to control optically and to alter electrically.

8.1.1 Polarized light

The various forms of polarization can be understood by considering a plane-polarized wave travelling in the z direction. It can be resolved into two

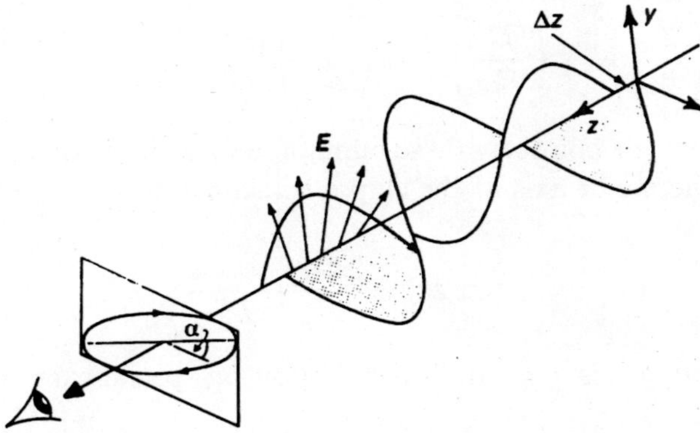

Fig. 8.2 The two components of a right elliptically polarized wave.

components, one polarized in the x–z plane and the other in the y–z plane, as shown in Fig. 8.2. A phase difference can be established between the two components, for instance by passage through a medium with an anisotropic refractive index such that the velocity of the y–z wave is greater than that of the x–z wave. This will cause the y–z wave to lead the x–z wave by a distance, say, Δz, i.e. by a phase angle $\delta = 2\pi\Delta z/\lambda$. δ will remain constant when the light emerges into an isotropic medium such as air.

Confining attention to the electric fields, we have for the x wave,

$$E_x = E_{0x}\sin(kz - \omega t) \tag{8.1}$$

and for the y wave

$$E_y = E_{0y}\sin(kz - \omega t + \delta) \tag{8.2}$$

in which E_{0x} and E_{0y} are the amplitudes, k is the propagation number ($k = 2\pi/\lambda$) and ω is the angular frequency. The locus of the tip of the resultant **E** can now be determined as the waves pass a plane $z =$ constant. If, for simplicity, we put $z = 0$ in Eqs (8.1) and (8.2) we obtain

$$E_x = -E_{0x}\sin(\omega t) = E_{0x}\sin(\omega t) \tag{8.3}$$

and

$$E_y = E_{0y}\sin(-\omega t + \delta) = -E_{0y}\sin(\omega t - \delta)$$

Then

$$E_y = -E_{0y}\{\sin(\omega t)\cos\delta - \cos(\omega t)\sin\delta\} \tag{8.4}$$

Substituting $\sin(\omega t) = -E_x/E_{0x}$ from Eq. (8.3) into (8.4) and rearranging yields

$$\left(\frac{E_x}{E_{0x}}\right)^2 + \left(\frac{E_y}{E_{0y}}\right)^2 - 2\left(\frac{E_x}{E_{0x}}\right)\left(\frac{E_y}{E_{0y}}\right)\cos\delta = \sin^2\delta \tag{8.5}$$

which represents an ellipse with semiminor and semimajor axes E_{0y} and E_{0x} respectively. The major axis of the ellipse is inclined to the x axis at an angle α, where

$$\tan(2\alpha) = \frac{2E_{0x}E_{0y}\cos\delta}{E_{0x}^2 - E_{0y}^2} \tag{8.6}$$

The above treatment is general, but the following special cases are important.

1. When $\delta = \pi/2$

$$\left(\frac{E_x}{E_{0x}}\right)^2 + \left(\frac{E_y}{E_{0y}}\right)^2 = 1 \tag{8.7}$$

and the tip of $\boldsymbol{E}$ describes an ellipse whose major and minor axes coincide with the x and y axes. When $E_{0x} = E_{0y}$ the ellipse becomes a circle.

2. When $\delta = 0$

$$\left(\frac{E_x}{E_{0x}}\right)^2 + \left(\frac{E_y}{E_{0y}}\right)^2 - 2\frac{E_x}{E_{0x}}\frac{E_y}{E_{0y}} = 0$$

i.e.

$$\left(\frac{E_x}{E_{0x}} - \frac{E_y}{E_{0y}}\right)^2 = 0$$

or

$$E_y = \frac{E_{0y}}{E_{0x}}E_x \tag{8.8}$$

The tip of $\boldsymbol{E}$ describes a straight line passing through the origin with slope E_{0y}/E_{0x}.

The paths described by the tip of $\boldsymbol{E}$ for the various values of δ are summarized in Fig. 8.3. It is apparent that plane-polarized light is a special case of elliptically polarized light, as indeed is circularly polarized light. The sense of the

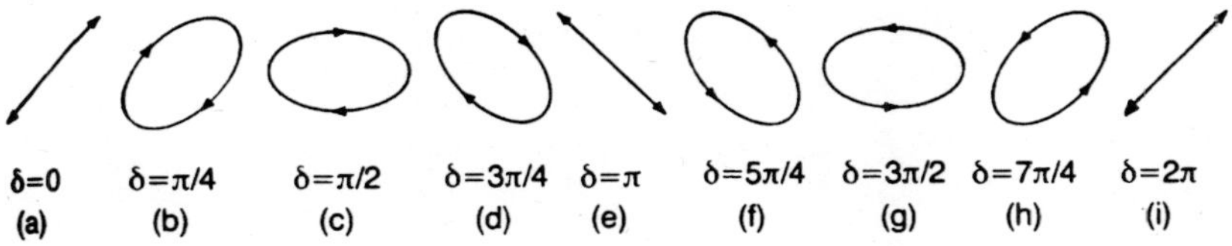

Fig. 8.3 How elliptically polarized light depends on the phase difference δ between plane-polarized components.

polarization is taken to be the sense of rotation of the $\boldsymbol{E}$ vector for oncoming light. For example (b) is described as 'right elliptically polarized light', (f) as 'left elliptically polarized light', and (a) and (e) simply as 'plane-polarized light'. This seems to be the most commonly adopted convention, although the opposite one might be encountered which is, in fact, more consistent with the definitions regarding the polarization and angular momentum of a photon.

There are materials, for example in the form of certain specially prepared polymer films, which, for light incident normal to the film, absorb to an extent dependent on the inclination of the plane of polarization to a unique axis in the plane of the film. Devices made from such films are termed 'polarizers': approximately 60% of the incident unpolarized light is absorbed, and that part transmitted is plane polarized. The $\boldsymbol{E}$ vectors for the transmitted light are perpendicular to the high-absorbance direction. If the incident light is plane polarized, the intensity transmitted depends on the orientation of the polarizer axis with respect to the plane of polarization of the light. A device used in this mode is usually referred to as an 'analyser'.

8.1.2 Double refraction

In the discussion of the piezoelectric effect in Chapter 6 the tensor character of the permittivity of a dielectric was recognized although attention was focused on the piezoelectric coefficients. Because the optical and electro-optical properties of dielectrics are determined by their refractive indices or, equivalently, by their permittivities (see Eq. (2.120)), it is now necessary to consider these parameters in some detail.

In an isotropic dielectric such as a normal glass, the induced electrical polarization is always parallel to the applied electric field and therefore the susceptibility is a scalar. In general this is not the case in anisotropic dielectrics when the polarization depends on both the direction and magnitude of the applied field. The three components of the polarization are written

$$P_x = \varepsilon_0(\chi_{11}E_x + \chi_{12}E_y + \chi_{13}E_z)$$
$$P_y = \varepsilon_0(\chi_{21}E_x + \chi_{22}E_y + \chi_{23}E_z) \quad (8.9)$$
$$P_z = \varepsilon_0(\chi_{31}E_x + \chi_{32}E_y + \chi_{33}E_z)$$

or

$$D_x = \varepsilon_{11}E_x + \varepsilon_{12}E_y + \varepsilon_{13}E_z$$
$$D_y = \varepsilon_{21}E_x + \varepsilon_{22}E_y + \varepsilon_{23}E_z \quad (8.10)$$
$$D_z = \varepsilon_{31}E_x + \varepsilon_{32}E_y + \varepsilon_{33}E_z$$

which can be abbreviated to

$$D_i = \varepsilon_{ij}E_j \quad (8.11)$$

with summation over repeated indices implied.

In an anisotropic dielectric the phase velocity of an electromagnetic wave generally depends on both its polarization and its direction of propagation. The solutions to Maxwell's electromagnetic wave equations for a plane wave show that it is the vectors $\boldsymbol{D}$ and $\boldsymbol{H}$ which are perpendicular to the wave propagation direction and that, in general, the direction of energy flow does not coincide with this.

The classical example of an anisotropic crystal is calcite ($CaCO_3$). This is the crystal with which the first recorded observation of 'double refraction' was made by Bartolinius in 1669. Because of the particular arrangement of atoms (and so of electric charges) in calcite, light generally propagates at a speed depending on the orientation of its plane of polarization relative to the crystal structure. For one particular direction, the *optic axis*, the speed of propagation is independent of the orientation of the plane of polarization. Because in calcite there is only one such axis the crystal is termed *uniaxial.*

Crystals may have two optic axes, not necessarily perpendicular, in which case they are termed *biaxial.* Orthorhombic, monoclinic and triclinic crystals are biaxial; hexagonal, tetragonal and trigonal crystals are uniaxial; cubic crystals are isotropic. In the following discussion attention is confined to uniaxial crystals.

The situation can be analysed more closely by considering a source of monochromatic light located at S in Fig. 8.4 from which two *wavefronts* propagate, one with a spherical surface and the other with an ellipsoidal surface. A wavefront is the locus of points of equal phase, i.e. the radii (e.g. SO and SE), which are proportional to the ray speeds and inversely proportional to the refractive indices. Figure 8.4 is a principal section of the wavefront surfaces because it contains the optic axis SY.

Those rays, e.g. SO, for which the electric displacement component of the wave vibrates at right angles to the principal section (indicated by the dots) travel at a constant speed irrespective of direction; they are the *ordinary* or o rays.

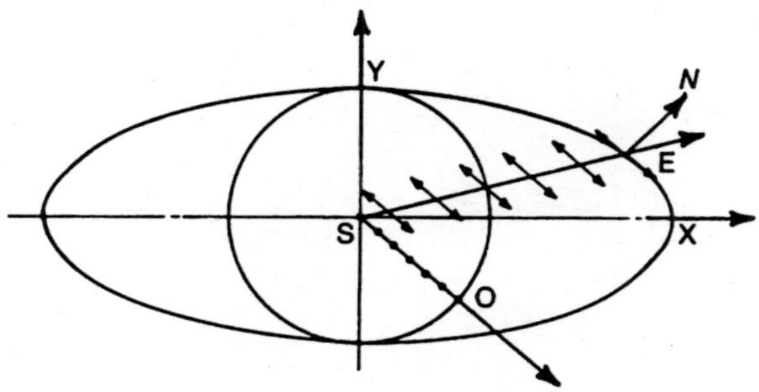

Fig. 8.4 A principal section of the wavefront surfaces for a uniaxial crystal; the dots on SO represent the $\boldsymbol{E}$ (and $\boldsymbol{D}$) vectors which are normal to the plane of the paper.

Those rays, e.g. SE, for which the electric displacement component lies in the principal section travel at a speed which depends on direction; they are the extraordinary or e rays. The refractive index n_e for an e ray propagating along SX is one of the two principal refractive indices of a uniaxial crystal; the other, n_o, refers to the o rays.

In Fig. 8.4 the velocity of the o ray is less than that of the e ray, except for propagation along the optic axis. It follows that $n_e - n_o < 0$, a situation defined as negative birefringence. Cases of both positive and negative birefringence occur: calcite and rutile (TiO_2) have values of -0.18 and $+0.29$ respectively, and both crystals are very birefringent. Values for the common ferroelectrics lie in the range -0.01 to -0.1.

It is customary to develop the optical theory of crystals in terms of the relative impermeability tensor $B_{ij} = (\varepsilon_r^{-1})_{ij}$, which is a second-rank tensor. It can be shown that B_{ij} is symmetrical, a property that requires $B_{ij} = B_{ji}$. It can also be shown that because B_{ij} is a symmetrical second-rank tensor it can be represented by

$$B_{ij}x_ix_j = 1 \tag{8.12}$$

which, when expanded, is a second-degree equation representing a surface. The notation in Eq. (8.12) implies summation over values of $j = 1, \ldots, 3$ for each value of $i = 1, \ldots, 3$. The equation is called the 'representation quadric' and possesses principal axes. The representation quadric for the relative impermeability tensor B_{ij}, with reference to its principal axes, reduces to

$$B_1x_1^2 + B_2x_2^2 + B_3x_3^2 = 1 \tag{8.13}$$

in which B_1 etc. are the principal relative impermeabilities. Under the assumption that the material is non-magnetic ($\mu_r = 1$) (see Eq. 2.120)

$$B_1 = (\varepsilon_r^{-1})_1 = \frac{1}{n_1^2} \text{ etc.} \tag{8.14}$$

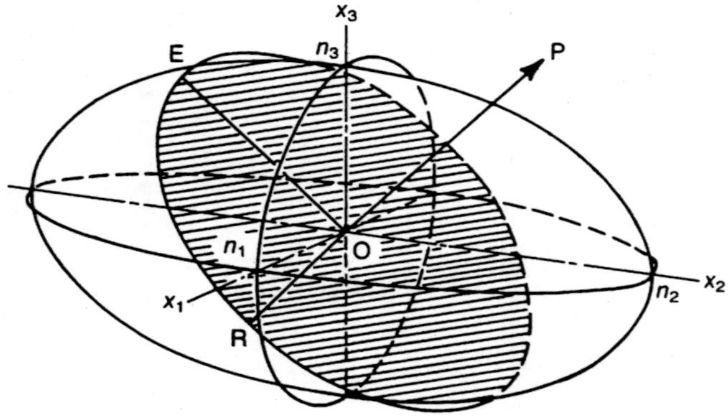

Fig. 8.5 The optical indicatrix.

where n_1 is the refractive index for light whose electric displacement is parallel to x_1. Eq. (8.13) can be written

$$\frac{x_1^2}{n_1^2}+\frac{x_2^2}{n_2^2}+\frac{x_3^2}{n_3^2}=1 \tag{8.15}$$

and is known as the *optical indicatrix*. This is an ellipsoidal surface, as shown in Fig. 8.5, and n_1, n_2 and n_3 are the principal refractive indices of the crystal. For biaxial crystals $n_1 \neq n_2 \neq n_3$.

If in Fig. 8.5 OP is an arbitrary direction, the semiminor and semimajor axes OR and OE of the shaded elliptical section normal to OP are the refractive indices of the two waves propagated with fronts normal to OP. For each of the two waves associated with a given wave normal, the electric displacement $\boldsymbol{D}$ vibrates parallel to the corresponding axis of the elliptical section.

In the special case when Ox_2 is the propagation direction the two waves have refractive indices n_1 and n_3; similarly, the two waves propagated along Ox_3 have refractive indices n_1 and n_2, and those propagated along Ox_1 have refractive indices n_2 and n_3.

For a uniaxial crystal the indicatrix is symmetrical about the principal symmetry axis of the crystal – the *optic axis*. If x_3 is the optic axis, the central section of the ellipsoid is a circle of radius n_o and the equation becomes

$$\frac{x_1^2}{n_o^2}+\frac{x_2^2}{n_o^2}+\frac{x_3^2}{n_e^2}=1 \tag{8.16}$$

8.1.3 The electro-optic effect

In the earlier discussion on dielectrics (see Section 2.7.1) a linear relationship between $\boldsymbol{P}$ and $\boldsymbol{E}$ was assumed. The justification for this assumption rests on

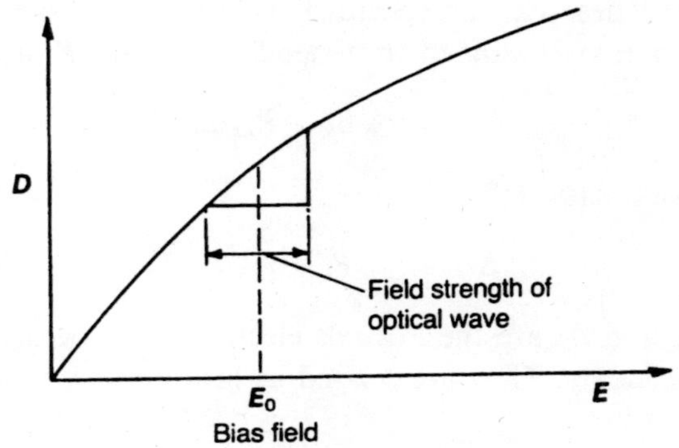

Fig. 8.6 Non-linearity in the ***D*** versus ***E*** relationship.

experiment, and the underlying cause is that the largest field strengths commonly encountered in practice (about $10^6\,\mathrm{V\,m^{-1}}$) are small compared with those that bind electrons in atoms (about $10^{11}\,\mathrm{V\,m^{-1}}$).

However, the linear response of a dielectric to an applied field is an approximation; the actual response is non-linear and is of the form indicated in Fig. 8.6. The electro-optic effect has its origins in this non-linearity, and the very large electric fields associated with high-intensity laser light lead to the non-linear optics technology discussed briefly in Section 8.1.4. Clearly the permittivity measured for small increments in field depends on the biasing field E_0, from which it follows that the refractive index also depends on E_0. The dependence can be expressed by the following polynomial:

$$n = n^0 + aE_0 + bE_0^2 \cdots \tag{8.17}$$

in which n^0 is the value measured under zero biasing field and a and b are constants. This dependence of n on E_0 is the electro-optic effect.

It is evident that if the material has a centre of symmetry or a random structure as in the case of a normal glass, reversal of E_0 will have no effect on n. This requires a to be zero, so that there is only a quadratic dependence of n on E_0 (and, possibly, a dependence on higher even powers). If, however, the crystal is non-centrosymmetric, reversal of E_0 may well affect n and so the linear term has to be retained.

In 1875 John Kerr carried out experiments on glass and detected electric-field-induced optical anisotropy. A quadratic dependence of n on E_0 is now known as the Kerr effect. In 1883 both Wilhelm Röntgen and August Kundt independently reported a linear electro-optic effect in quartz which was analysed by Pockels in 1893. The linear electro-optical effect is termed the Pockels effect.

The small changes in refractive index caused by the application of an electric field can be described by small changes in the shape, size and orientation of the optical indicatrix. These can be specified by changes in the coefficients of the indicatrix ΔB_{ij} which are assumed to depend on $\boldsymbol{E}$ (and $\boldsymbol{P}$) as follows:

$$\Delta B_{ij} = r_{ijk}E_k + R_{ijkl}E_kE_l \tag{8.18}$$

or, in terms of polarization $\boldsymbol{P}$,

$$\Delta B_{ij} = f_{ijk}P_k + g_{ijkl}P_kP_l \tag{8.19}$$

The constants r_{ijk} and f_{ijk} are the Pockels electro-optic coefficients, and R_{ijkl} and g_{ijkl} the Kerr coefficients. They are related as follows:

$$f_{ijk} = \frac{r_{ijk}}{\varepsilon_k - \varepsilon_0} \tag{8.20}$$

and

$$g_{ijkl} = \frac{R_{ijkl}}{(\varepsilon_k - \varepsilon_0)(\varepsilon_l - \varepsilon_0)} \tag{8.21}$$

where ε_k and ε_l are the permittivity values measured with $\boldsymbol{E}$ directed along k or l. The subscript on E_0 has now been dropped because the distinction between a bias field and the electric field component of an electromagnetic wave should be apparent from the context.

Whether or not the dependence is expressed in terms of $\boldsymbol{E}$ or $\boldsymbol{P}$ is a matter of choice; it seems customary in the literature relating to single crystals to use the r coefficient for the linear Pockels effect and g for the quadratic Kerr effect. In the case of electro-optic ceramics r and R are most commonly used.

The r and f are third-rank tensors and the R and g are fourth-rank tensors, and since they are both symmetrical the subscripts i and j can be interchanged, as also can k and l. Under these circumstances it is helpful to use a 'contracted' or 'reduced' notation such that $r_{ijk} \rightarrow r_{mk}$ and $R_{ijkl} \rightarrow R_{mn}$, where m and n run from 1 to 6. The $ij = m$ and $kl = n$ contractions are as follows: $11 \rightarrow 1, 22 \rightarrow 2, 33 \rightarrow 3, 23 \rightarrow 4, 13 \rightarrow 5$ and $12 \rightarrow 6$.

A complete description of the electro-optic effect for single crystals necessitates full account being taken of the tensorial character of the electro-optic coefficients. The complexity is reduced with increasing symmetry of the crystal structure when an increasing number of tensor components are zero and others are simply interrelated. The main interest here is confined to polycrystalline ceramics with a bias field applied, when the symmetry is high and equivalent to ∞mm (6 mm) and so the number of tensor components is a minimum. However, the approach to the description of their electro-optic properties is formally identical with that for the more complex lower-symmetry crystals where up to a maximum of 36 independent tensor components may be required to describe their electro-optic properties fully. The methods are illustrated below with reference to single-crystal $BaTiO_3$ and a polycrystalline electro-optic ceramic.

The Pockels electro-optic effect

Single-crystal $BaTiO_3$ $(T<T_c)$

At temperatures below T_c $BaTiO_3$ belongs to the tetragonal crystal class (symmetry group 4mm); it is optically uniaxial, and the optic axis is the x_3 axis ($n_o = 2.416, n_e = 2.364$). When an electric field is applied in an arbitrary direction the representation quadric for the relative impermeability is perturbed to

$$(B_{ij} + \Delta B_{ij})x_i x_j = 1 \tag{8.22}$$

where $\Delta B_{ij} = r_{ijk}E_k$. Employing the contracted notation and referring to the principal axes of the unperturbed ellipsoid (Eq. (8.12)) we can expand Eq. (8.22) to

$$\left(\frac{1}{n_1^2} + r_{1k}E_k\right)x_1^2 + \left(\frac{1}{n_1^2} + r_{2k}E_k\right)x_2^2 + \left(\frac{1}{n_3^2} + r_{3k}E_k\right)x_3^2 \\ + 2x_2x_3r_{4k}E_k + 2x_3x_1r_{5k}E_k + 2x_1x_2r_{6k}E_k = 1 \tag{8.23}$$

in which k takes the values 1, . . ., 3. From considerations of crystal symmetry it can be shown that $r_{23} = r_{13}$ and $r_{42} = r_{51}$, and that, except for r_{33}, the remaining 13 tensor components are zero. Furthermore if we assume that the electric field $\boldsymbol{E}$ is directed along the x_3 axis so that $E_1 = E_2 = 0$, Eq. (8.23) reduces to

$$\left(\frac{1}{n_o^2} + r_{13}E\right)x_1^2 + \left(\frac{1}{n_o^2} + r_{13}E\right)x_2^2 + \left(\frac{1}{n_e^2} + r_{33}E\right)x_3^2 = 1 \tag{8.24}$$

It follows that, in comparison with Eq. (8.22),

$$\Delta\left(\frac{1}{n_o^2}\right) = r_{13}E \qquad \text{and} \qquad \Delta\left(\frac{1}{n_e^2}\right) = r_{33}E$$

or, since $\Delta n_o \ll n_o$ and $\Delta n_e \ll n_e$,

$$\begin{aligned} \Delta n_o &= -\tfrac{1}{2}n_o^3 r_{13}E \\ \Delta n_e &= -\tfrac{1}{2}n_e^3 r_{33}E \end{aligned} \tag{8.25}$$

The induced birefringence is Δn, where

$$\Delta n = \Delta(n_e - n_o) = \Delta n_e - \Delta n_o = -\tfrac{1}{2}n_e^3\left(r_{33} - \frac{n_o^3}{n_e^3}r_{13}\right)E \tag{8.26}$$

or, simply,

$$\Delta n = -\tfrac{1}{2}n^3 r_c E$$

where

$$r_c = r_{33} - \frac{n_o^3}{n_e^3} r_{13} \tag{8.27}$$

$r_c \approx r_{33} - r_{13}$ if $n \approx n_o \approx n_e$.

Polycrystalline ceramic

The form of the electro-optic tensor for 6mm symmetry is identical with that for the 4mm symmetry dealt with above. Therefore the induced birefringence for a field directed along the x_3 axis is again

$$\Delta n = -\tfrac{1}{2} n^3 r_c E \tag{8.28}$$

where $r_c = r_{33} - r_{13}$.

A useful figure of merit for linear electro-optic materials is the half-wave voltage V_π which is the voltage that must be applied to opposite faces of a unit cube of material to induce a phase difference of π between the extraordinary and ordinary rays. It is discussed further in Section 8.2.2 (see Eq. (8.44)). Values for V_π are included in Table 8.1.

The Kerr quadratic electro-optic effect

Single-crystal $BaTiO_3$ ($T > T_c$)

At temperatures above 130 °C $BaTiO_3$ is cubic (symmetry group m3m); it is optically isotropic with $n = 2.42$. When an electric field is applied in an arbitrary direction the representation quadric is perturbed to

$$B_{ij} + \Delta B_{ij} = 1 \tag{8.29}$$

where $\Delta B_{ij} = R_{ijkl} E_k E_l$.

Table 8.1 Properties of some electro-optic materials

Material	r_c/pm V^{-1}	V_π/V	R/nm^2 V^{-2}	$\lvert g_{11} - g_{12}\rvert$/m^4 C^{-2}
Single crystals				
$BaTiO_3$ (room temperature)	80	500	–	–
$BaTiO_3$ (>130 °C)	–	–	–	0.13
$LiNbO_3$	17.5	3030	–	–
$LiTaO_3$	22	2800	–	–
$SrTiO_3$	–	–	–	0.14
$KTa_{0.65}Nb_{0.35}O_3$	–	–	–	0.17
Ceramics				
PLZT(8/65/35)[a]	520	400	–	–
PLZT(8/40/60)	100	≈200	–	–
PLZT(9/65/35)	–	–	380	0.018

[a]For explanation of the notation see Section 8.2.1.

Because the material is isotropic, the electric field can be directed along the x_3 axis without any loss in generality, and so $E_1 = E_2 = 0$. Symmetry requires $R_{11} = R_{22} = R_{33}$ and $R_{12} = R_{13} = R_{23} = R_{31} = R_{32}$, $R_{44} = R_{55} = R_{66}$, and the remaining components to be zero. Therefore the indicatrix can be written

$$x_1^2\left(\frac{1}{n^2} + R_{12}E^2\right) + x_2^2\left(\frac{1}{n^2} + R_{12}E^2\right) + x_3^2\left(\frac{1}{n^2} + R_{11}E^2\right) = 1 \tag{8.30}$$

Now

$$n_1 = n + \Delta n_1 = n - \tfrac{1}{2}n^3 R_{12}E^2 = n_2$$

and

$$n_3 = n + \Delta n_3 = n - \tfrac{1}{2}n^3 R_{11}E^2$$

The induced birefringence is given by

$$\Delta n = n_e - n_o = n_3 - n_1 = -\tfrac{1}{2}n^3(R_{11} - R_{12})E^2 \tag{8.31}$$

or, in terms of the polarization,

$$\begin{aligned}\Delta n &= -\frac{n^3}{2}\frac{(R_{11} - R_{12})P^2}{\varepsilon_0^2(\varepsilon_r - 1)^2} \\ &= -\tfrac{1}{2}n^3(g_{11} - g_{12})P^2\end{aligned} \tag{8.32}$$

Polycrystalline ceramic

In the case of polycrystalline ceramic (6mm) the form of the electro-optic tensor is the same as that for m3m symmetry except that $R_{66} = \frac{1}{2}(R_{11} - R_{12})$. Therefore, when a field is applied along the x_3 axis, the induced birefringence is again

$$\Delta n = -\tfrac{1}{2}n^3(R_{11} - R_{12})E^2 \tag{8.33}$$

or, in terms of the polarization,

$$\Delta n = -\tfrac{1}{2}n^3(g_{11} - g_{12})P^2 \tag{8.34}$$

The electro-optic properties of a few important materials are summarized in Table 8.1.

8.1.4 Non-linear optics

Second harmonic generation (SHG)

In the previous section the effects of a d.c. bias field on optical properties were discussed. The effects arise because the bias field is sufficiently strong ($\sim 10^6\,\mathrm{V\,m^{-1}}$) to significantly disturb the structure of an anisotropic crystal such as $LiNbO_3$, a disturbance described by changes in the optical indicatrix. The strength of the electric vector ($\sim 10^2\,\mathrm{V\,m^{-1}}$) of the light used in such experiments acts on the bound electrons in an essentially linear way. We now turn attention to laser light for which the electric vector can have magnitudes in the range 10^8–$10^{12}\,\mathrm{V\,m^{-1}}$. Under such intense fields the interaction with the electrons is non-linear.

Field strengths as high as this, if applied statically or at low frequency, are strong enough to detach electrons from atoms or ions and cause breakdown. As a component of light the peak fields persist for only very short times ($\sim 10^{-15}\,\mathrm{s}$) and so do not cause breakdown.

The non-linearity in the response can be expressed by

$$P = \varepsilon_0\,(\chi^{1)}E + \chi^{2)}E^2 + \chi^{3)}E^3 + \ldots) \tag{8.35}$$

Because the susceptibilities are tensors, for the linear dependence we have

$$P_i = \varepsilon_0 \chi_{ij}^{1)} E_j \tag{8.36}$$

and for the quadratic

$$P_i = \varepsilon_0 \chi_{ijk}^{2)} E_j E_k \tag{8.37}$$

In the literature relating to second harmonic generation d-coefficients are sometimes used rather than χs when, employing a contracted notation (jk = s (1,2, . . ., 6))

$$d_{is} = \chi_{is}^{2)}/2 \tag{8.38}$$

with 's' running from 1 to 6, the values of the permutated 'jk'.

If laser light of sufficient intensity is propagating through an optically non-linear material, and the time dependence of the electric vector component of the electromagnetic field is given by $E = E_0 \sin(\omega t)$, then from Eq. (8.35), the polarization is given by

$$\begin{aligned} P &= \varepsilon_0\{\chi^{1)}E_0 \sin(\omega t) + \chi^{2)}E_0^2 \sin^2(\omega t) + \chi^{3)}E_0^3 \sin^3(\omega t) + \ldots\} \\ &= \varepsilon_0 \chi^{1)} E_0 \sin(\omega t) + \tfrac{1}{2}\chi^{2)}E_0^2\{1 - \cos(2\omega t)\} \\ &\quad + \tfrac{1}{4}\chi^{3)}E_0^3\{3\sin(\omega t) - \sin(3\omega t)\} \end{aligned} \tag{8.39}$$

The term $\chi^{1)}E_0 \sin(\omega t)$ represents the response expected from a linear dielectric. The second term contains the components $\frac{1}{2}\chi^{2)}E_0^2$ and $-\frac{1}{2}\chi^{2)}E_0^2\cos(2\omega t)$: the first of these represents a constant polarization which would produce a voltage across the material, that is rectification; the second corresponds to a variation in polarization at *twice* the frequency of the primary wave. Thus the interaction of laser light of a single frequency with a suitable non-linear material leads to both frequency doubling (SHG) and rectification.

Whether or not second harmonics can be generated in a material depends upon the symmetry of its structure. For example, in the case of isotropic materials such as cubic crystals (e.g. calcite) and normal glasses, reversal of $\boldsymbol{E}$ simply reverses $\boldsymbol{P}$; therefore there can be no even-power terms in Eq. (8.35) so that $\chi^{2)}$ etc. are zero and SHG is not possible. Specially modified glasses can generate second harmonics, as discussed in Section 8.4.

A difficulty arises in the production of second harmonic radiation because of 'dispersion', that is the dependence of wave velocity on frequency. We consider the primary wave (angular frequency ω) interacting with a bound electron and producing a second harmonic (2ω) wave. At this point the two waves are in phase. As the 'ω-wave' propagates it generates new second harmonic, '2ω-waves' and, because of dispersion, the 'ω-wave' will, in general, travel with a different phase velocity than that of the '2ω-wave'. As a result a 'new' '2ω-wave' will interfere with those generated earlier, and only constructively if they have the required phase relationship.

For a plane wave travelling in a direction normal to the surfaces of a plate of thickness L the following proportionality can be derived [9] for the intensity I of the second harmonic wave leaving the plate:

$$I \propto \{\sin^2 2\pi(n_1 - n_2)L/\lambda_0\}/(n_1 - n_2)^2 \tag{8.40}$$

where n_1 and n_2 are the refractive indices for angular frequencies ω and 2ω respectively and λ_0 (or $2\pi c/\omega$), is the wavelength of the fundamental frequency wave travelling in a vacuum.

From Eq. (8.40), because the intensity is inversely proportional to $(n_1 - n_2)^2$, the efficient production of SHG radiation in a crystal of appreciable thickness requires the matching of n_1 and n_2. In certain birefringent materials it is possible to choose a direction of propagation such that the refractive indices of the ordinary and extraordinary rays are equal. This is possible in, for example, the crystal potassium dihydrogen phosphate (KDP).

Frequency mixing

It is also possible to 'mix' frequencies when light beams differing in frequency are made to follow the same path through a non-linear medium. For example, if two waves with electric field components $E_1 \sin(\omega_1 t)$ and $E_2 \sin(\omega_2 t)$ follow the same path through a dielectric with the characteristic

$$P = \varepsilon_0(\chi^{1)}E + \chi^{2)}E^2) \tag{8.41}$$

then the second-order contribution is given by

$$\varepsilon_0\chi^{2)}\{E_1^2 \sin^2(\omega_1 t) + E_2^2 \sin^2(\omega_2 t) + 2E_1 E_2 \sin(\omega_1 t)\sin(\omega_2 t)\} \tag{8.42}$$

The first two terms in braces describe SHG and the last term describes waves of frequency $\omega_1 + \omega_2$ and $\omega_1 - \omega_2$. The output of a wave of frequency $\omega_1 + \omega_2$ is known as 'up-conversion' and has been used in infrared imaging. For example,

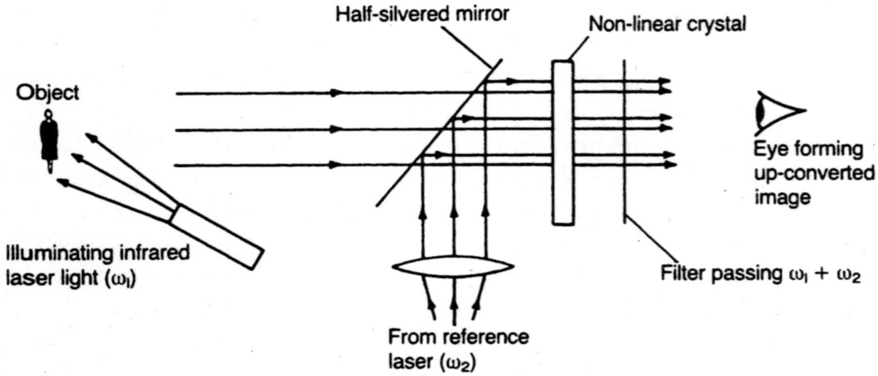

Fig. 8.7 Up-conversion of infrared to the visible frequency range.

as shown in Fig. 8.7, infrared laser light reflected from an object can be mixed with a suitably chosen infrared reference beam to yield an up-converted frequency in the visible spectrum. Also, the possibility of obtaining outputs corresponding to the sums and differences of the frequencies of two signals opens up the possibility of processing optical signals in similar ways to electronic signals. This is the subject of intensive research effort.

8.1.5 Transparent ceramics

For a ceramic to be useful as an electro-optic material it must, of course, be transparent. Ceramic dielectrics are mostly white and opaque, though thin sections are usually translucent. This is due to the scattering of incident light. Scattering occurs because of discontinuities in refractive index which will usually occur at phase boundaries (including porosity) and, if the major phase itself is optically anisotropic, at grain boundaries. For transparency, therefore, a ceramic should consist of a single-phase fully dense material which is cubic or amorphous; glasses usually fulfil these requirements.

It is instructive to examine Rayleigh's expression

$$\frac{I_\theta}{I_0} \propto \frac{1+\cos^2\theta}{x^2} r^2 \left(\frac{r}{\lambda}\right)^4 \left(\frac{n_p - n_m}{n_m}\right)^2 \qquad (8.43)$$

for the intensity I_θ of light scattered through an angle θ by a dispersion of particles of radius r and refractive index n_p in a matrix of refractive index n_m, where $n_p - n_m$ is small. The incident intensity is I_0, I_θ is measured at a distance x from the particles and λ is the wavelength of the scattered light ($r < \lambda$).

It should be noted that the scattering is proportional to $(r/\lambda)^4$ and so, for example, if r/λ is reduced from 10^{-1} to 10^{-2}, the scattering decreases by a factor of 10^4. This suggests that pores much smaller than the wavelength of the light will have only a minor scattering effect. Indeed, any change in refractive index that only persists for a fraction of a wavelength can be expected to have little effect. Thus dense ceramics composed of ultrafine particles of six 0.1 μm or less are expected to be transparent to visible light (λ(yellow)$\approx$0.5μm) even if the particles are optically anisotropic.

Because the scattering is proportional to $(n_p - n_m)^2/n_m^2$ it is expected to fall off rapidly as the refractive indices of the two phases become closer or as the birefringence of a single phase becomes less. However, if birefringence is introduced into an isotropic ceramic by the Kerr effect this may cause scattering in an otherwise transparent material.

The position is complicated in ferroelectric materials by the presence of domain walls which divide the crystals into small regions with differing refractive indices because of birefringence. Poling reduces the concentration of domain walls and may thereby reduce scattering in anisotropic ceramics. Crystal size within a ceramic affects the birefringence. Small crystals ($<2\,\mu$m) are under greater internal stress than larger crystals because they contain fewer domains and are less able to adjust to the essentially isotropic cavities in which they are embedded; consequently their optical anisotropies are reduced. The precise contributions to scattering from crystal size and domain structure have yet to be determined. However, it is found in practice that electrically controllable birefringence can be obtained in ceramics consisting of crystals with sizes below 2 μm and low birefringence, whilst controlled scattering can be obtained in ceramics with large crystals ($>5\,\mu$m).

8.2 Lanthanum-substituted Lead Zirconate Titanate

8.2.1 Structure and fabrication

Transparent single-crystal ferroelectrics, such as potassium dihydrogen phosphate (KDP), $BaTiO_3$ and $Gd(MoO_4)_3$, have long been recognized as useful electro-optic materials. However, the use of single crystals is limited by available size, cost and, in the case of KDP, susceptibility to moisture attack. In contrast, electro-optic ceramics do not suffer the same limitations, but prior to about 1960 transparent forms were not available. The 1960s saw the development of processing routes for the production of highly transparent ceramics in the $PbZrO_3$–$PbTiO_3$–La_2O_3 (PLZT) system which, as mentioned earlier (see Section 6.3.2), have the facility for changing their polar state under an applied field and have been exploited in a variety of successful electro-optic devices. The PLZT phase diagram is shown in Fig. 8.8; compositions are defined by $y/z/(1-z)$ in which y is the percentage of Pb sites occupied by La and $z/(1-z)$ is the Zr/Ti ratio.

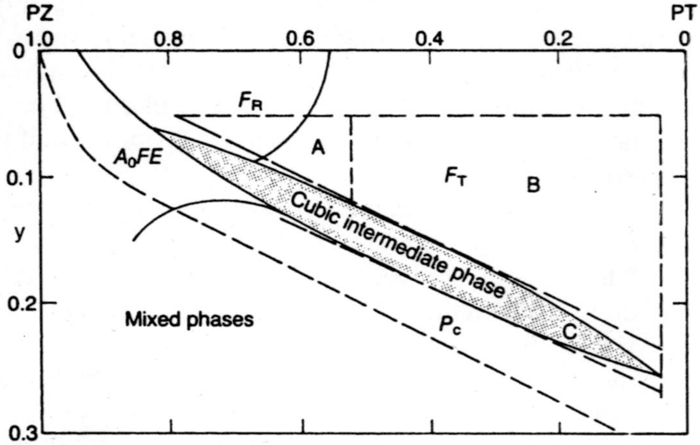

Fig. 8.8 Phase diagram of $Pb_{1-3y/2}La_y(Ti_{1-z}Zr_z)O_3$: A, memory; B, linear; C, quadratic.

Optimum homogeneity is needed in electro-optic materials both to avoid scattering due to local variations in composition and to ensure uniform electro-optic properties. Adequate mixing of the more slowly diffusing ions such as Zr, Ti and La is essential, and this calls for special chemical routes for the preparation of starting powders. As an example, the starting materials can be zirconium and titanium alkoxides mixed in predetermined ratios as liquids. An aqueous solution of lanthanum acetate is added and the resulting hydrolysis precipitates an intimate mixture of the three metal hydroxides. Lead oxide is milled with the precipitated material and the mixture is calcined at 900 °C. During the subsequent sintering stage the Pb^{2+} ions diffuse from the liquid phase and are distributed uniformly without difficulty. The slower-diffusing Ti^{4+}, Zr^{4+} and La^{3+} ions are already intimately mixed and have short paths to their final lattice sites.

Pore-free PLZT can be obtained by sintering under atmospheric pressure in oxygen or by applying a vacuum during the early stages of sintering. It is more usual to prepare cylindrical blocks of material by hot-pressing in Si_3N_4 or SiC dies as shown in Fig. 8.9. In both the pressureless sintering and hot-pressing routes the sample is surrounded by powders approximating in composition to that of the sample; the correct design of the powder is particularly critical if good quality material is to be made by the pressureless sintering route.

PLZT ceramic with a grain size less than 5 μm can be obtained by hot-pressing at 1200 °C and 40 MPa and holding for about 6 h. The crystals can be increased in size by a subsequent anneal in an atmosphere containing PbO vapour. Plates of large area for wide aperture devices require very careful annealing in order to eliminate variations in birefringence due to minor variations in density without undue crystal growth. Plates of the required thickness are cut from the hot-pressed pieces and the major surfaces are lapped and polished. If the plates are to

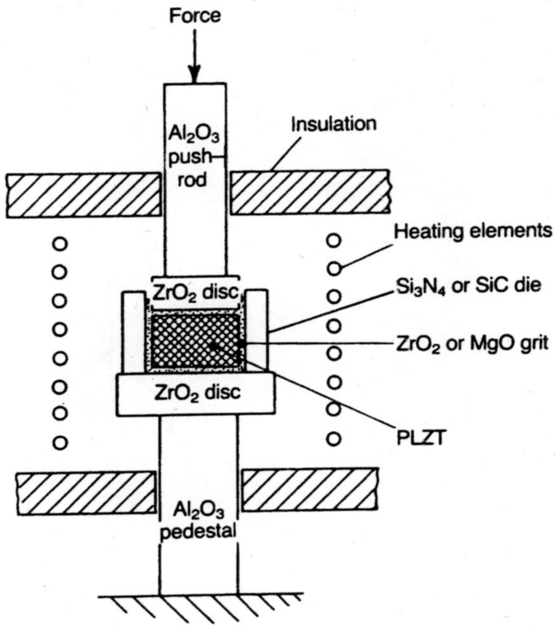

Fig. 8.9 Apparatus for hot-pressing PLZT.

be used exploiting a longitudinal electro-optic effect, where the electric field and light are parallel, transparent electrodes of indium tin oxide (ITO) can be applied to the major surfaces by sputtering and coated with blooming layers to reduce the loss of light by reflection. Where the field is to be at right angles to the direction of light propagation, grooves can be formed in the major surfaces and gold deposited in them to form an interdigitated electrode system. Blooming layers are again essential because the refractive indices may be about 2.5, so that approximately 40% of normally incident light would otherwise be reflected.

8.2.2 Measurement of electro-optic properties

In their ferroelectric state, the electro-optically useful PLZT compositions have an almost cubic structure, with the polar *c* axis being typically only about 1% longer than the *a* axes. Consequently the optical properties are almost isotropic and this, in part, is why high transparency can be achieved in the ceramic form. When an electric field is applied to the ceramic, domain alignment, or a field-enforced transition to the ferroelectric state, leads to the development of macroscopic polarization and so to uniaxial optical properties, i.e. the optic axis

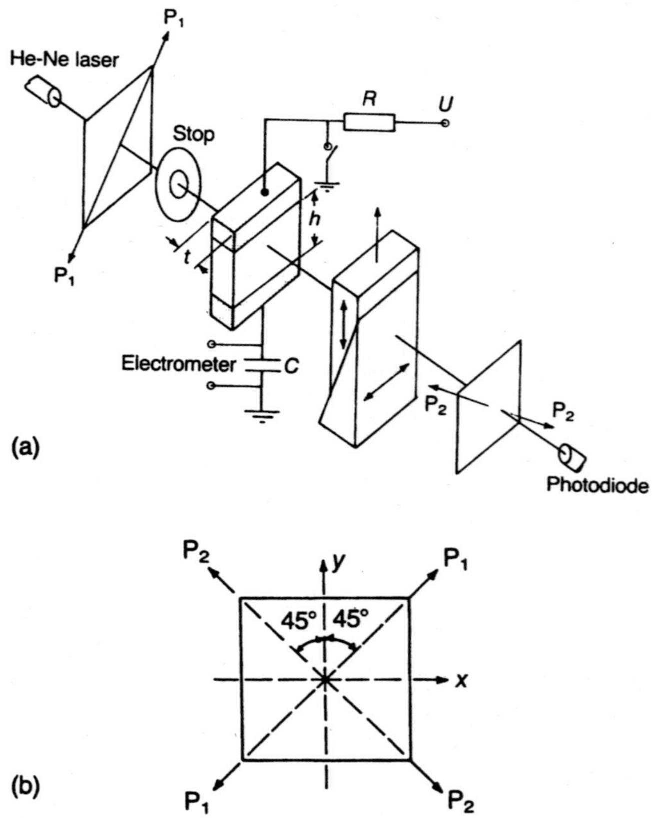

Fig. 8.10 (a) Optical system for measuring the electro-optic coefficient; (b) reference planes viewed along the optical axis from the photodiode.

coincides with the polarization. The effective negative birefringence is specified by $\overline{\Delta n}$, the magnitude of which depends on the applied field strength E.

The birefringence is measured using apparatus of the type shown in Fig. 8.10(a). He–Ne laser light ($\lambda = 0.633\,\mu$m) is passed through the polarizer P_1 and then through the electroded specimen. The specimen is in the form of a polished plate, of thickness t typically 250 μm, carrying gold electrodes with a gap of approximately 1 mm.

With no voltage applied to the electrodes an unpoled specimen is in the isotropic state and the light is passed on unchanged to the second polarizer P_2 where it is extinguished, as measured by the photodiode. Any field-induced relative retardation Γ between the horizontal and vertical components of the plane light that takes place during passage through the specimen leads to ellipticity and so a part of the light is transmitted by P_2.

The Babinet compensator is a calibrated optical device comprising two wedge-shaped pieces of quartz arranged so that their optic axes, and the light path, are mutually perpendicular. By moving one wedge relative to the other in the sense indicated, variable and known amounts of relative retardation can be introduced between the vertical and horizontal components of the plane-polarized light. Therefore the relative retardation introduced by the specimen can be measured directly from the adjustment of the calibrated compensator necessary to return the situation to complete extinction. It follows that

$$\Gamma = \overline{\Delta n}t = -\tfrac{1}{2}n^3 r_c E t = -\tfrac{1}{2}n^3 r_c \frac{U}{h}t \tag{8.44}$$

For $h = t$ and $\Gamma = \lambda/2$, $U = V_\pi = -\lambda/n^3 r_c$, i.e. the half-wave voltage referred to in Section 8.1.3 and Table 8.1.

An expression is now derived for the light intensity transmitted by P_2 in the case where the relative retardation introduced by the specimen is Γ and the compensator is removed. It is helpful to refer to Fig. 8.10(b) which represents the system viewed from the photodiode. If the amplitude transmitted by the polarizer P_1 is $a\sin(\omega t)$, then the x and y components incident on the PLZT are

$$x = \frac{a}{\sqrt{2}}\sin(\omega t) \tag{8.45}$$

$$y = \frac{a}{\sqrt{2}}\sin(\omega t) \tag{8.46}$$

On passing through the uniaxial negative PLZT the x component is retarded by an angle $\delta = (\Gamma/\lambda)2\pi$, where $\Gamma = t\overline{\Delta n}$. The amplitude a_T transmitted by P_2 is

$$\begin{aligned} a_T &= y\cos 45^\circ - x\cos 45^\circ \\ &= \frac{1}{\sqrt{2}}(y - x) \end{aligned}$$

i.e.

$$a_T = \frac{a}{2}\{\sin(\omega t) - \sin(\omega t - \delta)\}$$

which reduces to

$$a_T = a\sin\left(\frac{\delta}{2}\right)\cos\left(\omega t - \frac{\delta}{2}\right) \tag{8.47}$$

The time-averaged transmitted intensity I_T is given by

$$I_T = a^2\sin^2\left(\frac{\delta}{2}\right)\cos^2\left(\omega t - \frac{\delta}{2}\right)$$

i.e.

$$I_T = \frac{a^2}{2}\sin^2\left(\frac{\delta}{2}\right) = \frac{1}{4}I_0\sin^2\left(\frac{\Gamma\pi}{\lambda}\right) \tag{8.48}$$

where I_0 is the intensity incident on P_1. I_T is a maximum when $\Gamma = (2m+1)\lambda/2$, $\delta = (2m+1)\pi$ and a minimum when $\Gamma = m\lambda(\delta = 2m\pi)$, where m is an integer.

The total polarization in the specimen for a given applied field can be measured by the voltage developed across the series capacitor, because capacitors in series each carry the same charge. Closing the switch discharges the specimen and capacitor; opening it charges the specimen to a voltage close to U, provided that C is very much greater (at least a factor of 10^3) than the specimen capacitance. This allows the determination of hysteresis loops corresponding to the birefringence–field or birefringence–polarization relations.

8.2.3 Electro-optic characteristics

Depending on composition PLZT ceramics display one of three major types of electro-optic characteristic, i.e. 'memory', 'linear' or 'quadratic'. These are shown in Fig. 8.11, together with the corresponding hysteresis loops, and are discussed briefly below.

Memory

A typical PLZT exhibiting memory characteristics has the composition 8/65/35 and a grain size of about 2 μm. The general approach to obtaining a hysteresis loop is referred to above; the intermediate remanent polarization states (e.g. P_{r1} etc. in Fig. 8.12) necessary to generate the $\overline{\Delta n}$ versus normalized polarization plots (Fig. 8.11(a)(ii)) are obtained as follows.

The polarization is saturated by applying a field of approximately $3E_c$, where E_c is the coercive field. When the saturating field is removed, the polarization relaxes to the remanent states P_R or $-P_R$. Intermediate remanent states, such as P_{r1}, can be achieved by removing the field after bringing the electrical state of the material to C, applying an appropriate voltage for a limited time and controlling the charge movement by means of a high series resistance. There is a state D on Fig. 8.12 from which the polarization relaxes to zero. Intermediate states can be reached equally well from the positive E quadrants, and the right-hand side of Fig. 8.11(a)(ii) reflects the observed hysteresis between the two approaches. To achieve zero birefringence, coupled with zero polarization, it is usually necessary to depole a specimen thermally. Thermal depoling involves heating the specimen above the Curie temperature and cooling in the absence of a field, a procedure which randomizes the domain configuration.

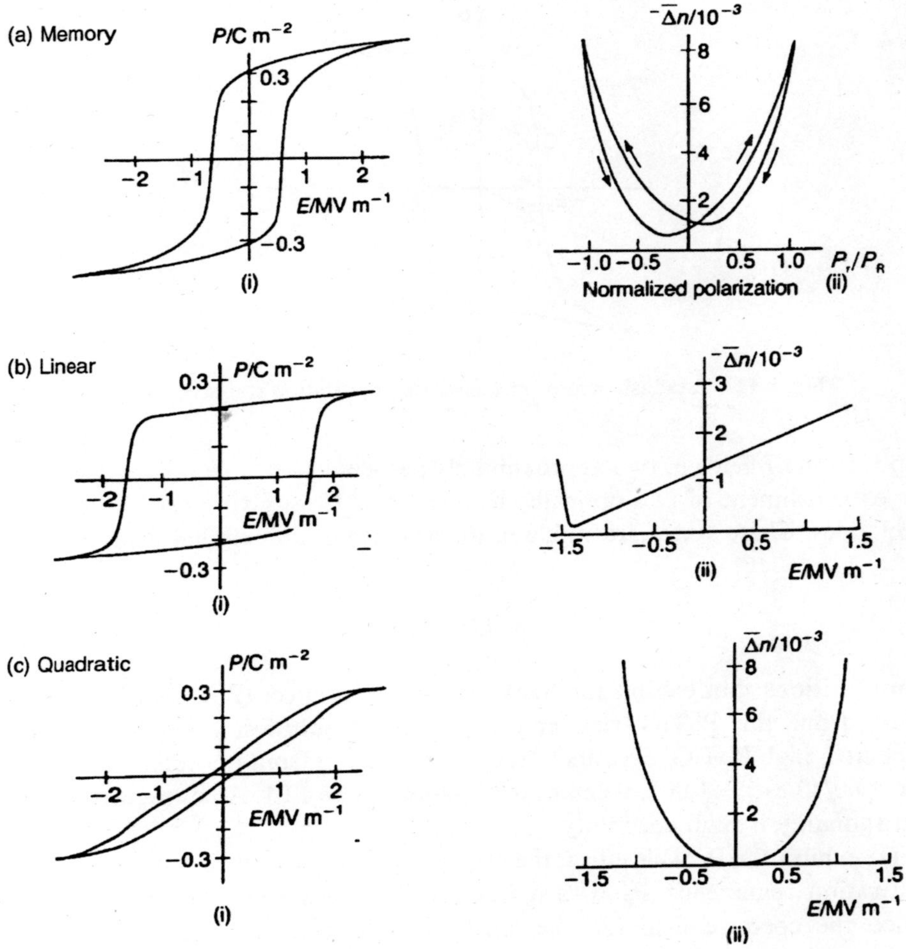

Fig. 8.11 Hysteresis and electro-optic characteristics of the three main types of PLZT: (a) memory; (b) linear; (c) quadratic (after G.H. Haertling (1971) *J. Am. Ceram. Soc.*, **54**, 303).

The reduced polarization states P_{r1}, P_{r2} etc. are due partly to 180° domain reversals, which do not affect the birefringence, and partly to changes in other domains (71° and 109° in rhombohedral crystals) which reduce both the polarization and the birefringence. The relation between birefringence and remanent polarization is therefore complex and, as indicated in Fig. 8.11, zero remanent polarization does not necessarily correspond to zero birefringence.

The existence of intermediate polarization states is not of great practical use because the effect of an applied field depends on both its exact mode of application and the previous treatment of a specimen. Reproducible results can be achieved by having an electrode and circuit arrangement that allows the

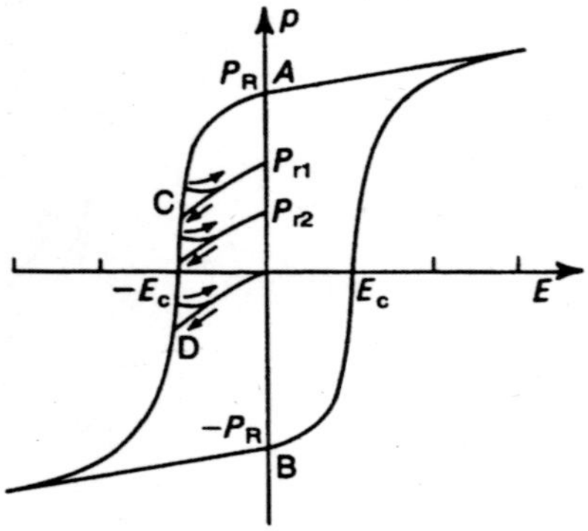

Fig. 8.12 Establishment of intermediate polarization states P_{r1}, P_{r2} etc.

application of fields in two approximately perpendicular directions. This permits the establishment of two optically distinct states by applying brief pulses to the electrodes. These states are stable in the absence of an external field.

Linear

Compositions that exhibit the linear electro-optic effect (Pockels effect) tend to come from the $PbTiO_3$-rich end of the solid solution range (Fig. 8.8). As expected, high $PbTiO_3$ favours tetragonal distortion from the cubic; for example, for 8/60/40 $c/a = 1.002$, whereas for 8/40/60 $c/a = 1.01$. A consequence of high tetragonality is high coercivity.

To exhibit the Pockels effect, the ceramic, e.g. 8/40/60, is first poled to positive saturation remanence P_R with a field of approximately $3\,MV\,m^{-1}$, i.e. about twice the coercive field (cf. the left-hand side of Fig. 8.11(b)(i)). $\overline{\Delta n}$ is then measured, at remanence, for positive and negative d.c. fields ranging from $-1.4\,MV\,m^{-1}$ to $2.0\,MV\,m^{-1}$. Over this range linearity is excellent (right-hand side of Fig. 8.11(b)(ii)) with a Pockels coefficient r_c of approximately $100\,pm\,V^{-1}$ (Table 8.1). Some PLZT compositions have higher r_c values, e.g. 8/65/35 (Table 8.1), but at the expense of departures from linearity for small fields. Grain size is known to have a very significant effect on the linearity of the $\overline{\Delta n}$–E characteristic.

Quadratic

Compositions that display the quadratic Kerr effect lie close to the ferroelectric rhombohedral–tetragonal boundary (Fig. 8.8); 9.5/65/35 is typical. At room temperature they are essentially cubic, but the application of a field enforces a transition to the rhombohedral or tetragonal ferroelectric phase, and the optical anisotropy increases with E^2. Because of their zero, or very low, remanence values, they are known as 'slim-loop' materials (left-hand side of Fig. 8.11(c)(i)).

Longitudinal effects

Strain bias

In the examples so far discussed, birefringence has been developed by electrically inducing a polarization in the plane of the PLZT plate, and this effect is the most widely exploited. However, it is possible to engineer a longitudinal effect in a number of ways, one being by strain biasing. A suitable strain can be introduced by cementing a PLZT plate to an inactive transparent plate, e.g. Perspex, and flexing the combination so that tensional strain of order 10^{-3} is induced in the PLZT, as illustrated in Fig. 8.17(a) below. This results in an orientation of the polar axes of the crystallites approximately in the direction of the strain and in an accompanying birefringence. The polar axes can now be switched through 90° by the application, via ITO transparent electrodes, of a field across the thickness of the plate, which reduces the birefringence to zero. The PLZT will revert to the birefringent state when the field is removed.

Scatter mode

An 8.2/70/30 composition is antiferroelectric at zero field but becomes ferroelectric when fields greater than $1\,\mathrm{MV\,m^{-1}}$ are applied. As a consequence the hysteresis loop has a narrow region at low fields which develops into a normal saturating characteristic at higher fields (Fig. 8.13). This behaviour is temperature sensitive but occurs to a sufficient extent for practical application between 0 and 40 °C.

The change to the ferroelectric state is accompanied by the development of birefringence ($\Delta n \approx -0.05$) and a marked scattering of light, particularly when the grain size lies in the 10–15 μm range. A ferroelectric domain structure only becomes apparent when the field exceeds about $2\,\mathrm{MV\,m^{-1}}$. There is uncertainty regarding the extents to which domain boundaries and grain boundaries contribute to the overall scattering effect.

Hot-pressed material has a grain size of about 2 μm which can be increased by annealing at 1200 °C in a PbO atmosphere provided by burying the specimen in powder of similar composition. Grain growth follows the relation

$$G \approx 2.0t^{0.4} \tag{8.49}$$

where G (in microns) and t (in hours) are the grain size and time respectively.

The scattering effect is exploited in the longitudinal mode, i.e. with the light and applied field directions collinear. The major faces of the plate are coated with ITO electrodes and antireflection layers. The arrangement of the device is shown in Fig. 8.14.

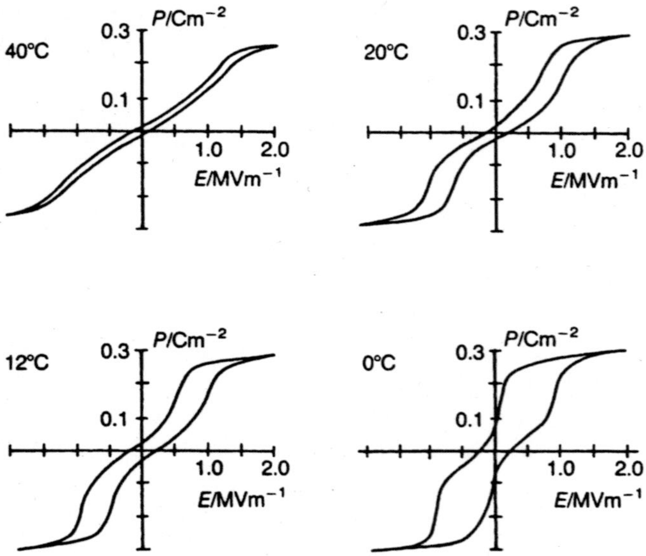

Fig. 8.13 Variation in P versus E hysteresis with temperature for 8.2/70/30 PLZT.

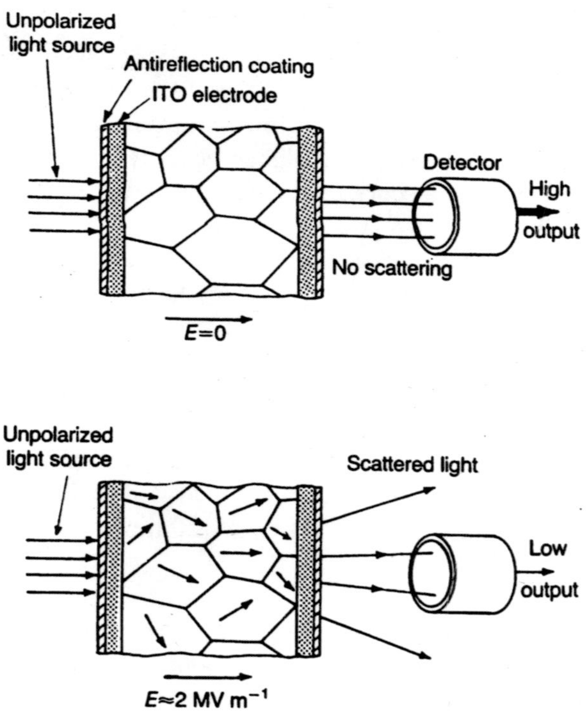

Fig. 8.14 PLZT used in the scattering mode.

Scattering has its maximum effect in reducing the light flux over a narrow angle, typically 3°; therefore it cannot be exploited in wide-angle devices. The 'scatter ratio', i.e. the ratio of the detected intensity in the transparent state to that in the scattering states, can be as high as 1000 for a plate 1 mm thick and an applied voltage of 2000 V.

As might be expected, scattering increases with specimen thickness since the number of scattering interfaces is thereby increased. However, for a given thickness there is a peak in scattering at a certain crystal size, since larger crystals, although having greater birefringence, lead to a smaller number of interfaces in a given thickness.

8.3 Applications

Brief descriptions of a selection of applications illustrating the various electro-optic effects and demonstrating ways in which they can be exploited are given below.

8.3.1 Flash goggles

Flash goggles designed to protect the eyes of pilots of military aircraft from the effects of nuclear flash are in production. Similar devices could also be of value to arc-welders, to those who suffer eye disorders and as part of television stereo-viewing systems.

The elements of flash goggles are a polarizer, a PLZT plate and a second polarizer arranged as shown in Fig. 8.15. The PLZT composition, 9.2/65/35, can be prepared by hot-pressing to a very pale yellow transparent ceramic with a grain size of about

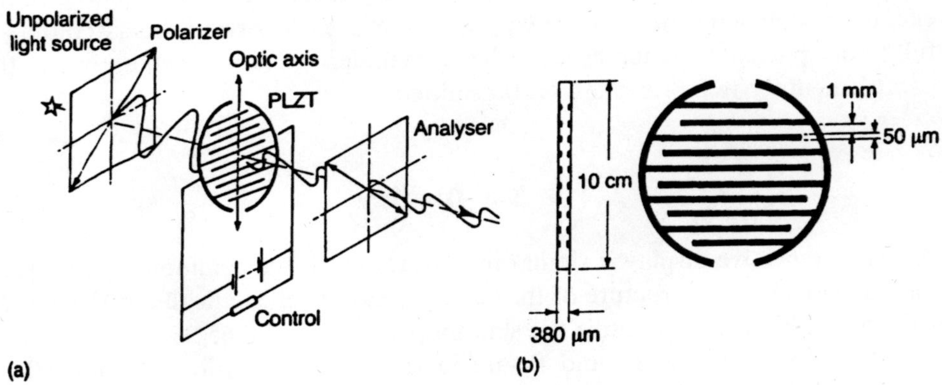

Fig. 8.15 (a) Arrangement for optical shutter; (b) detail of the electroding configuration.

2 μm. It is a slim-loop quadratic material with almost zero birefringence under zero applied field. The R value (Table 8.1) is about $4 \times 10^{-16}\,\mathrm{m^2\,V^{-2}}$.

For normal vision the light incident on the PLZT plate from a polarizer has its plane of polarization turned through 90° by the birefringence induced by a voltage (about 800 V) applied to the interdigitated electrode system (Fig. 8.15(b)). An analyser is so oriented as to allow the emergent light through to the wearer's eyes. A flash of intense light is detected by a photodiode which operates a switch that removes the voltage and closes the circuit between the electrodes. The plane of polarization is no longer rotated and the analyser now stops most of the light.

As finally designed, the PLZT is in the form of a plate 0.4 mm thick with corresponding grooves on both faces containing gold electrodes with gaps of about 1 mm between the interdigitated arrays. The embedded electrodes improve the field intensity distribution, and therefore the birefringence, across the thickness of the plate.

Switching times between the 'on' and 'off' states are typically 100 μs and the transmission ratios ~1000:1.

8.3.2 Colour filter

The optical shutter described above can function as a voltage-controlled colour filter. The PLZT is again of the slim-loop quadratic variety, and the polarizer–PLZT–analyser configuration as for the shutter.

When no voltage is applied, incident white light is extinguished. As the voltage is increased, the retardation reaches a value for which the mid-spectrum 'green' is retarded by $\lambda/2$ and so is fully transmitted; the remainder of the partially transmitted spectrum, along with the green, give the effect of approximately 'white' light. As the voltage is increased further, full-wave retardation for the shortest-wavelength primary colour, blue, is reached, and so that colour is extinguished. The other two primaries, red and green, are transmitted, to give the effect of yellow light. In the same way, as one of the other two primary colours is fully or partially extinguished, the remainder of it, together with the complementary two primaries, is transmitted.

8.3.3 Display

A PLZT reflective display is similar in appearance to the common liquid crystal display (LCD). The structure of the device is shown schematically in Fig. 8.16; a suitable PLZT composition is the slim-loop quadratic 9.5/65/35.

When the polarizers P_1 and P_2 are in the parallel position and no field is applied to the ITO electrode pattern, incident unpolarized light passes through P_1. The plane-polarized light then passes through the ITO electrode–PLZT

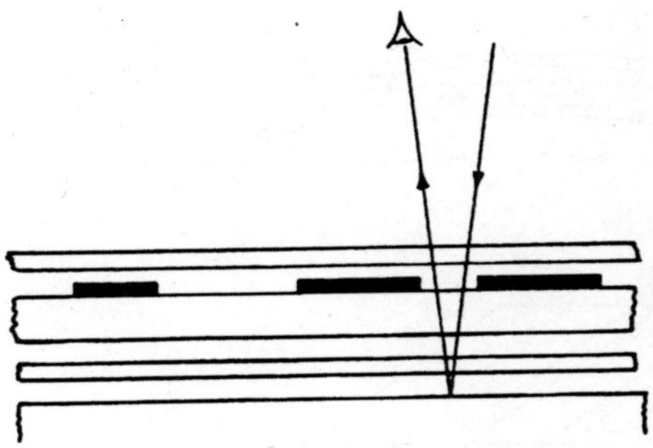

Fig. 8.16 Principle of the PLZT reflective display.

combination and the parallel P_2, and then, after reflection, passes back through the system to be observed. When a field is applied to the electrode pattern so as to induce a $\lambda/2$ retardation for a single passage through P_1, the light is extinguished at P_2. The activated segments then appear black against a light background. If P_1 and P_2 are arranged to be in the crossed position, the characters appear white against a dark background. The segmented alphanumeric electrode pattern is of the interdigitated form described above.

The disadvantages of the PLZT device, compared with the LCD, are high operating voltages (typically about 200 V) and high cost; an advantage is the fast switching times.

8.3.4 Image storage

PLZT can be exploited for image storage in a variety of ways: the first of the following two examples makes use of strain-induced birefringence and the second light scattering (Fig. 8.14).

A strain-biased antiparallel domain configuration is induced in the plane of a memory-type PLZT (7/65/35 with a grain size of 1.5 μm) plate approximately 60 μm thick, as shown in Fig. 8.17(a). A strain of 10^{-3} is achieved by bonding the PLZT to a thicker plate of transparent Perspex and bending the combination so that the ceramic is in tension in the convex surface. Combining a photoconductive layer, e.g. polyvinyl carbazole, CdS, ZnS or selenium, with the PLZT plate allows an optical pattern to be electrically stored, read and erased. Transparent ITO electrodes sandwich the photoconductor–PLZT combination.

When a light pattern illuminates the front face (Fig. 8.17(b)), the light passes through the transparent electrode to interact with the photoconductive layer, raising its conductivity. Simultaneously a voltage pulse (about 200 V) is applied across the electrodes, the major part of which, in the illuminated regions, drops across the ferroelectric. As a result the polarization is switched from the strain-biased configuration to a direction approximately normal to the plate. In this

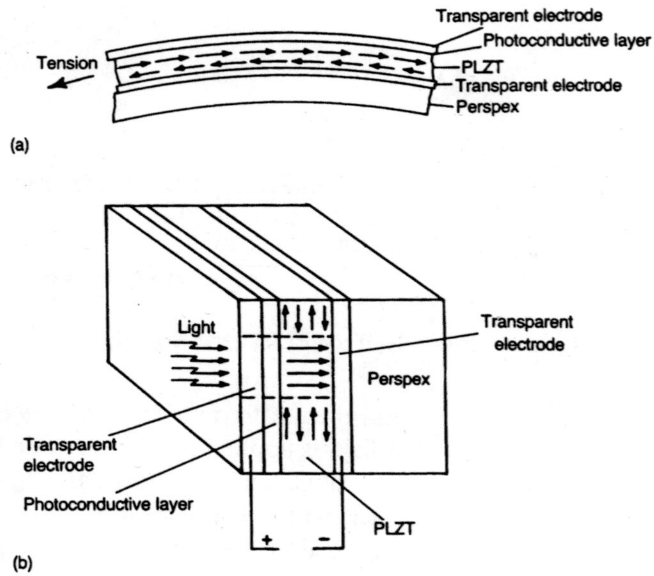

Fig. 8.17 Image storage device: (a) ferroelastic polarization; (b) field-induced polarization.

way the birefringence of the plate is modified across its area according to the incident light intensity distribution. The stored image can be read non-destructively with monochromatic light used in conjunction with the usual polarizer and analyser arrangement.

An improved display exploits scattering in a PLZT of similar composition to that described above but with the grain size increased to about 4.5 μm. This device has the advantage that it eliminates the need for the polarizer and analyser and the straining arrangement; also, white light can be used for both the 'write' and 'read' operations.

The ceramic plate, ITO electrodes and photoconductive layer are arranged as in the previous example. In its unpoled state the PLZT scatters light, probably at the 71° and 109° domain boundaries associated with the rhombohedral structure. When a voltage pulse is applied simultaneously with a light pattern incident on the plate, the polarization is switched to a direction normal to the plane of the plate and the scattering is reduced.

The photoconductive film is firstly uniformly illuminated so that the PLZT plate can be poled to saturation remanence. The image to be stored, for example from a transparent photographic 'positive', is then focused on the photoconductive film while a voltage of opposite polarity to that of the poling voltage is applied. This pulse is sufficient to bring the ferroelectric to the electrically depoled and optimum scattering state. Erasure of the stored image is achieved by uniformly illuminating the area and repoling the plate to saturation remanence.

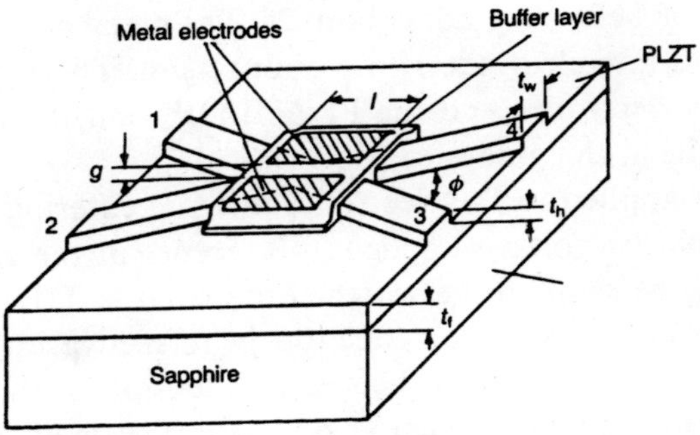

Fig. 8.18 Total internal reflection thin-film optical switch (after K. Wasa *et al.* (1984) *J. Lightwave Techn. IEEE.*, **LT-2**, 710–14).

The stored image can be read by passing collimated white light through the plate and then focusing the unscattered light to pass through a 2 mm aperture and onto a film. The PLZT plate is positioned in one focal plane of the focusing lens and the aperture in the other. Images can be formed with tones of grey and a resolution of approximately 80 lines mm^{-1}.

The techniques are capable of storing sequential information by using a raster-scanned laser beam with appropriately timed and sized voltage pulses.

8.3.5 PLZT films

PLZT in thin and thick film form can be exploited for a variety of applications including optical signal switching and modulation. An attraction of PLZT is its high electro-optic coefficient and in consequence low drive voltages to perform a particular function (c.f Table 8.1). The PLZT thin film switch described below is included to illustrate principles rather than because of very significant commercial importance.

There are various thin-film optical switch principles based on interference, diffraction or total internal reflection (TIR). The TIR switch, the essentials of which are shown in Fig. 8.18, depends on TIR of light passing from one medium to another of lower refractive index. TIR occurs for small differences in refractive indices when the angle of incidence approaches 90°, i.e. for light almost parallel to the surface of the material with the lower refractive index.

A suitable substrate is sapphire with the PLZT film deposited on an (0001) face. Because of the smallness of the mismatch (about 2%) between the O–O distances (287 pm in an (0001) face for sapphire and approximately 280 pm in a (111) face for PLZT) the PLZT grows epitaxially, with its (111) face registering with the sapphire (0001) face (*c* plane).

Typical dimensions of the sputtered PLZT film are as follows: thickness t_f about 0.35 μm; light guide step height t_h about 0.05 μm; light guide step width t_w

about 20 μm; angle ϕ between guides about 2°. The metal electrodes are of length l about 1.7 mm and interelectrode gap g about 4 μm. There is a buffer layer of Ta_2O_5 between the electrodes and the PLZT to prevent the metal affecting the propagation of light in the guide.

With no voltage applied to the electrodes a signal entering at 1 will exit at 3; with a sufficiently large voltage applied TIR occurs in the region between the electrodes, causing the signal to be switched to exit at 4. TIR occurs because the electro-optic effect gives rise to a reduction in refractive index for the guided light, whose $\boldsymbol{E}$ vector is parallel to the short dimension of the guide.

It is readily shown that the critical angle θ_c for TIR is given by

$$\theta_c = \sin^{-1}(1 - \tfrac{1}{2}n^2RE^2) \tag{8.50}$$

where the change in refractive index causing TIR is regarded as being due to the Kerr effect.

In the case of $LiNbO_3$ ($n = 2.29$), which exhibits the linear Pockels effect ($r_c = 17.5\,\text{pm V}^{-1}$), with an assumed θ_c of 88° (on the basis of the assumed 2° angle between the guides), E is calculated to be 3.3 MV m^{-1} which, for a gap of 4 μm, is an applied voltage of 13 V. If a 9/65/35 quadratic PLZT is assumed, then reasonable corresponding values are $R = 400\,\text{nm}^2\,\text{V}^{-2}$, $n = 2.5$ and a voltage of 1.4 V.

Thin PLZT films deposited by, for example, sputtering are too thin ($<1\,\mu$m) to achieve the necessary retardation for transverse mode optical devices. There is, however, potential for films having thicknesses in the range typically 2–25 μm. If these can be successfully and economically produced then they offer potential for a variety of devices including optical shutters, modulators and displays.

The substrate on which thick films are grown also has to be chosen with care since it influences grain orientation in the film, and particularly mechanical stresses arising because of thermal expansion mismatch. To avoid film cracking the film has to be maintained in compression requiring the substrate thermal expansivity to be greater than that of the PLZT.

The films can be grown by various methods (see Section 3.6.9). G.H. Haertling and G.C. Robinson [10] describe the production of 6 μm thick (9/65/35) PLZT films on sapphire substrates by an automated repeated dipping/firing process. The dipping solution consisted of the acetates of the metal ions, La, Zr, Ti and Pb. The films were the basis of a successful shutter and display based on the principles described above (see Sections 8.21 and 8.33).

8.4 Optical Non-linearity in Glass and Glass-ceramics

Glass in the form of the optical fibre plays an unrivalled passive role in communications' technology. Many messages, in the form of a string of light pulses generated by solid state (e.g. GaAs- or InP-based) lasers, are multiplexed and sent along a single fibre; at the receiving end there has to be an optical-to-electronic signal conversion using photodetectors and the messages demultiplexed by electronic switching.

At the present time the fastest electronic devices operate in the picosecond range ($\sim 10^{-12}$ s), slow in comparison to the duration of optical pulses which can be as short as a femtosecond (10^{-15} s). The communications systems engineer describes the use of electronics for signal processing as the 'electronics bottleneck'.

Optically non-linear crystals (e.g. $LiNbO_3$) provide a means of controlling optical signals but they have some disadvantages, including relatively high cost resulting from crystal growth, cutting and polishing operations, signal losses at interconnections with fibres caused by thermal expansion mismatch, and the high voltages needed to drive them. The possibility and attractiveness of exploiting glass fibres as 'active' devices for the control of signals, as well as for carrying them, has stimulated research interest.

Glasses have the advantage of high transparency over a wide frequency range, excellent 'forming' characteristics and, of course, total joining compatibility with the fibre network. However, the apparently macroscopically isotropic nature of glass would be expected to rule out the possibility of second harmonic generation, as discussed above. Third harmonic generation is expected, and second harmonic generation does occur in optical fibres irradiated with intense laser light. Non-linearity in glasses and the potential it has for optical modulators and switches is reviewed up to about 1990 by E.M. Vogel [11] and by E.M. Vogel *et al.* [12].

In 1991 R.A. Myers *et al.* [13] reported SHG in an electrically poled glass. A strong ($\sim 2.5\,MV\,m^{-1}$) electric poling field was applied across a 1.6 mm silica glass plate at $\sim 300\,°C$ for 15 min and maintained during cooling to room temperature. Second harmonic generation was observed with an efficiency of the same order as that achievable from $LiNbO_3$, the incident infrared laser light emerging with a green component. The second harmonic component was found to originate in a thin layer of the silica glass close to the anode. The observation led to a rapid growth of interest in poling glasses, particularly the chalcogenide* glasses, to induce non-linearity.

K. Tanaka *et al.* [14] found SHG in tellurium-containing glasses confirming that in this case the light originated throughout the volume of the glass and suggesting that the SHG is probably caused by local polarizations involving TeO_3 and TeO_4 structural units.

A further significant development is the observation by T. Fujiwara *et al.* [15] of enhanced SHG achieved through a combination of ultraviolet laser irradiation and electrical poling. A glass of composition $0.157GeO_2$–$0.843SiO_2$, poled at a field strength of $30\,MV\,m^{-1}$ and simultaneously irradiated with ultraviolet laser

* Glasses containing the chalcogenide elements, sulphur, selenium and tellurium.

light, developed strong non-linearity, the measured d-coefficient ($3.4\,\mathrm{pm\,V^{-1}}$) being larger than the corresponding value for $LiNbO_3$ ($\sim 2.6\,\mathrm{pm\,V^{-1}}$). J. Khaled *et al.* [16] observed similar effects in sputtered thin ($\sim 1\,\mu m$) glass (GeO_2:$SiO_2 = 1$:1) films. The origins of the non-linearities are not clear but are thought to involve defects and associated local electric fields. The effects decay with time but it seems likely that as understanding improves the decay rates will be reduced to acceptable values.

H.G. Kim *et al.* [17] observed SHG in a transparent tellurite-based glass-ceramic (see Section 3.7.4) and Y. Ding *et al.* [18] in thin ($\sim 0.35\,\mu m$) surface layers of $LiNbO_3$ having partially aligned crystallites and grown in a base glass of composition $0.35Li_2O$–$0.3Nb_2O_5$–$0.35SiO_2$. In addition to low voltage operation (for optical switches, for example) a further advantage offered by thin films is that the phase-matching requirement discussed above is relaxed.

The roles glass-based devices can play in photonics is well outside the scope of the present text and are dealt with comprehensively elsewhere [e.g. 11,12]. However, in addition to the effects mentioned above, the following two examples give some indication of the potential.

Second harmonic radiation has been generated along chemically modified (e.g. Ge- and P-doped) photo-irradiated silica fibres [12]. The high second harmonic conversions ($\sim 5\%$) along fibres as long as 30 cm require the SHG light to be phase-matched. The mechanism(s) by which phase-matched SHG can be generated in such long fibres is not fully understood but involves a periodic distribution of photo-induced defects (for example trapped electrons associated with dopant-induced oxygen vacancies) along the fibre length at just the periodicity demanded for phase-matching.

Very recently J.D. Mills *et al.* [19] have 'written' a nano-structured Bragg reflection grating within the volume of silica glass using focused infrared laser light. Such photo-induced structures have potential for a variety of photonics applications, including information storage.

There seems little doubt that the future will see glass playing an increasingly important 'active' role in communications and related technologies, in addition to its dominant role for carrying information along fibres.

Problems

1. A light wave of $\boldsymbol{E}$ vector amplitude E_0 is incident at an angle θ_i to the interface between two transparent media with refractive indices n_i and n_t. For $\boldsymbol{E}$ vectors in the plane of incidence the reflection coefficient $r_{\parallel}$ is

$$r_{\parallel} = \left(\frac{E_{0r}}{E_{0i}}\right)_{\parallel} = \frac{n_t \cos\theta_i - n_i \cos\theta_t}{n_i \cos\theta_t + n_t \cos\theta_i}$$

where θ_t is the angle of refraction. The corresponding coefficient when the $\boldsymbol{E}$ vector is perpendicular to the plane of incidence is

$$r_{\perp} = \left(\frac{E_{0r}}{E_{0i}}\right)_{\perp} = \frac{n_i \cos\theta_i - n_t \cos\theta_t}{n_i \cos\theta_i + n_t \cos\theta_t}$$

These are the Fresnel equations.

A parallel beam of light is incident normally on a thin slab of PLZT ($n = 2.5$). Ignoring absorption, what fraction of the incident intensity is transmitted through the slab? [Answer: 67%]

2. (a) The slab referred to in Question 1 has a thickness of 300 μm and an absorption coefficient at the particular wavelength (800 nm) of 65 m^{-1}. Calculate the total transmitted intensity taking absorption into account. (b) Calculate the thickness and refractive index of suitable antireflection coatings that might be applied to the PLZT. [Answer: 65%; 200 nm; 1.58]

3. A parallel-sided PLZT plate is positioned between two crossed Polaroid plates and the major faces of all three elements are normal to a parallel beam of monochromatic light ($\lambda = 750$ nm). The thickness of the PLZT plate is 1 mm and the electrodes, spaced 1 mm apart, are arranged such that a uniform electric field can be applied through the volume of the PLZT plate parallel to its major faces and at 45° to the transmission axis of each Polaroid plate.

   Calculate the voltage that must be applied between the electrodes to achieve maximum light transmittance through the system. Estimate the transmittance assuming 5% loss at each Polaroid surface and assuming that the PLZT element carries antireflection coatings and the material has an absorption coefficient of 65 m^{-1}. The PLZT is quadratic with $R = 4.0 \times 10^{-16}$ $m^2 V^{-1}$ and $n = 2.5$. [Answer: 346 V; 38% transmitted]

4. If in Problem 3 V_0 is the voltage necessary to produce maximum transmittance, draw the time variation of the average transmitted intensity when a voltage of $V_0 \sin(10\pi t)$ is applied.

5. Discuss the requirements for optical transparency in polycrystalline ceramics and outline a route designed to achieve transparency.

6. Explain why it is that a small addition of La_2O_3 to $BaTiO_3$ leads to semiconductivity whereas a similar addition to $PbTiO_3$ increases electrical resistivity.

7. Compile a list of the various methods for producing thin films of PLZT on a suitable substrate pointing out the merits or otherwise of each.

Bibliography

1. Haertling, G.H. (1988) Electro-optic ceramics and devices. In *Electronic Ceramics*, L.M. Levinson (ed.), Marcel Dekker, New York, pp. 371–92.
2. Herbert, J.M. (1982) *Ferroelectric Transducers and Sensors*, Gordon and Breach, London.
3. Aguilló-López, F. *et al.* (1994) *Electrooptics*, Academic Press, London.
4. Lines, M.E. and Glass, A.M. (1977) *Principles and Applications of Ferroelectrics and Related Materials*, Clarendon Press, Oxford.
5. Burfoot, J.C. and Taylor, G.W. (1979) *Polar Dielectrics and their Applications*, Macmillan, London.
6. Wood, R.M. (1993) *Optical materials*, The Institute of Materials, London.
7. Margulis, W. *et al.* (1998) Poled glasses, *Mater. Res. Soc. Bull.*, **November**, 30–5.
8. Yariv, A. and Yeh, P. (1984) *Optical Waves in Crystals*, John Wiley and Sons, Inc., New York.
9. Lengyel, B.A. (1966) *Introduction to Laser Physics*, John Wiley and Sons, Inc., New York.
10. Haertling, G.H. and Robinson, G.C. (1998) PZT and PLZT thick films on silver, sapphire and silicon. In *Integrated thin films and applications*. Pandey, R.K. *et al.* (eds.) *Ceramic Transactions* **86**, pp. 15–30.
11. Vogel, E.M. (1989) Glasses as non-linear photonic materials, *J. Am. Ceram. Soc.*, **72**, 719–24.
12. Vogel, E.M., Weber, M.J. and Krol, D.M. (1991) Nonlinear optical phenomena in glass, *Phys. Chem. Glasses*, **32**, 231–54.
13. Myers, R.A., Mukherjee, N. and Brueck, S.R.J. (1991) Large second order nonlinearity in poled fused silica, *Opt. Lett.*, **16**, 1732–4.
14. Tanaka, K. *et al.* (1996) Optical second harmonic generation in poled MgO–ZnO–TeO_2 and B_2O_3–TeO_2 glasses, *J. Non-Cryst. Solids*, **203**, 49–54.
15. Fujiwara, T. *et al.* (1997) Second-harmonic generation in germanosilicate glass poled with ArF laser irradiation, *Appl. Phys. Lett.*, **71**, 1032–4.
16. Khaled, J. *et al.* (2000) Generation of second harmonics in Ge-doped SiO_2 thin films by ultraviolet irradiation under poling electric field, *J. Appl. Phys.*, **87**, 2137–41.
17. Kim, H.G. *et al.* (1996) Transparent tellurite-based glass-ceramics with second harmonic generation, *J. Non-Cryst. Solids*, **208**, 303–7.
18. Ding, Y. *et al.* (1995) Second order optical non-linearity of surface crystallised glass with lithium niobate, *J. Appl. Phys.*, **77**, 2208–10.
19. Mills, J.D. *et al.* (2002) Embedded anisotropic microreflectors by femtosecond-laser nanomachining, *Appl. Phys. Lett.*, **81**, 196–8.

9

MAGNETIC CERAMICS

Over the past 60 years ceramic magnets have become firmly established as electrical and electronic engineering materials; most contain iron as a major constituent and are known collectively as 'ferrites'.

The science and technology of ferrites have their roots in the efforts of many in the early part of the last century. Significant contributors are S. Hilpert [1] in Germany, H. Forestier [2] in France and Y. Kato and T. Takei [3] in Japan. The remarkable discovery that the addition of non-magnetic zinc ferrite ($ZnFe_2O_4$) to a magnetic ferrite caused an increase in magnetism was explained in 1948 by L. Néel's [4] theory of uncompensated antiferromagnetism, or 'ferrimagnetism'.

During the 1930s the Philips researchers in The Netherlands made important contributions to the science and technology of ferrites, especially the pioneer, J.L. Snoek who, in the West, saw the industrial significance of ferrites as telecommunications was rapidly advancing. The development of the magnetically 'hard' hexagonal ferrite ('Ferroxdure') is also due to Philips [5].

The classic text by J. Smit and H.P.J. Wijn [6] reviews the subject up to the late 1950s and the authoritative paper by M. Sugimoto [7] surveys the past, present and future of ferrites from the perspective at the end of the 20th century. From the point of view of electrical properties they are semiconductors or insulators, in contrast to metallic magnetic materials which are electrical conductors. One consequence of this is that the eddy currents produced by the alternating magnetic fields which many devices generate are limited in ferrites by their high intrinsic resistivities. To keep eddy currents to a minimum becomes of paramount importance as the operating frequency increases and this has led to the widespread introduction of ferrites for high-frequency inductor and transformer cores for example. Laminated metal cores are most widely used for low-frequency transformers. This single example illustrates how metal and ceramic magnetic materials complement each other; often it is physical properties

Electroceramics: Materials, Properties, Applications. 2nd Edition. Edited by A. J. Moulson and J. M. Herbert.

which determine choice, but sometimes it is cost. Ferrites dominate the scene for microwave applications, and the transparency required for magneto-optical applications is offered only by them. Ferrites have also become firmly established as the 'hard' (or permanent) magnet materials used for high-fidelity speakers and small electric motors, to mention just two mass-produced components.

9.1 Magnetic Ceramics: Basic Concepts

9.1.1 Origins of magnetism in materials

Ampère, Biot, Savart and Oersted were among the first to demonstrate that conductors carrying currents produced magnetic fields and exerted 'Lorentz' forces on each other. They were also responsible for determining the laws governing the magnetic fields set up by currents. It was established that a small coil carrying a current behaved like a bar magnet, i.e. as a magnetic dipole with magnetic moment $\boldsymbol{\mu}$ (Fig. 9.1), and this led Ampère to suggest that the origin of the magnetic effect in materials lies in small circulating currents associated with each atom. These so-called amperian currents each possess a magnetic moment ($\mu = IA$), and the total moment of the material is the vector sum of all individual moments. The amperian currents are now identified with the motion of electrons in the atom.

Both the orbital and the effective spinning motions of the electron have associated angular moments quantized in units of $\hbar = 1.055 \times 10^{-34}$ J s. It is an elementary exercise in physics to show that the relationship between the magnetic dipole moment $\boldsymbol{\mu}$ and the angular momentum $\boldsymbol{L}$ for a moving particle of mass m and charge Q is

$$\boldsymbol{\mu} = \gamma \boldsymbol{L} \tag{9.1}$$

in which $\gamma = Q/2m$ is the gyromagnetic ratio. Therefore there is a unit of magnetic moment, corresponding to the quantum unit of angular momentum, in terms of which the moment of the spinning or orbiting electron is measured. This unit is the Bohr magneton ($\mu_B = (e/2m_e)\hbar$) which has the value 9.274×10^{-24} A m^2.

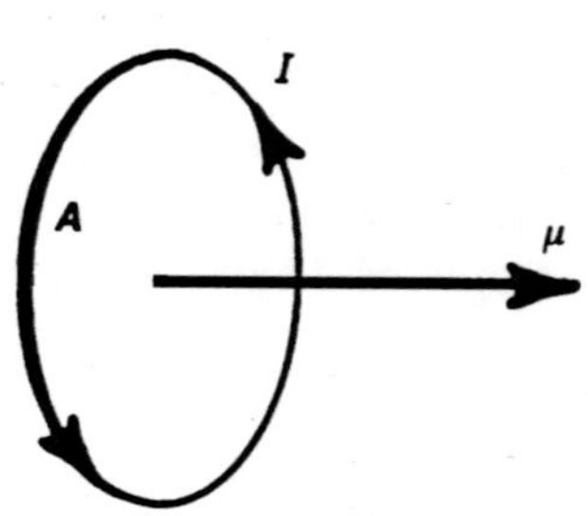

Fig. 9.1 The magnetic moment of a current loop.

It follows that once the total angular momentum of an ion, atom or molecule is known, so too is its magnetic moment. Most *free* atoms possess net angular momentum and therefore have magnetic moments, but when atoms combine to form molecules or solids, the electrons interact so that the resultant angular momentum is nearly always zero. Exceptions are atoms of the elements of the three transition series which, because of their incomplete inner electron shells, have a resultant magnetic moment.

There are two major contributions to the magnetic moment of an atom – from the orbiting electrons and from their spin. There is also a nuclear magnetic moment, but because it is of the order of 10^{-3} of that of the Bohr magneton, it is insignificant in the present context and will be disregarded.

The description of the orbital and spin states of an electron in terms of the quantum numbers n, l, m_l and s, and the calculation of the total angular momentum of a number of electrons in an atom in terms of J are outlined in Chapter 2. From that discussion, together with the above, it follows that an electron with orbital quantum number l possesses a total magnetic moment $\{l(l+1)\}^{1/2}\mu_B$, and this can be oriented with respect to a magnetic field in directions such that the component of magnetic moment in the field direction is also quantized with values $m_l\mu_B$.

The electron also behaves as if it had a component of magnetic moment of $1\mu_B$ in the direction of an applied field, resulting from its spin angular momentum s, and the total magnetic moment is $\boldsymbol{\mu} = 2\{s(s+1)\}^{1/2}\mu_B$. A consequence of the factor 2 is that the gyromagnetic ratio γ for a system of electrons possessing both orbital and spin angular momenta must be written $\gamma = g(e/2m_e)$, in which the Landé g factor is given by

$$g = \frac{3}{2} + \frac{S(S+1) - L(L+1)}{2J(J+1)} \tag{9.2}$$

and can take values from 1 to 2. For many ferrites g is close to 2, indicating that it is the total spin magnetic moment which makes the dominating contribution to magnetization; the orbital magnetization is said to be 'quenched'. The later discussions are concerned with the net magnetic moment of the partly filled 3d electron shell of atoms of the first transition series elements where, because of quenching, only the electron spin moments need be considered (see Chapter 2, p. 10).

9.1.2 Magnetization in matter from the macroscopic viewpoint

The field vector determining the Lorentz force on a current is the magnetic induction $\boldsymbol{B}$, which is measured in teslas. In principle, $\boldsymbol{B}$ at a point P can be

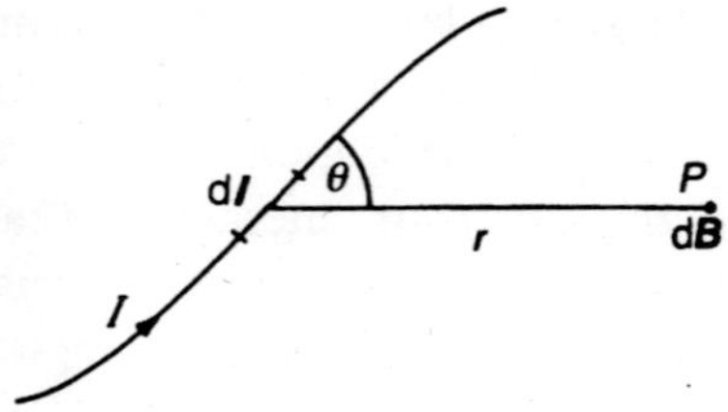

Fig. 9.2 Magnetic induction arising from a current element.

calculated for any system of currents *in vacuo* by the vector summation of induction elements d$\boldsymbol{B}$, arising from current elements Id$\boldsymbol{l}$ (Fig. 9.2), where

$$\mathrm{d}\boldsymbol{B} = \frac{\mu_0}{4\pi}\left(\frac{I\mathrm{d}\boldsymbol{l} \times \boldsymbol{r}}{r^3}\right) \tag{9.3}$$

The magnitude of d$\boldsymbol{B}$ is $(\mu_0/4\pi)(I\mathrm{d}l\sin\theta)/r^2$ in which μ_0, which is termed the permeability of a vacuum, has the value $4\pi \times 10^{-7}\,\mathrm{H\,m^{-1}}$.

An additional field vector, the magnetic field intensity $\boldsymbol{H}$ measured in ampères per metre, is defined so that in a vacuum $\boldsymbol{H} = \boldsymbol{B}/\mu_0$. Therefore, whereas $\boldsymbol{B}$ depends upon the medium surrounding the wire (a vacuum in the present case), $\boldsymbol{H}$ depends only upon the current.

In Chapter 2 (Eq. (2.74)) an important relation between $\boldsymbol{D}$, $\boldsymbol{E}$ and $\boldsymbol{P}$ was derived by considering the effects of polarization in the dielectric of a parallel-plate capacitor. An analogous relationship is now derived by considering the magnetization of a material of a wound toroid of cross-section A and mean circumference l (Fig. 9.3(a)). Because there is no break in the toroid, and so no free poles and consequent demagnetizing fields (see Section 9.1.3), $\boldsymbol{H}$ in the material must be due to the real currents only. The material becomes magnetized, i.e. it acquires a magnetic moment per unit volume or *magnetization* $\boldsymbol{M}$, and the

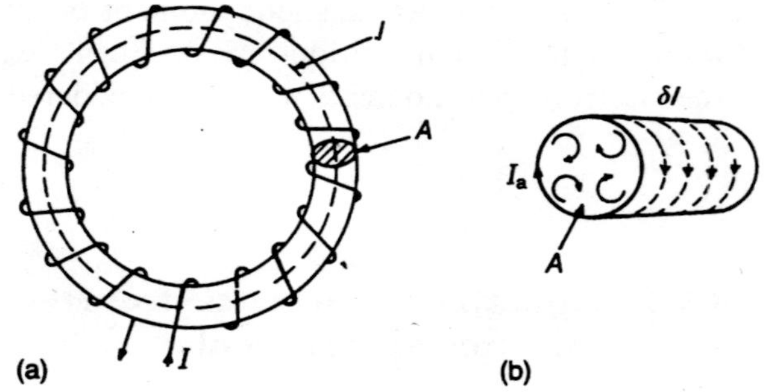

Fig. 9.3 Effects arising from the presence of a magnetic material; I_a is the amperian current per unit length of toroid.

magnetic moment of a volume element arises from microscopic currents, the 'amperian' currents.

Application of Ampère's circuit theorem $\oint \boldsymbol{B} \cdot \mathrm{d}\boldsymbol{l} = \mu_0 I_{\mathrm{total}}$ to the dotted path (Fig. 9.3(a)) around the coil wound on non-magnetic material gives $B = \mu_0 nI$, where I is the current and n is the number of turns per unit length. However, if the coil is wound on magnetic material then, because amperian currents are now also involved,

$$\boldsymbol{B} = \mu_0 nI + \mu_0 I_{\mathrm{a}} \tag{9.4}$$

Furthermore, from Fig. 9.3(b), because $I_{\mathrm{a}} \delta lA = M \delta lA$, it follows that

$$\boldsymbol{B} = \mu_0(\boldsymbol{H} + \boldsymbol{M}) \tag{9.5}$$

and, if the magnetization is assumed to be proportional to the magnetizing field,

$$\boldsymbol{B} = \mu_0(\boldsymbol{H} + \chi_{\mathrm{m}} \boldsymbol{H}) \tag{9.6}$$

where χ_{m} is the volume magnetic susceptibility. Therefore

$$\boldsymbol{B} = \mu_0(1 + \chi_{\mathrm{m}})\boldsymbol{H} = \mu_0 \mu_{\mathrm{r}} \boldsymbol{H} \quad \text{or} \quad \mu \boldsymbol{H} \tag{9.7}$$

where μ is the permeability and $\mu_{\mathrm{r}} = 1 + \chi_{\mathrm{m}}$ is the relative permeability of the material (the latter is a dimensionless quantity).

In general, the permeability and susceptibility are tensors; an implicit assumption made throughout the remainder of the chapter is that the vectors are collinear when the vector notation is dropped.

9.1.3 Shape anisotropy: demagnetization

Measurements of permeability and associated magnetic properties are usually made on toroids of uniform section when, to a close approximation, the flux density B is uniform throughout the material and lies entirely within it. In most practical applications the magnetic circuit is more complex, and variations in component section and permeability give rise to variations in flux density. Important effects arise from air gaps, which may be intentionally introduced or may be cracks or porosity.

The effect of the shape of a specimen on its magnetic behaviour, 'shape anisotropy', is expressed by a demagnetization factor N_{D}. A field H_{a} applied to a solid of arbitrary shape is reduced by a factor proportional to its magnetization M, so that the effective field H_{e} within the body is given by

$$H_{\mathrm{e}} = H_{\mathrm{a}} - N_{\mathrm{D}} M \tag{9.8}$$

It follows from Eq. (9.5) that

$$H_e = H_a - N_D\left(\frac{B}{\mu_0} - H_e\right)$$

Multiplying this equation through by μ_0/B and putting $\mu_{re} = B/\mu_0 H_a$, where μ_{re} is the effective relative permeability, gives

$$\frac{1}{\mu_{re}} = \frac{1}{\mu_r} + N_D\left(1 - \frac{1}{\mu_r}\right) \tag{9.9}$$

N_D can only be calculated simply if the field within the magnetized body is uniform, and it can be shown that this is the case only when the body is ellipsoidal. In ferromagnetic or ferrimagnetic bodies the permeability and magnetization vary markedly with field strength (cf. Section 9.1.10), so that M in Eq. (9.8) will vary in different parts of a body if the field is not uniform. For a sphere an exact calculation gives $N_D = 1/3$. A long rod with a length-to-diameter ratio of r can be approximated by a prolate ellipsoid with a large ratio r between the major and minor axes. Then

$$N_D \approx \frac{\ln(2r) - 1}{r^2} \tag{9.10}$$

If $r = 10$, $N_D = 0.02$, so that for $\mu_r = 1000$, $\mu_{re} \approx 50$, while for $r = 100$, $\mu_{re} \approx 700$.

The magnetic field which emanates from regions near the ends of magnets leads to the concept of magnetic poles which is useful in providing an approximate model in complex magnetic structures.

A thin flat disc magnetized normally to its plane will have a demagnetization factor close to unity so that its real permeability must be very high if its effective permeability is to be appreciable. Minor structural defects, such as fine cracks normal to the direction of magnetization, set up demagnetizing fields which can also markedly reduce effective permeability. The effect of such a crack can be calculated by considering a toroid of overall length l containing an air gap of length αl, as shown in Fig. 9.4; αl is assumed to be very small so that the effective cross-section A_g of the gap, is approximately equal to the cross-section A_m of the toroid. If the toroid carries nI ampère turns, then, by Ampère's theorem,

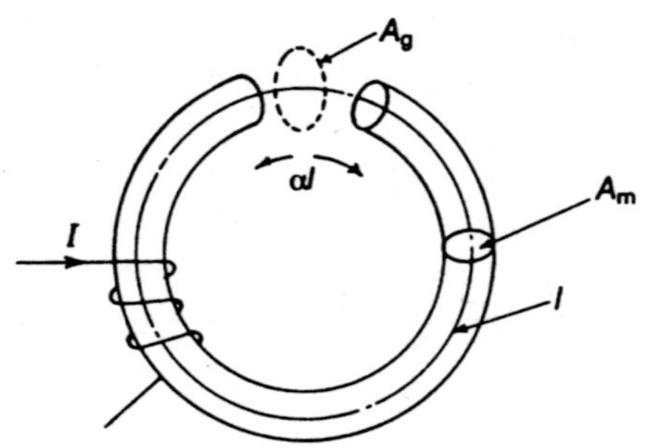

Fig. 9.4 The effect of an air gap in a toroid.

$$nI = \oint H\mathrm{d}l = H_m l(1-\alpha) + H_g \alpha l = H_e l \qquad (9.11)$$

where H_m and H_g are the fields in the material and the gap respectively and H_e is the effective field. Under the assumption that the total flux through the circuit remains constant, i.e. $B_m A_m = B_g A_g = B_e A_e$, and that $A_e = A_m = A_g$, where the subscripts have the same meanings as before, it follows that

$$\frac{1}{\mu_{re}} = \frac{1}{\mu_r} + \alpha\left(1 - \frac{1}{\mu_r}\right) \qquad (9.12)$$

From a comparison of Eqs (9.9) and (9.12) it can be seen that α corresponds to N_D and that, for μ_{re} to be close to μ_r α must be small compared with $1/\mu_r$. For example, suppose that, for a given material, $\mu_r = 1000$ and μ_{re} is required not to be less than $0.9\mu_r$. Then the maximum permitted crack width in a toroid of length 10 cm, say, is 10 μm. It is evident that if high-permeability ceramics are to be fully exploited they must be free from cracks normal to the intended flux direction.

9.1.4 Magnetic materials in alternating fields

The opening comments made in the corresponding section on dielectric materials (Section 2.7.2) apply equally here. Magnetic materials are used extensively as inductor and transformer cores where they are subjected to alternating magnetic fields. There are also many instances where an electromagnetic wave interacts with a magnetic material, as in microwave devices of various types. Under these conditions energy is dissipated in the material by various mechanisms and, analogously to the dielectric case, overall behaviour can be described with the help of a complex permeability $\mu^* = \mu' - \mathrm{j}\mu''$, in which μ' and μ'' are respectively the real and imaginary parts of μ^*. The complex relative permeability is $\mu_r^* = \mu_r' - \mathrm{j}\mu_r''$.

Important relationships can be derived by reference to the wound toroid (Fig. 9.3(a)) where the current I is now regarded as sinusoidal and represented by $I_0\exp(\mathrm{j}\omega t)$. The appropriate phasor diagram is shown in Fig. 9.5. If the instantaneous current is I, the corresponding field is nI and the flux linking the circuit is $\mu' n^2 IAl$. For a current varying sinusoidally the e.m.f. is

$$\begin{aligned} U &= \frac{\mathrm{d}}{\mathrm{d}t}(\mu^* n^2 AlI) \\ &= \mathrm{j}\omega\mu^* n^2 AlI \\ &= \mathrm{j}Aln^2\omega I\mu' + Aln^2\omega I\mu'' \end{aligned} \qquad (9.13)$$

Since μ'' is small, from Fig. 9.5 δ_m ($= \tan^{-1}(\mu''/\mu')$) is also small and therefore the voltage leads the current by nearly 90°.

The mean power $\bar{P}$ dissipated over a cycle of period τ is the mean value of the product of the current and the 'in-phase' voltage component U_i, i.e.

$$\bar{P} = \frac{1}{\tau}\int_0^{\tau} IU_i \mathrm{d}t = \tfrac{1}{2}I_0^2An^2l\omega\mu'' \tag{9.14}$$

or

$$\frac{\bar{P}}{V} = \tfrac{1}{2}I_0^2n^2\omega\mu'' = \tfrac{1}{2}H_0^2\omega\mu'' = \tfrac{1}{2}H_0^2\omega\mu' \tan\delta_m \tag{9.15}$$

where V is the volume of the toroid. $\mu' \tan\delta_m$ is a material constant proportional to the heat generated per unit volume in a magnetic body in an alternating field. It is not of great practical use unless the field at which it is measured is also specified, since both μ' and $\tan\delta_m$ are strongly field dependent except at very low fields. The behaviour is similar to that of ferroelectric dielectrics with respect to electric fields.

In the pot-cores used in inductors (see Section 9.5.1) for LC filter circuits, the fields are low enough for μ' and $\tan\delta_m$ to vary only gradually with the field and the ratio $(\tan\delta_m)/\mu'$ is sensibly constant. These cores are designed with small gaps of variable width so that their effective permeability can be adjusted in accordance with Eq. (9.12). If $\mu_r \gg 1$ Eq. (9.12) can be expressed in complex form to take account of losses and rearranged to give

$$\mu_{re}^* = \frac{\mu_r^*}{1 + \alpha\mu_r^*} \tag{9.16}$$

If μ_{re}^* and μ_r^* are put in the complex form

$$\mu_{re}^* = \mu_{re}'(1 - \mathrm{j}\tan\delta_{me}) \tag{9.17}$$

and correspondingly for μ_r^*, Eq. (9.16) becomes

$$\mu_{re}'(1 - \mathrm{j}\tan\delta_{me}) = \frac{\mu_r'\{1 + \alpha\mu_r'(1 + \tan^2\delta_m) - \mathrm{j}\tan\delta_m\}}{1 + 2\alpha\mu_r' + \alpha^2\mu_r'^2(1 + \tan^2\delta_m)} \tag{9.18}$$

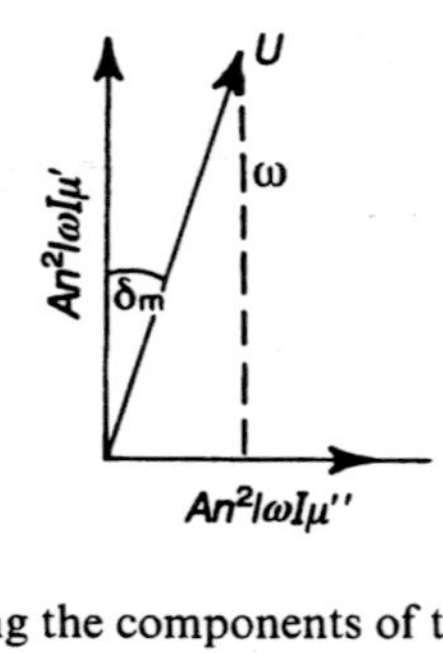

Fig. 9.5 Phasor diagram showing the components of the voltage U in phase and out of phase with the driving current I.

Since low loss is essential in pot-core applications, $\tan^2 \delta_m \ll 1$ and can be neglected. Separating real and imaginary parts then gives

$$\mu'_{re} = \frac{\mu'_r}{1 + \alpha\mu'_r}$$
$$\mu'_{re} \tan \delta_{me} = \frac{\mu'_r \tan \delta_m}{(1 + \alpha\mu'_r)^2} \quad (9.19)$$

so that

$$\frac{\tan \delta_{me}}{\mu'_{re}} = \frac{\tan \delta_m}{\mu'_r} \quad (9.20)$$

Eq. (9.20) shows that, for a small gap and at low field strengths, the ratio $(\tan \delta)/\mu$ of a high-permeability low-loss material is independent of the gap width and is therefore a useful material constant in the pot-core context. It is often referred to as the magnetic loss factor but is clearly quite different in character from the dielectric loss factor (see Section 2.7.2). It indicates that a reduction in permeability due to the introduction or enlargement of a gap is accompanied by a proportionate reduction in $\tan \delta_m$ (or increase in Q_m).

9.1.5 Classification of magnetic materials

There are various types of magnetic material classified by their magnetic susceptibilities χ_m. Most materials are *diamagnetic* and have very small negative susceptibilities (about 10^{-6}). Examples are the inert gases, hydrogen, many metals, most non-metals and many organic compounds. In these instances the electron motions are such that they produce zero net magnetic moment. When a magnetic field is applied to a diamagnetic substance the electron motions are modified and a small net magnetization is induced in a sense opposing the applied field. As already mentioned, the effect is very small and of no practical significance in the present context, and is therefore disregarded.

Paramagnetics are those materials in which the atoms have a permanent magnetic moment arising from spinning and orbiting electrons. An applied field tends to orient the moments and so a resultant is induced in the same sense as that of the applied field. The susceptibilities are therefore positive but again small, usually in the range 10^{-3}–10^{-6}. An important feature of many paramagnetics is that they obey Curie's law $\chi_m \propto 1/T$, reflecting the ordering effect of the applied field opposed by the disordering effect of thermal energy. The most strongly paramagnetic substances are compounds containing transition metal or rare earth ions and ferromagnetics and ferrites above their Curie temperatures.

Ferromagnetic materials are spontaneously magnetized below a temperature termed the Curie point or Curie temperature (the two terms are synonymous). The spontaneous magnetization is not apparent in materials which have not been exposed to an external field because of the formation of small volumes (domains) of material each having its own direction of magnetization (see Section 9.1.9). In their lowest energy state the domains are so arranged that their magnetizations cancel. When a field is applied the domains in which the magnetization is more nearly parallel to the field grow at the expense of those with more nearly antiparallel magnetizations. Since the spontaneous magnetization may be several orders of magnitude greater than the applied field, ferromagnetic materials have very high permeabilities. When the applied field is removed some part of the induced domain alignment remains so that the body is now a 'magnet' in the ordinary sense of the term. The overall relation between field strength and magnetization is the familiar hysteresis loop as illustrated in Fig. 9.10 below.

Spontaneous magnetization is due to the alignment of uncompensated electron spins by the strong quantum-mechanical 'exchange' forces. It is a relatively rare phenomenon confined to the elements iron, cobalt, nickel and gadolinium and certain alloys. One or two ferromagnetic oxides are known, in particular CrO_2 which is used in recording tapes. These ferromagnetic oxides show metallic-type conduction and the mechanism underlying their magnetic behaviour is probably similar to that of magnetic metals.

In *antiferromagnetic* materials the uncompensated electron spins associated with neighbouring cations orient themselves, below a temperature known as the Néel point, in such a way that their magnetizations neutralize one another so that the overall magnetization is zero. Metallic manganese and chromium and many transition metal oxides belong to this class. Their susceptibilities are low (about 10^{-3}) except when the temperature is close to the Néel point when the antiferromagnetic coupling breaks down and the materials become paramagnetic.

Finally, there are the important *ferrimagnetic* materials, the subject of much of this text. In these there is antiferromagnetic coupling between cations occupying crystallographically different sites, and the magnetization of one sublattice is antiparallel to that of another sublattice. Because the two magnetizations are of unequal strength there is a net spontaneous magnetization. As the temperature is increased from 0 K the magnetization decreases, reaching zero at what strictly speaking is the Néel point (see Fig. 9.7 below); however, it is commonly called the Curie point because of the generally close similarity of ferrimagnetic to ferromagnetic behaviour.

9.1.6 The paramagnetic effect and spontaneous magnetization

In the interests of simplicity, attention is confined to a crystal consisting of N atoms per unit volume for which the magnetic moment per atom arises from a

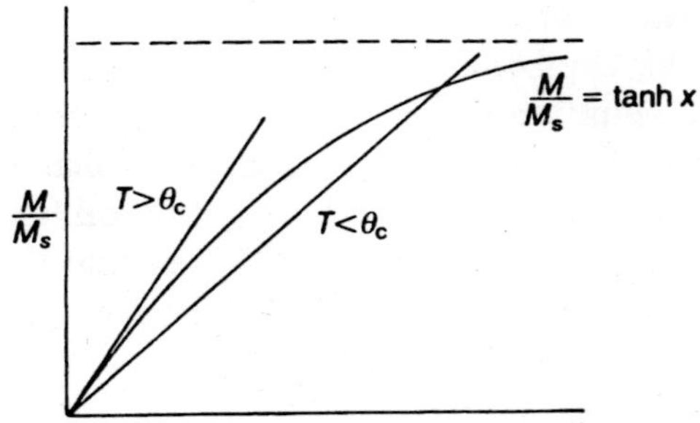

Fig. 9.6 Graphs illustrating the possibility of spontaneous magnetization below a critical temperature θ_c.

spinning electron only and the values resolved in the direction of an applied magnetic field are $\pm 1\,\mu_B$. Some of the moments will be directed parallel and some antiparallel to the field and, because there is an energy difference $2\mu_B B$ between the two states, the relative populations in them will depend upon temperature in accordance with Boltzmann statistics.

If $n\uparrow$ denotes the number of atoms per unit volume with moments in the direction of B and $n\downarrow$ the number in the antiparallel sense, the net induced magnetization is $M = (n\uparrow - n\downarrow)\mu_B$. It follows that

$$\frac{n\downarrow}{n\uparrow} = \exp\left(-\frac{2\mu_B B}{kT}\right) \tag{9.21}$$

and, since $N = n\uparrow + n\downarrow$, it is straightforward to show that

$$M = N\mu_B \tanh\left(\frac{\mu_B B}{kT}\right) \tag{9.22}$$

Under most practical conditions $\mu_B B \ll kT$, and then (using the approximation $e^x \approx 1 + x$ for $x \ll 1$) it follows that

$$M \approx \frac{N\mu_B^2 B}{kT} \quad \text{or} \quad \chi_m = \frac{N\mu_B^2\mu_0}{kT} = \frac{C}{T} \tag{9.23}$$

which is Curie's law. It should be noted that in the derivation B has been equated with $\mu_0 H$ since $\mu_r' \approx 1$.

In 1907 Weiss suggested that the magnetic field H_i 'seen' by an individual dipole in a solid was the macroscopic internal field H modified by the presence of dipoles in the neighbourhood of the individual. This idea was expressed by $H_i = H + wM$, (c.f. Eq. (2.84)) where w is known as the 'molecular field constant'. Under this assumption Eq. (9.22) is modified to

$$M = N\mu_B \tanh\left\{\frac{\mu_B\mu_0}{kT}(H + wM)\right\} \tag{9.24}$$

Spontaneous magnetization implies a value for M when $H = 0$, and the practical feasibility of this can be explored by putting $H = 0$ in Eq. (9.24) and then examining whether the equation can be satisfied by non-trivial values for M. Therefore, putting $N\mu_B = M_s$, where M_s is the saturation magnetization, we can write Eq. (9.24) as

$$\frac{M}{M_s} = \tanh\left(\frac{\mu_B \mu_0 w M}{kT}\right)$$

or, more conveniently,

$$\frac{M}{M_s} = \tanh x \tag{9.25}$$

where

$$x = \frac{\mu_B \mu_0 w M}{kT} \tag{9.26}$$

It follows from Eq. (9.26) that

$$\frac{M}{M_s} = \frac{kT}{\mu_B^2 \mu_0 w N} x = \frac{T}{\theta_c} x \tag{9.27}$$

in which θ_c has the dimensions of temperature and is the Curie temperature. Figure 9.6 shows the graphs of Eqs (9.25) and (9.27).

It is clear that, for spontaneous magnetization to be possible in principle, Eqs (9.25) and (9.27) must be consistent, i.e. both must be simultaneously satisfied for particular values of M/M_s and x. Also, for small values of x, $\tanh x \approx x$ and so, near to the origin, $M/M_s = x$; therefore for $T = \theta_c$ the equations have the same slope at the origin. This defines a temperature θ_c such that if $T > \theta_c$ (slope greater than unity) there is no intersection at non-zero values of x and, in consequence, no spontaneous magnetization is possible. For $T < \theta_c$ there is an intersection and spontaneous magnetization is a possibility. It follows that, for temperatures up

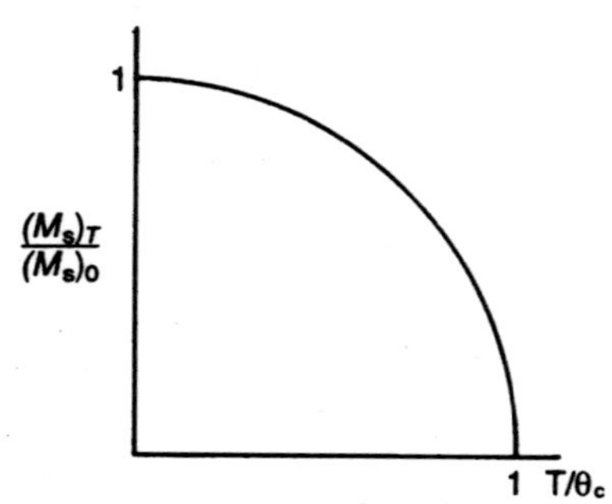

Fig. 9.7 The variation of the ratio of saturation magnetization at T K to that at 0 K with the ratio of the temperature to the Curie temperature.

to $T = \theta_c$, the M/M_s value corresponding to the intersection can be found, leading to Fig. 9.7 which shows good agreement with experiment for iron, cobalt and nickel.

According to the Weiss theory just outlined, ferromagnetism is caused by a strong internal magnetic field aligning the magnetic moments on individual ions. The physical origin of the internal field is now known to be quantum mechanical in nature and to involve the 'exchange forces' which determine the relative orientation of the spins on adjacent electrons. The exchange energy arising from exchange forces plays a dominating role in determining the nature of the important magnetic materials. In some instances, e.g. the metals iron, cobalt and nickel, the exchange energy is minimized for parallel spins; this is exceptional since, generally, the exchange energy is minimized when adjacent spins are antiparallel. In the non-metallic antiferromagnetics and ferrites the ordering is antiparallel through superexchange forces, so-called because they act between the spins of neighbouring cations with the involvement of an intermediate ion which, in the case of the ferrites, is oxygen.

9.1.7 Magnetocrystalline anisotrophy

A spinning electron, free from any restraints, can be aligned by an infinitely small field, implying an infinite permeability. A restraint leading to finite permeabilities in magnetic materials is caused by a coupling between the spins and the crystal lattice through the agency of the orbital motion of the electron. This spin–orbit lattice coupling results in orientation of the spins relative to the crystal lattice in a minimum energy direction, the so-called 'easy direction' of magnetization. Aligning the spins in any other direction leads to an increase in energy, the anisotropy energy $\mathcal{E}_K$. For a cubic lattice, such as a spinel, $\mathcal{E}_K$ is related to two anisotropy constants K_1 and K_2 by

$$\mathcal{E}_K = K_1(\alpha_1^2\alpha_2^2 + \alpha_2^2\alpha_3^2 + \alpha_3^2\alpha_1^2) + K_2\alpha_1^2\alpha_2^2\alpha_3^2 \cdots \qquad (9.28)$$

where α_1, α_2 and α_3 are the direction cosines of the magnetization vector relative to the crystallographic axes.

The approximate expression for $\mathcal{E}_K$ (Eq. (9.28)) leaves out all but two of the terms of an infinite power series, and even the term involving K_2 can often be safely neglected. If it is assumed that K_2 is negligible and K_1 is positive, then $\mathcal{E}_K$ has a minimum value of zero if any two of the direction cosines are zero, i.e. the anisotropy energy is a minimum along all three crystal axes and these are therefore the 'easy' directions. If K_1 is negative, the minimum occurs for $\alpha_1 = \alpha_2 = \alpha_3 = 1/\sqrt{3}$, i.e. the body diagonal is the 'easy' direction.

The anisotropy constants listed in Table 9.1, which anticipate the discussion of ferrites (Section 9.2 *et seq.*), indicate that the easy directions for cubic crystals are [111] except for those containing cobalt, in which case they are [100]. For

Table 9.1 Room temperature anisotropy constants of some important ferrites

Ferrite	K_1/kJ m^{-3}
Fe_3O_4	−11
$Mn_{0.98}Fe_{1.86}O_4$	−2.8
$Co_{0.8}Fe_{2.2}O_4$	+290
$NiFe_2O_4$	−6.2
$CuFe_2O_4$	−6.0
$Mn_{0.45}Zn_{0.55}Fe_2O_4$	−0.38
$Ni_{0.5}Zn_{0.5}Fe_2O_4$	−3
$BaFe_{12}O_{19}$	+330

hexagonal crystals the anisotropy energy $\mathcal{E}_{Kh}$ depends only upon the angle θ between the c axis and the magnetization vector, and is almost independent of direction in the basal plane. It can be expressed approximately as

$$\mathcal{E}_{Kh} = K_{1h}\sin^2\theta + K_{2h}\sin^4\theta \tag{9.29}$$

The energy minimum occurs for $\theta = 0$, making the c axis the 'easy' direction.

The concept of an anisotropy magnetic field H_A, which is referred to simply as the 'anisotropy field', is also introduced to describe magnetic anisotropy. The anisotropy energy is considered to be that of the saturation magnetization moment M_s in an induction $\mu_0 H_A$, i.e. $\mathcal{E}_K = -\mu_0 H_A M_s \cos\theta$, where θ is the angle between H_A and M_s.

9.1.8 Magnetostriction

Because of the spin–orbit lattice coupling referred to in the previous section, changes in the spin directions result in changes in the orientation of the orbits which, because they are restrained by the lattice, have the effect of slightly altering the lattice dimensions. This effect is known as magnetostriction.

The magnetostriction constant λ_m is defined as the strain induced by a saturating field; it is given a positive sign if the field causes an increase in dimensions in the field direction. For single crystals λ_m varies with the crystallographic direction, and so for the ceramic form it is an average of the single-crystal values. λ_m values for some polycrystalline ferrites are given in Table 9.2.

9.1.9 Weiss domains

The fact that spontaneous magnetization exists in, for example, a piece of iron, and yet the overall magnetization can be zero, is explained by the existence of

Table 9.2 Saturation magnetostriction constants for some polycrystalline ferrites

Composition	$\lambda_m/10^{-6}$
Fe_3O_4	+40
$MnFe_2O_4$	−5
$CoFe_2O_4$	−110
$NiFe_2O_4$	−26
$Ni_{0.56}Fe^{2+}_{0.44}Fe_2O_4$	0
$Ni_{0.5}Zn_{0.5}Fe_2O_4$	−11
$MgFe_2O_4$	−6
$TbFe_2$ a 'giant' magnetostrictor	$>10^3$*

*Included for comparison (H. Uchida *et al.* (2002) Progress in thin films of giant magnetostrictive alloys, *J. Mag. Mag. Mater.*, **239**, 540–5).

domains. Below its Curie temperature a ferromagnetic or ferrimagnetic body consists of a large number of small domains, each spontaneously magnetized to saturation. Each grain or crystallite in a polycrystalline magnetic ceramic may contain a number of domains, each differing from its neighbour only in the direction of magnetization.

A single crystal with uniform magnetization, i.e. a single-domain single crystal, has magnetostatic energy due to the external magnetic field which it generates. If the crystal is divided into oppositely oriented parallel domains, the energy will be greatly reduced since the flux can now pass from one to another of the closely adjacent domains (Fig. 9.8(a)). In cubic materials, such as spinels and garnets, zero magnetostatic energies are possible through the formation of closure domains (Fig. 9.8(b)) since the external flux is now close to zero. Even so, the system has magnetoelastic energy because magnetostriction leads to straining between the long and the triangular domains.

The boundary between two adjacent domains is known as a domain wall or Bloch wall – the latter term after F. Bloch who proved in 1932 that magnetization cannot change discontinuously at a domain boundary. A Bloch wall is the region between two domains in which the elementary spin moments change smoothly from one orientation to another. For example, in the case of the antiparallel domains the change in direction of the vectors in moving from one domain to an adjacent one would be as shown diagrammatically in Fig. 9.9. The walls have widths in the range 10–100 nm and an associated energy in the range $(1–10)\times10^{-4}\,J\,m^{-2}$.

An important property of a Bloch wall is its mobility. It can be seen from Fig. 9.9 that the application of $\boldsymbol{H}$ in the sense shown will cause the wall to move to the left by a series of minor rotations of the vectors. A more detailed consideration of wall displacements shows them to be of two types, reversible or irreversible; which one occurs depends upon the size of the displacement and the

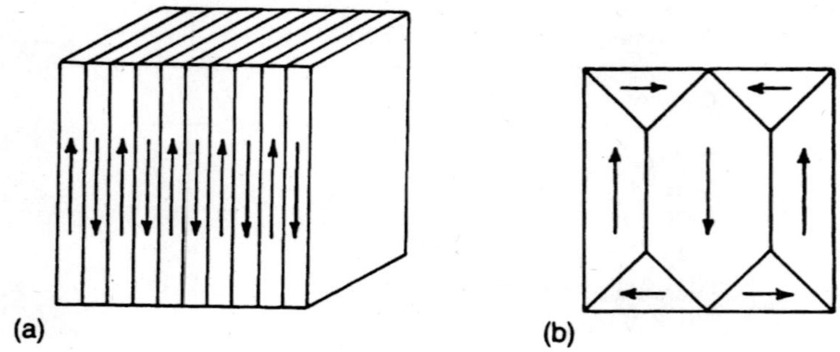

Fig. 9.8 Idealized magnetic domain configurations: (a) antiparallel domains; (b) flux closure domains.

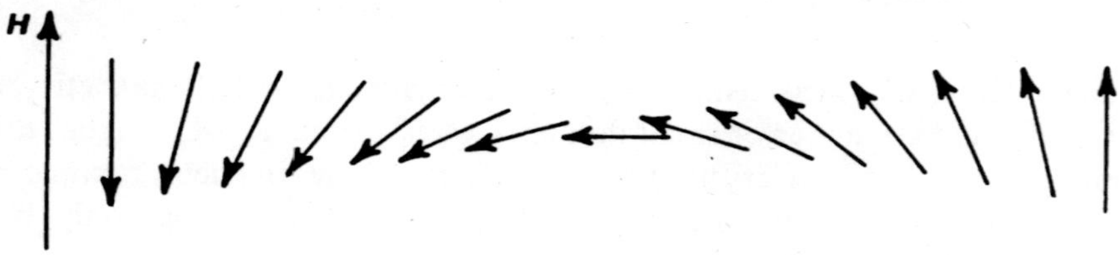

Fig. 9.9 The change in spin orientation across the width of a Bloch wall.

disposition of wall energy minima, which in turn depend upon crystal lattice defects and inhomogeneities of various types.

9.1.10 Magnetization in a multidomain crystal

The most characteristic feature of ferromagnetic or ferrimagnetic materials is the relationship between B and H (or M and H) – the hysteresis loop (Fig. 9.10) – which can now be considered in more detail in terms of the magnetization processes in a multidomain single crystal. The line deOba – the 'virgin curve' – represents the relationship determined experimentally when the specimen is demagnetized before each measurement of the induction for a given field. The change in B, very near to the origin, represents magnetization by reversible Bloch wall displacements, and the tangent OC to this initial magnetization curve is called the initial permeability μ_i. The steep rise in B represents magnetization by irreversible Bloch wall displacements as the walls break away from their pinning points, and the region ba represents magnetization by reversible and irreversible domain rotations from one easy direction to another more favourably aligned with the applied field. The latter process requires high field strengths because the magnetization within a domain is rotated against the anisotropy field.

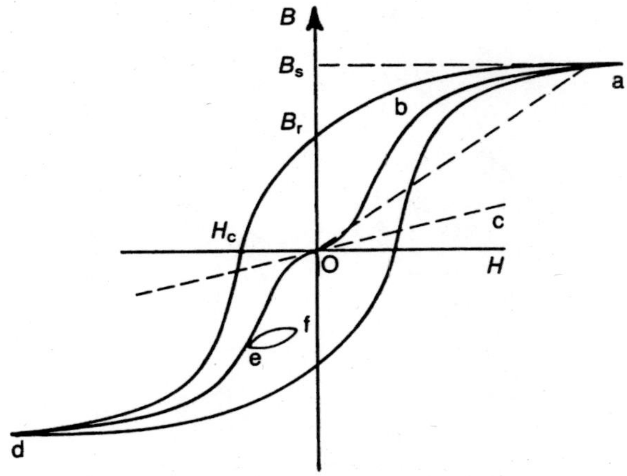

Fig. 9.10 Magnetic *B–H* hysteresis loop.

The slope of Oa, from the origin to the tip of the loop, gives the amplitude permeability μ_a which has a maximum value when the peak field corresponds to the point b on the virgin curve. If a relatively small alternating field is superimposed on a static field, a minor loop such as ef is obtained, and the amplitude permeability of this is known as the incremental permeability μ_Δ. As far as applications are concerned, μ_i is important for inductors where only small alternating fields are encountered, μ_a is important for power transformers when large alternating fields are involved and μ_Δ is important for transductors to which both alternating and static fields are applied.

If, after the material has been magnetically saturated to the value B_s, the field is reduced to zero, the magnetization vectors rotate out of line with the field towards the nearest preferred direction which is determined in part by magnetocrystalline anisotropy. The magnetization is thus prevented from complete relaxation to the 'virgin' curve and hence, for zero field, there is a remanent induction B_r. In order to reduce the induction to zero a reverse field H_c has to be applied. The coercive field or 'coercivity' H_c depends in part on crystalline anisotropy, as might be expected.

Because of hysteresis, energy is dissipated as heat in a magnetic material as it is taken round a complete *B–H* loop, and the hysteresis energy loss W_h per unit volume of material is

$$W_h = \oint B \mathrm{d}H \tag{9.30}$$

Magnetic materials are usually characterized as 'hard' or 'soft', depending on the magnitude of their coercivity and Figure 9.11 shows typical loops.

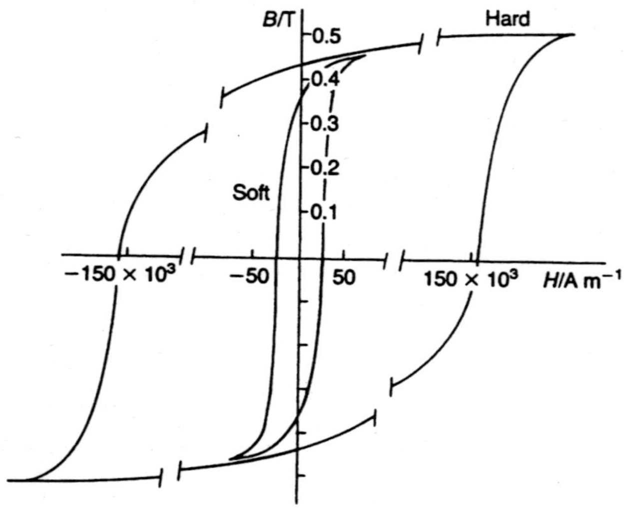

Fig. 9.11 Hysteresis loops illustrating the distinction between magnetically 'soft' and 'hard' materials.

9.2 Model Ferrites

9.2.1 Spinel ferrites: model $NiOFe_2O_3$

Magnetite (Fe_3O_4), a naturally occurring ferrite, is the earliest known magnetic material and was exploited as lodestone† many hundreds of years ago. Its composition can be written $FeOFe_2O_3$, when the structural relationship to the mineral spinel ($MgOAl_2O_3$) is apparent. There are many other possible compositions with the general formula $MeOFe_2O_3$, in which Me represents a divalent ion such as Mn^{2+}, Fe^{2+}, Co^{2+}, Ni^{2+}, Cu^{2+} or Zn^{2+}, or a combination of divalent ions with an average valence of 2. The substitutional possibilities are considerable, which has led to the extensive technological exploitation of ferrites.

In the spinel crystal structure the oxygen ions form a cubic close-packed array in which, as discussed in Chapter 2 (Fig. 2.1), two types of interstice occur, one coordinated tetrahedrally and the other octahedrally with oxygen ions. The cubic unit cell is large, comprising eight formula units and containing 64 tetrahedral and 32 octahedral sites, customarily designated A and B sites respectively; eight of the A sites and 16 of the B sites are occupied. The unit cell shown in Fig. 9.12 is seen to be made up of octants, four containing one type of structure (shaded) and four containing another (unshaded). In this representation some of the A-site cations lie at the corners and face-centre positions of the large cube; a

† The attractive power of magnetite was known to the Greeks (*c.* 100 BC); by the early 17th century – and probably much earlier – it was used by the Chinese to magnetize compass needles. Hence, 'lode' in 'lodestone' is synonomous with 'lode' in 'lodestar' (the Pole star).

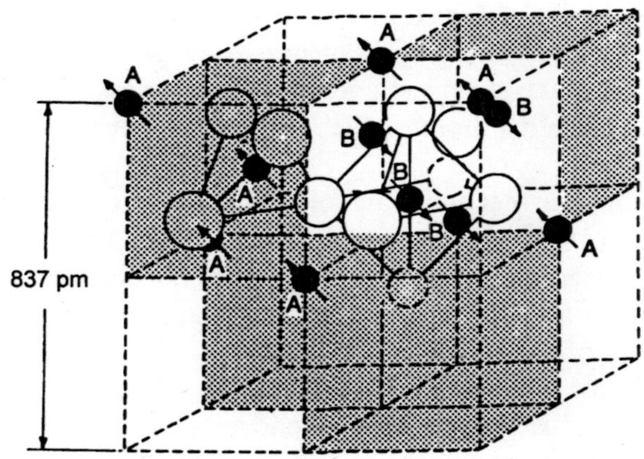

Fig. 9.12 The unit cell of a magnetic inverse spinel.

tetrahedral and an octahedral site are shown. The close-packed layers of the oxygen ion lattice lie at right angles to the body diagonals of the cube. The arrows on the ions, representing directions of magnetic moments, indicate that the B-site ions have their moments directed antiparallel to those of A-site ions, illustrating the antiferromagnetic coupling.

The occupancy of sites in the spinel-type minerals is conveniently represented with the help of the diagrams in Fig. 9.13. In the case of the mineral spinel the divalent ions (Mg) occupy the A sites and the trivalent ions (Al) the B sites (Fig. 9.13(a)). This is known as the *normal* spinel structure. In the spinel ferrites ($MeFe_2O_4$) different ions exhibit different site preferences. In nickel ferrite, for instance, the Ni^{2+} ions occupy B sites along with an equal number of randomly distributed Fe^{3+} ions, whilst the remaining Fe^{3+} ions occupy A sites. This is termed an *inverse* spinel structure (Fig. 9.13(b)). In practice, the structures of the spinel-type magnetic ceramics are a blend of the normal and inverse types, with the oxidation states and distribution of cations amongst the lattice sites depending upon thermal history, i.e. on firing schedules and kiln atmosphere.

Attention is now given to the ordering of the magnetic moments. The spin-ordering forces are of the 'superexchange' kind, so called because they act between the resultant spin on the cations through the agency of an intervening oxygen ion. Superexchange forces acting between A ions, between B ions and

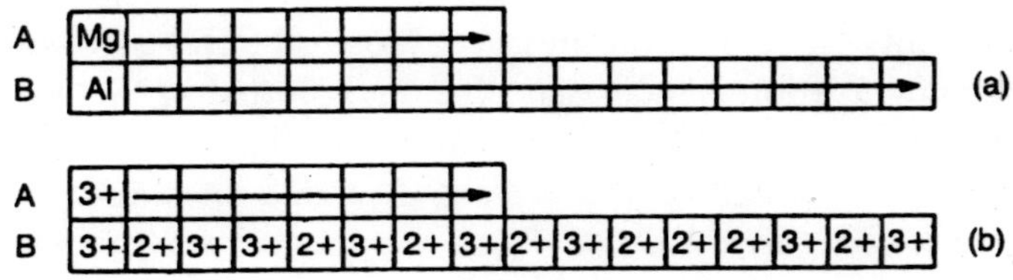

Fig. 9.13 Diagrammatic representation of site occupancy in (a) normal and (b) inverse spinels.

between A and B ions are referred to as AA, BB or AB interactions, and in each case the forces tend to hold the spins of nearest-neighbour cations antiparallel. The AB interaction is the strongest and in consequence moments on the A ions are directed antiparallel to those of the B ions. Néel introduced and developed this theory in 1948.

Before proceeding further with the analysis, the magnitude of the magnetic moments associated with the various cations must be known and, because of orbital moment quenching, only the spin moments need be considered when calculating the moment of the ions. For example, as a first step in calculating the magnetic moment of $NiOFe_2O_3$ it is readily shown that the resultant spin quantum numbers for Ni^{2+} and Fe^{3+} ions are 1 and 5/2 respectively. Because the spinning electron produces twice the magnetic moment expected from its angular momentum, the corresponding magnetic moments are $2\mu_B$ and $5\mu_B$. Also, because all the A moments are directed antiparallel to the B moments, it is evident from Fig. 9.13(b) that there is a resultant moment of $16\mu_B$ per unit cell which is due to the eight Ni^{2+} ions. The saturation magnetization can now be calculated from the cell dimensions:

$$M_s = \frac{16\mu_B}{\text{cell volume}} = \frac{16 \times 9.27 \times 10^{-24}}{(8.37 \times 10^{-10})^3} \approx 2.5 \times 10^5 \text{ A m}^{-1}$$

This compares well with the experimentally determined value (approximately 3×10^5 A m^{-1}). The discrepancy is probably due partly to the assumption that $NiOFe_2O_3$ has the ideal inverse spinel structure and partly to the incomplete quenching of the orbital moment.

An elegant confirmation of the essential correctness of the theory lies in the explanation of why the addition of a non-magnetic ion such as Zn to a spinel ferrite should lead to an *increase* in saturation magnetization. The magnetic moment per formula unit $MeFe_2O_4$, where Me represents Ni^{2+} or Zn^{2+}, is shown in Fig. 9.14 as a function of zinc substitution. Zinc ferrite is a normal spinel, indicating that Zn ions have a preference for the A sites, so that on substituting zinc for nickel the occupancy becomes $(Fe^{3+}_{1-\delta}Zn^{2+}_{\delta})(Fe^{3+}_{1+\delta}Ni^{2+}_{1-\delta})O_4$, in which the first and second brackets indicate occupancy of the A and B sublattices respectively. Thus the antiparallel coupling between moments on A and B sites is reduced because the occupancy of A sites by magnetic ions is reduced, and as a consequence the Curie point is lowered (see Fig. 9.18 below). However, the excess of moments on octahedral sites over those on tetrahedral sites is increased so that the magnetization is increased. The data plotted in Fig. 9.14 confirm this model up to $\delta \approx 0.4$. The fall in magnetization for higher values of δ is due to the reduced antiparallel coupling between the A and B sites, and it becomes zero when $\delta = 1$, i.e. $(Zn^{2+})(Fe^{3+}_2)O_4$.

Generally speaking, the spinel ferrites have low magnetic anisotropies and are magnetically 'soft'; exceptions are those containing Co^{2+} which is itself strongly

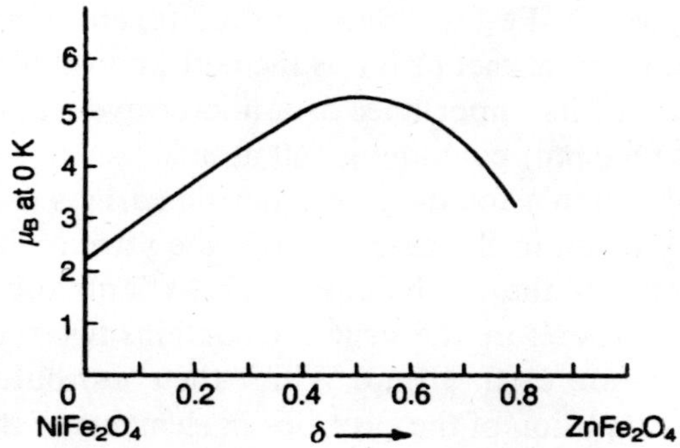

Fig. 9.14 Saturation magnetization per 'formula unit' for the ferrite $(Fe^{3+}_{1-\delta}Zn^{2+}_{\delta})(Fe^{3+}_{1+\delta}Ni^{2+}_{1-\delta})O_4$ as a function of δ.

magnetically anisotropic. Cobalt spinel ferrites can have coercivities approaching $10^5\,A\,m^{-1}$, placing them firmly in the 'hard' category.

9.2.2 Hexaferrites: model $BaFe_{12}O_{19}$

Barium hexaferrite ($BaFe_{12}O_{19}$) is the model for a family of 'M-type ferrites', so-called because they are based upon the hexagonal magnetoplumbite, or M-structure. Its crystal structure, though related to that of the spinels, is very much more complex. The large unit cell ($c = 2.32$ nm; $a = 0.588$ nm) contains two formula units, i.e. a total of 64 ions. The Ba^{2+} and O^{2-} ions together form a close-packed structure with some of the layers cubic close-packed and others hexagonal close-packed. The origins of the magnetic properties are basically the same as those already discussed and can be summarized as follows: of the 12 Fe^{3+} ions in a formula unit, nine are on octahedral sites two on tetrahedral sites and one on a five-coordinated site; seven of the ions on octahedral sites and the one on a five-coordinated site have their spins in one sense, and the remainder are oppositely directed. Thus there are four more ions with spins in the one sense than there are with spins in the other and, since there are five electrons with parallel spins in each Fe^{3+} ion, there are 20 unpaired spins per formula unit, leading to a saturation magnetization of $20\,\mu_B$ per cell volume. $BaFe_{12}O_{19}$ has a high magnetic anisotropy (Table 9.1) with its 'prefered direction of magnetization' ('easy' direction) along the c-axis. Various substitutions are made to tailor intrinsic magnetic properties, for example Sr for Ba, and partial substitution of Al for Fe to increase coercivity [8].

9.2.3 Garnets: models $Y_3Fe_5O_{12}$ (YIG)

'Garnet' is the name of a group of isostructural minerals with the general composition $3R'O \cdot R''_2O_3 \cdot 3SiO_2$. Examples are $3CaO \cdot Al_2O_3 \cdot 3SiO_2$

(grossularite), $3CaO \cdot Fe_2O_3 \cdot 3SiO_2$ (andradite) and $3MnO \cdot Al_2O_3 \cdot 3SiO_2$ (spessarite). Yttrium iron garnet (YIG) is the best known of a family of ferrimagnetic garnets because of its importance as a microwave material. It has a large cubic unit cell ($a \approx 1200$ pm) containing 160 atoms.

The general formula for the ferrimagnetic garnets is written $R_3Fe_5O_{12}$, where R stands for yttrium in the case of YIG; the yttrium can be totally or partially replaced by one of the lanthanides such as lanthanum, cerium, neodymium, gadolinium etc. Therefore the structure contains two types of magnetic ion, iron and one of the rare earth group. Whilst the contribution to the magnetization from the orbital motion of the electrons in elements of the first transition series is close to zero (quenching) because of the orbital–lattice coupling, that of the electrons in the lanthanide ions has a significant effect. The unpaired electrons of the first series elements are in the outermost 3d group and therefore are not shielded from the crystal field which is responsible for quenching. In the lanthanide ions the unpaired electrons in the 4f group are shielded by the 5s5p electrons and there is therefore an orbital contribution in addition to that of the unpaired spins. As a consequence the contribution of the lanthanide ions to the magnetization is somewhat greater than would be estimated from the simple rules governing the elements of the first transition series. A further consequence of this shielding is that the coupling of lanthanide ions to other magnetic ions is weaker than that between the ions of the first transition series.

The lattice site occupancy is conventionally represented by the formula $\{R_3\}_c[Fe_2]_a(Fe_3)_dO_{12}$, where $[\;\;]_a$ indicates ions on octahedral sites, $(\;\;)_d$ indicates ions on tetrahedral sites and $\{\;\;\}_c$ indicates ions on 12-coordinated sites. There is strong coupling with antiparallel spins between ions on the a and d sites, and thus, since all are Fe^{3+}, the net contribution is $5\mu_B$; the rare earth ions on the c sites have their unpaired spins coupled antiparallel to the Fe^{3+} on d sites and so contribute $-3\mu_R$, where μ_R is the strength of the moment of the R ion measured in Bohr magnetons. The resultant magnetization per formula unit, measured in Bohr magnetons, is therefore

$$M = 5 - 3\mu_R$$

in which μ_R is greater than 7 for gadolinium, terbium and dysprosium, and falls off to 3.5 for thulium, 2.7 for ytterbium, and zero for lutetium and yttrium. When $\mu_R > 5/3$, M is negative and its value at 0 K is dominated by the contribution from the rare earth ions. At higher temperatures the rare earth contribution decreases because of the weak coupling between $(Fe^{3+})_d$ and $\{R\}_c$, and thus the magnetization first falls to zero and then increases again. The zero magnetization point is known as the *compensation point*. Since at high temperatures magnetization in those rare earth garnet ferrites containing no substitute for Fe^{3+} depends mainly on the $(Fe^{3+})_d$–$[Fe^{3+}]_a$ coupling, they all have approximately the same Curie points.

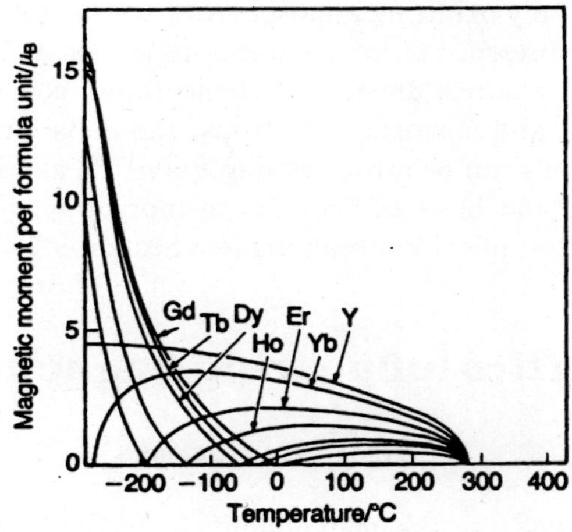

Fig. 9.15 Variation of saturation magnetization with temperature for various garnets.

The types of saturation magnetization–temperature characteristic for the various garnet ferrites are summarized in Fig. 9.15. Figure 9.16 illustrates for a series of the yttrium gadolinium iron garnets an important feature associated with a compensation point which is exploited in certain applications: the magnetization can be arranged to be almost independent of temperature over a chosen temperature range. For example, for the composition corresponding to x=0.6, i.e. $Y_{1.2}Gd_{1.8}Fe_5O_{12}$, the saturation magnetization is relatively stable over a wide temperature range centred around 50 °C.

A further important feature of the garnets is their high electrical resistivity resulting in a high Q suitable for certain microwave devices. This is because they

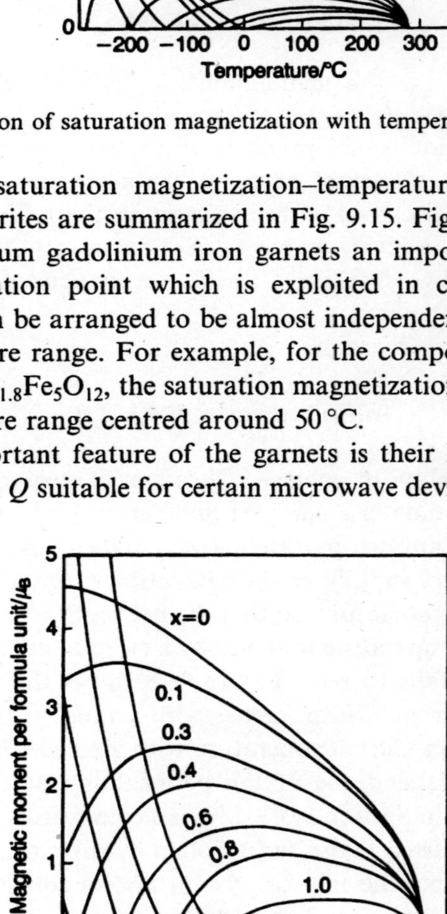

Fig. 9.16 Variation of saturation magnetization with temperature for $Y_{3(1-x)}Gd_{3x}Fe_5O_{12}$ for various x values.

can be sintered in fully oxidizing atmospheres since their magnetic properties do not depend on the presence of lower-valency ions such as Fe^{2+} or Mn^{2+}. Other properties, including lattice dimensions, temperature coefficients of magnetization and expansion and magnetic anisotropy, can be tailored by adjustments to composition. Garnets can be grown as single crystals, and thin films of complex composition, once the basis of 'bubble' memories, now attract considerable interest for magneto-optical applications (see Section 9.5.4).

9.3 Properties Influencing Magnetic Behaviour

9.3.1 Soft ferrites

Soft ferrites are used for the manufacture of inductor cores (pot cores) for telecommunications, low-power transformers and high-flux transformers such as television line output transformers, and as television tube scanning yokes (Fig. 9.17). The more important material characteristics for these and other applications are now discussed with emphasis on the influence of composition and microstructure. The review paper by A. Broese van Groenou *et al.* [9] and the monograph by E.C. Snelling [10] are recommended to supplement the discussion.

Initial permeability (μ_{ri})

High initial permeability is achieved through control of composition and microstructure. It depends in a complex manner on high saturation magnetization, low magnetic anisotropy and low magnetostriction. The magnetic anisotropy falls off very rapidly as the saturation magnetization falls to a low value near the Curie temperature, so that the net result is a peak in permeability just below the Curie temperature followed by a steep fall to a value close to unity as the magnetization falls to zero. Figure 9.18 shows the variation of μ_{ri} with temperature for MnZn and NiZn ferrites with a range of zinc contents; it also illustrates the change in Curie temperature with zinc content mentioned earlier (Section 9.2.1). The dependence of the permeability of the NiZn system on frequency is discussed in Section 9.3.1. Magnetic anisotropy is low in the MnZn ferrites and can be adjusted in the NiZn system by substituting small amounts of cobalt for the nickel, since the anisotropy constant for cobalt ferrite is opposite in sign to that of most ferrites (see Table (9.1).

Magnetostriction can be reduced by adjusting the sintering atmosphere during the application of the maximum temperature and afterwards so that a small amount of Fe^{2+} is formed, thus taking advantage of the opposite sign of the magnetostriction constant for Fe_3O_4 compared with that for most other ferrites

Fig. 9.17 Range of soft ferrite components: (i) TV scanning yoke (components kindly supplied by Philips Components Ltd.); (ii) UR core and TV line output transformer; (iii) E core for switched mode power supply; (iv) wide band transformer core; (v) core giving good magnetic shielding; (vi) high Q (adjustable) filter core (cf. Fig. 9.48); (vii) precision ferrite antenna for transponder; (viii) multilayer EMI suppressors; (ix) toroids for laser and radar pulse applications; (x) typical EMI shields for cables. ((ii)–(x) Courtesy of 'Ferroxcube UK'.)

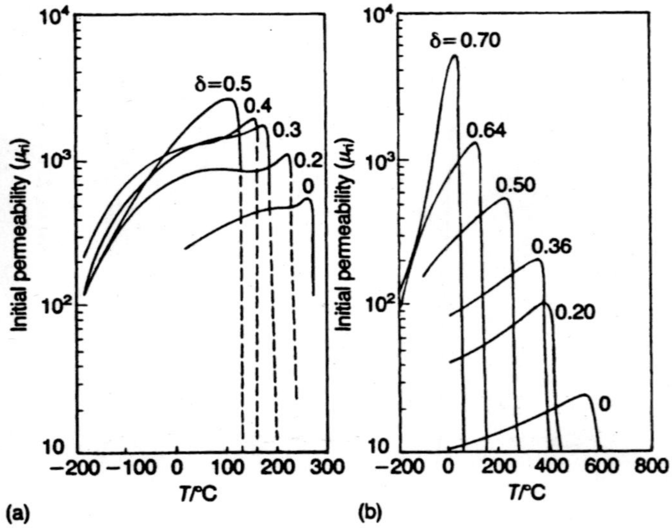

Fig. 9.18 The variation with temperature of the initial relative permeability μ_{ri} for different δ values in (a) $Mn_{1-\delta}Zn_\delta Fe_2O_4$ and (b) $Ni_{1-\delta}Zn_\delta Fe_2O_4$ (after [10]).

(see Table 9.2). The introduction of Fe^{2+} has the disadvantage of lowering the resistivity of the ferrite, but a high permeability can be obtained without excessive penalty in this respect.

Because a major contribution to μ_{ri} is from Bloch wall movements, microstructure has a significant influence. High magnetic anisotropy implies high-energy walls readily 'pinned' by microstructural defects. Thus, for a high-permeability polycrystalline ferrite, very mobile domain walls are required, demanding in turn large defect-free grains coupled with low magnetic anisotropy. Figures 9.19 and 9.20 illustrate the sensitivity of permeability to grain size, and Fig. 9.21 shows how porosity leads to reduced permeability, presumably because of domain wall pinning.

The loss factor $(\tan\delta)/\mu_{ri}$

The way in which magnetic loss in a material is expressed depends upon the particular application. For example, in the case of pot cores, when currents are small and hence the flux density is also small (typically less than 1 mT), the loss factor used is $(\tan\delta)/\mu_{ri}$ or its reciprocal $\mu_{ri}Q$ where Q is the quality factor.

The loss factor is usually expressed in terms of three main contributions, namely

$$\frac{\tan\delta}{\mu_{ri}} = \frac{\tan\delta_h}{\mu_{ri}} + \frac{\tan\delta_e}{\mu_{ri}} + \frac{\tan\delta_r}{\mu_{ri}} \qquad (9.31)$$

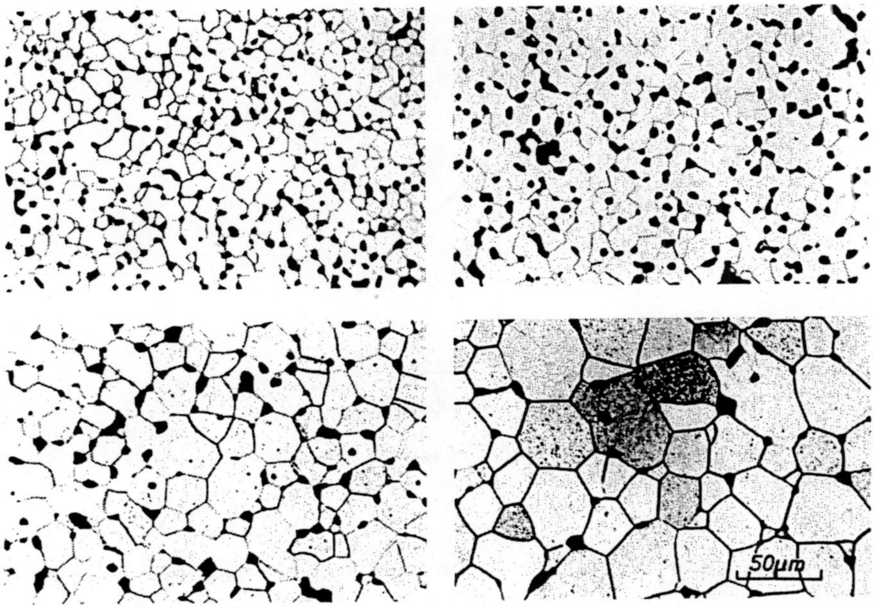

Fig. 9.19 Microstructure of high permeability Mn–Zn ferrites with a range of grain sizes: (a) $\mu_{ri} = 6500$; (b) $\mu_{ri} = 10\,000$; (c) $\mu_{ri} = 16\,000$; (d) $\mu_{ri} = 21\,500$. (Courtesy of Philips Technical Review.)

in which tan δ_h, tan δ_e and tan δ_r are the hysteresis, eddy current and 'residual' loss tangents. The processes contributing to residual loss are difficult to identify but include ferrimagnetic resonance and domain wall resonance (see below). In low-current high-frequency applications of high-resistivity ferrites, when hysteresis and eddy current losses are low, residual losses may dominate.

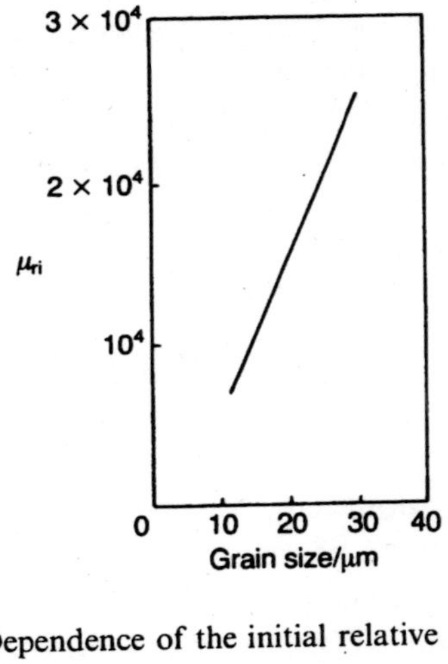

Fig. 9.20 Dependence of the initial relative permeability on grain size.

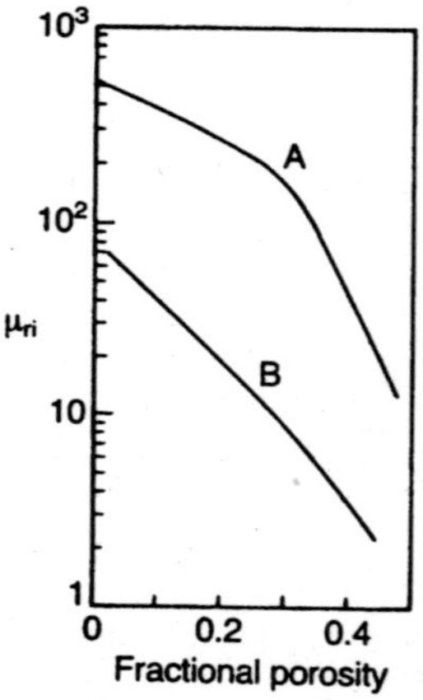

Fig. 9.21 Dependence of the initial relative permeability on porosity: curve A, $Ni_{0.5}Zn_{0.5}Fe_2O_4$; curve B, $NiFe_2O_4$ (after [6]).

P.J. van der Zaag [11] discusses loss mechanisms in soft ferrites and shows that a significant contribution to residual losses over the frequency range 500 kHz to 10 MHz originates from domain wall motion in multi-domain grains. This is evidenced by a predicted and observed decrease in 'intragranular' loss accompanying a transition from two-domain to mono-domain grains as the grain size is decreased in the range 0.2 to 5 μm).

It is shown (Section 9.1.4) that the loss factor is independent of gaps introduced in the magnetic circuit. In the case of pot cores (Fig. 9.48), where air gaps are introduced both intentionally and because of the imperfect join, the loss factor is a particularly useful parameter.

In the case of ferrites operating at high amplitudes (e.g. above 10 mT) and in the typical frequency range 15–100 kHz power is dissipated mainly through hysteresis, although eddy current loss may also contribute significantly. Losses in materials intended for such high-power applications are usually expressed in terms of power-loss density (Eq. (9.15)) measured directly from the alternating voltage across the winding on a toroid, the in-phase component of the current and the toroid volume.

Electrical resistivity

The resistivity ρ of an inductor core material is important because it determines eddy current losses. In general room temperature resistivities of ferrites lie in the range $10^{-1}-10^{6}\,\Omega$ m, many orders of magnitude higher than that of the most resistive of the ferromagnetic alloys (about $8\times10^{-7}\,\Omega$ m). Typical resistivity–temperature data for MnZn and NiZn ferrites are shown in Fig. 9.22. For both

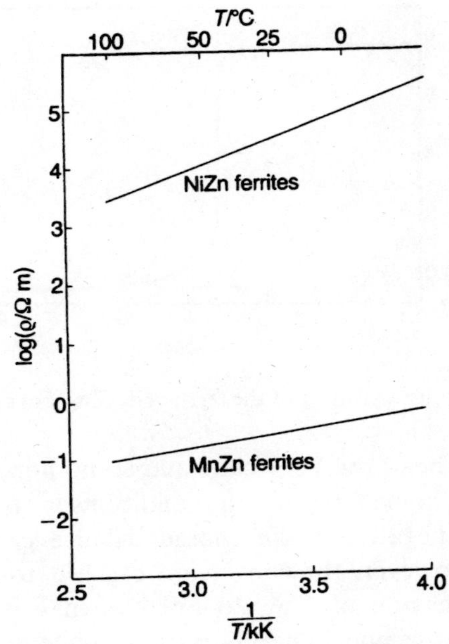

Fig. 9.22 Temperature dependence of the resistivity for NiZn and MnZn ferrites.

types the conductivity mechanism is believed to be electron hopping between ions of the same type on equivalent lattice sites, e.g. $Fe^{3+} \leftrightarrow Fe^{2+}$, $Mn^{3+} \leftrightarrow Mn^{2+}$ or $Ni^{3+} \leftrightarrow Ni^{2+}$ (see Section 2.6.2). In the ideal inverse spinel structures $NiFe_2O_4$ and $MnFe_2O_4$ there is no opportunity for hopping. However, as mentioned above, in practice the structures are not ideal and ions of the same type but differing oxidation states do occur on equivalent lattice sites. In the case of MnZn ferrite, for example, in order to maintain a major proportion of the Mn in the Mn^{2+} state so as to achieve the required magnetic properties, sintering is carried out in a slightly reducing atmosphere which carries with it the risk of converting some of the Fe^{3+} to Fe^{2+}; this must be carefully controlled for otherwise hopping conductivity could rise to an unacceptable level. In contrast, the NiZn ferrite family can be fired in air without oxidizing a significant proportion of Ni^{2+} to Ni^{3+} and ensuring that the concentration of Fe^{2+} is negligible. Therefore the opportunity for electron hopping is very much lower than in the case of the MnZn ferrites. As always, the chosen composition and fabrication steps are a compromise to achieve optimum properties for particular applications. The extent to which electrical conductivity is sensitive to composition, firing conditions and microstructure is illustrated by the following examples.

Figure 9.23 shows electrical resistivity data for $Ni_{0.3}Zn_{0.7}Fe_{2+\delta}O_{4-x}$ in which δ measures the amount of iron over the normal stoichiometric amount. For $\delta = 0$

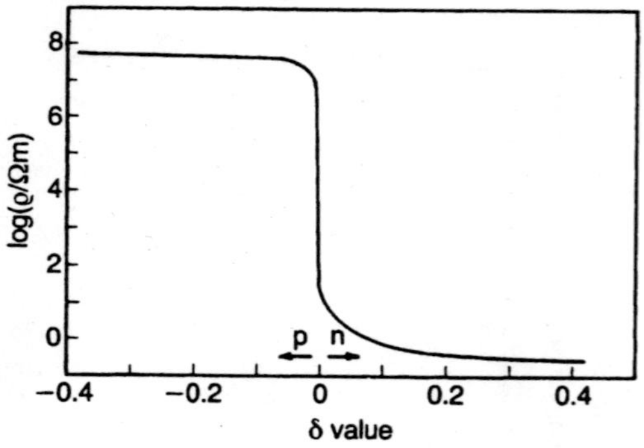

Fig. 9.23 Dependence of the resistivity of the ferrite $Ni_{0.3}Zn_{0.7}Fe_{2+\delta}O_{4-x}$ on the iron content.

all iron is assumed to be in the Fe^{3+} state and so no hopping occurs. Positive δ implies the presence of some Fe^{2+} and the conditions for n-type electron hopping ($Fe^{2+} \rightarrow Fe^{3+}$). With negative δ the charge balance can be restored by the formation of Ni^{3+}; however, the conversion of Ni^{2+} to Ni^{3+} requires greater energy than the conversion of Fe^{2+} to Fe^{3+} when δ is positive. The charge balance when δ is negative may in part be maintained by the formation of oxygen vacancies. In addition, the concentration of Ni ions is only one-seventh that of Fe ions. The net result is a very large increase in resistivity when the iron content is reduced (see Section 2.6.2).

The addition of cobalt to a nickel ferrite has been employed to increase resistivity, as illustrated in Fig. 9.24. For an explanation it is necessary to consider the third ionization potentials of chromium, iron, manganese, cobalt and nickel, which increase in this order. The addition of cobalt tends to maintain the iron in the Fe^{3+} state by virtue of the equilibrium

$$Fe^{2+} + Co^{3+} = Fe^{3+} + Co^{2+}$$

Similarly the presence of Ni^{3+} is discouraged because of the equilibrium

$$Ni^{3+} + Co^{2+} = Ni^{2+} + Co^{3+}$$

It appears that the small amounts of Co^{2+} and Co^{3+} which occur, probably on equivalent lattice sites, have little effect upon conductivity because of the large hopping distances.

The importance of the rate of cooling from the sintering temperature is apparent from the following data for a ferrite of composition $Ni_{0.4}Zn_{0.6}Fe_2O_4$. After sintering in air at 1300 °C followed by rapid cooling, the resistivity was 10 Ω m. Analysis showed the FeO content to be 0.42%. When the same sample was reheated and then cooled slowly, the FeO content was 0.07% and the resistivity was more than 1000 Ω m. The explanation is that at high temperatures Fe^{2+} is favoured and is quenched in by the rapid cooling. During slow cooling,

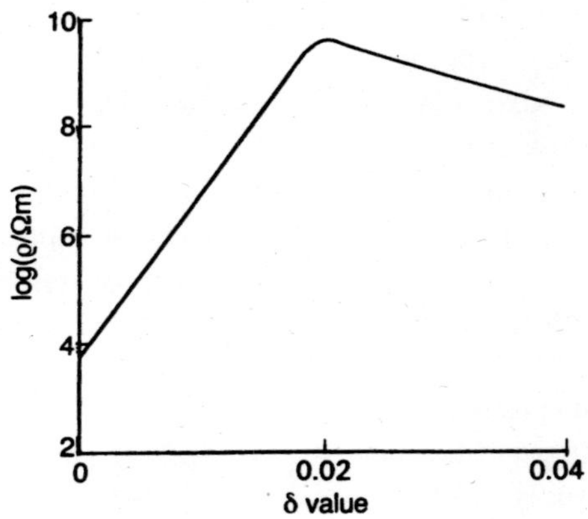

Fig. 9.24 Dependence of the resistivity of the ferrite $NiFe_{1.9}Co_{\delta}O_4$ on the cobalt content.

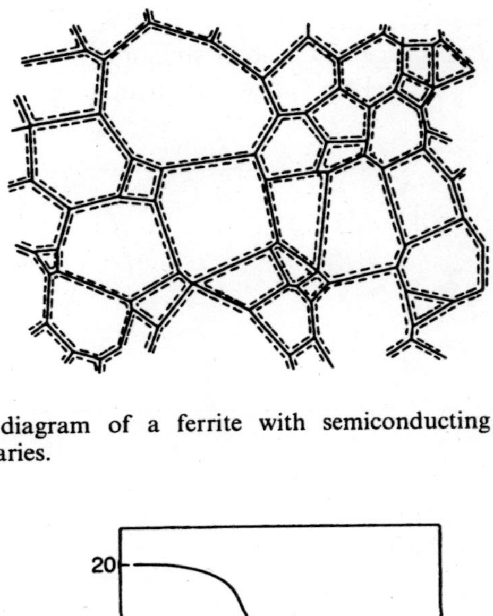

Fig. 9.25 Schematic diagram of a ferrite with semiconducting grains surrounded by insulating grain boundaries.

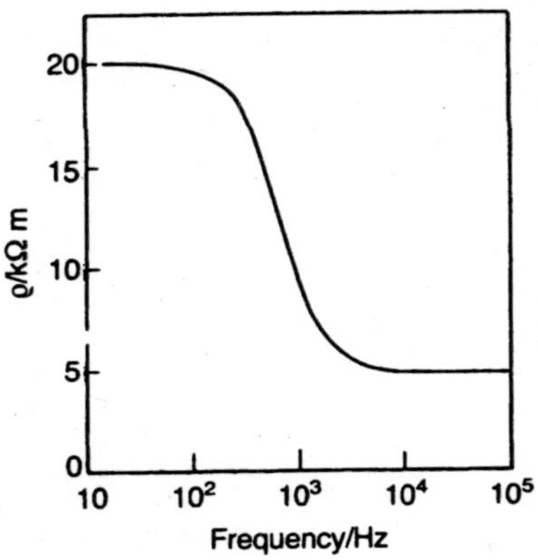

Fig. 9.26 Dependence of the resistivity of the ferrite $Ni_{0.4}Zn_{0.6}Fe_2O_4$ on frequency.

reoxidation can occur, and the resulting lower Fe^{2+} content restricts the opportunity for electron hopping.

Microstructure can also play an important role in determining electrical resistivity. If the cooling rate is arranged so that, because of the high oxygen diffusivity along grain boundaries compared with that in the single crystal, the grain boundaries oxidize preferentially, it results in each semiconducting grain being surrounded by a grain boundary region of relatively high resistance (Fig. 9.25). This configuration of zones of differing conductivities can be represented by an equivalent electrical circuit (see Section 2.7.5) comprising a series connection of shunted capacitors, where one capacitor type represents the grain and the other the grain boundary zone, and the shunts are the resistances of the two regions. An analysis of the model predicts frequency dispersion of both resistivity (Fig. 9.26) and permittivity (Fig. 9.27).

Another way of modifying the resistivity of the grain boundary region is to add small amounts of CaO and SiO_2. Figure 9.28 shows the effect of such additions on the resistivity and loss factor of a ferrite. Studies of the grain boundary show that the Si and Ca ions substitute in the ferrite crystal lattice close to the boundary, rendering the region insulating.

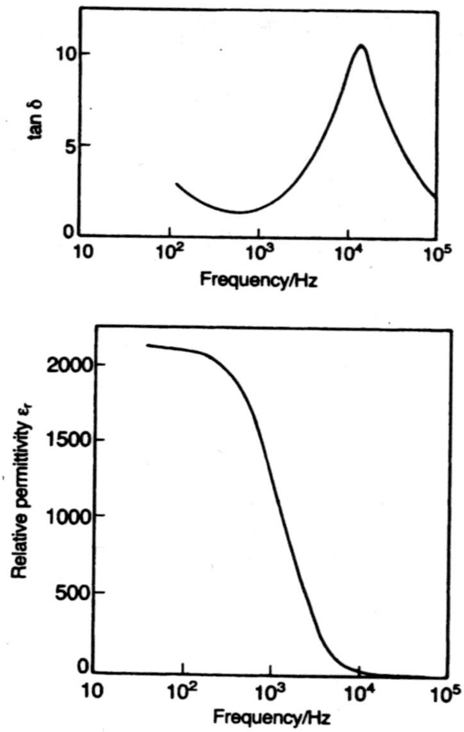

Fig. 9.27 Dependence of tan δ and ε_r on frequency for the ferrite $Ni_{0.4}Zn_{0.6}Fe_2O_4$.

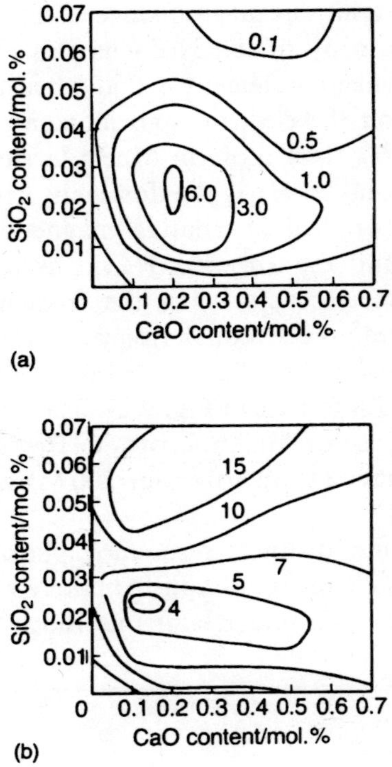

Fig. 9.28 Influence of SiO_2 and CaO additions to the ferrite $Mn_{0.68}Zn_{0.2}Fe_2O_4$ on (a) resistivity (Ω m) and (b) loss factor ($(\tan\delta)/\mu'_{ri} \times 10^{-6}$) at 100 kHz (after [9]).

Permittivity (ε_r)

Above a frequency of 1 GHz the permittivity of MnZn ferrites is around 10, but at 1 kHz it can reach values in the range 10^4–10^6. The dispersion is caused by the relatively insulating grain boundaries referred to above. The same effect occurs in the NiZn ferrites, although to a less marked extent, and Fig. 9.27 shows the frequency dispersion of ε_r and the dielectric loss tangent for the ferrite $Ni_{0.4}Zn_{0.6}Fe_2O_4$. If an equivalent circuit is used, and reasonable values for the resistivity, relative permittivity and dimensions are assigned to the grain and grain boundary phases, a relaxation frequency of approximately 20 kHz is estimated, which agrees quite well with experimental data.

Resonance effects

The replacement of the MnZn ferrite family by the NiZn ferrites as the application frequency rises is partly due to the higher resistivities of the NiZn ferrites, but there are also important resonance effects. Table 9.1 shows that the magnetocrystalline anisotropy of the NiZn ferrites is higher than that of the MnZn ferrites. An equivalent statement is that when the magnetization vectors are in the 'easy' direction the electron spin moments in NiZn ferrite are in a higher anisotropy field H_A than those in MnZn ferrite. In practice the applied fields are usually small, and so H_A is the dominant field.

When excited by an applied alternating magnetic field the magnetization vector will precess around the anisotropy field as discussed more fully later (Section 9.3.4). Resonance occurs when the frequency of the applied field coincides with the natural precessional frequency, i.e. the Larmor frequency $\omega_L = \gamma\mu_0 H_A$, with the result that the permeability falls and losses increase, as shown for a family of NiZn ferrites in Fig. 9.29. The onset of such 'ferrimagnetic resonances' restricts the use of MnZn ferrites to frequencies of less than about 2 MHz. At higher frequencies, up to about 200 MHz, compositions from the NiZn family are used.

There is another, quite distinct, resonance phenomenon concerned with domain wall movements occurring at approximately one-tenth of the ferrimagnetic resonance frequency. To understand this Bloch wall motion needs to be

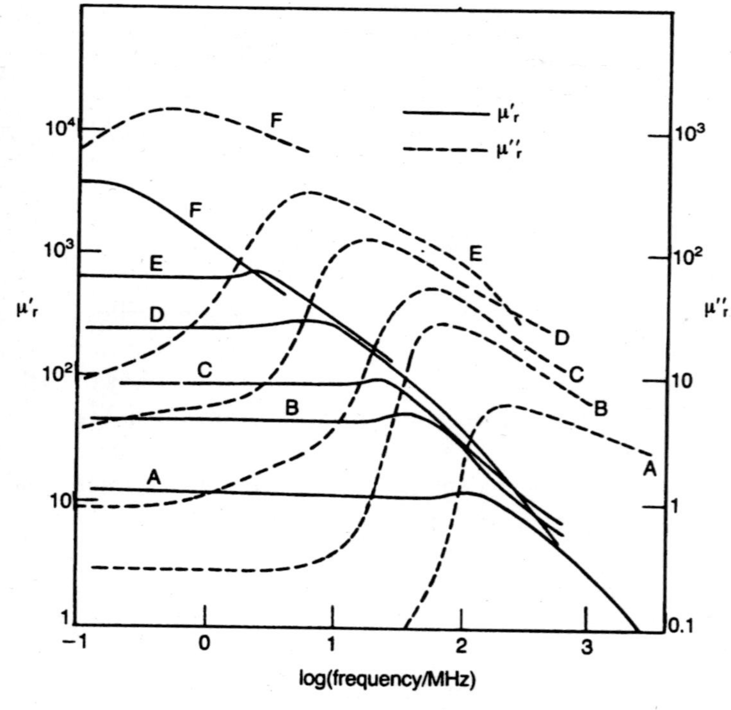

Fig. 9.29 Magnetic properties of polycrystalline $Ni_{1-\delta}Zn_\delta Fe_2O_4$ as functions of δ and frequency: curve A, $\delta = 0$: curve B, $\delta = 0.2$; curve C, $\delta = 0.36$; curve D, $\delta = 0.5$; curve E, $\delta = 0.64$; F, $\delta = 0.7$ (after [6]).

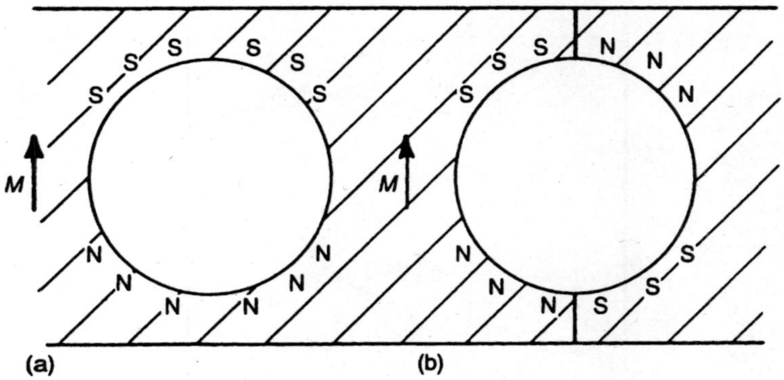

Fig. 9.30 Intersection of a Bloch wall by a pore.

considered in more detail. The initial part of the virgin magnetization curve is associated with 'reversible' Bloch wall movements; when larger fields are applied the movements are irreversible. This is best appreciated by considering the wall intersecting a second-phase inclusion, say a pore, as indicated in Fig. 9.30.

The pore in Fig. 9.30(a) possesses a high magnetostatic energy by virtue of the free poles at its surface. When a Bloch wall intersects the pore (Fig. 9.30(b)), this energy can be reduced. There is therefore a tendency for walls to intersect as many pores and other second-phase inclusions as possible. The wall is said to be 'pinned' at the inclusion, and a certain threshold field must be applied to free it. It is worth noting the close analogy that can be drawn between this situation and grain boundary movement during sintering.

When a small external field is applied the wall is displaced only slightly from the energy minimum position and, on removal of the field, the wall returns *nearly reversibly* to its original configuration. When larger fields are applied the wall may be displaced over a potential energy barrier to a new distant minimum. On removal of the field, the wall cannot return to its original position because of the intervening energy barrier. Such wall displacements are thus *irreversible*. Lattice strain, lattice defects of all types and second-phase inclusions affect the extent to which 'reversible' wall motion can occur and, in consequence, the magnetic hardness of the material.

From what has been said it follows that small displacements of a pinned domain wall introduce restoring forces and, since the wall has inertia and its movement is accompanied by energy dissipation, an equation of motion can be written for a sinusoidal applied field:

$$m\ddot{x} + \beta\dot{x} + cx = 2M_sB(t) \tag{9.32}$$

where x is the displacement normal to the wall, m represents the inertia, β is a damping coefficient and c is a stiffness coefficient. This equation describes damped forced harmonic motion so that a resonance effect will occur at a characteristic frequency $\omega = (c/m)^{1/2}$ if the damping is small.

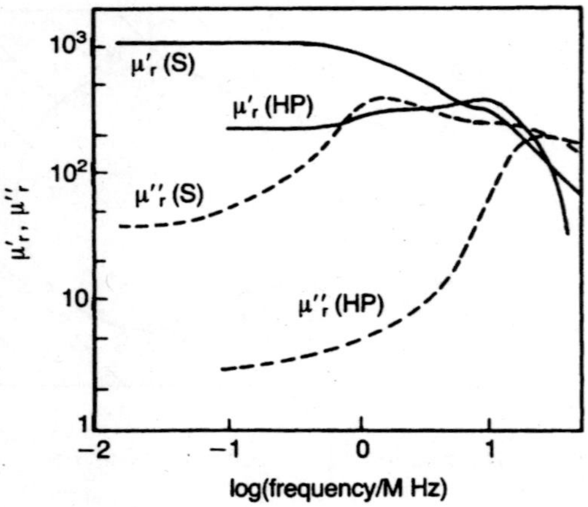

Fig. 9.31 Dependence of the permeability of the ferrite $Ni_{0.36}Zn_{0.64}Fe_2O_4$ on frequency and microstructure: S, normally sintered; HP, hot-pressed (after [9]).

Microstructure control can be used to reduce domain wall resonance effects. As the grain size is reduced the formation of domain walls becomes increasingly energetically unfavourable (see Section 9.3.2) so that their total effect at resonance is greatly diminished. Figure 9.31 shows the frequency dependence of the permeability for two NiZn ferrites with the same composition, resistivity and density but differing in grain size. The 'normally' sintered material has a grain size of approximately 35 μm and the loss component μ'' shows two maxima. The lower-frequency maximum is due to wall resonance while the higher-frequency maximum is a ferrimagnetic resonance effect. The hot-pressed material, which has a grain size of approximately 1 μm, shows no domain wall resonance, only evidence of the higher-frequency ferrimagnetic resonance. Although the fine-grained material has a lower permeability, the loss factor $(\tan \delta)/\mu_{ri}$ is improved over a wide frequency range.

A further resonance effect depends upon the dimensions of the specimen and so is referred to as 'dimensional resonance'. The explanation is as follows. In a loss-free medium the propagation velocity for electromagnetic waves is given by

$$v = (\mu\varepsilon)^{-1/2} = f\lambda \qquad (9.33)$$

where f is the frequency and λ is the wavelength. For example, a MnZn ferrite might have $\mu_r = 10^3$ and $\varepsilon_r = 5 \times 10^4$ and so, for a frequency of 1.5 MHz, $\lambda = 28$ mm. The consequence of this is that if certain dimensions of the specimen are close to $\lambda/2$ then a standing-wave system can be established (c.f. Section 5.6.5). Under this resonance condition the net flux in the specimen is zero and so the effective permeability falls to zero and, if the material is 'lossy' both μ'' and ε'' peak. The effect is shown in Fig. 9.32 for two different specimen sizes.

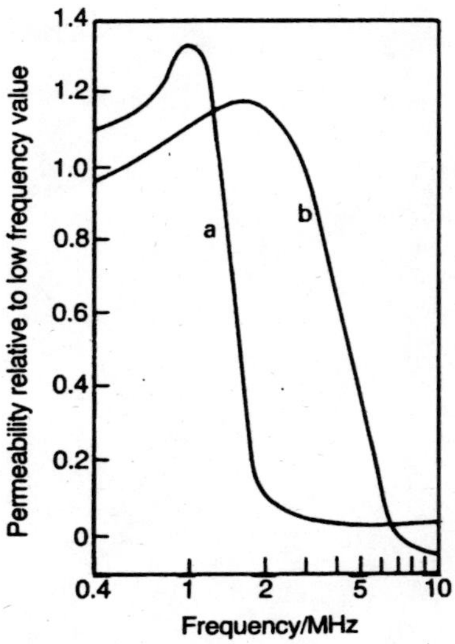

Fig. 9.32 The effect of dimensional resonance on permeability for two MnZn ferrite components of different sizes; (a) larger and (b) smaller. Permeability values are expressed relative to those measured at 1 kHz.

9.3.2 Hard ferrites

Permanent magnetic materials are distinguished from the 'soft' variety by their high coercivity H_c, typically above $150\,\mathrm{kA\,m^{-1}}$ (see Fig. 9.11). The most common engineering use to which permanent magnets are put is to establish a magnetic field in a particular region of space. The magnet must be permanent in the sense that it is capable of withstanding the demagnetizing effect both of its own internal field and from externally applied fields such as those encountered in d.c. motors.

A magnet is not a store of energy to be tapped as required until exhausted; rather, its function is to set up a magnetic potential field in much the same way as the earth sets up its gravitational field. Work is done as magnetic materials are moved around in the field, but the net energy change is zero.

The operating state of a permanent magnet lies in the second quadrant of a B–H hysteresis loop since the magnet is always subject to its own demagnetizing field. Apart from its coercivity and remanence, a permanent magnet material is rated by its 'maximum energy product' $(BH)_{max}$.

The various parameters are now considered separately by reference to the important hexaferrites.

Remanence (B_r)

Remanence is determined partly by the saturation magnetization M_s and partly by the extent to which domain development disorients the magnetization

Fig. 9.33 Microstructure of an oriented barium hexaferrite. (Courtesy of Philips Technical Review.)

vectors on removal of the saturating field. There are two general points to be made. First, because the magnetism in hexaferrites has its origin in ferrimagnetic coupling and because a large proportion of the ions in the crystal are non-magnetic (Ba, Sr, O), their saturation magnetization values (and hence, in general, their B_r values as well) are low compared with those of the metallic magnets. Second, the hexaferrites have high uniaxial anisotropy and thus B_r can be improved by tailoring the microstructure so that the crystallites are oriented to have their c axes ('easy' or 'preferred' direction) along a chosen direction; this is also the case with the well-known metallic cobalt alloys Columax, Alcomax, etc.

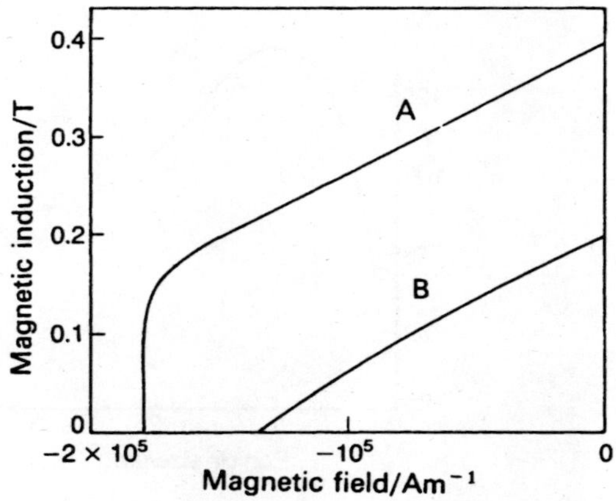

Fig. 9.34 Demagnetization curves for an oriented barium hexaferrite (curve A) and an isotropic hexaferrite (curve B).

Although a single hexaferrite crystal might be anisotropic, a ceramic comprising a very large number of randomly oriented microcrystals is normally isotropic. Thus the ceramic has a lower remanent magnetization than that offered by the intrinsic properties of the material. The situation can be significantly improved by lining up the microcrystals so that their *c* axes all point in the direction in which maximum magnetization is required. This is achieved by subjecting the ferrite particles, usually in the form of a slurry, to a strong magnetic field during the ceramic forming process. Figure 9.33 is a micrograph showing the alignment of particles and Fig. 9.34 shows the effect on remanence. As can be seen, it is necessary to distinguish between an *oriented anisotropic* and *an isotropic* hexaferrite ceramic, even though both consist of anisotropic crystallites.

Coercivity (H_c)

The high coercivity of the hexaferrites depends basically on their high magnetocrystalline anisotropy which results in anisotropy fields of approximately 1400 kA m^{-1}. However, the coercive fields achieved in practice are only a fraction of this value, because smaller fields can cause Bloch wall movements. It is found that the coercive field is a maximum when the average grain size is around 1 μm (Fig. 9.35). Reduction in particle size reduces the content of domain walls because their existence becomes energetically unfavourable when the volume they would occupy becomes an appreciable fraction of that of a particle (the width of a domain wall is about 0.1 μm). Also, the small grain size indicates an absence of grain growth during sintering, and it is during growth that the grains free themselves from defects such as lattice irregularities, second-phase

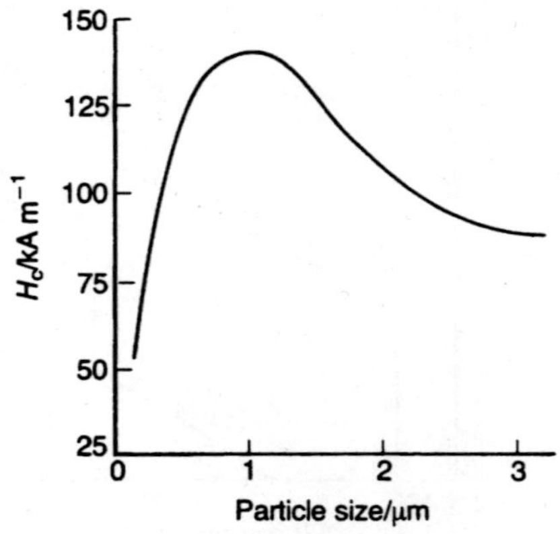

Fig. 9.35 The dependence of coercivity on the particle size of a barium hexaferrite powder.

inclusions and porosity. Small grains are therefore likely to be highly defective so that any domain walls are tightly pinned.

'Maximum energy product' $(BH)_{max}$

Consider a permanent magnet designed with the intention of setting up a magnetic field between its pole pieces (Fig. 9.36).

Applying Ampère's circuit theorem to the magnetic circuit and noting that the total flux remains constant leads to

$$(B_g H_g)V_g = -(B_m H_m)V_m \tag{9.34}$$

where B_g, B_m and H_g, H_m are respectively the fluxes and fields in the air gap and the magnetic material (H_m is the demagnetising field), and V_g and V_m are the gap and material volumes. Since the magnetic energy density is proportional to the product of B and H, equation (9.34) states that, for a magnet of given volume V_m, the magnetic energy in the gap is a maximum when the product $B_m H_m$ is a maximum. Therefore $(BH)_{max}$ is a figure of merit for a permanent magnet

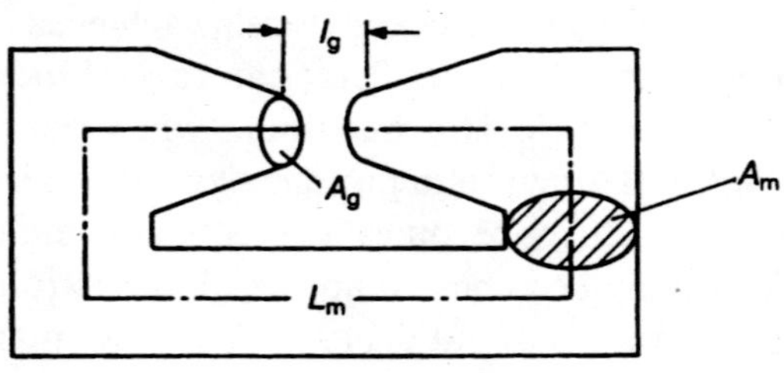

Fig. 9.36 Permanent magnet with an air gap.

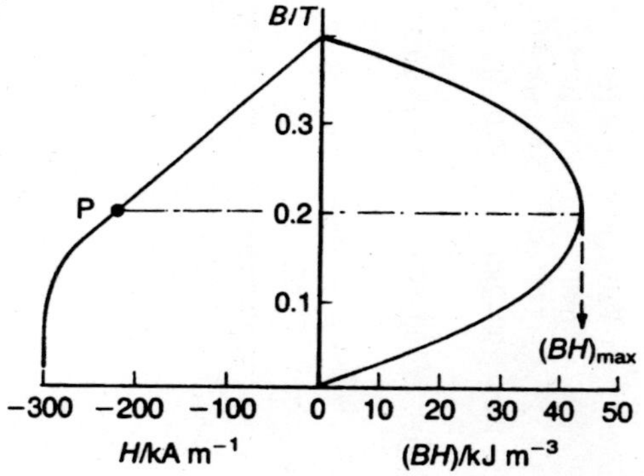

Fig. 9.37 Demagnetization and energy product curves.

material. As is evident from Fig. 9.34, the energy product is also significantly improved by the particle-orienting fabrication process.

Figure 9.37 shows a typical demagnetization curve for a hard ferrite together with the energy product curve. A horizontal line from the $(BH)_{max}$ point intersects the demagnetization curve to define the optimum working point P. The magnetic-circuit designer can arrange for the magnet to operate at P by appropriate design of magnet shape, since this determines the demagnetizing field.

Since at remanence $B_r = \mu_0 M_r$, where M_r is the remanent magnetization, the demagnetizing field is given by

$$H_D = -N_D M_r = -\frac{N_D B_r}{\mu_0} \tag{9.35}$$

If H_D exceeds a certain field strength H_i, it will reduce the remanent magnetization permanently. For the magnet to retain its full strength it is therefore necessary that $H_i > H_D$, i.e. $H_i > N_D B_r / \mu_0$ or $\mu_0 H_i / B_r > N_D$. This relation indicates that shapes with large demagnetizing factors may not retain their remanent magnetization. As a rough guide H_i can be replaced by H_c, and it is one of the advantages of ceramic permanent magnet materials that the value of $\mu_0 H_c / B_r$ is close to unity so that they can be used in shapes with high demagnetizing factors such as large thin discs. Ceramic magnets in the form of discs are most efficient when their thickness-to-diameter ratio is 1:2 since they then operate near the point P in Fig. 9.37, but they can also be used effectively with much larger aspect ratios (i.e. flatter discs). Ni–Al–Co magnets, however, have a $\mu_0 H_c / B_r$ ratio of about 0.1, so that if they are in the form of cubes ($N_D \approx 0.3$), for example, the remanent magnetization would not be retained. Such metal magnets operate most efficiently when in the form of rods with length-to-diameter ratios of about 5.

Table 9.3 Typical properties of some important magnetic materials

	μ_{ri}	$\mu_{r\,max}$	$(\tan\delta/\mu_{ri})/10^{-6}$	H_c/A m^{-1}	B_s/T	T_c/°C	ρ/Ω m
Soft materials							
Fe	150	5000	–	80	2.16	770	10×10^{-8}
Fe + 4%Si	2000	35 000	–	30	1.93	690	60×10^{-8}
Mumetal[a]	50 000	200 000	–	3.0	0.75	350	55×10^{-8}
MnZn ferrites	500–10 000	–	1 at 10 kHz to 50 at 1000 kHz	5–100	0.35–0.50	90–280	0.01–1
NiZn ferrites	10–2000	–	20 at 100 kHz to 200 at 100 MHz	15–1600	0.10–0.40	90–500	10^3–10^7
			$(BH)_{max}$/kJ m^{-3}	H_c/kA m^{-1}	B_r/T	T_c/°C	
Hard materials							
Carbon steel[b]			1.6	4.4	0.9	770	
Ticonal GX[c]			60	58	1.35	860	
$Nd_{15}Fe_{76}B_8$			360	800	1.4	310	
Isotropic barium ferrite			6.0	200	0.3	450	
Anisotropic (oriented) barium ferrite			35	320	0.4	450	

[a]77Ni–16Fe–5Cu–2Cr.
[b]Approximately 1% C.
[c]14Ni, 8Al, 24Co and 3Cu.

9.3.3 Summary of properties

Table 9.3 summarizes typical values of the relevant properties of some of the important magnetic material types.

9.3.4 Microwave ferrites

Conventional electronic circuits handling frequencies of up to, say, 300 MHz ($\lambda \approx 1$ m) usually comprise resistors, capacitors, inductors, diodes and transistors to control energy flow. As the frequency increases so the wavelengths approach the dimensions of the circuit and microwave techniques dominate. Although there is no clear defining lower-frequency limit for microwaves, typically they lie in the range 1–300 GHz (wavelengths in the range 30 cm to 1 mm). In the context of waveguide technology microwave frequency bands are designated code letters as listed in Table 9.4.

Microwaves can be propagated down a waveguide, which is simply a metal 'pipe', and since the advent of microwave technology materials have been introduced into waveguides to change their propagation characteristics; devices such as isolators, gyrators, phase shifters and circulators are based on this approach. Certain ferrites play an important role in these devices, principally because their high electrical resistivity coupled with low magnetic losses lead to what in this context is called 'low insertion loss'. Therefore microwaves can pass some considerable distance through the ferrites and, while doing so, are modified in a predetermined way by interaction between the magnetic and electric field components of the wave and the magnetic and dielectric properties of the material; when electromagnetic waves pass through non-magnetic dielectrics (cf. Chapter 8) it is only the interaction between the electric field component and the material which is significant.

When a magnetic field H is applied to a spinning electron the angular momentum vector is inclined at an angle to the field direction. Because a magnetic moment is associated with the angular momentum, the electron experiences a torque and precesses around the field direction with the Larmor angular frequency $\omega_L = 2\pi f_L = \gamma\mu_0 H$. This is analogous to the precessional motion of a spinning top with its axis of rotation inclined to the gravitational field.

Table 9.4 Microwave band designations

Frequency range/GHz	1.12–1.70	1.70–2.60	2.60–3.95	3.95–5.85	5.85–8.20	8.20–12.4	12.4–18.0	18.0–26.5	26.5–40.0	40.0–60.0	60.0–90.0	90.0–140.0	140.0–220.0
Designation	L	R	S	H	C	X	Ku	K	Ka	U	E	F	G

In the case of ferromagnetic and ferrimagnetic materials the spins are strongly coupled over the region of a domain, but when a sufficiently strong magnetic field is applied the domain structure disappears and the material possesses a magnetization vector M_s which precesses around H with the Larmor frequency. Because of the spin–orbit lattice coupling the precessional energy is steadily dissipated, and in consequence M_s gradually spirals in towards the H direction as indicated in Fig. 9.38(a). A measure of the time taken for the energy transfer to occur is the relaxation time τ.

The precessional motion can be maintained by a suitable radio frequency field superimposed on the steady field. For example, in Fig. 9.38(b), when a steady field H_z is applied along the z axis and a radiofrequency field H_{rf} is applied in the x–y plane and rotates in the same sense and at the same frequency as the precession, resonance occurs. 'Gyromagnetic resonance' as outlined above is in principle the same as 'ferrimagnetic resonance' referred to earlier (Section 9.3.1), except that in the former case the material is magnetically saturated by a strong applied field. In practice the steady field, which determines the Larmor frequency, is made up of the externally applied field, the demagnetizing field and the anisotropy field, and is termed the effective field H_e. Figure 9.39 shows the H_e values at which resonance occurs in some of the important communications and radar frequency bands.

At low applied fields, when the ferrite is magnetically unsaturated, broad-band losses occur, the so-called 'low-field losses', and the increasing loss as H_e decreases can be identified with that occurring on the high-frequency side of the peaks in Figs 9.29 and 9.31. Low-field losses arise because the radio-frequency field resonates with the magnetization in individual domains which precesses around the anisotropy field H_A. It can be shown that this resonance occurs over a range of frequencies between $\mu_0\gamma H_A$ and $\mu_0\gamma(H_A + M_s)$; thus, to

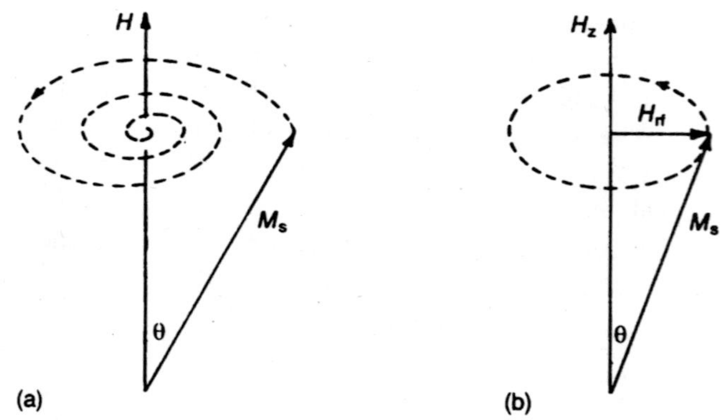

Fig. 9.38 Precessional motion of magnetization: (a) M_s spiralling into line with H as the precessional energy is dissipated; (b) precession maintained by an applied radio frequency field.

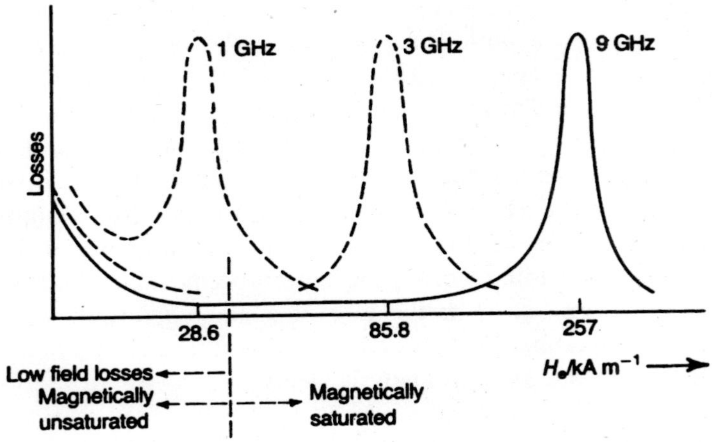

Fig. 9.39 Effective field values H_e at which resonance occurs in some important frequency bands.

avoid the onset of low-field losses in low-frequency applications, it is necessary for H_A and M_s to be as small as possible.

To understand the principle of operation of important 'non-reciprocal' (see below) microwave devices, consider what occurs when a plane-polarized microwave is propagated through a ferrite in the direction of a saturating field H_e. The wave can be resolved into two components of equal amplitude but circularly polarized in opposite senses, i.e. into a 'right-polarized' and a 'left-polarized' component. These two components interact very differently with the material, leading to different complex relative permeabilities $\mu^*_{r+} = \mu'_{r+} - j\mu''_{r+}$ and $\mu^*_{r-} = \mu'_{r-} - j\mu''_{r-}$, as shown in Fig. 9.40. Because of the

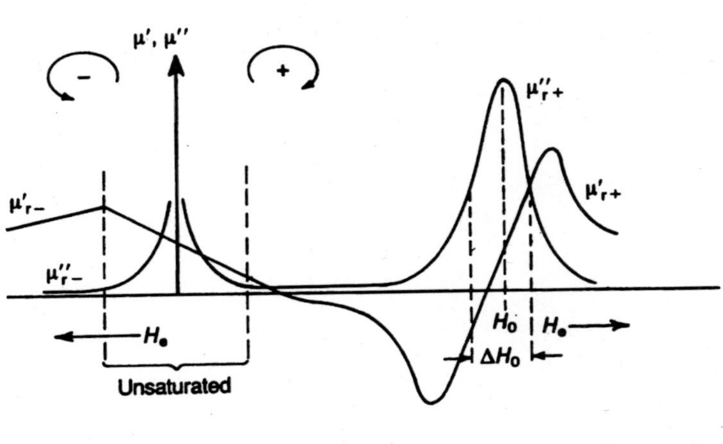

Fig. 9.40 Dependence of the permeability on H_e for 'right' and 'left' circularly polarized components of a plane-polarized microwave of frequency $\omega = \gamma\mu_0 H_0$.

different permeabilities, the phase velocities of the two components differ, so that the negative, or left, component is retarded relative to the positive, or right, component, causing a clockwise rotation of the plane of polarization. The sense of the rotation relative to H_e does not depend on the direction of propagation. Thus, if a plane wave is rotated 45° in the clockwise sense during its passage through a saturated ferrite, then if it were to be reflected back, it would suffer a further 45° rotation. Because of this the process is said to be 'non-reciprocal'.

A rigorous analysis of the interaction of the microwave field with the precessing magnetization involves the tensor permeability and is beyond the scope of the present text. For the case of a plane-polarized wave of angular frequency ω propagating through a magnetically saturated ferrite in the direction of the saturating field, such an analysis yields

$$\mu_{r+} = 1 + \frac{\omega_M}{\omega_0 - \omega} \tag{9.36}$$

and

$$\mu_{r-} = 1 + \frac{\omega_M}{\omega_0 + \omega} \tag{9.37}$$

in which μ_{r+} and μ_{r-} are the relative permeability values for the right and left circularly polarized components respectively, $\omega_M = \mu_0 \gamma M_s$ and $\omega_0 = \mu_0 \gamma H_0$, H_0 is the resonance field and M_s is the saturation magnetization.

Because the wave velocities v_+ and v_- of the two components differ, a phase difference ϕ develops between them as they propagate. The plane of polarization is thus rotated by an angle $\theta = \phi/2$. The angular rotation θ/L of the plane of polarization per unit path length can be estimated as follows. Suppose that the plane-polarized wave travels a distance L through a medium of relative permittivity ε_r and relative permeabilities μ_{r+} and μ_{r-}. If the average velocity is v, then the fast and slow components travel distances of $(L/v)v_+$ and $(L/v)v_-$ respectively, where v_+ and v_- are the fast and slow speeds. The path difference is

$$\Delta = \frac{L}{v}(v_+ - v_-) \tag{9.38}$$

and the phase difference is

$$\phi = \frac{L}{v}(v_+ - v_-)\frac{2\pi}{\lambda} \tag{9.39}$$

where λ is the average wavelength. The plane of polarization is therefore rotated through θ where $\theta = \phi/2$

$$\theta = \frac{L\pi}{v\lambda}(v_+ - v_-)$$

and

$$\frac{\theta}{L} = \frac{\pi}{\upsilon\lambda}(\upsilon_+ - \upsilon_-) \tag{9.40}$$

Since $\upsilon^2 \approx \upsilon_+\upsilon_-$

$$\frac{\theta}{L} = \frac{\pi\upsilon}{\lambda}\left(\frac{1}{\upsilon_-} - \frac{1}{\upsilon_+}\right)$$
$$= \pi f\left(\frac{1}{\upsilon_-} - \frac{1}{\upsilon_+}\right) \tag{9.41}$$

where f is the frequency of the wave. Because

$$\upsilon_- = \frac{c}{(\mu_{r-}\varepsilon_r)^{1/2}} \quad \text{and} \quad \upsilon_+ = \frac{c}{(\mu_{r+}\varepsilon_r)^{1/2}} \tag{9.42}$$

where c is the velocity of light in a vacuum,

$$\frac{\theta}{L} = \frac{\omega\varepsilon_r^{1/2}}{2c}(\mu_{r-}^{1/2} - \mu_{r+}^{1/2}) \tag{9.43}$$

For a device operating below resonance the applied field is small compared with H_0, although sufficiently strong for the ferrite to be magnetically saturated. Under these conditions $\omega_0 \ll \omega$ and $\omega_M < \omega$, and therefore it follows from Eqs (9.36) and (9.37) that

$$\mu_{r+}^{1/2} \approx 1 - \frac{1}{2}\frac{\omega_M}{\omega} \tag{9.44}$$

and

$$\mu_{r-}^{1/2} \approx 1 + \frac{1}{2}\frac{\omega_M}{\omega} \tag{9.45}$$

Therefore Eq. (9.43) becomes

$$\frac{\theta}{L} = \frac{\varepsilon_r^{1/2}}{2c}\omega_M = \frac{\varepsilon_r^{1/2}}{2c}\mu_0\gamma M_s \tag{9.46}$$

In a typical case where $\varepsilon_r = 10$ and $M_s = 170\,\text{kA}\,\text{m}^{-1}$, $\theta/L = 10°\,\text{mm}^{-1}$.

The width ΔH_0 of the resonance absorption curve measured at half peak power – the '3 dB resonance line width' – should in general be as small as possible since this implies a narrow range of frequencies over which strong interaction with the ferrite can occur; however, there are certain broad-band applications where this would not be the requirement. There are two main contributions to the linewidth:

1. the intrinsic linewidth characteristic of the single crystal and
2. additional effects arising in the case of polycrystalline materials and influenced by magnetic anisotropy and microstructural features.

The intrinsic single-crystal linewidth can be shown to be inversely proportional to the relaxation time.

In a polycrystalline material each crystallite experiences its own effective field, which is determined in part by magnetic anisotropy. Therefore, because of the different orientations of the crystallites, there will be a distribution of resonant frequencies; in consequence, narrow linewidths are associated with low anisotropies. Magnetic inhomogeneities – inclusions and pores – also disturb the local effective fields and so lead to line broadening. The effect of porosity on resonance linewidth can be expressed by the relationship

$$\Delta H_p = 1.5 M_s p(1 - p) \tag{9.47}$$

where p is the fractional volume porosity. Thus it is evident that a pore-free single-phase microstructure is necessary for minimum linewidth.

There is also a strong dependence of resonance linewidth on surface finish, which is exemplified by the following data for small (0.35 mm diameter) YIG spheres in a resonant cavity. The linewidths for surfaces finished with 15 μm grit, 5 μm grit and 'highly polished' were 560 A m^{-1}, 240 A m^{-1} and 40 A m^{-1} respectively.

At high microwave powers 'spin waves' may be excited rather than the uniform precession discussed so far. Power from the radiofrequency field is then coupled into the spin-wave system and the energy dissipation shows a higher than proportional increase with increasing power. The power threshold P_c at which spin-wave resonance sets in is measured in terms of a material parameter ΔH_k, the spin-wave linewidth, which is a measure of the damping of the spin waves. There are two approaches to increasing ΔH_k and thus the microwave power-handling capability of the ferrite – tailoring the composition or tailoring the microstructure. For example, the substitution of holmium and dysprosium for gadolinium in $Y_{2.9}Gd_{0.1}Fe_5O_{12}$ leads to an approximately tenfold increase in

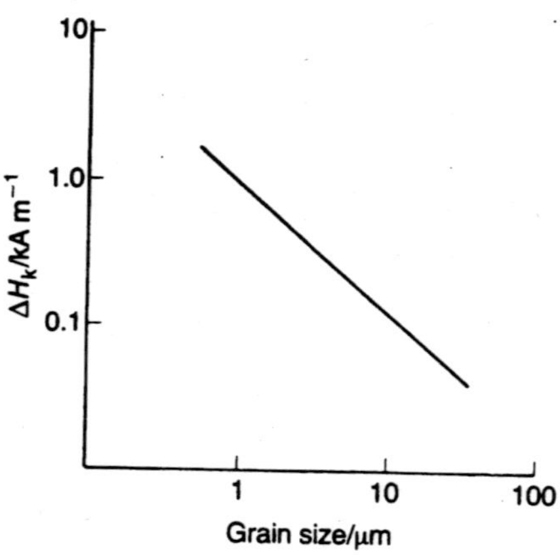

Fig. 9.41 Dependence of the spin-wave linewidth ΔH_k on grain size.

ΔH_k. In addition, the spin waves can be strongly damped if the grain size of a polycrystalline ceramic is kept small. For example, Fig. 9.41 shows that a reduction in grain size from 30 μm to 1 μm increases ΔH_k by a factor of approximately 30 and thus, because $P_c \propto \Delta H_k^2$, increases the power-handling capability by a few orders of magnitude. Therefore, although single crystals offer the optimum microwave properties at low fields, a fine-grained material performs more satisfactorily at high power levels.

9.4 Preparation of Ferrites

The choice of composition of a soft ferrite is made to achieve one or more of the following properties:

1. maximum permeability;
2. minimum losses;
3. temperature variation of properties between prescribed limits;
4. adequate saturation magnetization;
5. satisfactory behaviour over the required frequency band;
6. finally but by no means the least important, minimum cost.

The basic choice of composition for the great majority of applications lies between MnZn and NiZn ferrites. The former have a lower material cost and are superior in all the above properties except 5 and the latter are more suitable for operation at higher frequencies. $(MnZn)Fe_2O_4$ has the disadvantage of requiring a low oxygen pressure atmosphere during sintering so that its manufacture necessitates a special kiln with a supply of suitable gas for the atmosphere. $(NiZn)Fe_2O_4$ can be fired in air. Fabrication generally follows the lines described in Chapter 3. Features unique to ferrites are outlined below.

9.4.1 Raw materials

The raw materials can be minerals that have only been purified by mechanical methods or the purer products of chemical processes. Both haematite (Fe_2O_3) and magnetite (Fe_3O_4) occur in large deposits of better than 90% purity that, for some ferrites, require little more than grinding before use. A purer grade of Fe_2O_3 can be obtained from the pickle liquor that results from the removal of oxide crusts during steel fabrication. The liquor contains iron chlorides and can be converted into oxide by spraying into heated air in an apparatus similar to that used in spray drying but operated at a higher temperature.

High-purity oxides can be obtained by calcining oxalates in an atmosphere with a controlled oxygen content:

$$Fe(CO_2)_2 \rightarrow Fe + 2CO_2$$
$$2Fe + 3O_2 \rightarrow 2Fe_2O_3$$

The product consists of very small crystals that need little grinding. Co-precipitation of mixed oxalates can also give intimate mixing.

Most transition elements are available in a pure state as metals which can be dissolved in acids. A mixture of nitrates can be evaporated to dryness and calcined to form precursor oxide mixtures for the preparation of spinel and garnet ferrites. Alternatively, mixed oxides, carbonates or oxalates can be precipitated. Microwave ferrites that are required to be of high purity can be prepared by one of these chemical routes.

Barium ferrite adequate for certain permanent magnet applications can be prepared from crude mineral iron oxide. A past interest in the material for perpendicular magnetic recording stimulated studies of a variety of chemical routes, including hydrothermal synthesis, to produce powders [12].

9.4.2 Mixing, calcining and milling

Mixing is usually carried out in a ceramic ball mill with steel balls. The inevitable attrition of the mill balls leads to an addition of 0.5–1 wt% to the iron content for which allowance must be made. Care must be taken to maintain the quantity and size distribution of the mill balls and to remove those that have become so small that they cannot be separated from the slip or powder on sieving.

Calcination (typically at temperatures in the range 1000 °–1100 °C, depending on composition) is carried out in saggars or continuously in a rotating tube kiln. An air atmosphere is used even though this results in a high Mn^{3+} content in MnZn ferrites, but the state of oxidation is rapidly rectified during sintering. A low Mn^{3+} content can be a disadvantage during the early stages of sintering since, at about 430 °C, MnZn ferrites take up oxygen, converting Mn^{2+} to Mn^{3+} with a resulting lattice shrinkage. The temperature gradients within a furnace are often considerable at this stage in firing and the pressings are weak. One part of a large piece may be shrinking while another part is expanding, resulting in the formation of cracks. At temperatures above 700 °C the thermal gradients are greatly reduced because of the rapid interchange of radiant energy so that effects of this type are less likely to cause problems.

9.4.3 Sintering

If a low oxygen pressure is required during sintering nitrogen is injected at a point in the tunnel kiln where the temperature is approximately 1000 °C, and the air is effectively displaced at a position where the temperature is 1300 °C. The composition of the atmosphere is monitored by passing samples from the hot zone through a paramagnetic oxygen meter or by using a zirconia solid state electrolytic device (see Section 4.6.1) within the furnace. The low oxygen pressure (around 1 kPa (7.5 Torr)) is maintained until the pieces have cooled to around 500 °C. A controlled atmosphere can also be obtained by burning methane or town gas in a limited supply of air. The large-scale process in a continuous kiln does not allow full atmosphere control because the extent to which oxygen gains access to the hot zone is governed by so many variables, one of which is the packing of the work on the trolleys which inevitably varies with the sizes and shapes of the pieces concerned. Where close control is essential, as with pot cores, the pieces are fired in well sealed batch kilns. In this case the oxygen pressure can be programmed to correspond to the equilibrium pressure of the desired balance of oxidation states at both the maximum temperature and as it falls to around 900 °C. Below 900 °C the rate at which oxygen is exchanged between the ferrite and the atmosphere is sufficiently slow that, provided that the cooling is rapid, there is no need to maintain tight control. Means are often provided for moving the sintered material into a cooled compartment when the temperature has fallen sufficiently.

In the cases of $NiFe_2O_4$ and the rare earth garnets for microwave use, in which the losses due to conductivity must be minimized, an oxygen atmosphere may be required for sintering so that the concentration of Fe^{2+} ions is reduced to a very low level.

9.4.4 Single-crystal ferrites

The growth of single crystals is discussed in Section 3.10. Because of a combination of high permeability and abrasion-resistance single crystal Mn–Zn ferrites are used for the read-out heads in VCRs. The crystals can be grown by the Bridgman–Stockbarger method. Garnet ferrite crystals are required for many microwave applications and as thin layers for magneto-optical devices. Since these compounds melt incongruently they are grown from solutions based on lead oxide. The gadolinium gallium garnet crystals melt congruently and can be grown directly from the melt.

9.4.5 Magnets with oriented microstructures

For hard ferrites for use as permanent magnets relative permeability is unimportant and is usually near to unity in the magnetized state. The important parameters are a high coercive field, a high remanent magnetization and the maintenance of a suitable combination of these properties over the operational temperature range. Only ferrites with the hexagonal magneto-plumbite structure with compositions close to $MO{\cdot}6Fe_2O_3$, where M is Ba or Sr, are used. Somewhat better properties can be obtained with strontium, but the cost of the raw material usually determines which is used at a particular time. In what follows, barium ferrite only is referred to on the understanding that either barium or strontium may be involved.

Production processes are broadly similar to those used for soft ferrites, but various techniques can be introduced to orient the easy axes of magnetization of the crystals in the sintered ceramics. Since all the iron is in the Fe^{3+} state, sintering can be carried out in air. Simple shapes can be pressed from a mixture of Fe_2O_3 and $BaCO_3$ and sintered at 1300 °C to form permanent magnets of sufficient quality for some applications. More commonly a mixture of $BaCO_3$ and Fe_2O_3 is calcined and treated in the way described for soft ferrites. The milling of the calcine may be more prolonged since the coercive field is increased by having a smaller crystal size. The calcination is often carried out in rotary calciners (Fig. 3.1) on a continuous basis yielding the fully formed ferrite. When milled this calcine forms tabular particles with the easy direction of magnetization normal to their larger surfaces. These features permit some degree of particle orientation during pressing.

Orientation of the single-crystal powder particles by an applied field requires that they should be able to rotate relative to one another so that, on average, their easy directions of magnetization are aligned. Only a limited extent of alignment can be obtained using dry powders, and it is more effective to make the ferrite powder into a slurry with water which is removed during pressing through a specially designed die. Demagnetizing and static fields are applied in rapid succession from coils round the die, the water is sucked out of the slurry while the static field is still applied and pressure is exerted by a top punch (Fig. 9.42(a)). If suitable deflocculants are used, a high degree of dispersion can be obtained in the slurry so that the particles can be oriented effectively. The filter cake is subjected to moderate pressure (15 MPa, 1 tonf in^{-2}) to compact it and remove excess water. It is usually necessary to apply a demagnetizing field before the compact can be removed from the die without the risk of its breaking. The direction of magnetization required is usually normal to the major surfaces of the pressed piece so that the magnetizing coils can be placed round the die to give a field parallel to the motion of the punches. The process is slow compared with normal die pressing, and the consequent higher cost can only be justified when the improved energy product is needed in the intended application. A complex die for producing small motor magnets is shown in Fig. 9.42(b) and a production press in Fig. 9.43.

Some degree of anisotropy can also be achieved by applying a field during the extrusion of a paste formed from barium ferrite powder and a viscous fluid medium. In this case the plate-like shape of the particles also assists orientation

(a)

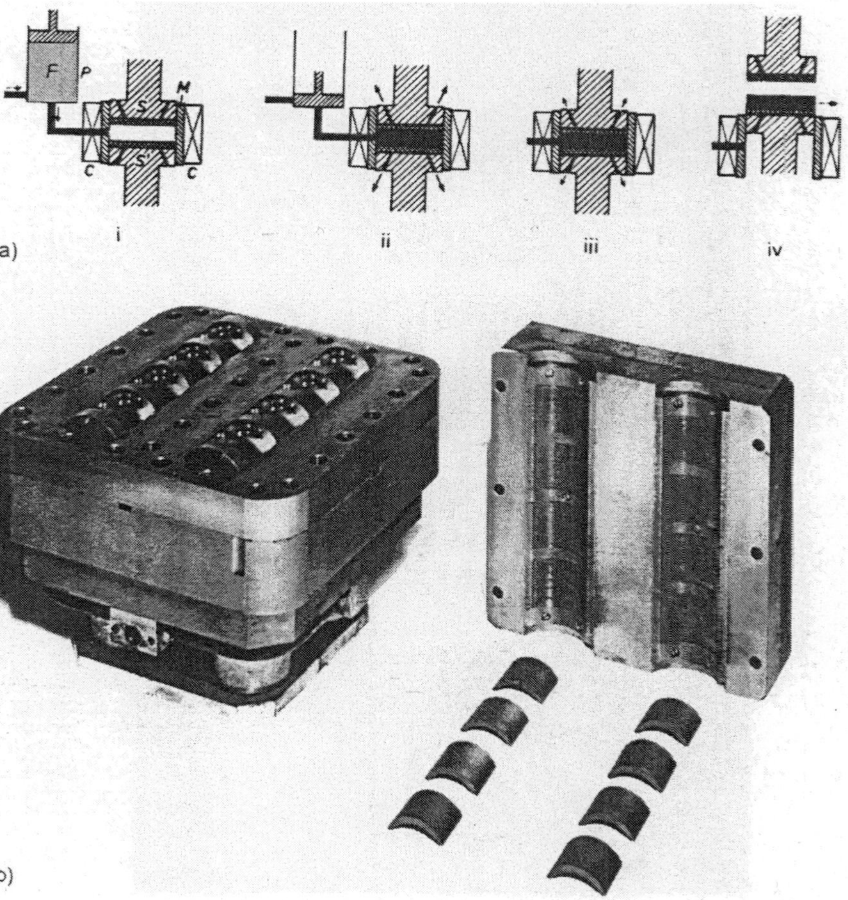

(b)

Fig. 9.42 (a) Diagram of the pressing cycle: M, Die; S,S′, punches; P, injection pump; F, suspension; C, magnet coil. The punches contain filters and drainage channels for the water. (i) Before and (ii) after the material is injected into the compression space; the slurry is densified from 40% to 13% water content, and the particles are oriented by the magnetic field; (iii) after densification by slight upward movement of the lower punch, the amount of the movement depending on the thickness of the product; demagnetization; (iv) removal of the compacted product. (b) Die for simultaneous pressing of eight oriented magnet segments. (Courtesy of Philips Technical Review.)

since the plane of the plates tends to align parallel to the direction of extrusion (Fig. 9.44).

Barium ferrite powder is also incorporated in rubber or polymer matrices to make, for instance, the familiar magnetic gaskets used to latch refrigerator doors. In this case the particle size must be adjusted to an optimum of around 1 μm since this gives maximum coercive field. The powder can be incorporated by

Fig. 9.43 Production press for oriented ferrite components. (Courtesy of Philips Technical Review.)

Table 9.5 Properties of barium ferrites made using different processes

Property	*Unit*	*Calendered with polymer*	*Sintered*		
				Anisotropic	
			Isotropic	*Extruded*	*Wet pressed*
Remanence	T	0.16	0.20	0.30	0.40
Coercivity	$kA\,m^{-1}$	112	140	150	180
$(BH)_{max}$	$kJ\,m^{-3}$	5	7	15	30
Density	$kg\,m^{-3}$	3400	4800	4800	4800

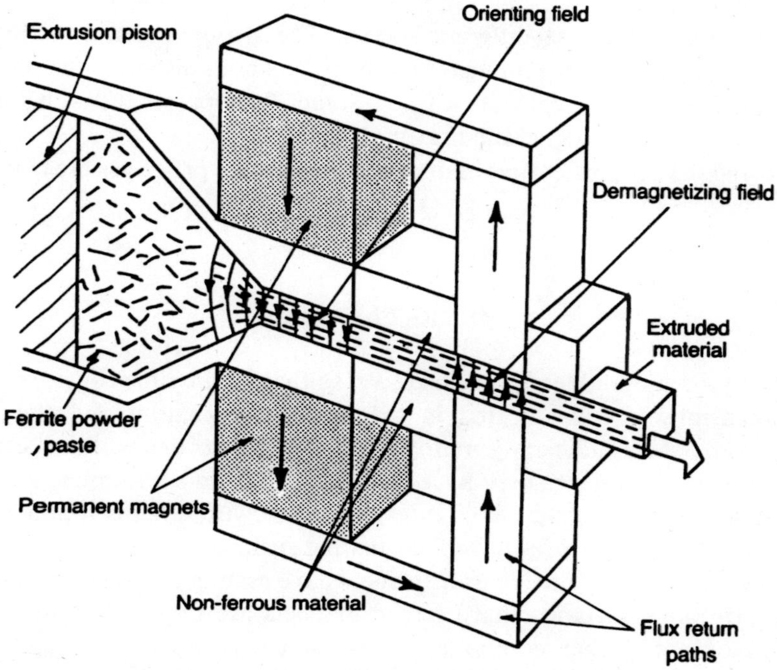

Fig. 9.44 Orientation of particles during extrusion.

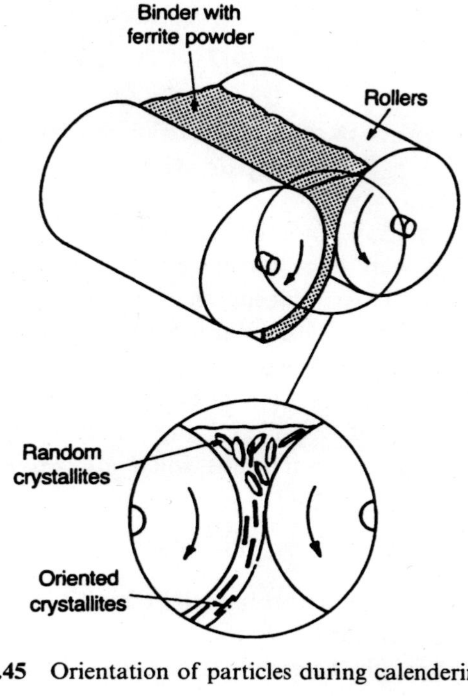

Fig. 9.45 Orientation of particles during calendering.

calendering, that is by passing a mixture of polymer and ferrite between closely spaced rollers rotating at different speeds. The strong shearing action as the mixture passes through the nip between the rollers both disperses the powder and orients the plate-like particles (Fig. 9.45). Some 20–30 passes through a nip 1 mm wide may be needed to give the maximum effect.

The principal properties obtainable with the various processes are compared in Table 9.5.

9.4.6 Finishing

Machining the sintered ferrite is expensive, so that all possible control over shape and dimensions is concentrated in the pressing operation and the sintering conditions must be sufficiently uniform to keep dimensions within specification. However, certain surfaces may have to be improved in smoothness and flatness so that they can form magnetic circuits by contact with other components with a minimum air gap. Pot-cores have their mating surfaces lapped, and the E, C and I cores used in high-frequency transformers have mating surfaces ground flat and parallel. In many instances hard ferrite magnets are required to fit into metal housings or onto other components, and some diamond grinding is usually required to achieve the required dimensional tolerances.

9.5 Applications

9.5.1 Inductors and transformers for small-signal applications

A major use of high-quality pot-core inductors is in combination with capacitors in filter circuits. They were once extensively used in telephone systems but solid state switched systems have replaced them. The soft MnZn and NiZn ferrite systems are dominant for pot-core manufacture, although metal 'dust cores' are used for certain applications. The important design principles can be understood by reference to a simple series LCR circuit to which a sinusoidal voltage U of angular frequency ω is applied (Fig. 9.46). The circuit impedance $Z = R + j(\omega L - 1/\omega C)$ is a minimum when $\omega L = 1/\omega C$. Thus the resonant frequency ω_0 is given by

$$\omega_0 = (LC)^{-1/2} \tag{9.48}$$

At this frequency the current through the circuit is a maximum and the total applied voltage U appears across R; the voltages across L and C are out of phase by π. The current through the circuit is U/R and the voltage across C is

$$U_c = \frac{U}{R}\frac{1}{j\omega_0 C}$$

It therefore follows from this and Eq. (9.48) that

$$\left|\frac{U_c}{U}\right| = \frac{1}{\omega_0 RC} = \frac{\omega_0 L}{R} = Q \tag{9.49}$$

where Q is the 'circuit magnification factor' or 'quality factor'. It can also be shown that

$$Q = \frac{\omega_0}{2\delta\omega} \quad \left(\text{or } \frac{f_0}{2\delta f}\right) \tag{9.50}$$

where $\delta\omega$ (or δf) is the change in ω (or f) from ω_0 (or f_0) required to reduce the power dissipated in the circuit to half the resonance value. Alternatively, in terms of voltages, $\delta\omega$ is the change from ω_0 required to reduce the voltage across C (or L) to $1/\sqrt{2}$ of its peak resonance value. The circuit is therefore selective in its response to signals of different frequency, and Q measures selectivity. The practical aim is usually to maximize the circuit Q, and from Eq. (9.49) this requires L and R to be maximized and minimized respectively. In order to achieve a high circuit Q the materials which are used in the circuit components must be chosen with care.

As discussed in Section 9.3.1, the quality of the ferrite is measured by its loss factor $(\tan\delta)/\mu_{ri}$ which should be as small as possible, and in the following discussion it is assumed that this is the case.

The inductance L of the core is given by

$$L = \mu_{re}\mu_0 n^2 D \tag{9.51}$$

where μ_{re} is the effective relative permeability, n is the number of turns and D is a constant depending upon core geometry. It follows that, for a given L, a higher μ_{re} will allow n to be reduced proportionately to $\mu_{re}^{1/2}$. A decrease in the number

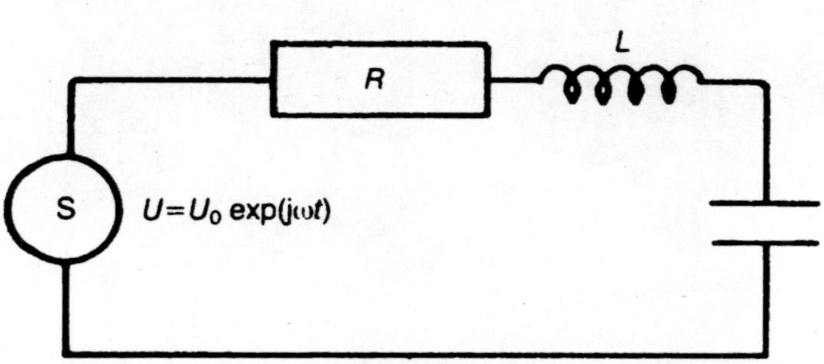

Fig. 9.46 Series *LCR* circuit.

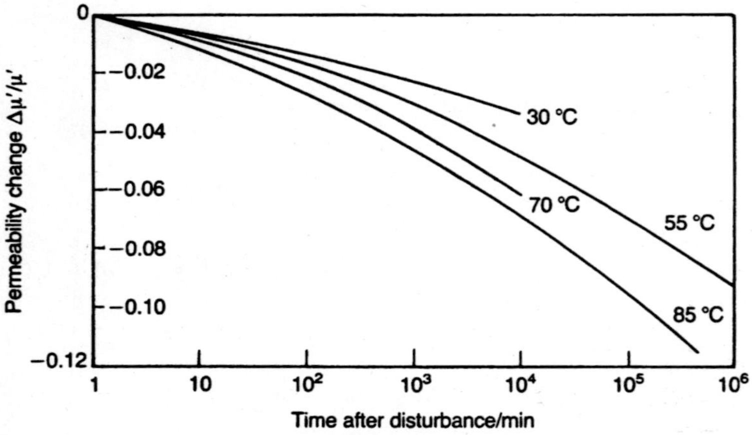

Fig. 9.47 Change in permeability with time or 'disaccommodation'.

of turns results in lower winding-resistance losses (the so-called 'copper losses') and hence higher Q; a high μ_{re} also enables the core size to be kept small.

There are two ferrite material properties which were not discussed in Section 9.3.1 but which are important in the inductor context: they are the temperature and time stabilities of the permeability which, of course, determine the stability of the inductance. The temperature coefficient of permeability must be low, and this has been achieved for certain MnZn ferrite formulations as indicated in Fig. 9.18. A small residual temperature coefficient of inductance can be compensated by a suitable coefficient of opposite sign in the capacitance of the resonant combination.

Fig. 9.48 Pot core: (a) complete unit; (b) schematic diagram of structure. (Courtesy of Philips Components Ltd.)

A time variation in inductance arises because the permeability diminishes after the establishment of a fresh magnetic state; this effect is known as 'disaccommodation' and is illustrated in Fig. 9.47.

The fall in permeability is roughly proportional to the logarithm of time. Therefore, if there is a loss of 1.5% between 1 min and 1 week (approximately 10^4 min), it will take 10^4 weeks (about 200 years) for the loss of a further 1.5% to occur! Disaccommodation is thought to be due to the directional ordering of ion pairs such as Mn^{2+} and Fe^{2+}, and it is increased by a high concentration of metal ion vacancies which favours ionic diffusion. The vacancy concentration can be kept low by minimizing the oxygen in the sintering atmosphere.

The permeability is also affected by stress: the induced change decays logarithmically with time as a form of disaccommodation. Inductors are usually made in two parts so that the winding can be introduced on a bobbin; the parts must then be joined closely together to prevent the permeability from being affected by an uncontrolled air gap. If a mechanical clamp is used, the force must not be excessive and must be the same in similar units. Its effect must be allowed to decay before the inductance value is finally adjusted.

A typical design of a pot core for an inductor is shown in Fig. 9.48. There is a gap in the central limb of the core when the surfaces of the outer parts are in contact. There is also a hole through the centre of the core into which a magnetizable rod can be inserted. The position of this rod relative to the gap can be altered so as to adjust the value of the inductance. Apart from allowing this adjustment, the gap reduces and controls the magnetic losses of the core (Eq. (9.12)). If μ_r is the relative permeability without a gap and μ_{re} is the relative permeability with a gap, the losses due to the core ($\tan \delta_m = 1/Q$), its temperature coefficient and changes due to disaccommodation are all reduced by the factor μ_{re}/μ_r (see Section 9.1.4). In many filter applications a constant bandwidth is required over a wide range of frequencies so that the ratio of bandwidth to centre frequency diminishes as the frequency increases. Q therefore needs to be higher and tolerance on inductance variations less at higher frequencies. This can be achieved, without altering either the core design or the ferrite composition, by increasing the gap width. Cores are manufactured with diameters ranging from 10 to 45 mm and inductance values ranging from the order of microhenries to 5 H. Various core geometries are available (Fig. 9.17) but no new principles are involved. Choice is determined mainly by considerations such as size economy and compatibility with printed-circuit-board technology.

In the use of ferrites as inductors for tuned circuits, the Q of the materials is of prime importance. Wide-band transformers are also used extensively in communications systems, for example to transform signal voltages and to provide impedance matching and d.c. isolation of one part of a circuit from another. Pulse transformers are of increasing importance because of the rapidly growing use of pulsed signals in communications technology. Since a pulse can be synthesized from its Fourier components, a pulse transformer needs to be, in effect, a wide-band transformer. An analysis of the equivalent circuit for a transformer shows that core losses have little effect on the transformer's efficiency, provided that they are acceptable at the lower end of the frequency band to be transmitted. For example, an MnZn ferrite is a perfectly acceptable material for a wide-band transformer for the frequency range 100 kHz to 50 MHz. The reason for this is that as the frequency increases the core losses (equivalent to a resistance shunting the transformer's mutual inductance) become less significant compared with the impedance of the transformer. For wide-band transformers it is the high initial permeability of the MnZn ferrites coupled with an acceptably low loss which makes them the favoured material. Because mating surfaces must inevitably lead to a reduction in effective permeability, the transformer cores are available in toroid as well as pot-core forms.

An important application of ferrites is for shielding sensitive equipment (e.g. data-processing, telecommunications and audio-visual equipment) from electromagnetic interference (EMI). Both NiZn- and MZn-based ferrites components are capable of suppressing interference up to the GHz frequency range by virtue of the high impedance they present to high frequency currents. The ferrite parts are made in a variety of shapes to enclose the leads to be shielded, as shown in Fig. 9.17.

Multilayer technology is exploited just as for multilayer capacitors (see Section 5.4.3) to EIA size specifications, and inductor chips can be bandoliered ready for surface mounting, typical inductance values lying in the range 1 nH to about 20 μH. They find important applications as EMI suppressors as well as for a wide range of applications as a lumped circuit component in equipment of all types – communications, 'entertainment', computers, etc. (see Fig. 9.17).

9.5.2 Transformers for power applications

There are many high-frequency (typically 1 kHz to 1 MHz) applications where a high saturation flux density is the prime requirement. Examples are power transformers to transmit a single frequency, those required to transmit over a narrow frequency band, as in the power unit of an ultrasonic generator, and wide-band matching transformers to feed transmitting aerials. Television line output transformers are, commercially speaking, important examples of power transformers and, although not a transformer, the beam deflection yoke (Fig. 9.17) falls into the same category.

The saturation magnetization of ferrites (about 0.5 T) is too low for them to compete with Si–Fe laminations (about 2.0 T) at power frequencies. At higher frequencies eddy current losses preclude the use of metallic ferromagnets, and ferrites are widely used.

The growth in the demand for switch mode power supplies has led to a corresponding demand for high-frequency ferrite-cored power transformers. The switch-mode principle is not new, and it is the availability of suitable power-switching transistors, rectifiers, capacitors and ferrite cores which has enabled full advantage to be taken of the small size and inherent efficiency of the type of supply. In a typical unit the mains supply is rectified, smoothed and chopped at a high frequency, usually in the range 25–100 kHz, by a power transistor. This chopped voltage is transformed to the desired voltage by a ferrite-cored transformer and finally rectified and smoothed. Transformers capable of handling up to 2 kW are readily available.

As far as material properties are concerned, the general requirements of ferrites for power applications are similar to those for pot cores, but the permeability can be allowed to vary more widely. As high fields are likely to be applied, the hysteresis losses will be important. They can be minimized by keeping the crystalline anisotropy to a low level, for instance by substituting cobalt for about 1% of the nickel in $(Ni,Zn)Fe_2O_4$ (cf. Table 9.1). The internal shape anisotropy due to inhomogeneities and porosity can be minimized by ensuring that mixing, pressing, and sintering conditions are such as to yield a homogeneous high-density body.

As expected, the NiZn ferrite system is available for the higher frequencies (up to approximately 5 MHz), whereas for frequencies up to about 100 kHz the MnZn ferrites are favoured because of their higher permeabilities.

The shapes used are mostly simple E or C cores and I bars (Fig. 9.49). These allow simple high-speed winding machines to be used either to put windings directly on the ferrites or to wind bobbins which can be slipped onto the accessible limbs. The C and E cores must have their magnetic circuits completed by securing I cores across the appropriate limbs. The mating surfaces must be ground so that the air gap is minimal. Alternatively, cores can be fabricated as closed magnetic circuits and fractured into parts onto which windings can be fitted. The ceramic texture must be such that the two parts can then be clamped together without any damage to the mating surfaces. The process of fracturing is assisted by pressing grooves into the 'green' shape and can be achieved either mechanically or through thermal stress by the local application of a flame or a hot wire. Careful production control is needed to give a good yield of suitable fractures.

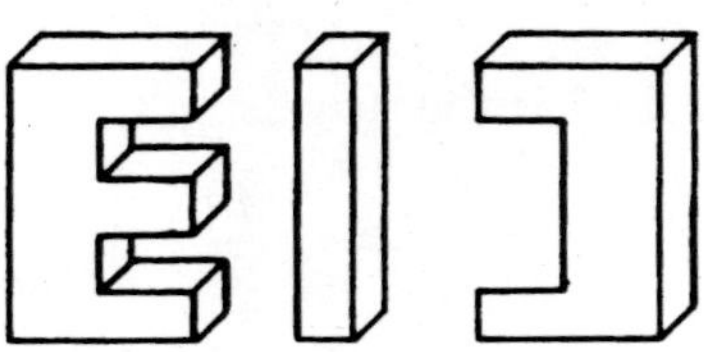

Fig. 9.49 Transformer core shapes.

Ferrites having a square hysteresis loop (e.g. (Mg, Mn, Zn) Fe_2O_4), developed for the now outdated computer core memories, have found applications in the form of a toroidal 'saturable inductor' for regulating the output currents in switched-mode power supplies.

9.5.3 Antennas

Small antennas comprising a suitable winding on a ferrite rod are needed for radios, pagers and for RF identification transponders. In the case of a transponder information is stored in a tag (an application for ferroelectric memories – see Section 5.7.5), which can be read by a signal from an interrogating unit. The cylindrical rods (Fig. 9.17) are precisely dimensioned with sizes down to 0.76 mm diam. by 4.82 mm. The tag carries no battery, sufficient power to transmit the stored information being derived from the interrogating signal. Applications include car key identification so that if the correct code is not detected then the steering wheel and fuel supply are blocked, airport baggage checks, control of access to buidings and traceability in food chains.

A short winding of n turns of area A will develop an e.m.f. of amplitude U if placed in an alternating magnetic field of angular frequency ω and amplitude H, where

$$U = \mu_0 \omega H A n \tag{9.52}$$

If a long rod of effective relative permeability μ_{re} is inserted in the coil the e.m.f. becomes

$$U = \mu_0 \mu_{re} \omega H A n \tag{9.53}$$

Because the magnitudes of the electric and magnetic vectors of an electromagnetic wave are related by $H = (\varepsilon_0/\mu_0)^{1/2} E$ and the wave velocity *in vacuo* is given by $c = (\varepsilon_0 \mu_0)^{-1/2}$, it follows that

$$U = \mu_{re} \omega \frac{An}{c} E$$

A figure of merit for an antenna is the effective height h_e, where

$$h_e = U/E \tag{9.54}$$

and, since $c = (\omega/2\pi)\lambda$ where λ is the wavelength,

$$h_e = 2\pi \mu_{re} \frac{An}{\lambda} \tag{9.55}$$

Therefore h_e will increase at shorter wavelengths and with increased μ_{re}, which is related to the material permeability μ_r by Eq. (9.9). Values of h_e for two materials

Table 9.6 Antenna rod dimensions and properties

μ_r	*Length*/mm	*Diameter*/mm	*Demagnetizing factor* N_D	μ_{re}	h_e/mm
175	200	9.5	0.0043	100	5.9
175	150	12.7	0.0112	59	6.3
500	200	9.5	0.0045	154	9.1
500	150	9.5	0.0073	108	6.4

Short central winding of 40 turns; 1 MHz ($\lambda = 300$ m).

are given in Table 9.6. The h_e values are very small compared with, for example, those for a correctly designed dipole interacting with the $\boldsymbol{E}$ vector of the electromagnetic field. By comparison with a dipole the ferrite-cored antenna is very inefficient, but very much more compact!

The signal power available to the first stage of a receiver is proportional to $h_e^2 Q_a$, where Q_a is the inverse of the loss tangent for the rod and coil combination together with any metal objects that may absorb or distort the radiation. It is therefore important that the ferrite should have low losses in the frequency range for which the antenna is designed. $(Ni,Zn)Fe_2O_4$ is used at frequencies above 1 MHz since its losses are less than those of $(Mn,Zn)Fe_2O_4$. Figure 9.29 gives the permeability and loss for some compositions of the former type as a function of frequency and for a range of zinc contents.

Since the ferrite is required in rod form it is usually made by extrusion. A simple circular section is sometimes replaced by one having ribs parallel to the axis of a central cylinder. This structure is intended to avoid the effects of dimensional resonance (see Section 9.3.1) which would cause the antenna to have poor response over a small part of its total frequency range. The effect of a ribbed structure is that only a fraction of the material will have a resonant dimension at any particular frequency so that, compared with an unribbed rod, there will be a minor drop in effective height over a wider frequency band. The relatively high losses at megahertz frequencies usually damp the resonance and their effects can be largely eliminated by suitable design.

9.5.4 Information storage and optical signal processing

Information storage

The continuously evolving technologies concerned with information storage are many and complex and involve metals, ceramics and polymers in active roles. The following discussion is concerned only with those aspects which involve ceramics and is therefore very restricted. Comprehensive descriptions of the technologies in general can be found elsewhere, for example in the monograph edited by W. Gerhartz [12] and in recent overviews [13,14].

During the few decades prior to 1980 the square magnetic hysteresis loop ferrite 'cores' dominated computer memory technology. The binary information was determined by the sense of the magetization around the core. They were non-volatile and intrinsically 'radiation hardened' and still used in 1990 for specialized applications, for example military, where those attributes were essential. They have been totally displaced by the semiconductor memory offering much higher storage density and speed. It is interesting that FeRAMS (see Section 5.7.5), the ferroelectric analogue offering high storage density, high access speed and non-volatility, are now manufactured commercially.

In the conventional tape-recorder the recording medium consists of small needle-shaped particles dispersed in an organic binder and supported on a flexible polymer film. The traditional recording technology is 'longitudinal', that is the induced changes in magnetization occur in the place of the film. Most tapes incorporate particles of γ-Fe_2O_3 (maghemite, which has a defect spinel structure) that are synthesised chemically in the form of needles typically 0.2 to 0.5 μm long with aspect ratios in the range 5 to 10. The needles are single domain and the magnetic anisotropy arises as a result of the different demagnetizing fields along the length and perpendicular to it ('shape anisotropy'). The spontaneous magnetization lies along the length of the needle and the coercive force is typically in the range 50 to 100 $kA\,m^{-1}$.

The record/read head, positioned close to or in contact with the recording surface, is essentially a gapped soft electromagnet (e.g. Mn-Zn ferrite) positioned so that the fringing field in the gap intersects the magnetic particles in the tape. The signal fed to the head coil modulates the field which, in turn, switches the magnetization in the particles. In play-back mode the modulated magnetization in the tape induces voltages in the coil as the tape moves relative to the head. The 'bit' length, the minimum distance between magnetization reversals, is about 1 μm, corresponding to one half wavelength in analogue recording.

Perpendicular recording, which involves magnetization changes normal to the plane of the tape or disc, has attracted considerable interest because of the higher storage density offered than with longitudinal recording. This is because the magnetic state of a particular bit is not subjected to the demagnetizing field from a neighbouring bit of opposite polarity, whereas in the case of longitudinal recording it is, placing a lower limit on bit spacing. Small (typically $\sim 0.08 \times 0.03 \times 0.1\,\mu m^3$) barium hexaferrite particles are arranged to lie with their c-axes (the preferred direction of magnetization, or 'easy' direction; see Section 9.2.2) perpendicular to the plane of the tape [12]. Although the technology has not been adopted the interest stimulated the development of routes to preparing suitable powders. These included both 'mixed oxide' followed by milling, and 'chemical', including 'hydrothermal' [15]. A rather novel route involved fusing the barium and iron oxides with boric oxide to form a glass from which the barium hexaferrite particles were precipitated, the glass being removed by acid dissolution.

Pure iron particles, coated to inhibit oxidation, are now used to an increasing extent for data storage and may ultimately totally displace the ferrites.

The hard disk of a computer exploits thin magnetic metal films with a present-day data storage capacity of about 10 Gb in^{-2} whereas in the case of flexible tape memory systems incorporating magnetic particles the capacity is about 100 times smaller.

Applications exploiting magneto-optical effects

Magneto-optical recording exploits the Kerr effect which describes rotation of the plane of polarization on reflection from a magnetic material. Both the Kerr and Faraday effects (the latter describing rotation of the polarization plane on passing through a transparent magnetic material) are exploited in optical signal processing.

If a thin transparent layer of anisotropic magnetic garnet is viewed through a crossed polarizer and analyser system in a microscope, a pattern of the type shown in Fig. 9.50(a) is seen. This pattern is due to the Faraday rotation effect which was discussed in Section 9.3.4. Domains magnetized in a direction normal to the film rotate the plane of polarization of incident light in one sense while those magnetized in the reverse direction rotate it in the opposite sense, so that the domain pattern is observable in the light emerging from the analyser. If a field is now applied normal to the film, one set of domains expands while the other shrinks as magnetization in a single direction becomes established (Fig. 9.50(b)). There is a stage before magnetization is complete when the remaining domains, which have their magnetizations in the opposite direction to

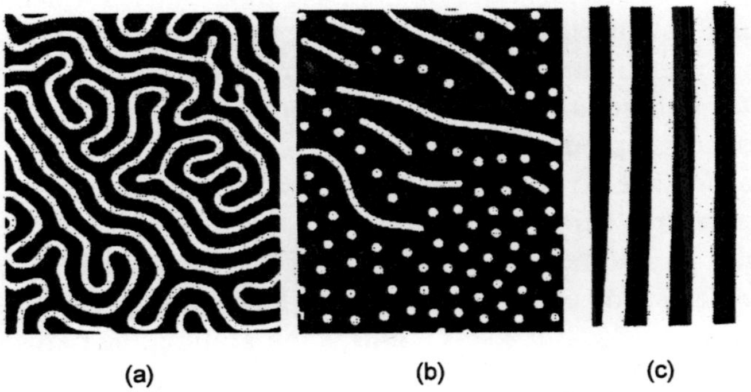

Fig. 9.50 The various domain structures observable in epitaxial iron garnet films: (a) maze, (b) 'bubbles', and (c) stripe.

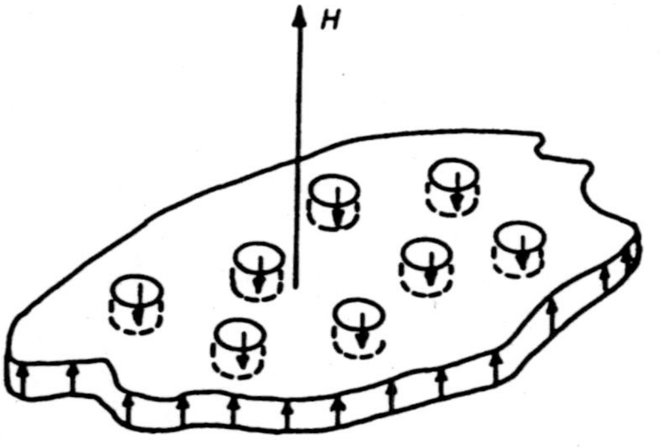

Fig. 9.51 Schematic diagram of bubbles showing directions of magnetization.

the field, are in the form of short circular cylinders with their axes normal to the film and appear under the microscope as circular ‘bubbles’ (Fig. 9.51).

These effects are shown by all magnetic materials that have uniaxial anisotropy, and the size of the stable bubbles is inversely proportional to the saturation magnetization; thus barium ferrite and metallic cobalt give bubbles of diameter about 0.05 μm and 0.01 μm respectively. Garnet is a cubic structure and therefore does not have uniaxial anisotropy in the massive form. When it is grown as a single-crystal epitaxial layer on a substrate of differing composition, strain is induced by the mismatch between the lattice dimensions of the magnetic garnet and the substrate. This is enhanced on cooling to room temperature by the difference in thermal expansion between the two garnets and results in an anisotropic axis normal to the film. The anisotropy and magnetization of the film, and consequently the size of the bubbles, can be adjusted by replacing yttrium by other rare earths and iron by other trivalent cations such as Ga and Al. $Y_{2.6}Sm_{0.4}Ga_{1.2}Fe_{3.8}O_{12}$ is a typical formulation giving bubbles of diameter 4–8 μm in films of similar thickness to the bubble diameter.

In 1990 there was strong interest in data storage devices which exploited trains of small ‘bubbles’ injected into a garnet film; the presence or absence of a bubble representing the binary-coded information. Bubble memories offered the advantage of non-volatility but are now commercially unimportant because of inadequate access speed.

It is also possible to develop ‘stripe’ domains as shown in Fig. 9.50(c), the domain geometry being determined by fields applied in the plane and perpendicular to it. The pattern as a whole can be rotated by rotating the in-plane field. The stripe pattern, arranged to have an appropriate spacing by the applied magnetic field, can act as a diffraction grating, the paths of light passing through it being controlled through corresponding changes in the applied magnetic field. The effect has the potential for switching signals in optical communications systems (cf. Sections 8.3.5 and 8.4).

An important application for garnets is for isolators in optical communications systems. Reflections from end-faces of components and interconnections in an optical transmission line have a destabilizing effect on the operation of the laser sources and have to be eliminated. This is achieved using an optical 'isolator', the optical analogue (λ typically in the range 1.30–1.55 μm) of the microwave isolator described in Section 9.5.5.

An optical isolator comprises a magnetic garnet crystal plate sandwiched between crossed polarizing elements, and a permanent magnet for applying the necessary field. The device allows the transmission of forward light whilst blocking any reflected light. Such isolators are incorporated in the semiconductor laser modules and so are miniaturized with overall dimensions typically 3×3 mm. Bi-substituted iron garnet is used for the active element because of its low saturation magnetization and large rotating power, and rare-earth magnets are employed to apply the appropriate magnetic field.

For one particular commercial isolator ($(YBi)_3(FeAl)_5O_{12}$; precise details not disclosed) an optical rotation of 1060 deg cm^{-1} is quoted for a wavelength of 1.55 μm which compares with 175 deg cm^{-1} for YIG ($Y_3Fe_5O_{12}$). An additional advantage offered by the substituted garnet is a smaller saturation magnetization of ~60 mT compared with ~178 mT for YIG.

The substituted garnet plates can be produced by slicing from a single crystal rod grown by the floating zone method or, more economically, from thick films grown onto a host substrate by liquid phase epitaxy (see Section 3.11). The plates are lapped, polished, coated with antireflection layers and then diced to produce the final elements (~2 mm side by ~100 μm thickness) ready for assembly into the isolator.

There are many potential applications for thin film magnetic garnets, including wave-guiding in the plane of films, and image storage and display technologies. The monograph by A.K. Zvezdin and V.A. Kotov [16] is recommended for a comprehensive treatment of the materials and magneto-optical applications.

9.5.5 Microwave devices

The elementary description of microwave propagation in magnetically saturated ferrites given in Section 9.3.4 provides a basis for an understanding of the principles of operation of devices such as circulators, isolators and phase shifters, all of which find extensive use in microwave engineering. For example, a circulator might be used in a radar system where a single antenna is employed for both transmitting microwave power and receiving a reflected signal, as shown schematically in Fig. 9.52(a).

An isolator is a device which allows microwave power to be transmitted in one direction with little attenuation while power flowing in the reverse direction is, ideally, completely absorbed. It is used to avoid interaction between the various

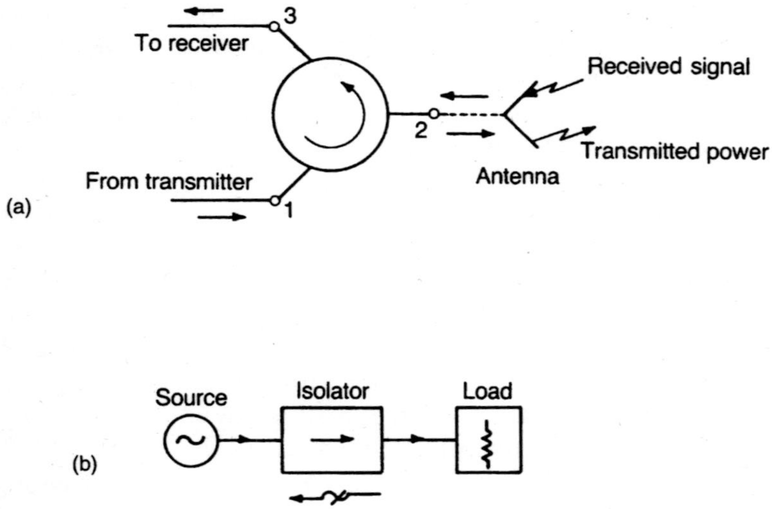

Fig. 9.52 (a) Schematic diagram of a circulator for a radar system; (b) an isolator.

parts of a microwave system. For example, in high-microwave-power industrial processing systems it is essential to prevent the high-power source – a magnetron for instance – from instability and destruction due to reflected power from the load. This function is illustrated diagrammatically in Fig. 9.52(b).

An example of the use of ferrite phase shifters is in a phased-array antenna. This comprises an array of coherently radiating dipoles, in which the microwaves from each are controlled in phase and amplitude. The main beam emitted is produced by interference between the radiation from the individual elements; the phase difference between the radiation from the individual dipoles determines the attitude of the beam in relation to the array. Stepwise changes in the phase differences thus make it possible to control the direction of the beam electronically. In fact the individual phase shifters can be controlled by a computer, thus enabling a single immobile array to perform inertialess search, acquisition and tracking functions simultaneously on a large number of targets. These devices are described in the following section, but first the Faraday rotation isolator is briefly discussed because it illustrates the basic ideas; at one time it was an important device but it has now been superseded, principally by the junction circulator.

Faraday rotation isolator

Figure 9.53 shows the essential parts of a Faraday rotation isolator. The electric field of the forward wave is perpendicular to the plane of the resistive card (a thin

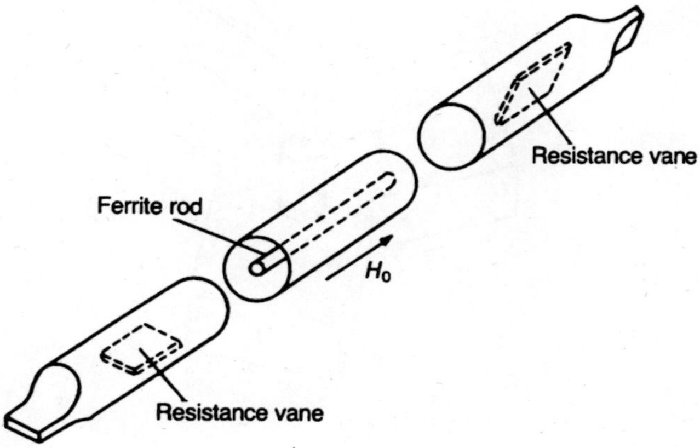

Fig. 9.53 Faraday rotation isolator.

vane of absorbing material such as a coating of nichrome deposited onto a thin glass strip) and so it passes it with little attenuation. The length and magnetization of the ferrite and the applied field H_e are chosen so that the plane of polarization of the radiation is rotated by 45° in the correct sense for it to pass into the right-hand guide without reflection. Radiation propagated in the reverse direction will also be rotated by 45° so that the plane of polarization is now at 90 ° to that of the forward wave. Therefore it will not propagate in the left-hand guide, and also the resistive card will be in the same plane as the electric vector so that the radiation is absorbed.

Although this type of isolator has largely been superseded by other types, it has advantages at very high frequencies over devices which necessitate the ferrite operating at resonance: the very high applied magnetic fields required for millimetre wave resonance isolators are avoided.

Y-junction circulator

The most frequently used ferrite device is a three-port circulator in waveguide or stripline configuration, as illustrated in Fig. 9.54. In the waveguide configuration a ferrite cylindrical disc is located symmetrically with respect to the three ports. In the stripline configuration two ferrite discs are placed on each side of the strips. In both cases a magnetic field is applied perpendicular to the planes of the discs.

Unlike the case of the Faraday rotation isolator, even an elementary description of the operational principle of the circulator is not easy, involving as it does the dimensional resonance of the microwave field within the ferrite cylinder. In this context the word 'resonance' does not signify 'gyromagnetic resonance' but a standing-wave resonance determined by the dimensions of the

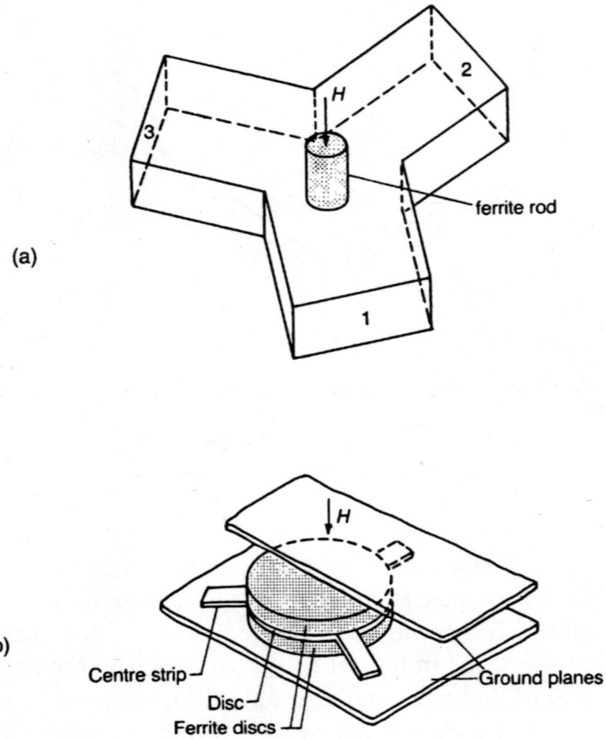

Fig. 9.54 Y circulators in (a) waveguide and (b) stripline configuration.

ferrite (cf. Section 9.3.1). In the particular resonance modes excited, the electric field vectors are perpendicular to the plane of the disc. The field configuration generated can be regarded as two counter-rotating patterns which, because of the applied field H, do not resonate at the same frequency. This is because of the different permeability values depending on whether the magnetic component of the microwave field is rotating in the same or the opposite sense to that of the electron precession. By appropriate choice of dimensions and biasing field, depending on the operating frequency, the positions of the maximum and minimum electric field can be made to coincide with the port positions. For example, for power entering port 1 (Fig. 9.54(a)) the electric vector $\boldsymbol{E}$ can be arranged to be zero opposite port 3, so that no voltage exists there and in consequence no power leaves; because $\boldsymbol{E}$ is a maximum opposite port 2, power passes through it. The device therefore acts as a transmission cavity with power entering at port 1 and leaving at port 2. Because the device is symmetrical, power entering port 2 will set up similar field configurations resulting in power leaving port 3 and port 1 will be isolated. Similarly, power entering port 3 is transmitted to port 1 and port 2 is isolated.

In addition to its use as a circulator, the device can be operated as an isolator, by terminating one of the ports with a matched load, or as a microwave power switch, by reversing the sense of the biasing field.

Phase shifters

The phase of a microwave passing along a guide can be shifted by inserting pieces of ferrite at appropriate positions. For example, in a rectangular guide along which the simplest microwave field configuration (TE_{10} mode) is propagating, the magnetic field component is circularly polarized in the plane of the broad face of the waveguide at a distance approximately a quarter of the way across the guide; on the opposite side of the guide the magnetic field component is circularly polarized in the opposite sense. This can be seen from Fig. 9.55 which shows the instantaneous magnetic field configuration looking down on the broad face of the guide. As the pattern propagates from left to right the field directions at A and B change as indicated, as the field configurations at points 1, 2, 3 and 4 arrive at A and B. Therefore if two slabs of ferrite are positioned in the guide and magnetic fields, well below resonance strength, are applied as shown in Fig. 9.56, the effective path length of the wave is altered and a phase shift is introduced.

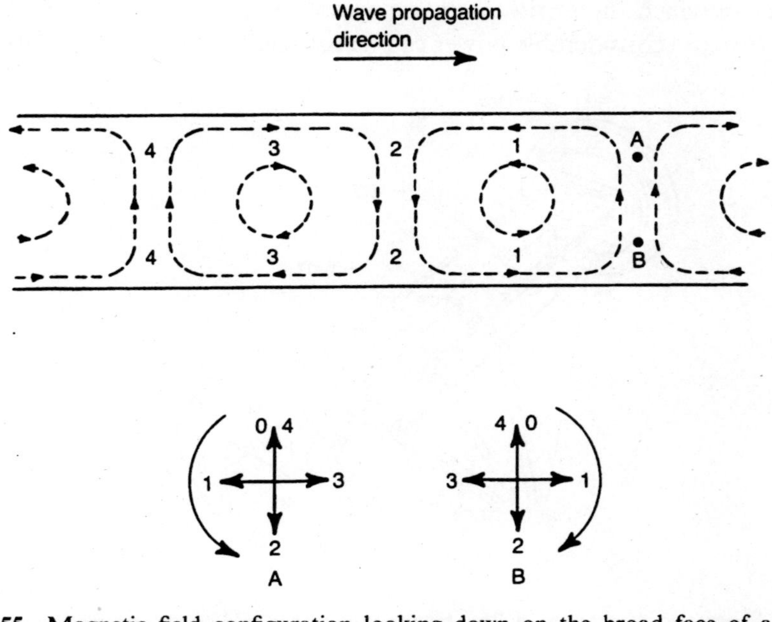

Fig. 9.55 Magnetic field configuration looking down on the broad face of a waveguide propagating the TE_{10} mode.

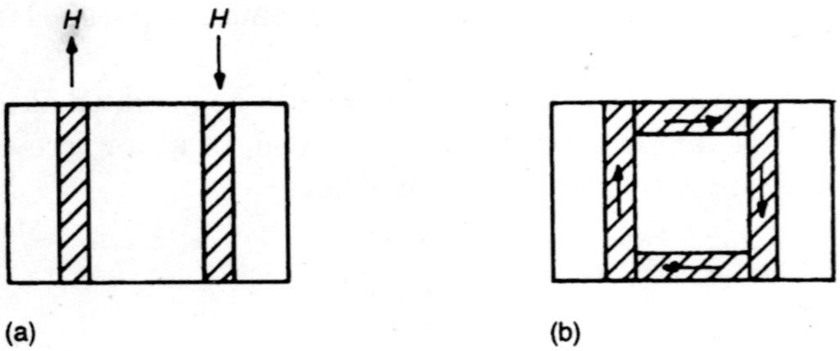

Fig. 9.56 Phase shifters: (a) non-reciprocal isolating arrangement; (b) latching.

It is worth noting that if the ferrites are magnetized to resonance and a forward wave is absorbed by the ferrite, it will be relatively unaltered in the reverse propagation direction. The device can therefore function as an isolator.

A development of the configuration of Fig. 9.56(a) is to complete the magnetic circuit using permanently magnetized ferrite pieces, as shown in Fig. 9.56(b); this is the basis of the 'latching ferrite phase shifter'. The static magnetization may now correspond to positive or negative remanence. Thus the phase can be shifted by a magnetizing pulse through wires threading the core. By using a series of such phase shifters in the feeder to a dipole of an antenna, the overall phase shift can be controlled over a range of steps as small as desired. The control is digital and can therefore be computer directed, forming one element of a 'phased-array antenna'.

The way in which the ferrite parts are arranged in the guide is illustrated in Fig. 9.57. Because considerable power can be handled by radar phase shifters the

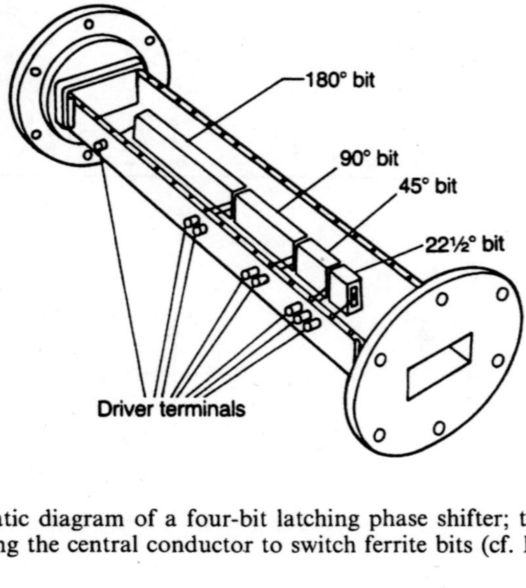

Fig. 9.57 Schematic diagram of a four-bit latching phase shifter; the driver terminals pass current pulses along the central conductor to switch ferrite bits (cf. Fig. 9.56(b)).

inevitable heat developed has to be removed. This can be achieved by ensuring good thermal contact between the ferrite and the metal parts. BeO and BN ceramics can be designed in as heat sink materials.

Microwave ferrite types

Perhaps more than in most other examples that can be cited of materials exploitation, designing a ferrite for a particular microwave application is a compromise: a desired value for a particular parameter is generally achieved at the expense of the optimum values of other parameters. Nevertheless, a select number of spinel and garnet compositions have become favoured for microwave applications.

As is evident from the discussion in Section 9.3.4, saturation magnetization is an important parameter. In the nickel spinel family M_s values ranging from 400–40 kA m^{-1} ($4\pi M_s = 5000$–500 G) can be tailored by substituting Zn^{2+} for Ni^{2+} to raise M_s or by introducing Al^{3+} to the B sites to lower M_s. Lowering M_s also increases the power-handling capability of the ferrite since it partly determines the power threshold P_c (see Section 9.3.4) of the microwave at which spin-wave resonance sets in. Nickel ferrites are used to best effect for applications at X-band frequencies and above and, because of their large spin-wave linewidth, for handling high peak power levels. Undesirable features of the nickel spinels are their magnetic and dielectric loss factors which are higher than acceptable for some applications.

The magnesium ferrites have M_s values ranging from 200 to 40 kA m^{-1} ($4\pi M_s = 2500$–500 G). Small additions of Mn^{2+} are found to suppress the formation of Fe^{2+} which would increase dielectric losses due to electron hopping. As in the case of nickel spinels, the addition of Al^{3+}, which goes mainly into the B sites, leads to a reduction in M_s.

The lithium spinels $Fe^{2+}(Li^{1+}_{0.5}\,Fe^{3+}_{2.5})O_4$ have M_s values ranging from 380 to 32 kA m^{-1} ($4\pi M_s = 4800$–400 G). They have exceptionally high Curie temperatures – up to about 670 °C – and so the M_s values are relatively insensitive to temperature under normal operating conditions. M_s values can be reduced by the substitution of Ti^{4+} into B sites, with additional Li^{1+} for charge compensation. Coercivity can be tailored by the addition of Zn onto A sites which reduces anisotropy; zinc addition also increases M_s. Addition of Co^{2+} increases the power-handling capability (see Sections 9.1.7 and 9.3.1). Care has to be exercised in the sintering to prevent the formation of Fe^{2+} and consequent poor dielectric properties; small quantities (about 1%) of Bi_2O_3 are added to reduce sintering temperatures and so alleviate the problem. Except for the low insertion loss, where the garnets have a clear advantage, the lithium spinels are regarded as having superior properties for most situations.

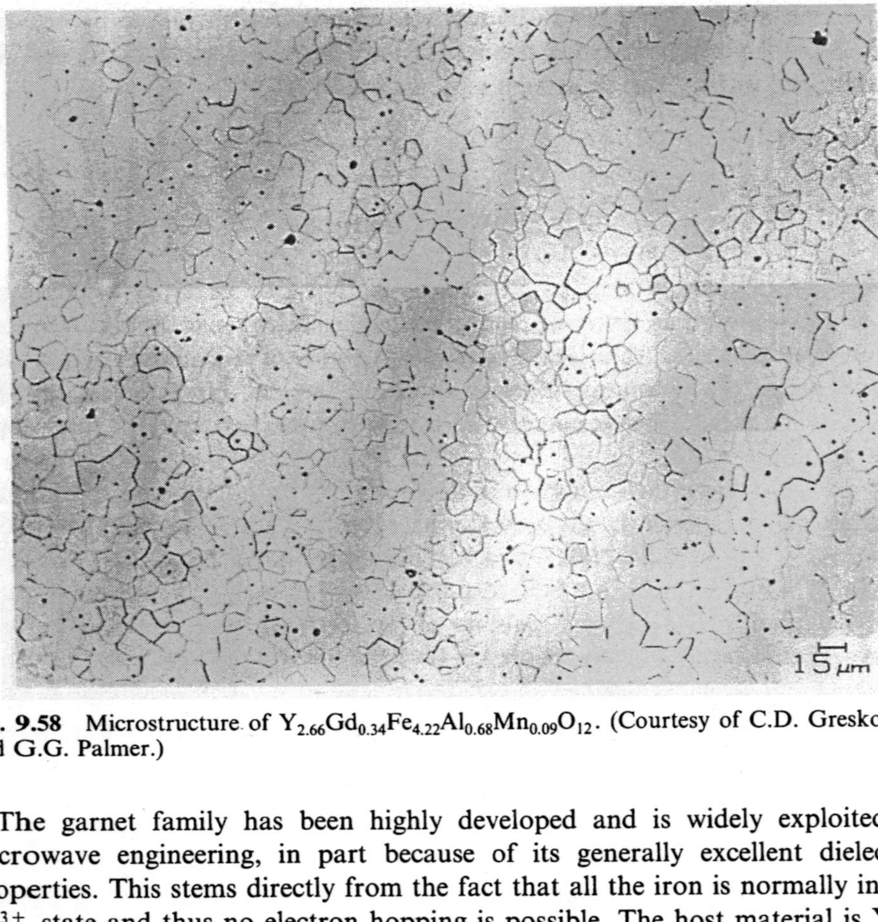

Fig. 9.58 Microstructure of $Y_{2.66}Gd_{0.34}Fe_{4.22}Al_{0.68}Mn_{0.09}O_{12}$. (Courtesy of C.D. Greskovich and G.G. Palmer.)

The garnet family has been highly developed and is widely exploited in microwave engineering, in part because of its generally excellent dielectric properties. This stems directly from the fact that all the iron is normally in the Fe^{3+} state and thus no electron hopping is possible. The host material is YIG which has M_s values in the range 140–8 kA m^{-1} ($4\pi M_s = 1800$–100 G). To improve the temperature insensitivity of M_s around room temperature Gd^{3+} ions are substituted for Y^{3+} (Fig. 9.16). The composition of garnets can be designed to enable them to handle high microwave power. For example an X-band three-port circulator operating at a peak power of about 200 kW (average power, 200 W) exploits the garnet $Y_{2.1}Gd_{0.9}Fe_5O_{12}$ ($M_s = 980$ kA m^{-1}; $4\pi M_s = 1230$ G).

Garnets are also exploited in the construction of latching ferrite phase shifters because they can be tailored to optimum performance with respect to low insertion loss, temperature-insensitive M_s, high remanence, low coercivity and high power-handling capability. A square hysteresis loop is a requirement, and this is influenced by microstructure as the optimum squareness ratio occurs for dense pore-free material with a narrow distribution of grain size. A garnet developed for a phased-array radar system has the composition $Y_{2.66}Gd_{0.34}Fe_{4.22}Al_{0.677}Mn_{0.09}O_{12}$ and a good microstructure, as shown in Fig. 9.58.

9.5.6 Permanent magnets

As discussed in Section 9.3.2, permanent magnet materials are required to give a maximum flux in the surrounding medium for a minimum volume of material and to be resistant to demagnetization. Suitability of a material for a particular application is judged by its B_r, H_c and $(BH)_{max}$ values, which are given in Table 9.3 for some of the principal hard magnetic materials. Permanent magnets can be found in a wide range of applications including consumer, industrial, electronic, automotive and medical equipment. The outstanding magnetic properties of the rare earth variety make it ideally suited to high performance applications, for example motors for computer disc drives and parts for medical technology where the high cost of the magnet material can readily be absorbed.

There are many applications for which the magnetic properties of the ferrites are quite adequate and this, coupled with their relatvely low cost, has assured them of a large and sustainable market, for the foreseeable future, at least. They have higher coercive fields than those for the mass-produced metal magnet types and this allows them to be made into shapes (e.g. d.c. motor field magnets) involving high demagnetizing fields.

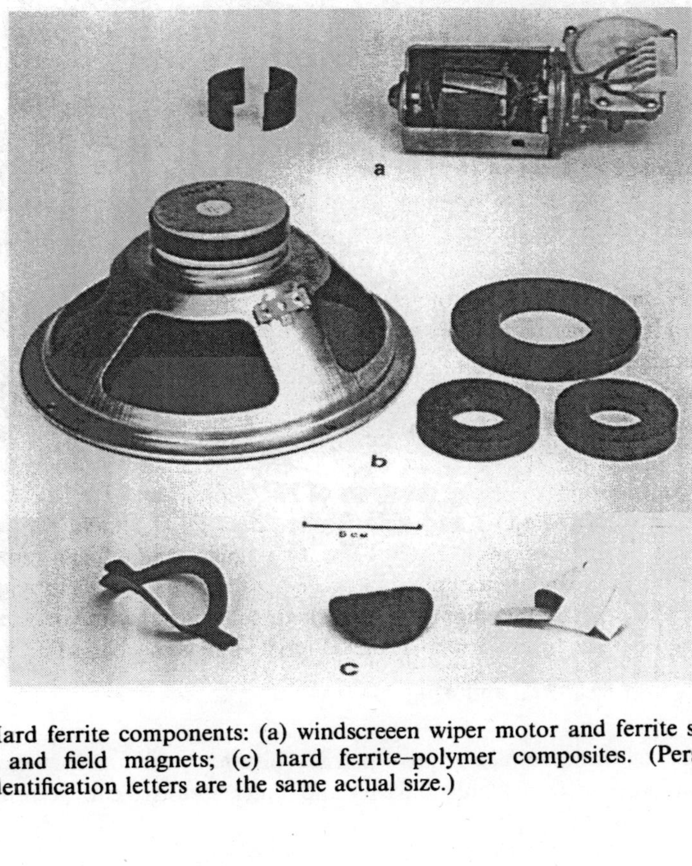

Fig. 9.59 Hard ferrite components: (a) windscreeen wiper motor and ferrite segments; (b) loudspeaker and field magnets; (c) hard ferrite–polymer composites. (Perspective size distortion; identification letters are the same actual size.)

An important market is for the field magnets in small d.c. motors for which the shape demands high coercivity. Automotive applications for small motors include the windscreen wiper (see Fig. 9.59(a)) and air-blower, window lift, and pumps for fuel, ABS and power-steering. Similar motors are used in washing machines and general kitchen appliances and also for portable tools. Another important application is for 'stepping' motors which convert electrical impulses into incremental mechanical movements for clocks, copying and fax machines and medical equipment.

Generally speaking the higher quality hard ferrites are based upon Sr-hexaferrite and those for less demanding applications on the cheaper Ba-hexaferrite.

Ferrite magnets are used in the form of flat toroids in moving coil loudspeakers (Fig. 9.59(b)), and large pieces for magnetic separators for mineral beneficiation and water filters.

Small pieces find many diverse applications, for example in toys, door latches, display boards, TV tube correction magnets, etc., and the plastic composite variety is best known as a refrigerator door seal (Fig. 9.59(c)).

Problems

1. Derive an expression for the Bohr magneton and calculate its value.

   A particular crystal has a cubic unit cell of side 1 nm. Each cell has an associated induced magnetic moment of $10^{-7}\,\mu_B$ for an applied field of $10^3\,\mathrm{A\,m^{-1}}$. Calculate the susceptibility and offer an opinion as to whether the material is diamagnetic, paramagnetic or ferromagnetic. [Answers: $\chi = \pm 9.3 \times 10^{-7}$; diamagnetic]

2. Calculate the magnetic moment of an isolated Fe^{2+} ion and also that associated with an Fe^{2+} ion in magnetite. Explain why they are different.

   The saturation magnetization of magnetite is $5.2\times10^5\,\mathrm{A\,m^{-1}}$ and the unit cell is of side 837 pm. Assuming the inverse spinel structure, estimate the magnetic moment (in Bohr magnetons) of the Fe^{2+} ion. [Answers: $6.7\,\mu_B$ and $4\,\mu_B$; $4.11\,\mu_B$]

3. Calculate the spin-only magnetic moments of Ni^{2+}, Zn^{2+} and Fe^{3+}.

   Nickel ferrite ($NiO{\cdot}Fe_2O_3$) and zinc ferrite ($ZnO{\cdot}Fe_2O_3$) have the inverse and normal spinel structures respectively. The two compounds form mixed ferrites. Assuming that the magnetic coupling between ions is the same as in magnetite and that the orbital angular momenta are quenched, calculate the magnetic moment (in Bohr magnetons) per formula unit of nickel ferrite and $Zn_{0.25}Ni_{0.75}Fe_2O_4$. [Answers: $2\,\mu_B$, $0\,\mu_B$ and $5\,\mu_B$; $2\,\mu_B$ and $4\,\mu_B$]

4. A soft ferrite with complex relative permeability $\mu_r^* = 2000 - 7j$ is in the form of a toroid of cross-sectional area $0.5\,\mathrm{cm}^2$ and inner radius 3 cm. A primary winding

comprising 200 turns and a secondary winding of 100 turns are wound uniformly on the toroid. Calculate the r.m.s. value of the e.m.f. generated across the secondary winding when a sinusoidally varying current of frequency 50 Hz and amplitude 2 A is passed through the primary.

Calculate the effective permeability and loss tangent of the ferrite and the e.m.f. when a 0.1 mm gap is introduced into the toroid.

Consider the implications of the results of the calculations for the manufacture of soft ferrites for high-quality inductors. [Answers: 5.23 V; 1033, 1.8×10^{-3} and 2.7 V]

5. A disc, 2 cm in diameter and 5 mm thick, magnetized normal to its faces and manufactured from a ferrite with $B_r = 0.3$ T and $H_c = 200$ kA m^{-1} is required. Comment on the practical feasibility of the requirement. Assuming the ferrite composition to be fixed, what steps might be taken to realize the requirement? [Answer: $H_D = 238$ kA m^{-1}]

6. Draw a block diagram illustrating the fabrication route for a ferrite-loaded plastic suitable for a refrigerator door seal.

7. Derive an expression for the angular frequency of the precessional motion of an electron situated in a magnetic field.

8. The ferrimagnetic resonance phenomenon in a ceramic ferrite in the form of a strip $70 \times 5 \times 2$ mm is to be exploited in a particular microwave device, with the static field applied along the length of the strip. If the design frequency is 2 GHz and the coercive field for the ferrite is 100 A m^{-1}, what limiting value is placed on the saturation magnetization?

   It is anticipated that, because the device is to be used at high microwave power levels, there is the risk of the onset of spin waves. How might the microstructure of the ferrite be designed to reduce this risk? Suggest a fabrication route suitable for achieving the objective. [Answer: $M_s < 56.7$ kA m^{-1}]

9. A 1 GHz plane-polarized electromagnetic wave passes down a microwave ferrite rod 50 mm long situated in a uniform magnetic field parallel to the axis of the rod. The magnetic field strength is adjusted so that the relative permeability values for the right and left circularly polarized components of the wave are 6 and 4 respectively. Calculate the angle through which the plane of polarization of the wave is rotated on travelling the length of the rod. The relative permittivity of the ferrite for both waves is 16 (velocity of light in a vacuum, 3×10^8 m s^{-1}). [Answer: 54°]

10. Explain why it is that magnetic domain structures in a thin magnetic garnet layer can be visualized in a polarizing microscope and how the various structures can be exploited. Describe two routes for fabricating such layers.

Bibliography

1. Hilpert, S. (1909) The structures and origins of the magnetic properties of ferrites and iron oxide, *Ber. Deutsch. Chem. Ges.*, **42**, 2248–61.
2. Forestier, H.M. (1928) Transformations magnétiques du sesquioxyde de fer, de ses solutions, et de ses combinaisons ferro-magnétiques, *Annales Chimie (Paris)*, **9**, 316–401.
3. Kato, Y. and Takei, T. (1933) Characteristics of metallic oxide magnet, *J. Inst. Elect. Eng. Japan*, **53**, 408–12 (in Japanese).
4. Néel, L. (1948) Propriétés magnétiques des ferrites: ferrimagnétisme et antiferromagnétisme, *Ann. de Phys.*, **3**, 137–98.
5. Went, J.J. *et al.* (1951/2) Ferroxdure, a class of new permanent magnet materials, *Philips Tech. Rev.*, **13**, 194–208.
6. Smit, J. and Wijn, H.P.J. (1959) Ferrites – physical properties of ferromagnetic oxides in relation to their technical applications, Philips Technical Library, The Netherlands.
7. Sugimoto, M. (1999) The past, present and future of ferrites, *J. Am. Ceram. Soc.*, **82**, 269–80.
8. Kools, F. (1991) Hard magnetic ferrites. In *Concise Encyclopedia of Advanced Ceramic Materials*, R.J. Brook (ed.), Pergamon Press, Oxford, pp. 200–6.
9. Broese van Groenou, A., Bongers, P.F. and Stuijts, A.L. (1968) Magnetism, microstructure and crystal chemistry of spinel ferrites, *Mater. Sci. Eng.*, **3**, 317–92.
10. Snelling, E.C. (1988) *Soft Ferrites*, Butterworths, London.
11. van der Zaag, P.J. (1999) New views on the dissipation in soft magnetic ferrites, *J. Mag. Mag. Mater.*, **196–197**, 315–19.
12. Gerhartz, W. (ed.) (1992) *Imaging and Information Storage Technology*, VCH, Cambridge, UK.
13. Li, Y. and Menon, A.K. (2001) Magnetic recording technologies: overview. *Encyclopedia of Materials: Science and Technology*, **5**, Elsevier, Amsterdam, pp. 4949–57.
14. Bissell, P.R. (2001) Magnetic recording: VHS tapes. *Encyclopedia of Materials: Science and Technology*, **5**, Elsevier, Amsterdam, pp. 4957–61.
15. Ataie, A. *et al.* (1995) Effect of hydrothermal synthesis environment on the particle morphology, chemistry and magnetic properties of barium hexaferrite, *J. Mater. Sci.*, **30**, 5600–6.
16. Zvezdin, A.K. and Kotov, V.A. (1997) *Modern Magnetooptics and Magnetooptical Materials*, Institute of Physics Publishing, Bristol, UK.

INDEX

Electroceramics: Materials, Properties, Applications. 2nd Edition. Edited by A. J. Moulson and J. M. Herbert.

 ISBN 0 471 49747 9 (hardback) 0 471 49748 7 (paperback)

Lightning Source UK Ltd.
Milton Keynes UK
UKOW01f0902240714

235627UK00002BA/3/P

9 780471 497486